2

WORLD EXPLORATION (OLD WORLD)

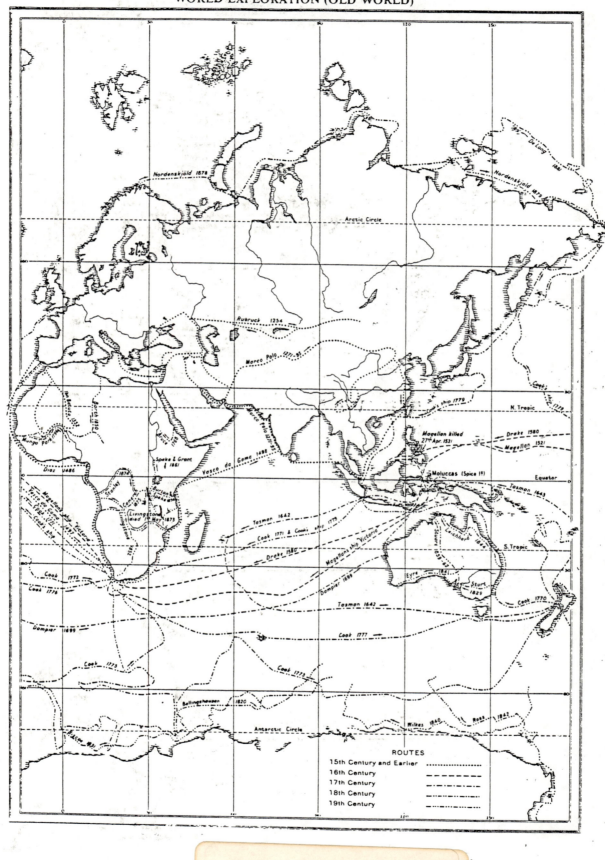

Arctic Circle

Cabot 1497
Frobisher 1576
Cabot 1498
Hudson 1610
Champlain 1613
Marquette & Jolliet 1673
Lewis & Clark
Cook 1778
Cook 1778
Columbus 1493
Drake 1577
De Soto 1539-1543
Coronado 1540
Columbus 1492
N. Tropic
Drake 1579
Captain Cook killed 14th Feb. 1779
Hawaii
Drake
Cardenas 1524
Vespucci 1499
Equator
Orellana 1541
Cook 1772
Tasman 1642
Cook 1774
Pinzon
Almagro 1535
Magellan 1521
S. Tropic
Drake 1577
Cook 1776
Cook 1769
Dampier 1699
Cook 1774
Biscoe 1831
Bellingshausen
Biscoe 1831
Antarctic Circle

OCEANOGRAPHY
Perspectives on
a Fluid Earth

OCEANOGRAPHY

*Perspectives on
a Fluid Earth*

STEVE NESHYBA
College of Oceanography
Oregon State University

JOHN WILEY & SONS
New York / Chichester / Brisbane / Toronto / Singapore

Cover Photo: © M. Zide/Woodfin Camp & Associates

Library of Congress Cataloging-in-Publication Data:

Neshyba, Steve.
    Oceanography: perspectives on a fluid earth.

    Bibliography: p.
    1. Oceanography.   2. Marine biology.   3. Submarine geology.   I. Title.

GC16.N47   1987   551.46   86-24597
ISBN 0-471-81761-9

Printed in the United States of America

10 9 8 7 6 5 4 3 2 1

Dedicated . . . . . .

first, to the curious students who seek to fashion their own perspectives of the sea, for whatever reasons . . .

second, to the sea itself, for all its wonders that have enchanted me . . .

and last, to the ephemeral kinship to the mother sea that we and its living creatures share.

PREFACE

I believe that a first study in oceanography offers a broad-base education all its own, for some outstanding reasons:

. . .Its scope of topics is so very wide that there is constant need to meld the views of biologists with physicists, chemists with geologists, and more. It is a multidisciplinary study without peer, the modern world's closest analogy to the "natural philosophy" of Darwin, Newton, and other early scientists.

. . .We deal constantly with the fluid properties of water, a common substance but one very uncommon in its behavior.

. . .The ocean is a living space with three dimensions. We gain much insight into life in the cold, dark ocean depths through its enormous difference from living on a well-lighted, planar surface.

This text introduces a world made of fluids. I designed it this way after reaching two conclusions about students enrolled in my Introductory Oceans class. First, most students have never been shown that fluids have *structure*, or behave according to physical–chemical laws, or "do" anything except run downhill, a precept that is conditionally wrong anyway. Second, paradoxically, they nurture ideas that there must be a definable order in the marine ecosystem—how else to explain the observed, orderly patterns in the life cycles of marine life forms? In this sense, we all seek to learn what our marine kin already know.

The text is aimed at the upper-division course level, for students in nonscience majors who, as a group, have diverse scholastic backgrounds. So, in place of the usual introduction, I construct an overview of oceanography in the first three chapters, Part I. With this the students quickly gain a synthesis of oceanography, from the reasons why we study oceans to descriptions of how we carry out studies, and thereby moves the entire class toward a common resource base for the rest of the course.

The text has seven parts. Parts I–V make up a basic, one-quarter course in introductory oceanography, organized according to the traditional subjects. Part II is the geological base: basin shapes, tectonics of plates and ridges, the continental margins and sediments. Part III looks at the physical properties, the characteristics, and the behavior of seawater: air–sea interaction, climate moderation, sea salts and their impact on ocean and marine life behavior. Part IV examines the dynamic world ocean, its currents and circulation. Part V treats marine biology in an ecosystem format: system behavior, alternate ways to organize marine biology, biogeography, and the fascinations of light in the sea. A special feature are two chapters that take students on simulated cruises. Chapter 8 uses winter and summer cruise data to show seasonal changes in open seas and coastal upwelling; Chapter 15 uses CALCOFI cruise data to describe the dominant plants and animals in the California Current food chains.

Placing *coastal oceanography* in a separate Part VI is a move for flexibility in course design. A distinct one-quarter course on coastal oceanography is had by substituting Part VI for Parts IV and V and adding Chapter 19, The Sea and Man. Instead of deep-ocean water mass behavior, one studies the unique processes that drive coastal currents, eddies, and upwelling. The movement of coastal sediment in response to tide currents and waves replaces basin sedimentation. Chapter 16 takes a special look at marine plants and animals in coastal habitats, the rocky shore, the sandy beach, and coral reefs. Chapter 18 describes the coastal zone. Chapter 19, The Sea and Man, useful in any introductory course, is especially applicable in the coast ocean context because it includes an extensive discussion of the Law of the Sea, fisheries, and pollution. These topics focus directly on coast ocean environments and make their study even more of an interdisciplinary course than the blue-water oceanography of the basic Parts I–V.

The text is equally flexible for semester course design. PARTS I–VI make a solid semester of introductory oceanography. Alternatively, adding the special topics of Part VII to the basic Parts I–V yields a comprehensive course in general oceanography suitable as an elective course for fishery, geography, and business majors. Chapter 19, in addition to its coast ocean aspects, treats the mineral resources of the deep-ocean floor. Chapter 20 focuses on the unique role that polar oceans play in global climate and economics. Because so many introductory courses are taught by instructors who have specialized in geology, Chapter 21 is included to give the history of global geological theories during the past half century.

Corvallis, Oregon   Steve Neshyba
September 1986

ACKNOWLEDGMENTS

It is a pleasure to acknowledge those persons who, in one way or another, have helped me in this work. I need first to acknowledge countless students whose questions have challenged me to acquire new insights and to practice how to teach them to others. Herbert F. Frolander, my colleague in introductory oceanography classes at Oregon State, was a constant source of encouragement. Charles B. Miller and Lawrence M. Small, biological oceanographers at Oregon State, patiently guided me in many aspects of their discipline and reviewed many drafts. Louis I. Gordon and Arthur Chen, both chemical oceanographers, were important resources for the interdisciplinary aspects of the chemistry in the text. I have drawn heavily from my polar research studies carried out with Victor T. Neal, who supplied a number of photographs as well. A general thanks goes to all my colleagues at Oregon State for support during this work.

Thanks go also to the late Sir Edward Bullard, whose seminar on the history of the development of plate tectonic theory led me to organize the chapter on this subject as a chronological sequence of newly discovered evidence. Ellen Drake introduced me to the history of Alfred Wegener's contributions to this global geology.

I thank the State of California for its support of the CALCOFI project, the reports of which were invaluable in organizing the chapter describing plants and animals of an eastern boundary community. John McGowan, Joe and Frieda Reid, and the late John Isaacs, all of Scripps Institution of Oceanography, were important resource persons.

Then there are those many reviewers of the manuscript, unknown to me personally, whose candor led me to see my work through the eyes of others, and whose suggestions helped to shape the text format. Special thanks go to T. M. Hammond, whose suggestion led to the inclusion of the coral reef section, the research for which was a delightful experience, and to Mark Hixon for its review.

Johanna Neshyba produced the graphical art for figures and graphs, assisted by Linda Henselman. Drawings of biological specimens were done by Robert Rose. And Irene, my wife, together with an IBM portable computer, produced the original manuscript submitted to the publisher. To all of them, my sincere appreciation.

S. N.

CONTENTS

PART VII   SPECIAL TOPICS IN MARINE
STUDIES   *365*
*By which we acknowledge that the reader is now able and
entitled to examine some of the most complex of issues and
topics in marine sciences.*

PART I

(Jeanette Thomas/Visuals Unlimited.)

PERSPECTIVES IN MARINE SCIENCE

In which we draw a broad picture of the fluid marine environment, its behavior, and the nature of problems that oceanographers study

---

CHAPTER 1   OUTLINE OF PERSPECTIVES FOR OCEAN STUDY

We explore the many reasons why we must study the vast bodies of fluid water that fill the basins in the earth's crust.

CHAPTER 2   THE NATURE OF THE FLUID OCEAN

We gain new insight into how the fluid world behaves by contrasting its characteristics with those of our more familiar terrestrial world.

CHAPTER 3   MARINE SPECIALISTS AND INTERDISCIPLINE STUDIES

The intricate details of fluid environments are studied best through specialization, but solutions to real problems often require a mix of many experts, an interdiscipline approach.

CHAPTER 1

OUTLINE OF PERSPECTIVES FOR OCEAN STUDY

We begin by examining the question "Why are you here?" Why do you, the reader, choose to invest a part of your science education in the basics of the complex field we call oceanography? These questions are philosophical in nature and require judgmental answers. A better opening would use questions such as "What is oceanography?" "What is achieved by investing time to learn how the ocean systems operate?" Such questions are pragmatic, serving to focus the reader on reality. One reality is that we know more about outer space than inner space; we have mapped the surface of the moon with much more detail than the floors of our oceans. And though we compose songs about the stars and moonlight, we sustain our physical and mental strength munching fish sandwiches and drinking fluids laced with evaporated seawater. This reality is more difficult to understand than science itself.

The real reasons for studying oceanography have to be based in *natural* philosophy. Nothing is more natural than our inherent curiosity about the how and why and where of the universe about us. This chapter sets out to define some whys and wherefores.

Perspectives

It is worthwhile at the outset to outline major reasons why oceans should be studied. These provide focal points around which are organized the detailed analyses of the nature of the ocean system, chapter by chapter. Clearly, the list is not exhaustive because the body of knowledge about the oceans is constantly evolving. Today such a list would include the following.

• The oceans exist—our curiosity overwhelms all other reasons to study them.

• The oceans provide us food—which acknowledges the history of our dependency on and exploitation of the oceans for their food potential.

• The fluid oceans and the atmosphere both support modes of transportation superior to land modes in many ways, but their effective use requires extensive study of currents and winds.

• The oceans are an important source of minerals, from salt to more exotic elements such as magnesium, and from phosphate fertilizer to the prosaic sands.

• Ocean water, in all its phases—liquid, solid, and vapor—is the principal medium for the distribution of heat energy across the planet. In this way the study of weather and climate is intimately linked with oceans.

• Ocean water, because of its ability to dissociate complex molecular structures, contains virtually all known elements. Yet it chemically stabilizes itself so that it never becomes either too acid or too alkaline. This self-stabilizing behavior has a profound impact on the life-sustaining quality of seawater. Indeed, it is widely believed that oceans fostered the evolution of living molecules on earth.

• Ocean water, because of its absorptive properties, removes and exchanges gases with the atmosphere; thereby it enters indirectly into the control of radiant-energy transfer between earth and outer space.

• Oceans, because they cover more than 70% of the earth's surface, evaporate more water than is precipitated upon them, and thus drive the hydrologic cycle on which terrestrial life is absolutely dependent. The ocean, whether tropical or polar, is heated or cooled from the "top" downward; its heat budget is almost entirely controlled by processes acting at its surface alone. In contrast, because evaporated seawater enters the atmosphere at the base of its air column, the circulation of the amosphere is driven from its "bottom" upward.

• The oceans at any instant contain the major fraction of all kinetic energy that earth derives from the sun. Stated another way, the amount of solar-derived energy stored in a unit area water column far exceeds the amount of such energy contained in equal area columns of land or atmosphere. The oceans thus have a key role in our efforts to develop alternative energy sources to fossil fuels.

• Oceans and lands are not symmetrically distributed over the earth's surface. This fact, a result of the complex geological history of the earth, is central to the dynamic motions of both ocean and atmosphere; it has also governed the development of human society.

• The oceans, in terms of ability to sustain living communities, offer almost 80 times as much living space as the terrestrial world. However, because the fluid medium in ocean basins is so well mixed in time and space, the number of different kinds of life forms in ocean space is much less than the number on land space.

• Seawater, through its property of high specific heat, maintains a relatively constant temperature in spite of its exposure to a wide range of conditions, from tropical zones of excess solar heating by radiation to polar zones of excess cooling, also by radiation. This constancy has profound control over the life styles of oceanic organisms, making them very different from those of terrestrial species.

• The ocean fluid, being a thousand times denser than the fluid air in which most terrestrial life exists, sustains life forms in sizes that *on the average* are substantially smaller than those found on land. The often-stated adage "in the scheme of life it pays to be small" is particularly appropriate to life in the sea. Yet oceans also sustain the largest of all animals that has ever lived on earth, the blue whale.

• The margins of ocean basins where land and sea meet are among the most organically productive of all earth areas. They are productive because these are zones of convergence of energy and mass: the oceans deliver to their shores the energy of waves gathered from the vast reaches of wind-exposed surface water, whereas rivers deliver the raw chemical materials that nurture life.

• Humans have joined in the convergence to the margin zones of ocean basins, not only by settling there in large numbers, but also by transporting into the occupied coastal zones the major fraction of all organic material produced by agriculture, mining, and manufacture in the interior of continents.

• Polar oceans, a last frontier, may be the most important of all to our survival. The control of the earth's climate hinges uniquely on the energy of transitions between liquid and solid phases of water, and on the albedo (solar reflectance) of ice-covered oceans.

Within this rationale for ocean studies is a myriad of complex processes: physical, biological, chemical, geological, meteorological, and more. Humans are interwoven into these matrices. It is our task to unravel the separate elements of such matrices, to describe each in qualitative and quantitative terms, and then to recombine them to gain an understanding of ocean behavior. This is what "marine science is all about."

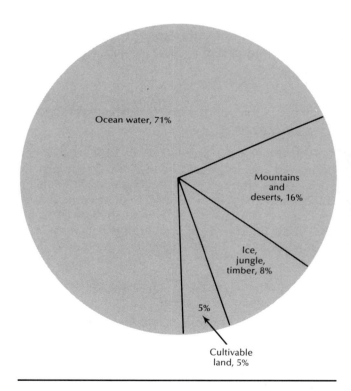

Figure 1.1. A schematic pie that shows how the earth's surface divides into distinct geographical domains. As of late 1979, worlds population was about 4.34 billion persons. The total area of the earth is 510 million km², the area of land (29.22%) is 149 million km², and the area of oceans and marginal seas (70.78%) is 361 million km². Translated into individual allotments, each person's share of cultivable land was 1.44 acres (about 0.58 hectare) and of ocean 20.4 acres (8.3 hectares).

In a larger sense, however, much of our "perspective" of oceans is controlled by our need to know more of how oceans impact on us and vice versa. The following sections are designed to amplify such perspectives.

Oceans and Food Production

When I began ocean studies in 1962, crude calculations showed that for each person then on earth there were 2.7 acres of arable land. An updated analysis in the late 1970s showed that the allocation had diminished to 1.44 acres per person. The division of earth into geographical domains is diagramed in Figure 1.1. Although 1.44 acres of arable land is more than adequate to sustain the health of one person, whether it remains arable over long periods depends on climate variation; we shall see that the oceans play a strong role in climate changes.

The space pie of Figure 1.1. also assigns to each of us about 20 acres of ocean. To evaluate the relative ability of our land and ocean allotments to provide food, let us examine the estimated plant productivity of lands and seas in Table 1.1. The entire world's *net*

primary production for the total land, (109×10^9 dry tons/yr), compared with that for the total ocean, (55×10^9), has a land-over-ocean ratio of about 2 to 1, in spite of the fact that ocean area exceeds land area by a ratio again of 2 to 1. Stated another way, in terms of plant production, acre for acre, lands exceed the oceans by about 4 to 1. The initial conclusion is that our 20-acre ocean allotment is equivalent to plant production on 5 acres of land, bringing our "resource base" to roughly 6.4 acres of equivalent land. But there is more to this exercise that must be taken into account.

Our ability to harvest the actual production of plants from both land and ocean allotments is critical to this comparison. Here other numbers from Table 1.1 are used. Note that in terms of *biomass*, defined as the mean quantity of dry plant mass per unit area at any instant, lands carry 12.5 kg of dry plant material per square meter. In contrast, oceans carry only 0.009 kg/m². These values have the ratio 1389 to 1, lands to oceans. What this means is that, at any instant in time, the average land unit is carrying over a thousand times more plant vegetation than the average sea unit.

What is the cause of this great difference between

Table 1.1 A comparison of plant production for various different land and oceanic areas. (After Whittaker and Likens, 1975.)

	Area 10^6 km²	Net Primary Productivity, Per Unit Area, dry g/(m²)(yr), normal range	World Net Primary Production, 10^9 dry tons/yr	Biomass per Unit Area, dry kg/m², mean
Lake and stream	2	100–1500	1.0	0.02
Swamp and marsh	2	800–4000	4.0	12
Tropical forest	20	1000–5000	40.0	45
Temperate forest	18	600–3000	23.4	30
Boreal forest	12	400–2000	9.6	20
Woodland and shrubland	7	200–1200	4.2	6
Savanna	15	200–2000	10.5	4
Temperate grassland	9	150–1500	4.5	1.5
Tundra and alpine	8	10–400	1.1	0.6
Desert scrub	18	10–250	1.3	0.7
Extreme desert, rock, and ice	24	0–10	0.07	0.02
Agricultural land	14	100–4000	9.1	1
Total land	149		109	12.5
Open ocean	332	2–400	41.5	0.003
Continental shelf	27	200–600	9.5	0.01
Attached algae and estuaries	2	500–4000	4.0	1
Total ocean	361		55	0.009
Total for earth	510		164	3.6

the land–ocean production ratio, (4:1), and the biomass ratio, (1389:1)? Seemingly, these numerical ratios should be the same. The answers to this question are given in Chapter 3. Here we state only that because marine plants are microscopic in size, reproduce very quickly, and are each "completely" consumed when grazed by an animal, the amount of plant material in the water at any instant is quite small. Stated another way, biomass is small, so that in economic terms direct harvesting of microscopic marine plants is prohibitive.

If the ocean is to feed us, it must do so from its production at the first animal level, that is, the grazing herbivore. But if marine plants are microscopic in size, it follows that the herbivores must also be quite small; the principal marine grazers, called copepods, average something like 2 mm in length. Even though these marine animals carry out the grazing function equivalent to that of terrestrial cattle, their small size prevents our harvesting them economically. In fact, we begin to harvest ocean production only at a still higher level in the process called the "food chain," namely at the level of fish such as herring, anchovy, or menhaden. All these fishes are of the *Clupeoidea* suborder; they aggregate in large schools, (which makes them economical to harvest), and feed by filtering seawater, (which allows them to feed directly on tiny copepods and even the very small plants).

What is the potential production of fish in the oceans? Ryther (1969) estimated both primary productivity and fish production for three categories of ocean waters and made the startling projections of Table 1.2. Overall, Ryther claims that the total ocean fish production is 241.6 million wet tons/yr, of which 99% occurs in the coastal ocean and its included upwelling zones. We define coastal waters as those lying over the shelflike, submerged margins of continents, in all about 10% of total ocean area. Upwelling zones are specific parts of coastal oceans, usually adjacent to the shorelines where deeper waters are forced upward into the sunlit surface layer. These make up only about 1% of the coastal waters, or only 0.1% of total ocean area, but yield as much as one-half of total ocean fish production. In contrast, the 90% of total ocean area classed as open ocean produces only about 0.7% of all fish. Why is the open ocean so barren in fish production? Does this mean that primary productivity is also low in open oceans, supporting only a weak overall food chain? Answers to these questions are yours to glean from the chapters to follow.

The numbers in Table 1.2 are for *total* fish production. This is not the *maximum sustainable yield* of fish from the sea. Currently, the world harvests about 70 million wet tons of fish per year. In comparison, Ryther estimates that only about *100* of the 240 million tons of total fish production can be harvested for human use on a sustained basis. The remainder must be left for reproduction as well as to predation by other carnivores such as guano birds, seals, penguins, salmon, or whales.

Fishing with nets and small boats. Artesenal fishing is a stable occupation and food source the world over. (Steve W. Ross/Visuals Unlimited.)

Table 1.2 Estimates of primary productivity and fish production by major ocean provinces. (After Ryther, 1969.)

Provinces	Area, %	Primary Productivity		Fish Production, wet tons/yr
		carbon/(m²)(yr)	tons of carbon/yr	
Open ocean	90.0	50	16.3×10^9	1.6×10^6
Coastal ocean*	9.9	100	3.6×10^9	120×10^6
Upwelling zones	0.1	300	0.1×10^9	120×10^6
			Total fish production	241.6 million tons/yr

*Excluding upwelling zones.

We need to carry this discussion one step further. In 1970 the Food and Agriculture Organization of the United Nations estimated that the annual worldwide deficit in protein food for humans was about 10 million tons. If we reckon that only 10% by weight of fresh fish is protein, a maximum sustainable yearly catch of 100 million tons yields only 10 million tons of protein, that is, just enough to satisfy the world protein deficits for 1970. Clearly, food from the oceans is not a panacea to humankind's hunger. The present imbalance is in fact even aggravated because some 40% of the world fish catch is processed into fishmeal, which is used to fatten chickens, hogs, and more exotic or marketable products such as salmon (northern United States), catfish (southern United States), and carp (Orient).

The numbers just analyzed are for fish produced ''naturally'' and harvested by ''hunting'' them. What is the existing and potential role of mariculture, that is, the culture of fish and shellfish in seawater? It is often said that mariculture is a means to major new food sources that could offset protein deficits. Reality, however, is something else. Almost without exception, existing mariculture grows products aimed at middle- and upper-class clientele, with markets among restaurants and fancy-food producers. Salmon and trout grown in raceways to one-pound size are aimed exclusively at the restaurant trade, as is the culture of prawns in tropical coast areas. Recent experiments in St. Croix, Virgin Islands have pumped nutrient-rich water from deeper depths into surface ponds where algae are grown and then fed to oysters or other filter-feeding shellfish. These operations are designed as an industry for developing nations that lack other resources to sell. Marketing these high-priced fish and shellfish will increase their export income. But mariculture schemes offer little hope for offsetting the protein deficiency of a nation's own hungry people.

Even so, there is great potential for mariculture. I have visited deep into the almost uninhabited Archipielago de Los Chonos of southern Chile to witness a pilot project for growing mussels *(Mytilus edulis)* to market size in just nine months! But the translation of this potential into actual production is a formidable social and engineering task. A few years ago a major United States cereal company sponsored a feasibility study on the construction and operation of *kelp* farms offshore. Conceived as a free-floating, 100,000-acre unit, each farm would be self-sustaining in energy and could produce a wide variety of foods and chemicals. Many other studies of kelp farming have been initiated, but as yet no farm has actually been established.

In coming decades we will increasingly be called on to *evaluate* all types of food production from the sea, both in the ''natural'' sense of fishing and in that of ''farming'' the sea. The scope of oceanography in this text is designed to provide needed perspectives by which evaluations are logically made.

Oceans and Weather

The oceans play a key role in shaping the global patterns of winds and weather, which constitute climate. This is a topic of great current interest, and from a number of valid views. One reason to advance our knowledge of *how* the oceans influence weather and climate is related to the subject of the previous section, namely, the ability to forecast food production through long-range forecasts of growing conditions.

Our knowledge of how ocean and atmosphere interact has grown greatly during the past decade. Marine meteorologists are now reasonably accurate in predicting weather patterns for periods from a few days up to a few months, and marine geophysicists have now constructed estimates of long-ago climates and the behavior of ancient oceans, based on analyses of deep-core samples extracted from marine sediments. On the time scale of 2 to 3 years, however, weather prediction is still very inadequate.

An example of the coupling between ocean and atmosphere on the short time scale is contained in Figure 1.2. The contours of the chart mark the *difference* between the actual temperature of the sea surface during September 1978 and its 20-year historical mean temperature; these data are called temperature "anomalies." The surface waters in the Gulf of Alaska were anomalously cold. A description of *how* and *why* cold ocean surface water affects changes in the atmosphere is given in Part III of the text. Here we give only the effect: a high-pressure ridge devel-

oped in the atmosphere over the colder water, causing a diversion of winds from a "normal" westerly path to a northerly flow that carried moist marine air far northward over the interior of Alaska. Fairbanks, Alaska then enjoyed relatively mild air temperatures and greater-than-normal rainfall as the moist air masses cooled. On the eastern flank of the high-pressure ridge, the air masses moved southward, deep into the interior of the western United States. Phoenix, Arizona received these air masses, by now quite dry and cold, and experienced *lower* daily temperatures than did Fairbanks, even though it lies 30° of latitude farther south.

What causes such anomalous sea surface temperatures and atmospheric circulation? How long do these conditions persist? Are the anomalies trans-

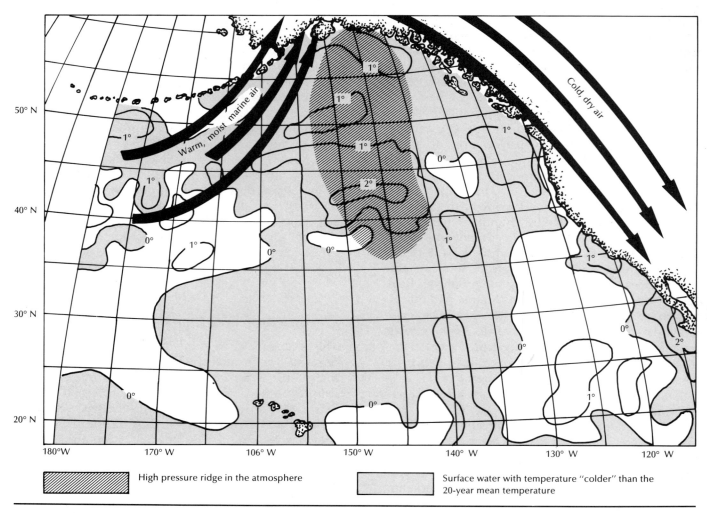

Figure 1.2 Sea Surface Temperature Anomalies in the North Pacific. The sea surface temperature for September 1 through 20, 1978, deviated from the 20-year mean for the years 1948 through 1967. (From *Fishing Information*, 1978.)

A view of cloud patterns over the Northern Hemisphere. Photographs like these are routinely taken by weather satellites that occupy geostationary orbits, meaning that the satellite location is fixed above a spot on earth. GOES 2 satellite is positioned permanently 22,500 miles above the earth at a point on the equator and 75°W longitude. This photograph is selected to show Hurricane David well formed off the tip of Florida; Hurricane Frederic is growing in intensity to the southeast of David.

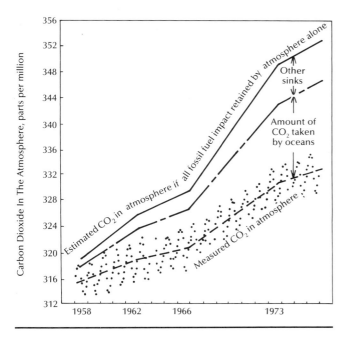

Figure 1.3 From 1958 to 1976 the measured value of carbon dioxide in the atmosphere at the Mauna Loa Observatory in Hawaii showed a secular increase. The data also indicate a definite annual cycle. Some years show a sudden change in the *rate* of CO_2 increase (as in 1973, presumably because of the oil embargo). (From various sources.)

ported by the mean ocean circulation to alter weather patterns elsewhere? There are questions designed to shape the perspective with which the reader will enter later sections of the text.

Oceans and World Climate—Long Term

With the exponential rise in the rate of burning fossil fuel since the beginning of the industrial revolution has come a steady rise in the level of carbon dioxide (CO_2) gas in the atmosphere. The trend over recent years is reflected (Figure 1.3) in the rise in concentration of CO_2 measured in the air at the Mauna Loa Observatory, Hawaii from 1958 to the present. It is rising at the rate of one part per million per year.

This measured concentration of CO_2 (dashed line in Figure 1.3) accounts for only about 52% of the total that has been released by burning fossil fuel (solid line). The oceans take up 35% of the remaining 48% (broken line). Other sinks, including increases in the growth rates of vegetation and forests—because more CO_2 is available—account for the remaining 13% (Woodwell, 1978; Broecker et al., 1979).

The principal long-term reservoir for atmospheric CO_2 is as dissolved gas in ocean water. Its permanent removal from the ocean is through the carbonate sediments on the ocean floor. At present the atmosphere holds about 700 billion metric tons of carbon in gaseous CO_2 form; in contrast, the oceans are believed to be a reservoir of some 40,000 billion metric tons as dissolved CO_2 gas or as carbonates.

The *rate* of exchange of CO_2 across the sea surface is quite slow, however. The deep waters must be exposed to the air interface to absorb new CO_2 gas, and the time required for turnover of the deep ocean layers is quite long. We estimate that 700 years are needed to recycle the deep Atlantic waters to the surface; for the Pacific basin the estimated time is much longer, about 2500 years.

What are some possible consequences of this trend? The most commonly discussed aspect of a rise in the level of CO_2 is the increased "greenhouse effect." In this process, the CO_2 gas acts to trap the long-wave radiation that carries energy away from the earth's surface; the result is a rise in the mean surface temperature of earth. Some believe that even small increases in the average earth temperature will cause substantially faster melt rates of polar ice caps. This could force an increase in sea level and possible flooding of major transportation and cultural centers in low-lying coastal zones. Others point to the recently noted rise in the concentration of airborne aerosols (suspended particles) as a potential cause of a reverse effect. An increased "dustiness" of the atmosphere will allow less incoming solar energy to penetrate to the earth's surface, which will tend to cool the earth.

Will the opposition of those two effects, both caused by humans, result in the net temperature of the earth remaining nearly the same? An answer is possible only with a thorough understanding of the dynamic interactions between oceans and atmosphere. Whether or not the oceans can relieve us of our excess CO_2, in time to avoid the trauma of major climatic change, is a problem for the future.

Exploitation of Oceanic Mineral Resources

Salt is absolutely necessary for our physiological well-being. Moreover, our dependence on salt from seawater is at least as old as known history. The use of sea salt continues to this day. Recently, Israel began constructing a combined canal-tunnel to divert Mediterranean Sea water into the Dead Sea, the Dead Sea lying some 240 meters *below* the level of the Mediterranean. A prime objective of the project is to raise the level of water in the Dead Sea to com-

pensate for the diversion of Jordan River water into irrigation projects. The history of salt production from evaporation pans located at the southern end of the Dead Sea is as old as the biblical literature of the region and is intimately entwined with the changes in its ancient cultures. Loss of inlet water to irrigation has caused major problems in these salt works as the level of the Dead Sea fell.

What are the physical properties of water that impact so greatly on our history and behavior? Water is a highly active solvent, for its dissociative properties allow it to contain heavy concentrations of most elements. During World War II we constructed industrial plants to extract magnesium from seawater; the lightweight metal was essential to the war effort. Recent discoveries of extensive pools of hot brine on the floor of the Red Sea pave the way for industries to extract minerals of several types. As discussed later, such brine pools are the result of seawater percolating through fissures in new magmatic rock that is exposed by the geological process of sea floor spreading.

After salt, the most widespread exploitation of mineral resources from the oceans is that of offshore petroleum deposits. A rather recent development, oil production from the continental shelf, has created a host of problems related to pollution of prime fishing areas. The chemical and physical processes involved in breaking down or removing concentrated oil spills are very complex, and the biological aspects are even more so. Treating oil spills with chemicals often results in a "cure" more biologically damaging than the oil itself. Because most oil spills occur over the continental shelf, a knowledge of coastal currents is essential in deciding how to treat each spill. Truly, these problems take on a complex interdisciplinary nature; hence, the perspective of interdiscipline studies in marine science is essential.

Recently, an explosive expansion in worldwide agriculture has placed heavy pressure on the fertilizer industry. The shallow waters around some continents, where deposits of phosphorite have formed by precipitating out of solution, are a prime source of phosphates. As yet these deposits are largely untapped; the industry is still mining ancient shallow-sea deposits that are now uplifted and accessible on land. Moreover, the mining of underwater deposits near shore might cause pollution by releasing large quantities of suspended sediments over productive fishing grounds. What are the behavior patterns of animal and plant species that depend on nearshore waters at some point in their life cycles and that would be impacted by offshore mining?

Still more recently, we have discovered large

Guajiro Indians harvesting salt evaporated from shallow coastal ponds on the Guajira Peninsula of Columbia, South America. (Victor Englebert/Photo Researchers.)

quantities of mineral nodules lying distributed across vast areas of the deep ocean. These nodules, like the phosphorite, are also hydrogenous sediments, that is, sediments accumulated by the precipitation of minerals out of the water solution. The nodules contain manganese, iron, nickel, cobalt, and copper in varying degrees and ratios, but in very high concentrations compared to those of land-based ores. They are an important world resource.

Which nations shall claim ownership and jurisdiction over offshore mineral resources? How are potential pollution problems to be handled when one nation's activity impacts on another nation's resources? These and many other problems demand new perspectives, which can be gained only through study of the oceans.

The Marine Biosphere

How is ocean life organized? In what ways does the marine biosphere differ from our terrestrial sphere? In terms of volume of living space, the marine sphere is 80 times larger than the terrestrial sphere (Figure 1.4). Is the large volume of hydrospace an important factor in how marine communities are organized? What are the properties of a *fluid domain* that shape the nature of its biosphere?

Biologists estimate that marine *pelagic* plant species number about 10,000. *Pelagic* species occupy the interior ocean waters; *benthic* species live in close association with the ocean bottom. In contrast, the number of species of seed-bearing terrestrial plants alone (Spermatophyta) is about 200,000. What are the fundamental reasons that the ratio of marine pelagic to terrestrial plant species is so low, about 1:20? How must this information influence the perspective with which the reader enters into studies in marine science?

A comparison of animal species yields an even more striking contrast: the pelagic ocean sustains an estimated 10,000 to 15,000 species compared to some 1.5 million species of recent land animals. Here the ratio of species, ocean to land, is of the order of 1:100 to perhaps 1:150!

What are the special facets of the marine biosphere that explain why 98% of the livable volume of the earth sustains only some 1% of the total number of species found on earth? Stated the other way around, why should the terrestrial biosphere with only 2% total livable space so surpass the marine biosphere in terms of diversity of species?

Does the answer lie in the fundamental differences between fluid and solid life-supporting media? If so, we will gain much help in fashioning ocean perspectives by comparing the two environments. Does the answer lie in the fact that the ocean medium is three dimensional whereas the land world is predominantly a two-dimensioned living space? If so, our perspective should expand to include a study

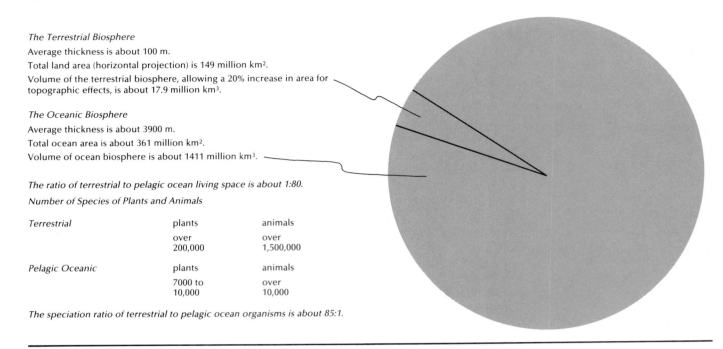

The Terrestrial Biosphere
Average thickness is about 100 m.
Total land area (horizontal projection) is 149 million km².
Volume of the terrestrial biosphere, allowing a 20% increase in area for topographic effects, is about 17.9 million km³.

The Oceanic Biosphere
Average thickness is about 3900 m.
Total ocean area is about 361 million km².
Volume of ocean biosphere is about 1411 million km³.

The ratio of terrestrial to pelagic ocean living space is about 1:80.

Number of Species of Plants and Animals

Terrestrial	plants	animals
	over 200,000	over 1,500,000
Pelagic Oceanic	plants	animals
	7000 to 10,000	over 10,000

The speciation ratio of terrestrial to pelagic ocean organisms is about 85:1.

Figure 1.4 A comparison of the terrestrial and oceanic biospheres in terms of the total volumes of living space in each and the approximate numbers of plant and animal species in each. (Compiled from various sources; pelagic ocean plant numbers courtesy Greta Fryxell, Texas A&M University.)

of the vertical structure of the living space in the ocean. Is the answer obtained through a comparison of successful strategies for the survival of species, ocean versus land? Or is the key perspective tied to the fact that the fluid medium for marine life is itself in constant motion? Or is the small number of known marine species just a reflection of the fact that much of the marine world is still unexplored?

What factors are common to both the marine and terrestrial biospheres? Are not the principles of ecology, the laws that govern living things, universal? If they are, the perspectives we have each already acquired by just living for a score or more years will help us gain perspective on the marine biosphere.

Summary

There the vast differences between the ocean and terrestrial worlds. If we consider all of earth to be fluidlike, the nature and behavior of the liquid ocean are still far different from those of the plastic earth and of the gaseous atmosphere. Marine science treats the fluid ocean, its shape, origin, properties, behavior, and the life that it sustains.

Marine life forms, their shapes, behavior, and life cycles, are also much different from terrestrial counterparts. In large measure, such differences stem from the constant motion of the ocean; life cycles and behavior must be closely tuned to the movement of the supporting medium itself. The limited diversity in oceanic species is also explained by ocean movement. And because oceanic life forms are on the average small, the first two levels in the pyramid that we call the food chain, the plants and the grazers are essentially free-drifting plankton. The body sizes of marine species are substantially smaller than those of species found on land because water is very dense, a thousand times denser than the fluid air in which we exist. A thorough familiarity with the basic physical properties of water is essential in studying marine science.

The tools of analyses of the marine scientist are much different from those used by terrestrial investigators, not only because the fluid ocean is in constant motion but also because there is so much of it. Overwhelming limitations to the progress of marine science are the labor and high cost of extracting information from a three-dimensional biosphere some 80 times larger than our own.

Nevertheless, we have progressed rapidly in describing the role of oceans on planet earth, and it is precisely because of past progress that enough new knowledge of oceans is now in hand that we can begin to form *perspectives* in marine science. Beyond such perspectives lies the ultimate goal—understanding.

Study Questions

1. Using the allotments of cultivable land, 1.44 acres, and ocean, 20.4 acres, that are due to each of us, what is the *maximum* amount of dry plant material that we could possibly get off each allotment per year? Use data from Table 1.1.

2. But suppose I ask you just to compare the amounts of dry plant material that I could expect to find on my allotments of land and ocean at any one instant in time (this is the definition of standing biomass). Could you do this using the data of Table 1.1?

3. Why is there so much difference between the numbers comparing land and ocean in question 1 and those comparing the same land and ocean in question 2?

4. Compare the *sizes* and *predation techniques* of the principal herbivores of land to those of the principal herbivores of the oceans.

5. Here we play a suppose game: Suppose that the *primary production* of the oceans were identical to that of lands. Could an acre of ocean feed the same *maximum* number of people as could an acre of land? Explain.

6. What is the potential production of fish from all the ocean that could be harvested on a sustainable basis, according to Ryther?

7. Argue against the following statement. If we harvest the oceans for their maximum sustainable fish production, we would be able to solve all our human starvation problems.

8. What is one basic, essential difference between a coastal ocean and an open ocean that could account for the vast difference in their relative fish productions?

9. In the short-term sense, how do the ocean and atmosphere communicate with one another in a sufficiently vigorous way that the condition of the sea surface can actually influence weather patterns of the atmosphere?

10. In the long-term sense, the continued buildup of higher levels of CO_2 gas in the atmosphere may lead to extensive changes in the climate of earth. Explain.

11. What will the role of the oceans be in the projected climate changes from higher atmospheric levels of CO_2?

12. This is a true statement. Much of the new carbon that today enters the oceans via the air–sea interface is moved rapidly into the sediments collecting on the ocean floor. Explain one mechanism responsible for this rapid removal.

13. What are some important chemicals or minerals that might be extracted from the oceans, either from the water directly or from the sediments on the ocean floor?

14. Almost everything we learn about the ocean would lead us to speculate that the *number of species of living things* in the oceans should be much larger than the corresponding number for land. There is twice as much ocean area, there is 80 times as much "living-space volume," and, what's more, the plants and herbivores at the base of the food chains are very tiny! So, based on what you have learned of the oceans to this point, why are there far fewer species in the fluid world?

CHAPTER 2

THE NATURE OF THE FLUID OCEAN

In our language we often communicate ideas and perspectives through the use of metaphors. For example, "dry as a bone" takes the place of a more definite statement that "moisture content is less than, say, 8%." Or, snow "blankets" the earth, a statement that has many scientific connotations. A layer of snow shields the earth from view, which tells about the optical transmission of snow; it insulates earth, which communicates the idea that snow is a relatively poor conductor of heat; or fresh-fallen snow has a physical texture somewhat like a fluffy wool blanket. The richness of language gives us an ability to choose metaphors that distinguish between a "host" of "shades" of meaning, from the grandoise to the minuscule, from major differences in scales to subtle, minute variations.

As far as we can tell, snow also falls upon the oceans. Yet we would not choose the blanket metaphor in this case. But which metaphor to use? In making that selection, we come face to face with the "differences" between fluid ocean space and our own thoroughly familiar and comfortable terrestrial space at the bottom of the fluid atmosphere. The purpose of this chapter is to describe these differences and, in so doing, increase our ability to understand the marine science described later.

Contrasts in Livable Space— The Third Dimension

What are the characteristics that we perceive about our terrestrial environment? How thick is it, that is, what is its vertical extent? Does its thickness change if we alter our elevation relative to sea level? Which characteristics of the environment do change as we shift to a different elevation, and which stay the same?

In the vertical direction, our habitat or livable space is about 2 m thick, which is our typical body height. If we choose as our measure the height of the tallest living thing, the California redwood tree, the thickness increases to about 100 m. In either case, our habitat is extremely thin compared to the marine world whose thickness varies from that of shallow tide flats to the more than 11,000-m depth of the Marianas Trench, the deepest known point of the ocean basins (see Figure 4.7 for its location).

We are keenly aware of the topography of land space. Art and literature abound with illustrations that are keyed to topography, and our language is rich in topographic terms. Peculiarly, however, we generally consider the ocean to be a flat surface. In fact, the "levelness" of the ocean surface makes it possible to observe the "sphericity" of earth in a way not possible on land. We seldom see a painting of the "sea" that does not include some land form, or ship, or birds, or perhaps cloud forms, each of which serves the purpose of focusing the viewer's eyes away from the otherwise featureless horizon. Literature is filled with descriptions of the conditions of the sea surface—of waves and storms, spray and whitecaps, or the changing colors of the sea—but it is almost completely bereft of metaphors for the interior of the ocean.

Contrasts in Classification— A "Geography" of Oceans

The ocean interior is no less rich in differences of quality, texture, and topography to marine life than *terra firma* is to us. In terrestrial geography we classify land forms in many ways to distinguish soil types, climate types, roughness, vegetation, and the like. Oceanographers do the same for oceans. We classify the ocean world into provinces, but now the third dimension is needed. We select boundaries for different life-support zones based on scientific analyses of their differences.

In Figure 2.1*a* the division into life zones is made on the basis of depth to the bottom and on the shape and location of features of the ocean basin. In this classification, used mostly by biologists, *neritic* waters are those over the shallow margins of continents, the zones we call the continental shelves. *Pelagic* waters are the thin surface layers of the open ocean, and *bathypelagic* describes the deep, (and dark), interior waters; some texts subdivide the interior waters into mesopelagic, bathypelagic, and even hadalpelagic. Biologists also differentiate zones for the ocean floor itself, the *benthic* zones. For example, the coastal zone between high- and low-tide levels is called the *littoral,* and the shallow floor of the rest of the continental shelf is called the *sublittoral. Bathyal, abyssal,* and *hadal* refer to benthic zones at still greater depths.

Figure 2.1*b* presents an alternate division into life-support provinces, one that is based on the characteristics of the fluid domain itself.

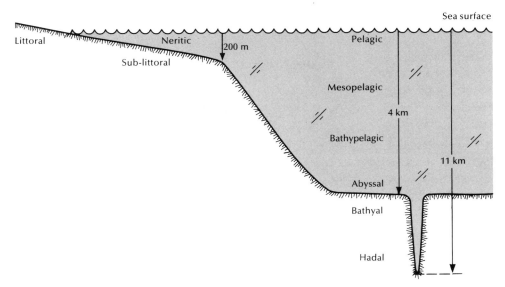

(a) Division of the ocean into principal life zones based on the depth of water and associated characteristics of the ocean floor. Littoral zones are those associated with shorelines, sublittoral those associate with shallow ocean floor. Neritic refers to the shallow waters above continental shelves; pelagic means open-ocean, upper-layer waters; bathyal, like littoral, refers to the substrate of the ocean floor, hence to a province suited for benthic life forms. Bathypelagic refers to life forms in the interior ocean.

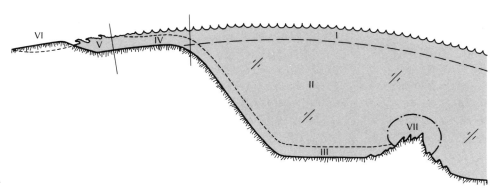

(b) Division of the ocean into life provinces based on the nature of the fluid domain.
  I. Oceanic photic zone.
  II. Oceanic interior sector.
  III. Oceanic abyssal zone.
  IV. Coastal margin zone.
  V. Coastal energy dissipation zone.
  VI. Estuarine zone.
  VII. Submarine hydrothermal vent zone.

Figure 2.1 Dividing ocean space into life-support zones.

Province I. The vertical dimension of this open-ocean photic zone is limited to the depth to which more than 1% of available sunlight penetrates, about 100 m. All photosynthesis is limited to this thin layer, and virtually all organic material in the oceans originates here. It is also the zone of air–sea exchange.

Province II. The vast interior waters have neither light nor substrate: there is neither photosynthesis nor any place for organic material to collect. The food base for the community of organisms living here is the myriad of small particles of organic debris, called *detritus,* that settle slowly out of the productive Province I enroute to a final settling place on the bottom. We probably know less about this zone than any other, partly because it is the largest by far of all the provinces, and partly because we can "see" so little of it during any one expedition. The living community has strange forms, from tiny animals that spin large bubbles of mucus with which to "catch" falling food particles to small carnivores that flash their bioluminescent "lures" to attract their prey. (Readers are encouraged to read William Beebe's account of his observations in a bathyscaphe, *The Arcturus Adventure,* 1926.)

Province III. Much of the organic detritus that settles out of Province I is consumed or dissolved in Province II, but some collects as an organic film on the ocean floor. Here it supports a pyramid-like food chain, one whose community of living forms is completely different and distinct from that in the upper provinces. This community is dominated by benthic forms that crawl over the floor or burrow in it or swim over it.

Province IV. The shallow waters over continental shelves differ from open-ocean waters. They receive large supplies of dissolved nutrients from coastal rivers. The sediment load is generally high, and the clarity of coastal waters is less than that of open-ocean water. The presence of sunlight at all depths, including the bottom, strongly influences the types and behavior of marine organisms. It is a zone of confluence of Provinces I and III.

Province V. Waters of the "edge zone" boundary between land and sea differ from other coastal shelf waters in that large amounts of wave energy are dissipated here. Water level adjacent to the shore is constantly changing with tides and storms. It is a highly specialized living space. Humans have a significant impact on this community.

Province VI. The estuarine zone is distinguished primarily by rapid changes of salt content of water, a feature that gives rise to unique adaptations by marine life. Tide wave energy dissipation is a strong factor. Humans have their greatest impact on this province.

Province VII. The interaction of bottom water and hot magma creates submarine geysers. A specialized living community has evolved around these hydrothermal vents; its organic base is the local production of bacteria that derive energy from chemical reactions involving sulfur compounds in the vent waters.

Contrasts in Structural Adaptation— A Cost-of-Living Approach

Simply stated, the terrestrial organism pays a much higher "price" for the privilege of living in the gaseous atmosphere fluid than that paid by the marine counterpart living in liquid fluid. The buoyancy offered to the marine organism by the dense fluid ocean, which reduces the "effective" force of gravity, and the relative constancy of the marine environment lower its price for living.

Buoyancy Effects

Both plants and animals in the terrestrial sphere invest a large fraction of their net caloric intake in the building of cellulose or bone body parts to support their body weight against the pull of gravity. Terrestrial plants also use their cellulose stalks in a competitive effort to overshadow a neighboring plant and to gain more sunlight. Plants also invest heavily in roots, for physical stability as well as to draw on the nutrients dissolved in ground water.

In contrast, the sea plant has need for neither root nor superstructure. It is completely immersed within the liquid soup from which all nutrients and water are absorbed directly through the exposed membranes of individual cells. It is easy to understand why most sea plants are unicellular, are microscopic in size, and reproduce by the simple method of dividing the nucleus. There is no need to provide for reproduction by seed. These statements hold for the several thousands of species of diatoms, dinoflagellates, and coccolithophores that comprise the principal families of plants in the sea.

But the cycle of life for the marine plant is not a total bed of rosewater. Most plants construct a protective but porous shell of material that is relatively dense compared to the density of seawater. Diatoms construct a hard shell of silica, a material like sand

(SiO_2). Dinoflagellates construct an outer membrane of chitin, a material similar to cellulose or to human fingernails. Coccolithorphores cover their cell bodies with coccoliths, tiny platelets that they construct out of calcium carbonate ($CaCO_3$). Even though most of the volume of a plant cell is in its protoplasm, living body tissue which has about the same density as seawater, the organism is slightly negatively buoyant and tends to sink slowly. A healthy plant cell may use some specific methods to counteract its lack of buoyancy: it can incorporate fats and lipids, which are less dense than seawater, or it may construct its shell in special shapes to reduce its rate of sinking (see Figure 9.9).

Marine animals also benefit from being immersed in a medium whose density is almost equal to that of living protoplasm. Marine species invest comparatively little in bone and superstructure because of the reduced effective gravity factor. Most true fishes incorporate a gas bladder to help control buoyancy; they control the volume of gases inside the bladder by diffusing the gases into or out of the blood-stream. For fishes that move cyclically and vertically in the water column, controlling buoyancy decreases the amount of energy used in up–down movement. (See the discussion of myctophid fish in the following section.)

The main function of the skeletal structure in marine animals is as a frame to which muscles are attached and against which muscles can lever body shape to gain locomotion. Many marine forms build an exoskeleton with the muscle and body parts internal; in these animals the exoskeleton also serves other purposes, such as protecting them from predators (for example, the crab species) or protecting them against desiccation if exposed to air (for example, the barnacles during low tides on rocky shores). Exoskeleton animals do not gain such protection free of charge: during periods of growth they must periodically molt off the old shells in order to construct newer, more voluminous ones. While in shell-less transition, they are highly vulnerable to predation.

Constancy of Environment

Terrestrial animals invest heavily in form, structure, and function in their effort to combat extreme ranges of climate factors. Many have evolved with thermoregulatory systems that maintain their body parts at relatively constant temperature; warm-blooded species pay a continuous caloric "tax" to meet this requirement. Others incorporate an ability to alter shape as a method of reducing skin exposure to cold. Still others use scales or feathers as insulation.

In contrast, marine animals live in a fluid whose temperature extremes are very mild. They have little need for complex thermoregulatory systems; most marine species are cold-blooded. There are, however, two sides to the temperature-control argument: without the ability to control body temperature, marine animals are highly susceptible to *small* thermal changes in their water habitats. For some, particularly open-ocean species, a desired water temperature is had simply by changing depth; water temperature decreases with depth. Migratory species, on the other hand, can face very real barriers wherever thermal changes are encountered. For example, it is thought that the salmon species evolved in the Northern Hemisphere but were prevented from dispersing into the Southern Hemisphere by the warm surface waters of the equatorial oceans. These waters acted as a thermal barrier. We do not find indigenous salmonids in the high-latitude rivers of South America.

A biological limitation to cold-blooded animals is that metabolism rate decreases with body temperature; their "available" energy *rate* is therefore less than that for warm-blooded species. Some species of fish have evolved with a combined cold–warm blood tissue system. Tuna are an example. They are highly streamlined, fast-moving species in constant motion for feeding and migration, with energy demands that could not be met by a cold-blood system alone; consequently, tuna body temperature is several degrees higher than that of the ambient water.

Motion Contrasts

To "go down to the sea in ships," to sample and study the moving ocean and its living communities, is the life's work of the oceanographer. But we are always glad when the cruise is ending and the ship approaches port, for us a haven of solid, unmoving cover. We, together with other terrestrial animals, feel secure only when we are in charge of our body movements. When we move, it is always along the horizontal. Both territory and direction within our domains are extremely important to us, and we cue unconsciously to the shape and texture of the land we "see" to the limits of line-of-sight vision.

No such comforts are available to marine life in the open oceans. Here the environment is itself in unceasing motion. We can from this singular contrast develop a logic with which to study observed patterns in the behavior of marine organisms.

Patterns of Motion

The marine animal may move itself relative to the fluid that sustains it, or it may choose not to move at all. If it chooses motion, it has the further choice of horizontal or vertical direction or both. In oceans living space takes on three dimensions, but the concept of "territory" vanishes because the surroundings lack visual, textural contrasts. If we accept this postulate, we take an important step toward understanding the behavior of marine life—that marine life seeks out, adapts to, and cues to environmental differences *in all three directions.*

One of the inherent physical properties of water (Chapter 6) is that it strongly absorbs and scatters light rays. This explains why the photic zone is so thin, a mere 100 m. It also forces a drastic "shrinking" of the visual world of a marine animal compared to that available to the terrestrial animal. For example, even in clear oceanic water the total "field of view" for a marine animal with eyes is a sphere of no more than about 30-m radius. In turbid coastal waters it is much smaller. Further, water itself "looks" the same in all directions. For the marine animal, there is no visual cue on which to base a decision to move in a specific horizontal direction; however, the vertical direction is a clear choice— move toward the light or away from it toward the dark waters below the photic zone.

Other factors that influence an animal's choice of vertical movement are *pressure,* which increases at the rate of one atmosphere for each 10-m increase in depth; *temperature,* which decreases rapidly with depth (see Figure 8.3 for a typical profile); and *movement* of the fluid itself. Later chapters describe in detail how the ocean is stratified into layers, each layer moving independently. The uppermost layer is driven in motion by the surface winds; it is coincident with the photic layer. Deeper waters do not feel the driving force of winds, so there is differential motion between the upper zone, which alone sustains marine plants, and the deeper water. Could we then postulate that the principal grazers of marine plants might evolve with the ability to migrate ver-

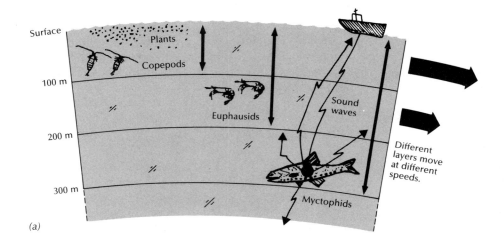

(a)

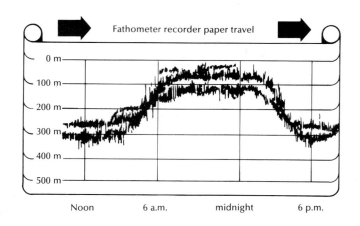

(b)

Figure 2.2 (a) A cross section of the upper ocean in which the principal participants of the deep scattering layer phenomenon are shown. Fathometers are instruments that transmit sound waves to measure the vertical distance between ship and ocean floor.

(b) Some sound energy is backscattered from the density discontinuity of the gas bladders in myctophid fish; this reflected energy is recorded as a heavy trace on the ship's depth recorder. The fathometer recording is a time history of the vertical migration of fish. This migration appears to follow the vertical migration of prey species such as the euphausids, which in turn follows the vertical migration of copepod grazers. The banded structure in the recorded echograph suggests that the fish may congregate into specific depth zones.

tically *between* such layers, and if so, to what advantage?

A clue that supports the postulate comes from examining the size of marine plants. The majority are microscopic-size, unicellular plant individuals. By deduction, we conclude that the grazer of such tiny plants must itself be small. The principal marine grazer is the group of animals called copepods. With an average length of some 2 mm, they are capable of swimming but at very slow rates compared to the speeds of ocean currents; they are therefore classed as plankton, which drift with fluid motion. Copepods are suspension feeders in general, capturing the microscopic plant cells that are in virtually complete suspension. The implication in this predator–prey setup is that the grazer population might rapidly deplete its environs of plants, for unlike its terrestrial counterpart, the cow, it cannot move to greener pastures by horizontal swimming.

But if the copepod puts its meager swimming ability into up–down motion, this predator–prey picture changes. By sinking, the grazer moves into a deeper layer whose speed is different from that of the surface water; therefore, its lateral displacement differs from that of the pasture in which it formerly grazed. By reversing its motion to upward swimming, the grazer finds itself in a new pasture on arriving back into the near-surface layer. Thus the postulate becomes logical, but do marine animals actually behave in this way? The answer is yes, as we next describe.

Deep Scattering Layer

During World War II, with its demands for increased marine transport, the Navy began receiving many reports of shallow waters and reefs in locations where no such features had been reported before. Moreover, the reports were almost always made during nighttime sailing in open-ocean waters. Investigation revealed that the ships' fathometers (sonic devices that measure bottom depth) were recording sound energy reflected from something other than the ocean floor. The source of reflection appeared at depths of 300 m or so during daylight hours but surfaced to within a few meters at night (Figure 2.2). Thus was born the biological research on what has come to be called the *deep scattering layer*.

Biologists found that the sound reflectors were the gas-filled buoyancy bladders of myctophid fishes, a type found over extensive areas of the world ocean. Reasoning that these fish migrated vertically in response to a similar up–down movement of prey, for their field sampling biologists used towing nets of various mesh size at different depths and times of day. Results shows that small shrimplike animals such as the euphausids also migrated vertically, as did the smaller copepods. The entire food chain moved vertically, apparently a response to the pattern set by copepods grazing on the fields of microscopic plants (Figure 2.2).

Why do the organisms surface during night hours—why not the reverse pattern? Clearly, this type of pattern correlates well with diurnal changes in light levels (called phototaxis), but is light intensity the real trigger? Are there different distributions of plant cells during day and night hours, making the grazing of these by copepods more efficient at night? Or is the security of the copepod from predation by the euphausids better during dark hours? The phenomenon is still being studied. In recent years has come the discovery that plant populations are indeed grouped into "patches," a pattern that would counter the vertical search efficiency of the copepod by reducing its success rate in finding *greener* pastures at *every* daily try. What we do not yet know is the precise mechanism or mechanisms that cause patchiness in plant concentration; the water itself may have nonuniform patches of "turbulent" motion and by this means play a key part in this aggregation of plants.

Size Scales and Size Specificity

Marine plants of the open ocean are small. They can be studied in detail only with the aid of microscopes. (We exclude here the very large plants such as kelp, which are found along continental margins.) Most are unicellular and, as indicated in an earlier section, thus force "smallness" on their grazers. Indeed, because plants are the foundation of all food chains, the entire marine food web has evolved to its present state as a direct result of its dependency on tiny plants. This fact has several important ramifications to marine science.

The Dispersion Effect

First, the fact of microscopic plant size together with the motion of the fluid means that plant production is extremely well dispersed throughout the ocean's photic zone. An important consequence follows immediately: the geographical area occupied by a single community, that is, a well-defined succession of marine life forms organized into a food web, must also be extremely large. This is a distinct contrast to the often small geographical areas of terrestrial communities. (See Chapter 13 for a discussion of biogeographical domains.)

But even though marine plants are very small, is it mandatory that the grazers also be small? For example, could not the concentration of plants per volume of seawater be great enough that grazers could literally "strain" the medium for food, which means the grazer might be quite large in physical size? In Chapter 1 we learned that the concentration of plant material at any one instant in time (biomass) is quite low. The basic causes of low plant concentration are twofold.

1. Individual plant cells are always sinking, albeit slowly, downward through the water column.

They must reproduce before exiting from the sunlit photic zone.

2. The supply of nutrients to plants is limited. In most oceanic photic zones the major fraction of the existing plant nutrients–the phosphates, PO_4, and the nitrates, NO_3–is "tied up" in the *living* plants and not available for the growth of new cells.

It follows then that the grazer population must be similarly dispersed. Moreover, the individual herbivore must be small enough to prey on individual plant cells.

The theme continues up the food chain. The small size and dispersion of grazers require that the juvenile stages of carnivores also be *of small size* but large in number to be able to crop efficiently. In summary, the sizes and numbers of individuals at all levels of the food chains are prescribed. Large carnivores must produce large quantities of small offspring, and this forces marine animals to reproduce by means of eggs.

Size Specificity

The dead bodies of marine plants and animals, whole or in parts, slowly sink in the ocean fluid. As a general statement, this means that there are no "leftovers" from partial consumption of prey. Unlike their terrestrial counterparts, marine predators ingest the entire prey at once. This is the basis of the *concept of specific size relationships between predator and prey*. The dependency of young animal larvae on food of the right size being available at the right time is well known, for example, by aquaculturists who raise oyster spat commercially. The young spat must have available a succession of algae of increasing cell size during the critical first few days after hatching. An error in the size and time matching of spat to algae can be disastrous to oyster survival. Figure 2.3 diagrams to *correct size scale* some of the main members of an open-ocean community off California. A comparison of sizes among predator and prey clearly demonstrates the bite-size principle; more than this, however, it demonstrates that the principle holds all the way down to the microscopic domain!

A second concept emerges from this discussion of size-specific relations. It is to the advantage of larvae to grow as rapidly as possible, for two reasons. First, from the view of being preyed *upon*, the best strategy is to move into and out of a given size category rapidly. Second, from the view of a predator, the system should avoid leaving a predator with the same prey-size dependency too long to avoid overharvest. Clearly, the statistical elements of *chance encounter* govern the behavior patterns and life cycles of marine organisms.

Organic Composition Contrast— Protein or Carbohydrate?

We terrestrial animals, to which territory is so important in survival strategy, need to expend large quantities of energy for movement. The expenditure is needed to meet two demands: first, a strong gravity force makes all movement a constant battle against friction and, second, active predation requires that the *rate* of energy use be very rapid during the "chase" time. We have thus evolved into a chemical machine in which the release of muscular energy depends on the storage and conversion of the energy-rich carbohydrates. (If this discussion is becoming a bit "heavy" for you, try munching a bar of chocolate.)

The marine animal, immersed in its buoyant environment, has much less need for such rapid or sustained energy delivery. (There are exceptions, such as the salmon, which must make sustained migrations to spawning grounds upriver.) Generally, *the "chase" in the marine world is notably short in both space and time*. This conforms directly with a previously developed idea that visual distance is restricted; a protracted evasion maneuver to escape one predator is equally likely to bring the prey into

Figure 2.3 Comparative sizes of dominant organisms in the California Current food chain (drawn to scale).

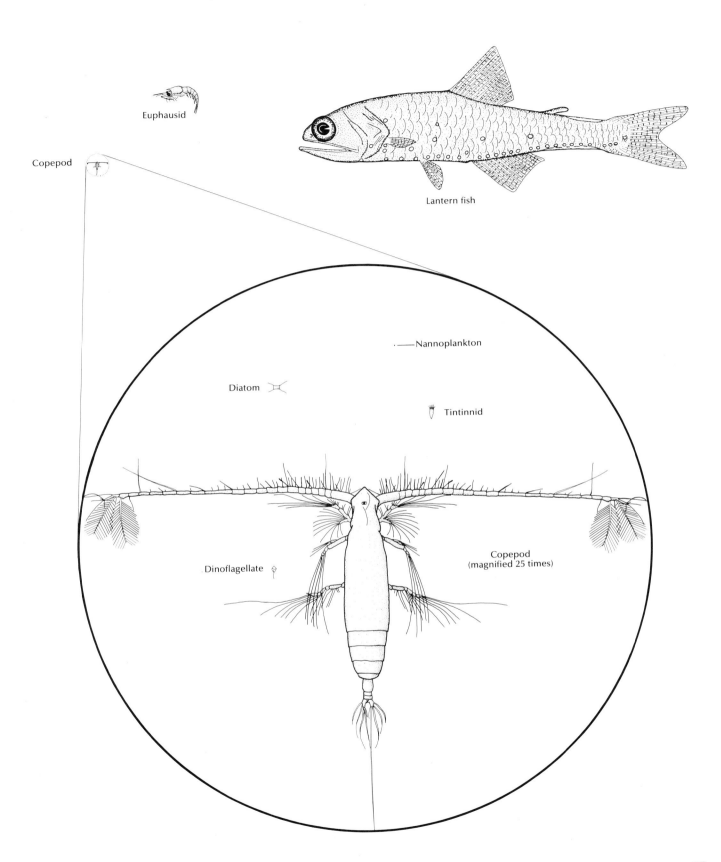

Euphausid

Copepod

Lantern fish

Nannoplankton

Diatom

Tintinnid

Dinoflagellate

Copepod
(magnified 25 times)

23

visual proximity of another predator. Once again we see the strong role of the *chance encounter* in governing survival strategy.

But the marine animal does rely to a considerable extent on proteins. The reasons are primarily two. First, the game of chance encounter that the marine biosphere requires its member to play forces the animal to reproduce in large quantities. The most efficient method evolved by nature for this is the egg. Second, the fact that the ocean is constantly moving means that the egg phase, and larval stages as well, is subject to widespread dispersion. Dispersion by itself would not necessarily demand that the number of eggs released per generation be large; if oceanic conditions in which larvae could survive were uniform, the survival rate of larvae would be independent of the dispersal limits. This is often not the case. Patchiness of marine plants alone would dictate that some larvae would die from lack of nutrition. Furthermore, there can be large changes in water conditions from point to point, particularly in coastal oceans, and in the type and amount of particulate matter in suspension. Therefore, to ensure species survival against the element of dispersion, the marine animal must generate large quantities of eggs.

Eggs must be rich in proteins. It follows that marine animals have evolved with primary reliance on protein chemistry, just as terrestrial counterparts have come to rely on carbohydrates. We put fish in our own diet for *protein*.

Contrasts in How Nutrients Recycle

This concept is best approached by organizing nutrient cycling, terrestrial versus oceanic, into *comparative scales of time and space*. For purposes of definition, nutrients are the chemical elements or ions that the living *plant* cell requires in its construction of shell and tissue. In the marine world, these nutrients are dissolved in water as ions of salts or other complexes; principal units are the nitrates (NO_3), the phosphates (PO_4), carbon and carbonates (which are derived for the most part from dissolved carbon dioxide CO_2), and silicate ($SiOH_4$).

The Terrestrial Case

In the typical terrestrial field or meadow, nutrients taken up by plants are returned to the soil within one year, that is, the annual growing cycle. There

are exceptions: tropical forests recycle fallen foliage more rapidly, whereas temperate forests tend to accumulate humus (dead organic material) over longer periods of time. From the space scale view, the terrestrial system returns nutrients to the immediate locale from which these are originally taken up.

The Marine Case

In contrast, the marine system for nutrient cycling is controlled by mechanisms of much longer duration in time, hundreds of years, and of much larger scope, global in scale. The keys to understanding these great contrasts are that (1) the ocean is forced to behave as a two-layered fluid system and (2) the mechanism that mixes the two layers operates on a size scale that is global. Figure 2.4 diagrams how the marine system works.

The upper layer is warmed by sunlight; as expected, this warming is greater in the equatorial regions than in the polar areas. Together with the patterns of surface winds, solar heating creates an upper ocean layer that is about 300 m thick near the equator, increases to perhaps 600 or 800 m of thickness in the central areas of ocean basins, and then diminishes to zero in polar zones, where solar energy is also diminished. All this is shown in Figure 2.4.

A thick layer of cold, dense seawater underlies the warm surface layer. In thickness it reaches to the deepest parts of the basins, and its temperatures are for the most part less than 4°C; in fact, the majority of all ocean water is at a temperature less than 2°C! It follows that we can locate the zone that separates the two layers by measuring just the temperature at successive depths below the surface of a column of seawater. When the temperature probe senses a rapid change to colder temperatures, it has encountered the separation zone; we call the zone a *thermocline* (for example, see Figure 8.3).

Another way to locate the boundary between the two ocean layers is to probe with a device that measures the density of seawater. In fact, it is the *difference in densities* of the two layers that (1) prevents these from being easily mixed together and therefore (2) sets the stage to explain the long time period involved in marine nutrient cycling.

Figure 2.4 indicates the way that the two-layered ocean overturns and eventually mixes. Warm surface waters are forced poleward carrying surplus solar heat to offset the deficit heat of polar zones. Near Greenland in the north and Antarctica in the south,

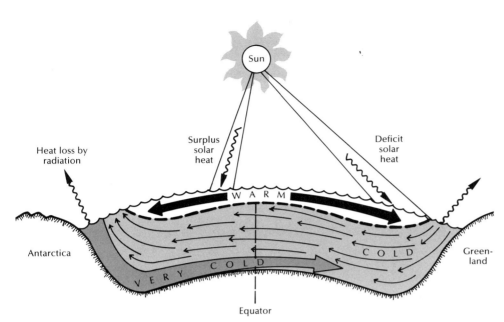

Figure 2.4 A cross section of the Atlantic Ocean, north to south, showing the two-layer structure typical of open oceans. The heavy dashed line marks the separation between the two layers. Organic particles that sink out of the warm surface layer are dissolved in the deep layer and require hundreds of years to recycle back to the surface.

surface waters are cooled to such a great degree that they become denser than other ocean waters and sink into the interior ocean to become the masses of water of the *deep* layer. These masses spread across the basins, filling the entire interior domains of all basins. Eventually, deep water is returned to the surface, particularly in zones called *upwelling* zones. The cycle is thus completed. In the Atlantic, a cycle requires about 700 years; in the Pacific about 2500 years.

How does this long time cycle impact nutrient cycling? We note again that all protoplasm, especially when combined with skeletal material, is heavier than seawater. All dead organisms must therefore sink toward the ocean floor. If the dead material is not recycled before it sinks out of the upper layer, it then enters the deep layer and so automatically becomes a part of the long-term cycle period of overturn. It takes hundreds of years before the nitrogen and phosphorus that dissolve into the deep-layer water are returned to the photic zone to become nutrients again for plants in the warm surface layer.

In terms of space scales, marine nutrient cycling is a global cycle, a result that is consistent with two facts, (1) that the medium is a fluid in constant motion and (2) that the deep ocean currents span the entire oceans. For example, organic material that sank to deep levels off Greenland at the time of the Crusades is today surfacing off Antarctica where deep Atlantic waters are upwelled. Our terrestrial experience in nutrient cycling does not prepare us well for studies of the marine system.

Summary—A Marine Strategy for Survival

In the marine world, what is the successful strategy for survival of life forms? The marine biological system we sample today is by definition a successful one because it exists. Figure 2.5 summarizes how the numerous factors described in this chapter interact to compose successful strategy for marine communities.

The great buoyancy of the liquid medium plus its dispersed and dissolved nutrients are key factors in nature's selection of single-cell plants as the successful marine adaptation. The size of plants then influences the behavior of the total food chain, as does the fact that plant production is generally nutrient-limited.

Dispersion plus the limited visual-contact field introduce the element of chance encounter that then influences the reproductive scheme. Species survival is ensured not by cover or camouflage or evasion maneuvers, although all of these are present, but rather by the production of astounding numbers of offspring. Multiple reproduction imposes stringent demands for high protein production to sustain egg development. Many eggs means small-sized larvae, which fits with the size limitation imposed by unicellular plants upon the grazer species. Large quantities of eggs also maximize chances for species survival; some offspring will survive the predation and some will survive the dispersion by ocean currents.

Although the reduced gravity factor eliminates ex-

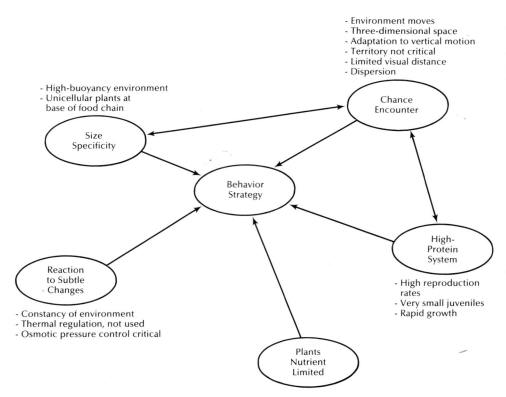

- Environment moves
- Three-dimensional space
- Adaptation to vertical motion
- Territory not critical
- Limited visual distance
- Dispersion

- High-buoyancy environment
- Unicellular plants at
  base of food chain

- High reproduction
  rates
- Very small juveniles
- Rapid growth

- Constancy of environment
- Thermal regulation, not used
- Osmotic pressure control critical

Figure 2.5 A strategy of behavior for marine animals, the consequence of the interaction of physiological factors and physical features of the fluid environment.

cessive body structure demands, there remains the slight negative buoyancy of living protoplasm. Successful adaptations are small size, control of body shape, and the use of flotation devices. Eggs are produced with sufficient fats and lipids to buoy them to the surface where food of correct size is available and the juvenile does not compete with the adult for the same food supply. Buoyancy control enables the individual to migrate vertically, and such behavior couples into the differential motion of ocean layers to ensure dispersion, and so on and so on.

We witness a marvelous world of adaptation to a moving, fluid medium. So unlike our own terrestrial system, it is equally successful *so far.* In the final analysis the protein deficits of humankind force us to ever greater reliance on the protein-rich marine system.

Study Questions

1. The traditional division of the ocean fluid into different "living spaces" (Figure 2.1*a*) is on the basis of depth of water, with an additional category(s) for those living spaces directly involving the bottom substrate. The alternate division scheme (Figure 2.1*b*) is based on specific physical characteristics of the ocean fluid, by which natural zones and realms are identified. Explain one advantage and one disadvantage to each method.

2. *One* physical property of water—*density*—explains the great differences in structural adapta-

tions of marine life relative to those for terrestrial life. Explain.

3. *One* physical characteristic of the seawater—the constancy of its thermal state—explains other major differences between terrestrial and marine life forms. Explain.

4. If most marine species have evolved without the ability to control internal body temperature, does it follow that they are *insensitive* to changes in the temperature of ocean water? Why do fishermen often tow their nets *along* the zones where waters of *different* temperatures come together?

5. Stated another way, are marine forms more highly sensitive to small changes in the temperature of seawater than terrestrial forms to changes in air temperature?

6. Which single property of seawater, more than any other, is responsible for the lesser degree of *territorial behavior* of marine pelagic species than of terrestrial species?

7. *If* territory is not a characteristic of pelagic species behavior, would you expect speciation in pelagic species to be greater or lesser than that for benthic species?

8. The single characteristic of the ocean world that describes it as overwhelmingly different from the terrestrial is that it moves. Describe some ocean consequences.

9. Contrast the sizes of the principal terrestrial herbivores with those of marine herbivores.

10. Your expertise regarding the moving ocean can be evaluated in another way. *If* the ocean fluid is in constant motion, why do so many species reproduce by means of eggs that are subject to widespread *dispersion?*

11. *If* the primary food supply on which all food chains rely—the plants—consists of such small individuals in the oceanic biosphere, how would you choose to reproduce if you were a marine organism, by live young or eggs? If by eggs, what size eggs?

12. The explanation given for the existence of the *deep scattering layer* phenomenon is that the copepod evolved a behavior pattern of using the differential surface currents as a way of *moving* from one pasture to another, even though its swimming ability is very limited. Would it be logical to explain the vertical migration of *all* the carnivores using the same "hunting ground" argument?

13. The majority of terrestrial herbivores *graze* during daylight hours. Why don't marine grazers do likewise? Is it necessary for the marine grazer to "see" the grass? Does marine grass taste better during the dark of the moon?

14. Describe how your mother explained to you that seafood was a high-protein food. Do you think you can devise a better explanation?

15. Write a short essay in which you describe *how* you would teach a fourth grade student the concept of *size specificity* as a controlling factor in how the marine biosphere is organized.

16. Write a short essay in which you explain to an eighth grade student that the limited distance marine species can see makes the "chase" (*a*) quite short and (*b*) subject to the statistics of *chance encounter*.

CHAPTER 3

MARINE SPECIALISTS AND INTERDISCIPLINE STUDIES

Until the mid-eighteenth century the number of published works in science were so few that an individual could, in one lifetime, study all existing information in all aspects of science—chemistry, biology, the mathematical arts, and so on. No longer. The world of science today belongs to the specialist; but what defines the limits of specialization is still much the same—it is the scope and breadth of literature that an individual can assimilate into his or her own studies.

In spite of the pressures for specialization, oceanography retains much of the flavor of "natural philosophy" in that each of us is trained to a high degree of proficiency in all the major disciplines. In many universities, this multidiscipline aspect is contained in a *core curriculum* in which the student oceanographer masters the fundamentals of physics, biology, chemistry, and geology as applied to oceans. This does not mean that career specialization does not occur in marine science—it does, and ever increasingly so. But our specialization tends to be "study of specific marine problems from a joint, multidiscipline view."

Introduction

In marine science we have learned that new insight is often gained only by a *joint* application of expert knowledge from several disciplines. For example, the biologist attempts to explain how the feeding appendages on the copepod actually work (see Figure 13.1); but until the explanation includes the effect of the water's viscosity, a fundamental physical property, it will not be an adequate one. Any hypothesis put forth to explain why such appendages are *successful* adaptations in the sense of evolution must also satisfy physical constraints. For example, the patterns of motion of the water medium itself, the buoyancy factor, and the vertical mixing of near-surface layers by winds and waves all affect the distribution of plants, which are the food to be gathered by the copepod.

The chemist works at defining the chemical pathways by which elements enter the fluid ocean from rivers and atmosphere and eventually exit by being taken up in the sediments. En route, these elements also participate in a complex of biological pathways, processes that also impact the *rates* of solution, dissolution, and sedimentation. Virtually half of the volume of marine sediments are biogenic in origin, so the geologist must know marine biology and chemistry to interpret the history of sediments. Conversely, feedback from marine geology is of fundamental importance to the biologist who studies the life cycles of organisms and to the chemist who studies reactions in seawater.

It is worthwhile at this point to examine the ocean world through the eyes of specialists in each of the classical sciences. The objective is first to identify the nature of specific problems of interest to each discipline and, second, to point out areas where an exchange of views between disciplines is necessary. For continuity through these discussions, we use a common base chart of earth on which are annotated a number of problems and topics. Many are common to two or more disciplines and are focal points for interdisciplinary studies.

The View from Physics

Physical oceanographers specialize in studies of the behavior of the fluid world. The physicist states that the motions of fluid ocean and atmosphere are a consequence of unequal heating of the earth's surface by the sun. In Figure 3.1 equal cross sections of radiance paths from sun to earth are shown to illuminate less surface area near the equator than in po-

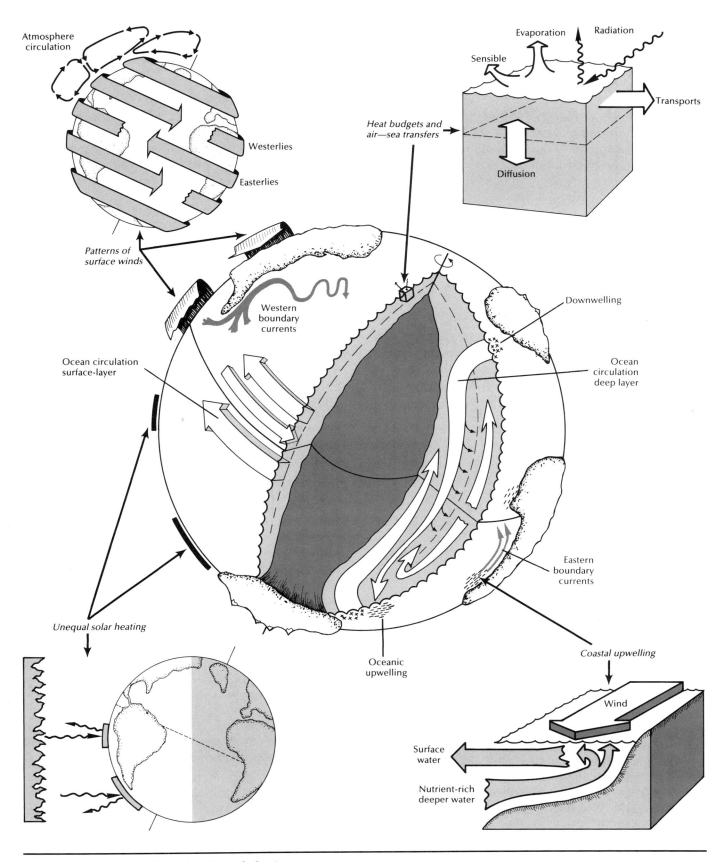

Atmosphere circulation

Westerlies

Easterlies

Evaporation

Radiation

Sensible

Transports

Heat budgets and air—sea transfers

Diffusion

Patterns of surface winds

Western boundary currents

Downwelling

Ocean circulation surface-layer

Ocean circulation deep layer

Eastern boundary currents

Unequal solar heating

Oceanic upwelling

Coastal upwelling

Wind

Surface water

Nutrient-rich deeper water

Figure 3.1 The oceans from the view of physics.

lar zones; equatorial zones receive more calories per square meter per day. This explains the unequal heating of the earth's surface and is the physical foundation for studies of the so-called *heat engine* of earth.

Unequal heating produces a difference in temperature between equatorial and polar zones. This is the fundamental physical factor that forces the atmosphere and ocean into motion. Motions of fluid earth are reactions to this forcing, motions that try to balance the unequal heat distribution by transporting heat from equatorial regions of surplus to polar regions of deficit heating. Descriptions of atmospheric winds and ocean currents are basic inputs to studies of the earth's heat budget. We estimate that the atmosphere and ocean share about equally in the process of redistributing heat over the global earth.

The atmosphere participates in the heat engine through a complex series of cellular motions called Hadley cells. These are sketched in cross section in Figure 3.1 (and a more detailed description is given in Figure 10.4).

Oceanographers are more interested in *surface* winds because these interact directly with the ocean. Figure 3.1 shows the pattern of surface winds over the global ocean; they exist in definite "belts" called *easterlies* or *westerlies*. (Meterologists describe winds in terms of the directions *from* which air flows, whereas oceanographers describe currents in terms of where the flow goes.) In general, these belts girdle the globe. Between the easterly winds along the equator and the westerly winds at latitudes 30° to 50° are zones of *tradewinds,* and on the poleward sides of the westerlies are another set of easterlies.

Surface winds generate motions in the surface layer of the oceans. Indeed, Figure 3.1 shows that surface winds are correlated with the patterns of major surface currents. Although we have names for specific currents, such as the particular western and eastern boundary currents, the fact is that wind-driven currents are organized into definite large-scale circulation patterns called gyres; gyre sizes are typically of the same scale as the main ocean basins, for example, the North Pacific or the North Atlantic gyres.

But because of an additional force generated by the spin of earth called the Coriolis force (see Figure 10.3), the pattern of flow is divided into separate gyres north and south of the equator. These gyres rotate in opposite directions. In fact, hemispherical patterns of motion in both ocean and atmosphere are "mirror images" across the equator.

Moreover, as a result of unequal heating by the sun, surface waters along the equator are warmer and hence less dense than polar surface waters. In polar regions, on the other hand, strong cooling and evaporation cause surface waters to become dense enough to sink deep below the surface. Together, these factors create and maintain a vertical motion pattern, called *thermohaline* circulation, by which upper-layer, warm subtropical waters move poleward while deep-layer, colder waters move toward the equator. From this comes the point of view by which we "model" the *ocean as a two-layered system* (Figure 3.1). Within the upper layer the time scale of wind-driven circulation around the gyres is of the order of 2 to 3 years; however, the time scale of deep circulation is hundreds of years.

Some mechanism must exist so that water that downwelled, sank out of the surface layer into the deep layer in polar zones, is upwelled, returned back to the surface. We find specific regions where such *upwelling* occurs. For example, in narrow zones along coastlines where winds are strong and directed parallel to the shore, deeper water is forced to the surface; other upwelling zones occur in narrow strips along the equator, particularly in the Pacific Ocean, and extensive upwelling occurs in the Southern Ocean around Antarctica. Upwelling zones are of great interest to biologists too, mostly because of the nutrient recycling, and to chemists who study how chemicals move from the surface to deep waters and back again.

Some recent events have catapulted physical oceanography onto the world stage along with meteorology. We have discovered that the equatorial easterlies in the Pacific "pile up" surface waters against the Asian continent; but when these winds relax in strength, as they do periodically, the surface waters move back across the Pacific in such volume and strength as to create major climate changes along the Pacific shores of South and North America. This is the El Niño phenomenon, so-called because it usually appeared off Ecuador and Peru during the Christmas season. These climate studies have now expanded into major global science missions such as TOGA (Tropical Ocean-Global Atmosphere), a worldwide effort to find the key "connections" between oceans and atmosphere that can be used to *predict* future climate changes.

Study Areas

Currents. What is the speed at which ocean fluid moves? With the development of a reliable ship's chronometer (timepiece, clock) in the early eighteenth century, navigators were able for the first time to measure their *longitude* for precise position at sea. An immediate result was that the "set" of a ship caused by surface currents could be measured. *Set* is the difference between where a ship is and where it should be. In this way the major currents were quickly charted. By the mid-nineteenth century Matthew Maury published compilations of currents inferred from these data. The first such navigation aid, it documented the intense north-flowing current, the Gulf Stream, along the Atlantic seaboard of the United States. Since then scientists have sought the physical basis by which an ocean develops a more intense current on its western boundary than on the eastern side. (An explanation was not produced until 1948; see Chapter 11.) Today, the Gulf Stream is studied more intensely than ever before, on the thesis that its dynamic behavior is the key to weather changes across the Northern Hemisphere.

Only in the latter half of this century has technology produced machines for measuring ocean currents accurately. Expensive and difficult to deploy, these machines yield only a fraction of the amount of direct measurements we shall need to describe ocean motion adequately; some sections of the ocean have yet to be sampled. A substantial part of ocean motion appears to be *turbulent*, that is, by definition unpredictable. Fortunately, remote sensors on satellites now hold promise of gathering the data from which we will be able to chart the ocean currents.

Waves. Fluid masses are unique in their ability to support wavelike motions. The oceans support a multitude of wave *types*, from the tiny "capillary" waves that give its surface a "riffled" appearance to waves of planetary scales, including tide waves (Chapter 17). Waves transport energy from one ocean region to another, just as currents do, but unlike currents waves transport very little mass. Wave motion is an integral part of how and where oceans obtain and dissipate motion energy, so the physics of wave generation continues to be studied.

Air–Sea Connections. The interface between the two fluids ocean and air is the focus of major study by physicists as well as chemists. Through this interface the winds pass the energy that drives currents. It is the zone of exchange of gases such as oxygen and carbon dioxide. From a physicist's view, the most critical of all materials that cross this interface is water vapor; it is the heat energy latent in the molecule of vapor that drives the major weather patterns and thus controls "weather" (an example was

shown in Chapter 1). We are measuring the *rates* of energy exchange across the air–sea interface, and the role of oceans in tranporting energy stored in one region to a release point in another region.

From the point of view of biology, the most critical element that diffuses across the interface is carbon dioxide, CO_2. The rate at which carbon is fixed into organic protoplasm by marine life determines in part the rate at which new CO_2 enters the oceans from the gaseous atmosphere. The chemist is also studying such rates, as is the meteorologist, because the removal of CO_2 from the atmosphere is a vital part of the problem created by humans in the burning of fossil fuel.

Equipment. Historically, the physical scientist first "probed" the oceans with mathematics. The equations that describe fluid behavior were developed over a period of 150 years, from about 1750 to 1900. Only in the last half century have scientists begun to deploy machines to obtain the data to verify the motion predicted by these equations. Figure 3.2 illustrates some of the machines that are basic to exploration of the physics of the oceans. Beginning with the historic HMS *Challenger* voyage of 1872 to 1876, physicists measured the temperature and the salinity—the concentration of dissolved salts—of ocean water and calculated its density. From these data they deduced the strengths of currents and launched the science of physical oceanography.

With advances in electronics after World War II came new ways to measure temperature, salinity and other physical characteristics *in situ*, here meaning *in the water*. For example, we now measure the speed of sound in the ocean with ever-increasing accuracy and at ever-decreasing intervals of space and time. This allows us to study how and how fast the energy of motion is finally dissipated into thermal energy or heat. We need these data to build accurate models of how friction modifies the dynamics of the ocean–atmosphere system.

The mainstay of equipment for the study of ocean physics has always been the ship, a mobile platform from which all sorts of measurements are taken (see, for example, Chapter 8). But even this dependency on ships is changing. We now depend on unmanned buoys to gather data; some buoys are anchored and others drift with winds and currents. The most recent advances, however, have been in the use of satellite-mounted remote sensors. By 1990 we shall have daily reports of global winds over the sea surface, together with wave data and even the color of the sea.

Key equipment for physical oceanographers. These oceanographers are bringing abroad a state-of-art electronic device for sensing water temperature and conductivity. The unit transmits to deck recorders through a conducting cable. Computers plot traces of temperature and salinity continuously as the sensor is lowered. Mounted above the sensor cage is a rosette of twelve sampling bottles; each can be separately triggered to capture a sample. (Courtesy Bob Collier, Oregon State University.)

Mooring current meters in the deep sea. Physical oceanographers install and retrieve many arrays of instruments that stay in the sea for periods up to one year, recording data internally and automatically. Here the crew is installing a current meter array. The first item over the side is the surface marker buoy, to mark the array so that it can be recovered later. Next is a series of eight flotation buoys, followed by the cylindrical current meter itself. To the right is a set of steel railcar wheels used as the anchor for the array; it is the last item lowered over the side. (Courtesy Bob Still, Oregon State University.)

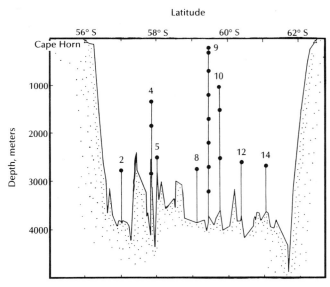

Strings of moored current meters. In 1975 an array of eight strings of current meters with as many as eight meters on a single string were installed across the Drake Passage as part of the International Southern Ocean Study. These meters remained *in situ* for a full year before being retrieved and their data analyzed. Ten of the nineteen instruments produced useful data for most of the year. (From Sciremammano and Pillsbury, 1980.)

Figure 3.2 Equipment and techniques used by physical oceanographers.

The Broad View from Biology

The work of biological oceanographers is probably the most complex of all the marine sciences. Theirs is the work of unraveling the mysteries of life cycles intricately tied to the motions of the life-sustaining fluid ocean itself. Figure 3.3 show this as the theme of the upper layer studies in energetics of food chains, population dynamics, and biogeography of open-ocean provinces. A similar theme directs much of the biologist's work in coastal oceans.

The biological oceanographer seeks to understand the schemes of adaptation. How do different organisms live and persist in the ocean habitat? In some schemes, as in the migration of eels from freshwater European rivers to the saltwater mid-Atlantic during spawning (Figure 3.4), how do organisms solve the problem of controlling body fluid salt content? Beyond this, how do migrating animals even know where they are going? What clues do they take from the surrounding medium?

Other questions arise from studying the *scales* of time and space involved with marine life. How does the distribution of plant species interact with the large-scale ocean circulation? Similar questions apply to animal distributions, as implied in Figure 3.3 in the sketch of oceanic areas where certain fish populations and stock are found.

The process of photosynthesis in the ocean environment is itself wrapped in unknowns. How do plants survive when the intensity of light photons decays so rapidly with depth in the water column? How does plant production depend on the physical factors such as vertical mixing induced by waves and turbulence? How does the biological oceanographer even *know* when enough samples have been taken to describe accurately the photosynthesis process in the open sea?

In seeking answers to questions such as these, the biological oceanographer tends to view the marine world through the eyes and other sensors of the organisms themselves (Chapter 12). In part, this is

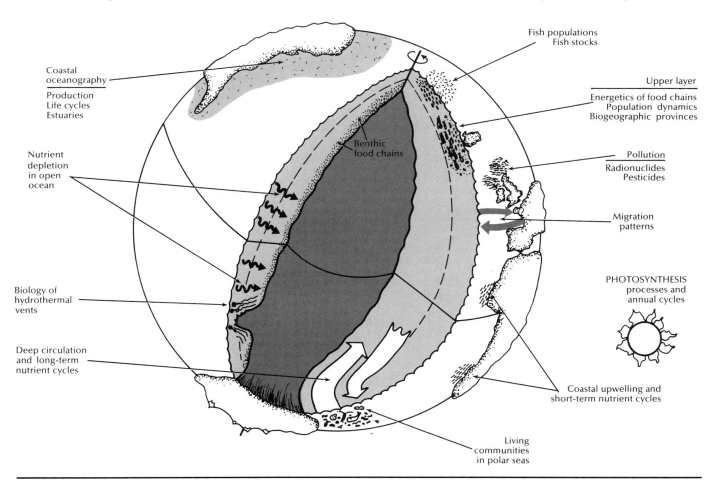

Figure 3.3   The oceans from the view of biology.

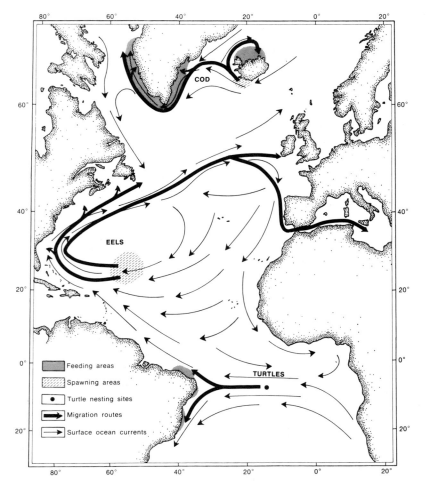

Figure 3.4
Migration routes of some marine animals, together with prevailing surface currents. (From Sumich, 1980.)

European eels. Adult eels are found in the rivers of Europe and in the Mediterranean, but spawn in the central part of the North Atlantic, the Sargasso Sea. Routes taken during migration to and from the spawning region follow main currents of the North Atlantic.

Cod. The fish migrate between spawning areas just south of Iceland and feeding areas along the route of the East Greenland Current.

Turtles. Notable among the migratory animals are turtles, which hatch at Ascension Island but feed over wide ocean areas that to date are not really very well known.

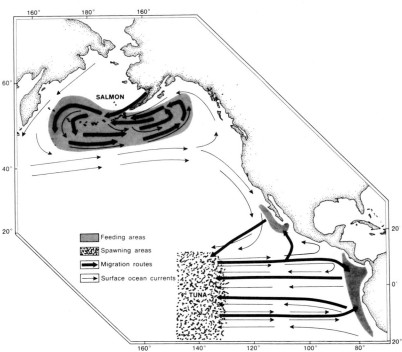

Salmon. Migration routes of Pacific salmon are well known. The feeding areas are for the most part in the Gulf of Alaska and in the Bering Sea; spawning is in the interior rivers of the Pacific Northwest.

Tuna. The fast-swimming, open-ocean tuna conduct one of the most complex migration patterns, coupling their travel between feeding and spawning regions to the east-west pattern of Pacific Equatorial currents.

why the field rapidly divides into areas of specialization. The coastal planktologist faces a vastly different ecosystem than does the open-ocean plankton investigator; for instance, the coastal ocean is much more variable, so its inhabitants must react differently to stimuli than do their oceanic cousins.

We distinguish between the biological oceanographer and the marine biologist, although both types of specialists often work in the same research group. Biologists focus on differences in the physiology and biochemistry of marine life and rarely go to sea; much of their work is done in laboratories. For example, studies in both the energetics of food chains and photosynthesis can be carried out in laboratories; both of these subjects are listed in Figure 3.3. The point is that biological oceanographers use the results of their laboratory research in planning their work at sea, particularly in deciding the how, when, and where and how repetitive should be the biological sampling carried out from ships at sea. If X copepods consume Y plant cells of species Z in the laboratory, can the same relationships hold for the real ocean?

Biological oceanographers must interact with specialists in all other marine disciplines. For instance, it is one study to sample the types and species of animals that make up the benthic food webs, but another study to find the numbers that relate the various species to an ecological framework. Here the biologists seek information from others. What is the *rate* of deposition of organic material on the ocean floor (chemistry)? How much does a community in one region depend on deep ocean circulation to bring organic food from distant provinces (physics)?

There are two oceanic regions where the interdiscipline dialogue between biological oceanographers and others is particularly important. These are (1) the submarine vent systems found along seismically active sea floor spreading ridges and (2) the polar oceans in general. Both are included in the biologist's view (Figure 3.3). Geologists and geochemists supply valuable insight to the biologist studying vent communities; physicists and meteorologists are important advisors to the polar biologists.

Study Areas

Populations of Organisms. The biological oceanographer focuses first on the dynamics of populations of organisms. He or she attempts to gather enough data from which to evaluate the energetics of a selected species, first, and later that of an entire food chain.

Populations per se are difficult to identify. Literally, a population means a collection of individuals of a species whose behavior is distinct from that of other collections of the same species. For example, populations of a herring species may mix on feeding grounds, but they separate into distinctly different populations during spawning (see figure 18.17).

The environment is an integral part of population study. There are vast biophysical and biochemical differences in populations of tropical and polar regions. Tropical waters enjoy ample sunlight throughout the year, are vertically stable because warm surface waters do not mix readily with deeper cold waters, and carry a wide diversity of species; populations tend to be small in size. Polar regions receive seasonal exposure to sun, are colder, hence force organisms to lower metabolic rates, and carry fewer species; populations tend to be large in size. Major fisheries occur in higher latitudes.

Vertical Distribution and Behavior of Marine Life. Depth (pressure) effects are still other mechanisms that control how populations are distributed. Life cycles of mid-depth organisms and those of the benthic communities on the ocean floor require different approaches of specialized study. Recent work has disclosed that the enzymes of many marine species function over limited depth ranges of about 1000 to 1500 m; populations at distinct levels are therefore constrained to those levels. If a mid-depth population also reproduces by eggs, and the eggs float to the surface layer for larval hatching and feeding, the organisms must somehow supply offspring with keys to two sets of enzymes, each appropriate to the pressure range they will occupy at different stages of the life cycle.

Migrations. Studies of animal migrations are an integral part of biological oceanography. Vertical migrations were discussed in Chapter 2, but there are well-known horizontal migrations of many species. Many migration routes correlate well with the patterns of ocean currents measured by physical oceanographers (see Figure 3.4). Eels that inhabit the rivers and estuaries of Europe as adults spawn in the central North Atlantic; the juveniles follow the major clockwise gyre of circulation of surface waters in returning to European rivers. The anadromous salmon of the Pacific follow an inverse behavior, returning from a nomadic feeding pattern in the central and northeast Pacific to spawn up the rivers of the Pacific Northwest. Within the complex of currents and countercurrents parallel to the equator in the Pacific, the rapid-swimming tuna execute repeated traverses between feeding and spawning grounds. What are the signals that trigger these massive migrations? And how do eels know to select east or west currents as migration routes?

Coastal Biological Studies. Readers will discover (Chapters 17 and 18) that the shallow waters of coastal oceans behave quite differently from those of the open ocean. The biosphere in these waters is also very different. There are more nutrients, there is much more "seasonality" as measured in everything from sunlight to freshwater runoff to pulses of animal migration, and there are human beings to introduce all sorts of complexities. It is a marine zone for which the number of specialty "sciences" grows daily. Most recently, because of the 1982 Law of the Sea Convention, economics and law have become entwined with basic science of the coastal biosphere.

Biological Sampling

The task of collecting an adequate suite of biological specimens is much more formidable for the biological oceanographer than for his terrestrial counterpart.

Seasonal Changes versus Ship Schedules. Open-ocean biospheres in surface waters undergo seasonal changes, linked of course to the annual cycle of sunlight intensity. But to take samples of marine life in different seasons means scheduling expensive research ships. The cost factor in ocean biology is orders of magnitude greater than that for land studies; a ship capable of sustained voyages in open oceans and equipped with machines to sample at all depth levels costs upward from $10,000 per day. Moreover, schedules for major cruises today require lead times of up to three years.

Sampling Time and Sampling Intervals. A common device used in biological sampling is the plankton tow net (Figure 3.5). Individual plankton tows near the surface can be done in short periods of time, say, one hour; if enough ship time is allotted, the investigator can make repeated hourly tows throughout the daily cycle. To avoid contaminating a sample taken at one specific depth with specimens at other depths, the biologist uses nets that open and close on command from the deck. A multiple net is deployed to study the horizontal distribution of organisms; the opening and closing of successive sections of the complex net are programmed during a single cast and tow.

Where mid-depth samples are needed, the device known as the Isaacs–Kidd trawl is deployed. Such tows require many hours, not only for casting and retrieving but also for the tow itself. Special propeller-driven logging devices monitor the amount of water "filtered" by the net so that estimates of population density are possible (Figure 3.5).

Benthic biologists extend the trawling technique to the ocean floor. Here even more time per cast is required, as much as 14 hours to gather a single sample. Devices other than nets and trawls are used to sample the organisms that burrow into the sediments and cannot be captured in trawls; generally these devices recover a "piece" of the ocean floor, mud and all.

For the capture of microscopic organisms, particularly small plankton, researchers take samples of the water itself and later filter the sample through micropore-size filters in the ship's laboratories.

The Moving-Ocean Problem. Clearly, a broad limitation to biological sampling is the inability to sample the *same* blob of water at different intervals of time—we simply do not know where the water column sampled yesterday is today. Single samples are inadequate. We must take multiple and repeated samples just to *estimate* a real-ocean biological situation.

The View from Chemistry

The famous oceanographic cruise of the HMS *Challenger* heralded the opening of the oceans to systematic sampling, analyses, and description. Sponsored by the British Admiralty, this cruise carried a team of scientists over the Atlantic, Pacific, and Southern Oceans to gather all manner of physical, chemical, biological, geological, and meteorological data. Out of this came some 50 volumes of analyses and descriptions.

It was the first time that samples of seawater from diverse oceans were gathered together and analyzed for similarities, both in chemistry and in indigenous specimens of marine life. We learned that the oceans were indeed *well mixed*, that although the ratio of H_2O to total dissolved salts may vary widely over oceanic regions, the *ratios of concentrations* of the different salts remained virtually the same. Two major principles emerged.

1. That the oceans were well mixed in salts. This led to the deduction that the steady-state concentra-

Bongo nets for sampling phytoplankton. Usually deployed in pairs to sample the plankton at two different depths, these nets are remotely opened and closed to avoid contamination of the sample by plankton at other than the selected tow depths. The "flying wing," here held by the crewmen, used to depress the nets down to selected depths during the tow. (Courtesy David Stein, Oregon State University.)

An Isaccs-Kidd mid-water trawl. A long, tapered sleevelike net, this IKMT has a mouth opening 2 m × 2 m. The V-shaped depressor wings along its lower edge allow the net to be towed horizontally but at depths of thousands of meters. The device mounted in the center of the mouth measures the volume of water that passes through the net during the tow. (Courtesy David Stein, Oregon State University.)

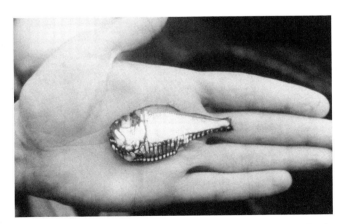

A specimen of hatchetfish, taken by mid-water trawl at 800-m depth. Fish taken from the deep interior of the sea are usually quite small, testimony to the relative scarcity of food there. (Courtesy David Stein, Oregon State University.)

A beam trawl for benthic organisms. This sledlike trawl is dragged along the bottom, to sample life forms on the ocean floor. Cameras and strobe lights mounted on the crossbar have two purposes: to photograph the organisms in their natural positions just before being scooped up, and to learn what type of creatures were able to evade or escape the net itself. (Courtesy A. G. Carey, Jr., Oregon State University.)

Figure 3.5 Equipment and techniques used by biological oceanographers.

tion of total salts is *large* compared with the rate that new salts are added, and that the different oceans "mix" relatively quickly.

2. That the observed differences in the concentrations of salt in different oceans had to be interpreted as variations in the concentration of *pure water* in seawater. Since there were no known sources or sinks of pure water in the interior of oceans, such variations were explainable only in terms of the net gain or loss of H_2O through exposure to the atmosphere which is a major source of water to the oceans, through precipitation, or a major sink through evaporation.

These were indeed important principles because, from that time on, all new hypotheses about ocean behavior were constrained to fit them. Now we had bounds within which to estimate the time needed for waters to circulate inside the major ocean basins. We also had the imperative to study processes that force their *vertical* circulation, and to find ways to measure such processes. Chemical oceanography was born, and from the outset it was coupled tightly with the physics of oceans and atmospheres (Figure 3.6).

The chemical oceanographer sees the ocean as a complex, multicomponent solution. The first task is to identify these components, organic and inorganic, and to measure precisely their concentrations (see Table 7.1). The second task then follows along three main routes.

1. How do the components *enter* the ocean fluid—what are the sources of chemicals?

2. How do the components *stay* dissolved in seawater—what are the reaction processes?

3. Where do the components eventually go—which are the chemical sinks? Not only must processes be identified, but also their *rates* must be measured.

We distinguish between marine chemists and chemical oceanographers. Marine chemists study the properties of seawater in laboratories and establish the chemistry and rates of processes. Chemical oceanographers apply known chemistry to the oceans and identify the scope of what is unknown.

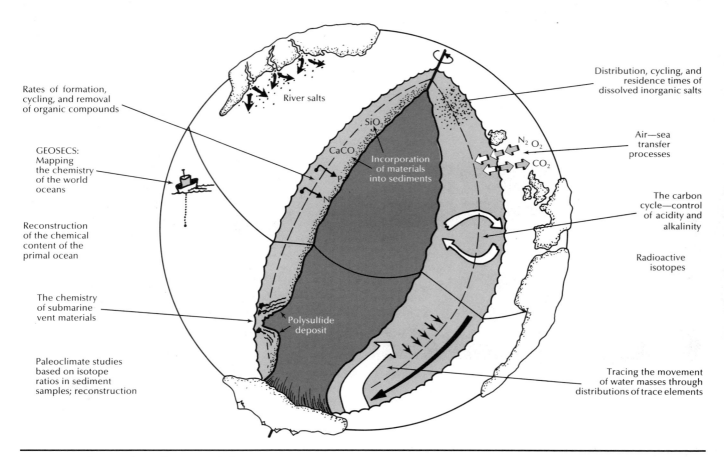

Figure 3.6 Oceans from the view of chemistry.

Study Areas

The Carbon Cycle—Control of Acidity–Alkalinity.

We found early on that the relative acidity of seawater samples from widely different locations *varied over only a narrow range.* On the pH scale from 1, highly acidic, to 14, highly alkaline, ocean values rarely are less than 7.8 or more than 8.3. Ocean water is thus slightly alkaline (7 is neutral) and stays that way. What chemical processes work to maintain this nearly neutral pH state of ocean fluid? Such processes are called "buffering" processes. That the oceans are buffered is of great consequence to the maintenance of life, which would not flourish in a soup of either highly acid or highly alkaline quality.

Chemical oceanographers now identify two main buffering mechanisms, each with a different time scale. The diffusion of CO_2 gas from atmosphere to ocean, together with dissociated salts such as calcium carbonate ($CaCO_3$), drives the *short-term mechanism.* On entering seawater, the CO_2 molecule readily combines with the water molecule to form HCO_3, which leaves one free H^+ ion, or CO_3, which leaves two free H^+ ions. Excess H^+ ions drive the fluid toward a higher-acidity state, but in their presence more calcium carbonate will dissociate and thus free more CO_3 . . . which then combines with the excess H^+ ions to form more HCO_3 . . . which reduces the number of free H^+ ions and drives the acidity state back toward neutral.

Measuring the *rates* of these chemical processes has been an ongoing task for chemical oceanographers. This task is made complex by the role played by marine plants and small animals that build shells of calcium carbonate. Their activity removes both calcium and carbonate ions, and thus alters the balance of these ions in the buffering mechanism. Therefore, chemical oceanographers and biological oceanographers must jointly research the pH-control processes.

Over much *longer time scales,* the buffering processes in the ocean surface layer are influenced also by the *rate* at which the calcium carbonate shells and bones are actually incorporated into the sediment. Some of this biological material dissolves to reenter the fluid ocean; such dissolution increases with ambient pressure, so that the deeper parts of the ocean basins never accumulate $CaCO_3$. Instead, the deep waters dissolve this material out of the detritus that falls from above. Deep water that is eventually upwelled, therefore, carries with it a resupply of calcium carbonate for the benefit of organic production in the photic zone, but this cycle requires hundred of years.

The Carbon Cycle and the Excess CO_2 Problem.

It now appears that, over time scales of hundreds of years or more, the oceans will be able to absorb and eliminate, via sediments, all the excess CO_2 that humankind is introducing into the atmosphere by burning fossil fuels. In this respect, the chemical oceanographer studies the rates of material transfer between air and sea, between ocean water and bottom sediments, and between ocean water and the biota that live in it. In essence, the oceans as an ultimate "sink" will be defined only after such process *rates* are well described.

Pollutant Chemistry.

During recent decades chemical oceanographers have measured concentrations of lead in ocean surface waters that are many times greater than that in deep waters. The source is mainly the lead additives to gasoline. After combustion the lead particulates become airborne, are distributed over the oceans mostly along the westerly wind belts, and eventually cross the air–sea interface into seawater. Methane and freon cycles are also being studied. Methane in air has properties like the greenhouse effect of CO_2 (Chapter 1). Freon is thought to interact with ozone in the upper atmosphere and thereby alter the intensity of ultraviolet light that reaches the sea surface (Chapter 14).

Amounts and types of new chemicals that humankind is releasing into coastal waters and coastal atmospheres are increasing. How these will impact the ocean as a life-support system is of great concern to us (see Chapter 19). These are multidisciplinary problems of a very complex nature. They require physical oceanographers to chart the currents and flushing times of coastal areas. Geologists enter these studies because clay particles carried in suspension in river outflows are integral parts of the chemical reactions that extract pollutants and concentrate them in coastal sediment. Biologists then study how the pollutants alter life cycles of marine organisms.

GEOSECS.

In 1969, following the discovery that pollutants from atmospheric nuclear tests were entering the deep ocean, chemical oceanographers and geochemists organized a project to map the entire deep ocean hydrosphere for its chemistry. The rationale was straightforward: we needed to acquire a chemical description of the "prehuman" deep ocean. Otherwise we would never know what the deep oceans were like, chemically, before the impact of humankind. The project was called GEOSECS (Geochemical Ocean Sections Study), and it carried out a number of cruises north to south and east to west across all the major ocean basins, covering 160,000 km of track with 316 stations sampled all the way to the ocean floor (Figure 3.7). By the end of the project each station represented an investment of $100,000 (Craig and Turekian, 1980).

An example of the new knowledge that came out of GEOSECS is the chart of concentration of radium-226 in the deep waters of the Pacific (Figure 3.7). Radium-226 is a naturally occurring, radioactive element that enters the ocean *from* the bottom as a

product of chemical transformation within the sediments. Scientists interpret the contours of radium-226 concentrations to show that the "oldest" water in all the ocean basins is the deep water of the Northeast Pacific.

Radiochemistry. Radiochemistry is increasing in importance in oceanography. It is used in analyses of *stable* isotopes such as oxygen-18 or *radioactive* isotopes like carbon-14 or strontium-95. Stable isotopes can be used as tracers to describe the movement of

Figure 3.7 Cruise tracks of the GEOSECS expeditions. The bottom map of the Pacific shows the distribution of the radium-226 isotope across the deep waters of the basin, both north and south. The highest concentrations are in the Northeast Pacific, suggesting that these are the oldest of all ocean waters. (Tracks from GEOSECS Atlas; radium-226 isotope data from Chung and Craig, 1980; 1 unit = 10^{-14}g ^{226}Ra/kg.)

water masses. For example, the ratio of ^{18}O to ^{16}O atoms in well-mixed seawater is constant; when seawater evaporates, the initial water vapor carries these in the same ratio as that of liquid water. Because ^{18}O is the heavier of the two states, however, it gets "rained out" faster. Then, a marine air mass that moves inland produces rainwater in which the $^{18}O/^{16}O$ ratio becomes smaller as the distance inland increases. For example, snowfall in the interior of Antarctica is extremely low in this isotope ratio, and this ratio is locked in the snow and ice that flow back to the sea in the form of melting icebergs. Therefore, finding a lower-than-average ratio anywhere can be interpreted as a higher-than-average local freshwater input. Scientists use this technique to estimate that the "flushing" time of the water beneath the floating Ross Ice Shelf of Antarctica is about 6 years; knowing this, we can then estimate the rate at which the ocean currents *deliver heat toward* Antarctica. This is extremely useful in the study of how Antarctica impacts upon global climate.

Tritium, ^{3}H, a radioactive isotope of hydrogen that has three times the mass of ordinary hydrogen, was produced in the atmosphere in large quantities during the high-altitude tests of nuclear explosives in the 1950s and is now found in seawater. With a half-life of 12.3 years (see Figure 19.23 for an explanation of half-life), tritium has now become a useful tracer of the *rate* at which water masses form and of their movement. Figure 3.8 shows how tritium was distributed in the North Atlantic in 1972. The contours are unmistakable evidence that surface waters (into which tritium first diffused from the atmosphere) are sinking off Greenland and slowly moving along deep levels toward the Southern Ocean (see also Figure 11.3).

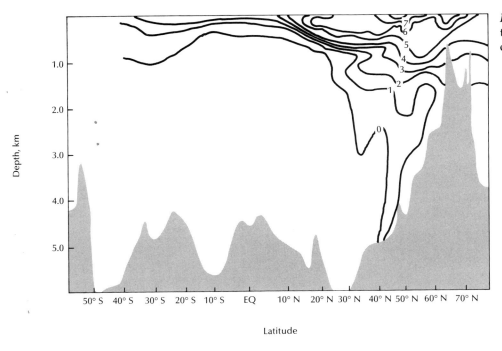

Figure 3.8 Use of radioactive trace elements in the study of ocean motion.

The distribution of tritium ^{3}H was sampled along a cruise track that followed the major basins of the western portions of the North and South Atlantic. These data are from 1972–1973. In the early 1950s oceanographers measured only about one atom of tritium for every billion billion (10^{18}) atoms of hydrogen, a very small concentration which is also used as a standard unit called 1 tritium unit (TU).

After the massive introduction of tritium as a by-product of the testing of nuclear weapons during the late 1950s, oceanographers found the isotope in ever-increasing concentrations. This graph shows clearly that by 1972–1973 the amount of tritium in surface waters had increased to more than ten times the level in the earlier measurements.

Tritium has a half-life of 12.3 years and will eventually disappear. In the meantime, its presence offers the potential of tracing the *rates* at which deep water masses move. The graph shows clearly that surface waters in the North Atlantic are sinking to depths of 5000 m. This is how we may interpret the deepening contours of tritium beginning at about 65°N latitude, in the vicinity of Iceland and Greenland. (From Hammond, 1977)

A crowded but highly automated chemistry laboratory on a modern research vessel. (Courtesy Bob Collier, Oregon State University.)

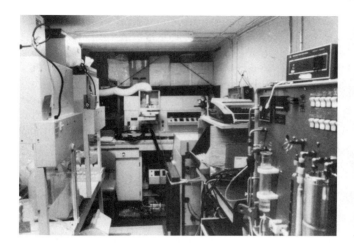

Taking large-volume water samples. For some chemical analyses, for example, measuring the concentration of radon gas in seawater near submarine hydrothermal geysers, relatively large volumes of sample are needed. (Courtesy L. I. Gordon, Oregon State University.)

A nighttime station. Recent expeditions have focused on chemical analyses of the particulates suspended in the water as well as of the seawater itself. So large a volume of seawater must be filtered to obtain an adequate sample of the particles that an elaborate filter–pump assembly is itself lowered deep into the water column. It filters hundreds of liters of sample *in situ.* (Courtesy Bob Collier, Oregon State University.)

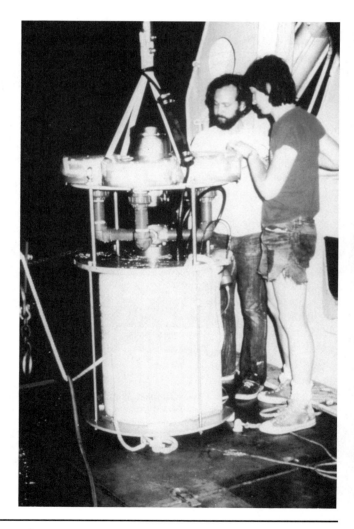

Figure 3.9 Equipment and techniques used by chemical oceanographers.

Sampling in Chemical Oceanography

The mainstay of sampling methods has been to collect water samples for later analysis in the laboratory. Testing for *salinity* now requires only about 50 cc (cubic centimeters) of seawater; this analysis is a measurement of the electric conductivity of the sample. But testing for concentrations of radioactive elements requires the collection of large quantities of water, up to several metric tons. This restricts severely the number of samples that can be taken and processed on a given expedition.

Chemical oceanographers also measure concentrations of the dissolved nutrients. These data are extremely useful to biological studies. Generally, we are mostly interested in the nutrient content of near-surface waters, and chemists have been able to develop autoanalyzer equipment that monitors nutrients and dissolved gases on a continuous basis as water is pumped directly from the sea into the ship's laboratory (Figure 3.9).

The View from Geology and Geophysics

The study of marine geology is quite "young" but deals with subjects that are very old. Early geologists must have known intuitively that the ocean floors and continents were interconnected in a geologic sense. Because so little was established about the ocean floors, and with so many theories to be worked out on continent structure and processes, they concentrated their efforts on land geology. Great strides were made in the nineteenth century by Charles Lyell, who systemized geology into a new science, and Charles Darwin, who combined new knowledge of "geologic" time with his theory of the natural selection process as the basis for the origin of species. But global theories of geology were slow in coming. The HMS *Challenger* oceanography expeditions carried no geologists, and although a number of samples of material were gathered from the ocean floor, the geology simply remained "undone" in the excitement created by the new vistas of zoology opened up through the collection of newly discovered marine species. Nor were geologists interested in topographic studies of the ocean floor made possible by the early soundings; submarine canyons (see Figure 5.2) were known to a few as early as 1903, but the geology community in general did not accept these discoveries until after World War I, when the development of the sonic fathometer provided undisputed evidence that canyons ex-

ist. A significant breakthrough came with the publication by B. C. Heezen and his colleagues (for example, Heezen et al., 1959) of physiograpic diagrams of the Atlantic Ocean and other ocean basins (see Figure 4.1.).

Marine geology is by nature a field activity based on the gathering of samples of rocks and sediments. Unlike terrestrial geology, where entire geologic vistas could be taken in at once simply by looking, early marine geology probed almost blindly along the ocean floors. Systematic cruises dedicated to collecting basin samples did not begin until the 1930s. Initial results led to the charting of ocean floors by the types of different sediments found (see Figure 5.11) and to developing composite maps of the morphology of the ocean floor.

Today marine geology is less of a separate discipline and is more interactive with geochemistry and geophysics. The fundamental reason for such interaction is the development during the past two decades of the theories of sea floor spreading and plate tectonics, both stemming from earlier hypotheses on the drift of continents. Figure 3.10 sketches the major thrusts of marine geology and geophysics today.

Study Areas

Morphology of Ocean Basins. Study continues in the traditional area of describing the "shape" of the ocean basins. The reader should be aware that vast areas of the ocean floor are still untouched by any probe yet devised. What little we know of the shape of the basins comes from the precision fathometer records gathered on various scientific cruises. Still, many of the large-scale features like the extensive trenches, the wide abyssal plains, and the extraordinary features like the mid-ocean ridges are now mapped.

The Plate Tectonic Revolution. Without doubt, one of the most earth-shaking developments in modern science has been the confirmation that the crust of planet earth moves across its surface as a series of solid "plates." These plates move against each other, slide by each other, or slide over and under each other in a constant process we now call "plate tectonics." Within this phenomenon we now recognize that the material of the ocean floor is distinct from the material making up the continents: it is heavier, and because it sinks lower into the plastic mantle of the earth, basins are created with continental rims and are now filled with the oceans.

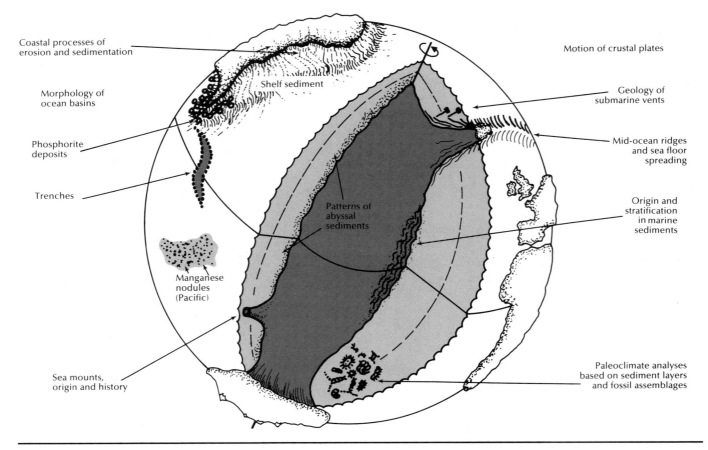

Figure 3.10 The oceans from the view of geology.

So it is that many of the topics outlined in Figure 3.10 as major geology perspectives of the oceans are related to this revolution in global geology. The subject is vast; the literature is already voluminous. The new knowledge that is being gained almost daily is expanding its influence into all the other sciences. No oceanography subject can be explored today without touching in some degree on this global phenomenon. In this text, the subject is first introduced in Chapter 4 and later Chapter 21 presents more details for the interested reader.

Coastal Processes of Sedimentation and Erosion. Overall, the sediments of the continental margins are the most intensely studied to date for several reasons. The push to develop the petroleum resources of these regions demands extensive studies of the *processes* of coastal sedimentation, the *rates* of accumulation, and, equally important, the *stratigraphy* of consolidated sediments. Geophysicists probe the topography of subfloor sedimentary rocks by using a point source of high-energy sound activated

near the surface and then recording on sensitive geophones the sound waves reflected from the various layers of subfloor. The technique is called *seismic refraction*. Marine geologists analyze core samples from holes bored through the layered formations to classify the structures in terms of fossil species, age, porosity, and so on, all of which are useful in determining petroleum reserves.

A second reason that marine geologists study continental margins is the impact of dense human populations in the coastal zones. Port facilities are continuously impacted by the shifting sands along coastlines, and many ports and channels are regularly dredged of sands that arrive either directly from rivers or by the alongshore littoral drift. Wave energy dissipation along shorelines erodes the beaches and creates new barrier islands. Coastal geology studies also include the description of relict beaches, evidence of where the ancient shorelines existed. Coastal oceans are useful dump sites. A study of the distribution of sizes of sediment particles can give clues to the "sorting" action by waves and currents. Coastal currents are a needed input into decisions of where waste and industrial effluents are to be dumped.

Sediments of the Deep Ocean. The distribution of types of sediments over the ocean basins has been mapped (see Figure 5.11). Lithogenous sediments, those derived from the erosion of land, cover a large portion. In general, these deep-sea muds cover the deepest parts of ocean basins. Here the mechanisms of sediment delivery are the turbidity currents (described in Chapter 5) and the prevailing belts of westerly winds that carry terrigenous dusts far out to sea. Glaciers that empty into the sea often carry substantial loads of terrigenous material. Icebergs that break free of glaciers and float out to sea disperse this material over large areas.

The greater part of the ocean floor is covered by sediments derived from the shells of plants and animals, called biogenous sediment. "Diatomaceous" sediments are built from the silica shells of diatom plants and radiolarian protozoa (see Chapter 15). We know that diatoms tend to be the dominant plant types in colder and nutrient-rich waters, and thus diatomaceous sediments dominate in the high latitudes around Antarctica and in the North Pacific. In the more temperate oceans the biogenous sediment is dominated by fossil shells of the order Foraminifera (see Chapter 15). These animals construct shells of calcium carbonate. Recall, however, that $CaCO_3$ tends to dissolve in the very deep parts of ocean basins (depths greater than 4500 m), and this is why much of the basin floor in Figure 5.11 is covered by deep-sea muds even in temperate latitudes where the overlying waters produce large amounts of foraminiferal shells.

In recent years marine geologists have developed methods of extracting long cores from the sediments of the ocean floor. The JOIDES program has carried out core sampling in all of the world basins, much as the geochemists have sampled the deep ocean fluid above these basin floors. But although the water geochemistry tells us very little of the *history* of oceans, the core data are a historical record of great utility. We can "read" the layered sediments like a historical atlas. As outlined in Figure 3.10, research in the origins and stratification of marine sediments lead to valuable insights on paleoclimates, the patterns of behavior of ancient marine life, and even such events as ancient volcanic eruptions. For example, geologists have discovered that one animal species in particular, *Globigerina pachiderma*, builds a spiral shell that "coils to the left" in warm surface waters, but reverses the spiral direction in colder waters. By counting the relative proportions of the two types of fossils in successive layers of sediment cores extracted from the sea floor, geologists can reconstruct the ocean's surface water temperatures during past geologic epochs.

Sediments of Economic Value. More recently, marine geology has virtually "exploded" in a flurry of studies associated with the effects of sea floor spreading. We are studying intensely the geothermal reactions where seawater interacts with rising magma along the active mid-ocean ridges (see Chapters 19 and 21). Rich deposits of polymetallic sulfides are found in these areas. Spurred by the 1982 Law of the Sea Convention, which grants economic control of coastal zones to the coastal state, many nations are surveying their continental margins for mineral and petroleum resources.

Other sediments with economic potential are the *manganese nodules* and *phosphorite* deposits. Both are hydrogenous sediments, forming by precipitation out of the seawater solution. Manganese nodules are rich in manganese, nickel, iron, and copper and occur as accretions around a core material (often a shark tooth or the earbone of whale), reaching sizes up to several centimeters. In some areas, particularly the Central Pacific, these nodules occur in such density as to be a potential mineral resource. Phosphorite is deposited in shallow water, also by precipitation out of solution. The accretions are irregular in shape but occur in sufficient quantity to constitute a valuable potential resource (phosphorous is a necessary fertilizer for agriculture). In warmer tropical waters, calcium carbonate can precipitate to form the white sand beaches of islands in shallow waters, such as those of the Florida Keys.

Sedimentation Rates. An important part of geological oceanography is the measurement of *rates* at which sediments accumulate. The sands and silts on continental shelves build at rates of over 10 cm/1000 yr. Some marginal seas such as the Gulf of California (into which empties the Colorado River) build sediment at up to 1 m/1000 yr. Deep ocean basins accumulate biogenous sediment at rates of about 1 cm/1000 yr, but the terrigenous muds and clays of ocean basins form at only about one-tenth this rate.

Sampling

The geological oceanographer's task is straightforward in principle: take samples of the ocean floor.

A box corer for sampling the floor sediments. When the lowest part of the frame touches the bottom, the square box is driven down into the sediment by heavy attached lead weights. Before retrieval, the triangle-shaped metal lid is levered down to close the bottom of the box, trapping the sample and protecting it from washout during the long haul up to the surface. Box corers sample only the topmost 30 to 50 cm of sediment. (Courtesy Bob Collier, Oregon State University.)

A piston corer for extracting long cores of sediments. The dartlike device hanging from the boom is a weighted steel tube that is dropped into the sediment, collecting a core sample inside that is about 2 m long. Within the derricklike structure is another piston corer, this one over 30 m in length, so large that special equipment, and ships, are needed just to deploy and retrieve the long core samples. (Courtesy Bob Collier, Oregon State University.)

Core samples. After retrieval, the cores are split and stored within plastic tubes under controlled humidity conditions. These samples are in effect a very valuable "library" of information. Scientists can retrieve pieces of ocean floor that date back millions of years in time. Notice the layered structure of this sample. (Courtesy L. D. Kulm, Oregon State University.)

Figure 3.11 Equipment and techniques used in marine geology and geochemistry.

But the execution is formidable. An early technique relied on the grab sampler, which was simply dropped to the bottom. Although rapid to deploy, the device does not protect the sample against washing by water during the time the sample is raised to the deck of the ship. It is still used for rapid initial surveys. An advanced mechanism, called the "box corer," (Figure 3.11), retrieves a protected sample of floor material; with this, the geological oceanographer can study the layered structure as well as the distribution of burrowing organisms within the sample.

A piston corer is used to extract samples deep in the basin floor. The device is lowered to just above the ocean floor, at which time a release is activated and the heavy device plunges the remaining distance. The core itself is retained within a plastic tube liner; cores up to 20 m long are routinely taken where the sediment allows such penetration. For cores of even greater penetration depth, a modified petroleum drilling rig is used. The most famous of such rigs is the *Glomar Challenger* used in the world-wide drilling program JOIDES. This ship-mounted rig is capable of entering holes in 4000 m of water to secure drill cores up to 1000 m long.

Underwater photography has been a very useful tool for geological studies of the surface of the ocean floor. Photographs often reveal ripple marks in the sediment, which suggest a winnowing action by bottom currents. They also show the marks of burrowing animals. Photography is the chief method used in surveying for deposits of manganese nodules.

With the development of self-contained submersible diving platforms, geologists now accompany the sampling instrument to the very ocean bottom. Here the operator can view the bottom and select samples directly as well as photograph the samples *in situ*.

Summary

The objectives of this chapter are two: first, to outline *how* specialists in each of the major science disciplines have defined their own areas of interest and, second, to point out *when* and *where* their work may be joined for mutual progress in understanding complex ocean processes. These outlines are presented on the same cutaway global section chart for each discipline, so that readers can find a number of regions or topics or processes that are common to two or more disciplines. For the five following examples dialogue among disciplines is not only probable but necessary in solving problems.

1. *Modeling the ocean as a two-layer system.* Sooner or later every oceanographer finds a need to treat the ocean as if it were composed of two layers: a warm, thin surface layer that feels the sun and wind and a deep, thick layer of cold water. Physicists use this model in describing circulation. But circulation theories must also fit the distributions of living organisms (biology) as well as patterns of bottom sediments whose distribution is affected by circulation (geology); they must also conform to the charts of "age of water" (chemists).

2. *The all-important topic of air–sea transfer processes.* Many of the characteristics of vast ocean space are controlled by processes that transfer material or energy across the air–sea interface. Chemists study the gas exchanges, including the key gas CO_2, which supplies carbon to organic life (biology), as well as carbonate ions, from which the biota construct the shells of $CaCO_3$ and SiO_2 that settle into the bottom sediments (geology). All disciplines need the models of heat budget for the ocean involving solar radiation and evaporation (physics).

3. *The story of sediments.* Sediment studies provide the best historical data for the planet earth for the past 200 million years. Sediments of ancient seabeds now exposed on continents extend this history much farther into the past (geology). Geochemists construct models of ancient climates that biologists use to explain the fossil record of evolution and extinction. Sediments are the main "economic" resource of the oceans, far outstripping food in dollar value.

4. *Land–ocean boundary zone processes.* Humans congregate near the ocean edge, drawn by its resources and transportation. Rivers converge to the sea carrying water and materials. Marine life abounds in the shallow shelf waters. Here all discipline studies support each other: tides, waves, and circulation (physics) control the dispersion of nutrients and eggs (chemistry and biology) and provide the energy that drives coastal erosion and deposition (geology).

5. *Deep–sea hydrothermal vent research.* Geologists and geochemists work out the chemistry of the hot submarine vent waters, and biologists devise models of the life-support system that the chemistry drives.

Study Questions

1. Chapter 2 contrasted the cycling of plant nutrients in the marine and the terrestrial biospheres, paying particular attention to the *space* and *time* scales involved. Discuss the contribution given in this chapter that *physical*, *biological*, and *chemical* oceanographers each make to this study.

2. Consider that you are a biologist and that you have been taking samples of surface water phytoplankton in coastal waters *for years and years.* Then along comes a physicist to explain how upwelling occurs. Having learned the physics of upwelling, how might you reorganize your biological data?

3. What is it that physical oceanographers *do?*

4. What is it that biological oceanographers *do?*

5. What is it that chemical oceanographers *do?*

6. What is it that geological oceanographers *do?*

7. What might be the impact on the biological oceanographer of learning that the time scale for the vertical circulation of the major oceans is *hundreds* of years?

8. Is there any connection between the migration routes of marine animals and the system of ocean currents?

9. If your answer to question 8 is yes, explain *how* a fish or eel would *know* when it is swimming with the current or against it? If you find it difficult to speak for a fish, how would *you* know whether your little yellow submarine is moving with the current or against it? Do you even care, as long as no one else knows how either?

10. Both chemical and physical oceanographers can *trace* the routes taken by different water masses moving through the fluid oceans, but they do this in different ways. Explain.

11. After analyzing water samples from many different parts of the world oceans, oceanographers of the *Challenger* era concluded that the oceans *were well mixed.* What was the basis for this conclusion? How can this information be useful?

12. With the discovery that the alkalinity of ocean fluid stays within a very narrow range, chemists began a search for the mechanism that maintains seawater in this state. What is the mechanism? Why should biological oceanographers be concerned?

13. What was the GEOSECS project and why did it come into existence?

14. A principal objective of geological oceanography is understanding how sediments accumulate on the ocean floor. Give reasons why this information is useful to (*a*) biologists, (*b*) physicists, and (*c*) chemists.

15. Most recently, geologists have discovered certain types of floor sediment that may have potential value as a source of metals. In what forms are these deposits found and which are the metals found in them?

16. Write a short essay to develop the idea that the two-layer *behavior* of the ocean is essential to understanding the limits to plant production in the ocean.

17. Show that the same two-layer model explains how chemicals that enter the ocean surface can eventually reach the deep ocean.

PART II

Using satellites to outline the shape of ocean basins. This image, taken by infrared devices aboard the METEOSAT satellite on March 6, 1978, paints the African continent dark and outlines the eastern margins of both the North and South Atlantic ocean basins. (Courtesy W. V. Burt, Oregon State University.)

SHAPES AND SEDIMENTS OF OCEAN BASINS

In which we examine the morphology of typical basins, study the processes that shape the margin zones between continents and basins, and explore how basin shape influences the behavior of liquid oceans.

CHAPTER 4 SHAPES OF OCEAN BASINS

What are the shapes of the basins in earth's crust in which the fluid ocean is collected? How did the basins originate and how do they change with time? Can we relate basin geology to fluid ocean behavior during past epochs?

CHAPTER 5 CONTINENTAL MARGINS AND BASIN SEDIMENTS

We study the unique zones where continents and oceans meet. Much of the history of ecosystem earth is recorded in the sediments of today's ocean basins.

CHAPTER 4

SHAPES OF OCEAN BASINS

All science begins with description. How do we describe the *shape* of a liquid mass? Our language contains virtually no metaphors by which we convey specific information about large bodies of fluids. Even the few adjectives in common use are misleading when applied to large volumes of fluid: for example, a *thick* fluid describes one with the consistency of molasses, not one having a large dimensional thickness.

There are two principal factors that determine the shape of fluid oceans:

1. The *shape of the basins* that contain them, described in this chapter.

2. Their *dynamic motion*, which alters their shape from what it is when they are at rest; this is described in Part IV.

Overview of Ocean Basin Morphology

If there is to be an "eighth" wonder of the world, surely it must be the publication just a few years ago of the first comprehensive topographic charts of the ocean basins (Figure 4.1). The first portions were produced in 1959 by Heezen and his colleagues, who brought together all available bathymetric data from the world oceans. For the first time we can see the relief of the entire global earth crust. For the first time we can place in perspective the features of the ocean basins together with those of continental land masses.

The chart is still not complete. It is based primarily on precise measurements of ocean depths along the tracks of many scientific ocean expeditions, and large areas of the ocean basin are not yet surveyed. The reader is cautioned that some detail is artistic interpretation, but the general, large-scale features are known, and these are discussed in following sections.

The reader is further advised that the chart of Figure 4.1 contains large distortions of the relative areas of lands and oceans. This is a consequence of the type of global projection used to map from a spherical earth onto a flat sheet. Here the projection is the Mercator, which distends the high (polar) latitudes; for example, Greenland appears larger than the entire continent of South America, which we know to be false. For a more detailed discussion of maps and cartography in general, the reader is referred to Appendix 1.

Area Contrast: Land versus Ocean

We gain two more perspectives of the global view when the *relative* areas of lands and oceans are compared. In Figure 4.2 a relation is drawn between land and sea area for each 10° band of latitudes, from pole to pole.

First, notice that the land masses occur mostly in the Northern Hemisphere, about 68% of total land. Stated another way, the Northern Hemisphere has twice as much land mass as the Southern. The reader will discover further along in the text that this unbalanced land location has a profound influence on global climate patterns.

For another perspective, contrast the land–ocean areas of the two hemispheres in the latitude sectors 30 to 60°N and 30 to 60°S. In the north sector at least one-half of the globe is land; this is the temperate zone, which gave birth to the history of *Homo sapeins*. In the south sector, however, there is virtually no land; this is the domain of the Southern Ocean and its globe-girdling Antarctic Circumpolar Current.

Last, bear in mind that the total ocean area covers 71% of the earth in contrast to 29% for land. As Figure 4.1 also shows, the Pacific is by far the largest of oceans, with almost 50% of all ocean area. The Atlantic and Indian Oceans divide the remainder about equally.

Elevation Contrast: Land versus Ocean

The most striking comparisons are made between the elevations of lands and seas. The artistry of the

THE FLOOR OF THE OCEANS

Based on Bathymetric studies by
Bruce C. Heezen and Marie Tharp
of the Lamont Doherty Geological Observatory
Columbia University Palisades, New York, 10964
SUPPORTED BY THE UNITED STATES NAVY
OFFICE OF NAVAL RESEARCH

Figure 4.1 A physiographic diagram of the surface of the earth's crust, made up of continents and ocean basins. (Heezen and Tharp, copyright Marie Tharp.)

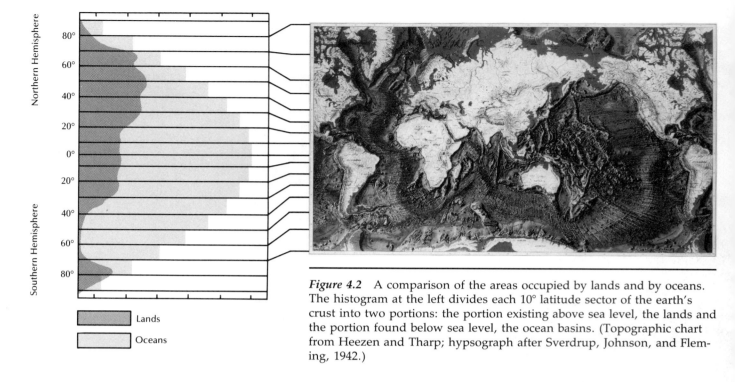

Figure 4.2 A comparison of the areas occupied by lands and by oceans. The histogram at the left divides each 10° latitude sector of the earth's crust into two portions: the portion existing above sea level, the lands and the portion found below sea level, the ocean basins. (Topographic chart from Heezen and Tharp; hypsograph after Sverdrup, Johnson, and Fleming, 1942.)

Heezen and Tharp chart of Figure 4.1 shows the continents as raised platforms above the general expanse of ocean floor. Further, the chart shows sharp changes in elevation around the margins of the continents, evidence that there is a distinct *edge* to the ocean basins.

How deep are the oceans and what is their average depth compared to the average elevation of lands? Answers can be had from Figure 4.3. The average elevation of land is 840 m above sea level, and the average depth for the ocean floor is about 3900 m below sea level. Also of interest is the range of elevations: the deepest ocean depth discovered so far is a sounding of over 11,000 m in the Marianas Trench just east of the Philippine Islands; contrast this with the height of Mt. Everest at 8880 m above sea level.

Figure 4.3 contrasts the difference in elevation of continents and ocean basins in still a different way, by using a bar chart with elevations lumped into 1000-m-thick categories. In this method two of the bars stand out in sharp distinction from the rest: the 0- to 1000-m elevation category and the 4000- to 5000-m depth category. Each represents over 20% of the globe's surface. Statisticians use the term *bimodal* to describe a distribution with two peaks, with the idea that there is a fundamental explanation. Geo-

physicists explain it this way: the material that makes up the ocean floor is denser than that comprising land masses, and therefore it "sinks" deeper into the crust of the earth than the land does. This is the principle of "isostasy," to which we return later in this chapter.

Still another way to contrast elevations is with a *cumulative* graph of the elevation of earth, also shown in Figure 4.3. The graph is constructed this way: labeling the highest elevation as 0%, simply add the percentage of earth covered by each successive depth interval to the previous total, step by step, until the entire earth surface is counted. Notice that the curve crosses 0 elevation at 29%, that is, the curve has summed up all those parts of earth covered by land, 29%. Another number of interest is that only 58% of the earth's surface lies above the mean ocean depth level; 42% of the earth's surface is "deeper" than 3900 m below our present sea level!

Typical Topographic Section across the Ocean Floor

There are two typical cross sections of the ocean floor that show the nature of its topographic relief. Section *A–A* in Figure 4.4 crosses the Atlantic Ocean from Recife, Brazil to Freetown in Sierra Leone, Af-

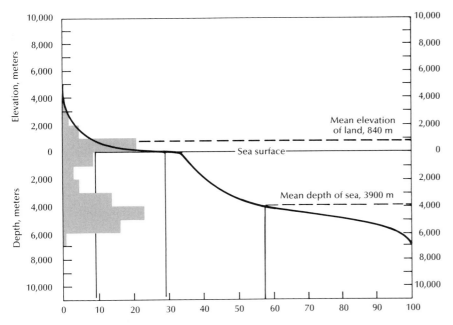

Elevation, meters

Depth, meters

Mean elevation of land, 840 m

Sea surface

Mean depth of sea, 3900 m

Percentage of the Earth's surface

Figure 4.3 A comparison of the relative elevations of the earth's crust. The shaded bars at the left represent the percentage of the earth's surface in each 1000-m interval of elevation above and below the sea surface. The curved line is a cumulative graph. For any selected elevation, the corresponding point on the graph gives the percentage of all the earth's surface that exists at elevations higher than the one selected. If we select the sea surface level, for example, we find that 29 + % of the earth's crust is at higher elevations; this represents the land area. (Based on Sverdrup, Johnson, and Fleming, 1942.)

rica. Section *B–B*, taken from the SCORPIO Expedition, crosses the entire Pacific Ocean along the 28°S latitude line. Compare the location sketches of sections *A–A* and *B–B* with the overall relief chart of Figure 4.1.

These two profiles of the ocean floor illustrate features that are common to all ocean basins as well as major differences between Pacific and Atlantic Ocean basin topography. *Notice* that these profiles are *not* to true scale. We are compelled to exaggerate the vertical dimension in sketches like these. The "true" scale of oceans over the earth can be modeled by picking up a soccer ball from a wet playing field; after the ball drips for a few seconds, the thickness of the remaining water film approximates the scale size of oceans on earth.

Section *A–A*. Section *A–A* is chosen because it shows very clearly why the word "basin" is chosen for the depressions in the earth's crust now filled with ocean water. Away from either shoreline the ocean floor drops rapidly to a comparatively smooth floor called the abyssal plains. These are at average depths of about 5000 m. The central part, however, contains a broken topographic feature that rises more than 1000 m above the level of the abyssal plains on either side. This is the Mid-Atlantic Ridge; reference to Figure 4.1 shows that the ridge extends along the entire length of the Atlantic Ocean, north

to south. Further, notice that the ridge is located about midway between the American and the Euro-African continents. In fact, the Atlantic Ridge is but part of an extensive ridge system that traverses all the major ocean basins. These are the spreading centers from which new ocean floor material is continuously being generated.

Section *B–B*. Section *B–B* across the South Pacific has two features in common with the Atlantic section. First, the ocean floor drops sharply away from the continents to the level of the abyssal plains, showing again the basinlike pattern. Second, the Pacific also has a ridge system, as can be seen in Figure 4.1, even though its ridge is not centrally located as is that of the Atlantic.

Section *B–B* also reveals major distinctions between the Pacific and Atlantic floors. First, there are the major trenches: one is found immediately adjacent to the South American continent, the Peru–Chile Trench, and the other is located to the northeast of New Zealand and is called the Tonga–Kermadec Trench (see Figure 4.7). The origin of these trenches is closely linked to the origin of the ridge-like spreading centers, as will be described in a later section.

Second, the Pacific basin floor topography is much more irregular than that of the Atlantic. We see numerous hills, ridges, and undersea mountain fea-

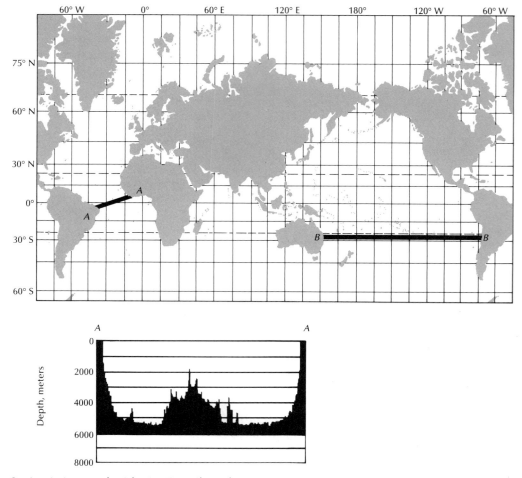

Section A–A across the Atlantic. (From Shepard, 1963.)

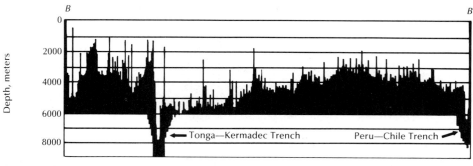

Section B–B across the Pacific. (From *SCORPIO* expedition, 1967.)

Figure 4.4 Typical cross sections of the ocean basins.

tures in the Pacific. The global relief chart, Figure 4.1, supports this view that the Pacific basin, particularly in its western part, is much more mountainous. The many tropical island groups of the Western Pacific, for example, the Samoan, Marianas, and others, are volcanic mountains large enough to penetrate the sea surface. The Hawaiian group is seen as a "chain" of such mountains originated through a series of volcanic eruptions.

Mid-Ocean Ridge Features—A Connected System?

The mid-ocean ridges traversed in both sections in Figure 4.4 are but part of an extensive network of such ridges found in the ocean basins. In length, the network contains over 65,000 km of interconnected ridges (Figure 4.5). In width, these irregular ridges extend 500 to 1000 km to either side of the ridgeline before merging into the abyssal plains. In height, they rise typically to heights of 1 to 2 km above the adjacent plains. In some areas, notably the Iceland Island group in the North Atlantic, the ridge actually penetrates the surface to form groups of connected islands.

A Science Basis for Basin Morphology

By now it is becoming clear to the reader that mere descriptions of the topography of ocean basins are inadequate to explain *why* the oceans have their present shapes. Why is the Pacific basin so much more broken and irregular than the Atlantic (Figure 4.4)? Conversely, why does the Atlantic basin contain a continuous ridge that not only is located in *mid*-ocean (Figure 4.5) but also meanders along

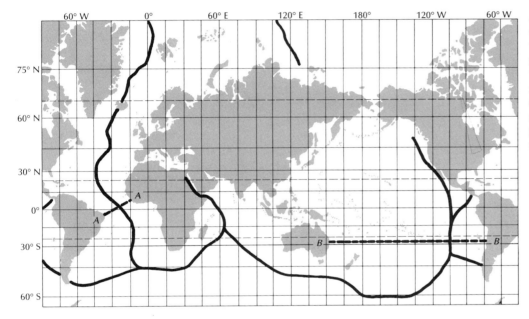

Figure 4.5 The distribution over the earth's surface of the center lines of mid-ocean ridges. Called "rift valleys," the centers of the ridges denote areas where the ocean floor is diverging, moving away from the center lines. Sections A–A and B–B are redrawn from Figure 4.4.

Birth of a new island in the Iceland group. Iceland is comprised of volcanic material ejected from the Mid-Atlantic Ridge, an active sea floor spreading zone that rifts the entire length of the Atlantic basin. This volcano, named Surtsey, created a new island in 1966.

paths identical with the shores of the American continents to the west and the Euro-African continents to the east? And why are the elevations of the earth's surface distributed in a bimodal pattern (bar chart Figure 4.3)?

Answers to these questions are found in a consistent way within the global geology theory called *plate tectonics*. It is a new science, no more than about a quarter century old. It has produced a virtual "storm" of new ideas about the oceans and lands, how they formed and how they move, appear, and disappear. It has already integrated two earlier themes on global geology, so a brief description of these will also serve to describe the tectonic envelope. But a caution to the reader—the focus of this text continues to be on the fluid ocean itself. We need here the outlines of global geology to continue an orderly construction of introductory oceanography. The fascinating development of plate tectonic theory itself is covered as a special topic in Chapter 21.

Continental Drift—A Step toward Plate Theory

In the fifteenth century exploration of the oceans increased dramatically as greed and curiosity drove Western adventurers to discover new shores. But the *mapping* of these new shores remained crude until the eighteenth-century invention of the ship's chronometer allowed mariners to take accurate navigational fixes of longitude. With new and accurate maps came the startling revelation that indeed the coastlines of the Americas, north to south, "fitted" very well with those of Africa and parts of Europe. This was the observational basis that drove the world of science to put together the evidence that could prove or disprove that these continents might once have been joined together.

Alfred Wegener (1880–1930) was an early and persistent proponent of continental drift theory. He amassed evidence of various kinds to support the theory (see page 437) and combined them to produce his version of how today's continents were once a large supercontinent called Pangea (Figure 21.1). According to Wegener, a major breakup of Pangea began about 200 million years ago. As the pieces drifted apart, the continents as we know them today began to take shape and the Atlantic Ocean was formed. But the Wegener hypothesis spawned others: for example, if the Atlantic basin was young, should the rocks and sediments of its floor not also be "young"? It also generated counterquestions. For example, unless an energy source could be found that was both plausible to science

and large enough to push continents around en masse, how could the drift hypothesis be taken seriously?

Age of Basin Floors: A Geological Clue. One clue that supported the idea of continental drift, and with it the shape of ocean basins, is that nowhere from any ocean basin have geologists obtained floor samples that are older than about 200 *million* years. (In contrast, rocks from some land masses, for example, East Greenland, are dated at about 3.8 *billion* years.) These data also give an estimate of the rate at which the continents drift apart: taking the width of the Atlantic as 7000 km and dividing by 200 million years yields a *rate* of expansion of 3.5 cm/yr. This is about the rate at which human fingernails grow.

The Mid-Ocean Ridges: Younger or Older? If there is a limit to the age of the ocean floors, another question comes to mind. Where is the basin floor oldest and where youngest? It is an important question, and answers came quickly. Figure 21.6, for example, demonstrates that the oldest floor in the Pacific is farthest from that basin's ridge feature, the East Pacific Rise. Similar age patterns were found for other ridges in other basins. Ridges, then, are the *youngest* parts of the ocean basins.

The Energy Question. The energy question had been settled even before geologists had put together the clues just advanced. Physicists and geophysicists had proved that the heat energy liberated by radioactive decay of materials deep in the earth could provide the large amounts of energy needed to "move continents." The next question was to describe the movement mechanism itself. Because the ridge features were the youngest parts of the basin floors, the mechanism had to "be" there, and this led us right into the next theory.

Sea Floor Spreading Theory

If the mid-ocean ridges are "sources" of new material for the basin floors, the material *must come from deeper levels in the earth crust*—there is no other source. We know from other evidence that the earth crust consists of a solid layer, the *lithosphere*, which carries our continents and ocean basins, and a deeper, plastic layer called the *asthenosphere*. New basin material must therefore come from the asthenosphere as a highly viscous, molten rock, called *magma*. Figure 4.6 shows how the new material enters the basin floor structure.

Magma from the asthenosphere rises upward into

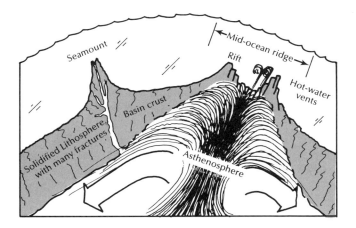

Figure 4.6. A mechanism that injects new material into the basin floor in the vicinity of the rift valleys of mid-ocean ridges. The source of the new material is the magmalike asthenosphere layer of the earth. The solidified lithosphere forming the new ocean floor is moved laterally away from the rift; hence the term "sea floor spreading."

A community of tube worms on the Galapagos Rise, a sector of the spreading sea floor rift in the East Pacific basin photographed from the research submersible ALVIN. These worms feed by filtering bacteria from the seawater around the hydrothermal vents in the rift zone. At the 3000-m level of this living community, bacteria extract energy for subsistence by reducing chemical compounds released from the magma itself. Other community members are mussels, also filter feeders, and scavenging crabs.

The research submersible ALVIN enroute to survey a sector of the East Pacific Rise, a sea floor spreading zone along the Pacific Ocean basin.

the basin floor itself. Contact with the overlying ocean cools the outer surface of the magma pool to the point that it solidifies. Cooling also causes the now-solid skin to contract, creating many fissures

through which seawater percolates downward to cool the material even deeper. This action occurs within the *rifting* zone. Scientists in submersibles have begun to survey these rift zones. They find numerous geysers of superheated water spewing from the fissures, clear evidence of the presence near the surface of hot magma material.

For consistency with the age pattern of the basin floor, the solidified crust newly injected in the ridge zones must move *away* from the ridge, as indicated in Figure 4.6. This behavior of the basin floor is presently under intense research by geophysicists. Two alternative mechanisms for such movement are proposed. The first states that the viscous material in the asthenosphere is itself organized into huge *convection* cells, and that the solidified basin crust is simply being "dragged along"; the open arrow in Figure 4.6 represents this mechanism. A second concept states that the reason a ridge extends far above the surrounding ocean abyss is just a consequence of the greater volume that the mass of hot, molten magma occupies, compared with its volume after cooling and solidification. Therefore, the cooled and highly fissured surface layer over the magma dome is just "sliding down" and away from the high dome.

For completeness, the spreading basin model must also include a process that *removes old ocean floor material*. That aspect is covered in a later section.

Self-consistency of the Basin Spreading Theory. Both of the mechanisms just described move the new, solid lithosphere away from the ridges. This clearly fits the observations that the basin floor ages with distance from the ridge. Both mechanisms also explain why the ridges are higher in elevation than the surrounding abyssal floor. We also find that the thickness of sediment increases with distance from the ridges, a result consistent with basin aging.

Pacific versus Atlantic Basin Spreading. Is there any difference in the spreading phenomena of the Pacific and Atlantic basins? One notable difference is in the rate. Earlier we gave a rough estimate of Atlantic spreading of 3.5 cm/yr, continent to continent, or one-half this rate for the movement away from its ridge. From Figure 21.6, a similar calculation for the Pacific basin yields a much higher rate: dividing the distance from Baja, California (on the East Pacific Rise) to the Marianas Trench, about 12,000 km, by the age difference across this span of the basin, about 136 million years, yields a rate of about 9 cm/yr, that is, five times the Atlantic basin rate.

Could this large difference in rate of spreading also explain why the Pacific basin has so many more seamounts than the Atlantic? Answers are not yet available. Nevertheless, a logical deduction would be that the more vigorous the mechanics of sea floor spreading, the more likely that fissures in the crust would exist to channel magma into the building of more volcanoes, hence more seamounts.

Elevated, ancient marine terraces on the west slopes of San Clemente Island, California. For such terraces to appear, either the sea level must fall or the land itself must be uplifted. When terraces appear in sequences, as seen here, the land was likely uplifted in a series of tectonic, seismic events.

Fracture Zones. The physiographic diagram of the ocean floors (Figure 4.1) shows that a large number of breaks occur along the entire ridge system. On an average, the breaks appear at intervals of about 100 km. Generally, they occur in pairs, that is, a break is continuous from one flank of the ridge across the top and down the flank on the other side. We now know that these breaks are actually fractures in the basin floor, along which one piece of basin floor moves away from the central ridge at a rate different from the adjacent block; hence a *fracture zone* is created. (The evidence for this description is another of the fascinating parts of the tectonic revolution in global geology; for details, the reader may slip to Figures 21.7 and 21.8.)

Fracture zones are strong evidence that new basin material is created at different rates within different sections of the ridge system. As long as the new material divides equally to each side of the ridge, the ridgeline itself is not changed. But in some places such division is unequal and the center line of the ridge itself shifts laterally along a fracture zone. This is clearly seen (Figure 4.1) in the South Pacific basin at about 50°S latitude, where the East Pacific Rise shifts sideways for several hundreds of kilometers along a succession of two to three adjacent fractures. From the view of ocean currents, "gaps" created in ridges by such lateral displacements permit the fluids in the deep basins to "communicate."

Some fracture zones also involve a *vertical* shift between adjacent basin blocks. One of the most noteworthy is the Mendocino fracture, which trends westward away from the Pacific coast of North America at Cape Mendocino, California. Called the Mendocino Escarpment because of its vertical shape, the basin floor on the north side of the fracture is more than 1000 m higher than the adjacent block to the south. This vertical wall in the basin floor extends for thousands of kilometers westward. It has a pronounced influence on the movement of deep waters in this part of the Pacific.

Nevertheless, the reader is cautioned that not all the fracture zones shown in the Heezen and Tharp diagram are documented by actual depth measurements. The artistic license included here, especially for the South Pacific, has to date been poorly sampled.

Sial, Sima, Isostasy, and Archimedes. New basin material formed at the spreading centers is different from the rock types that make up the continents. Continental rocks are composed of minerals rich in silicates of aluminum and are called *sialic* rocks. In contrast, basin material contains minerals that are rich in silicates of magnesium and is called *simatic* material; it is denser than continental rock.

As indicated in Figure 4.9, both the sialic continent and simatic basin crusts "float" on top of the viscous asthenosphere. The basin crust, because it is denser; will float *deeper* in the mantle than the continents. This concept, termed isostatic equilibrium or just *isostasy* by earth scientists, explains why the basin surfaces are depressed and the land surfaces are elevated relative to each other. The fluid oceans collect in the depressions between these two different materials.

This concept of isostasy, as we have indicated, explains the bimodal nature of the bar chart for surface elevations (Figure 4.3). It simply says that the combined weights of basin crust and of its overlying seawater force the basin floor to its present average depth of about 3900 m. The major parts of lands, on the other hand, ride in the mantle so that the average elevation of their surfaces is about 840 m higher than the surface of the oceans.

It is an interesting exercise to ask what Archimedes would have deduced had he been shown the bar chart of Figure 4.3. This early scientist developed the principle that a rigid body will sink into a fluid until the mass of displaced fluid just equals the mass of the total body itself. I cannot help but believe that had Archimedes known of the bimodal nature of earth surface elevation he would have deduced (1) that the earth's interior was fluidic and (2) that the stuff of ocean basins is different from land stuff, that it is denser and hence rides "lower in the fluid earth."

Plate Tectonics Concepts

The two trenches located in the topographic section B–B of Figure 4.4 are parts of a complex matrix of trenches found in all the ocean basins. Most trenches, however, are located around the rim of the Pacific Ocean (Figure 4.7). The existence of trenches has been known for over half a century, but their origins were explained only during the last two decades with the development of plate tectonics. The new plate theory developed because geophysicists asked the question "If new basin material is continuously being added along the ridgelines and is spreading across the basin, where is the old basin floor stuff?" Where has it gone and how? We know from other evidence that the ocean water itself has existed for much longer than the 200 million-year age of the Atlantic basin (see Chapter 7 on the origin

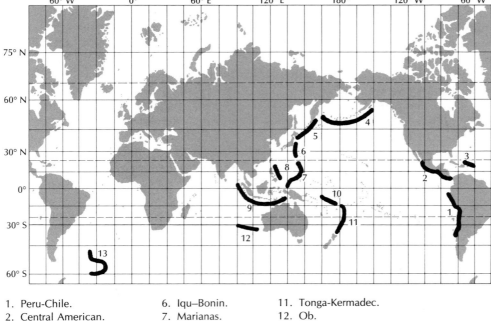

Figure 4.7 Locations of the major trenches of ocean basins

1. Peru-Chile.
2. Central American.
3. Puerto Rican.
4. Aleutian.
5. Kurile.
6. Iqu–Bonin.
7. Marianas.
8. Philippine.
9. Java.
10. Vityaz.
11. Tonga-Kermadec.
12. Ob.
13. Sandwich.

of sea salts and seawater). There must then be a process by which old basin floor vanishes.

The new theory states that the entire crust of the earth is segmented into a small number of integral plates. The actual number of plates is not known with any certainty. There seems to be about 13 major segments, and other smaller segments are being identified as more data accumulate. Figure 4.8 depicts the boundaries and names given to the major plates today.

As described in an earlier section on basin-spreading theory, the forces that move the plates across the globe are thought to result from friction drag between the lithosphere plates and the slow motions within the viscous asthenosphere. With each plate responding to different forcing, the boundaries where they contact, colliding or sliding against each other, must be zones of strong seismic activity. (In fact, geophysicists learn where the plate boundaries are by plotting epicenters of earthquakes, as in Figure 21.9.)

Plate Boundary Types. Plate boundaries fall into one of three categories: divergent, convergent, and slip. The mid-ocean ridges are examples of *divergent* boundaries: the newly formed basin material spreads away from the ridgeline in opposite directions. At the second type of boundary, called a *convergent* boundary, two plates collide. In some places two plates may just move laterally along their common boundary, with neither divergent nor convergent motion; these can be called *slip* boundaries. All three types of boundaries are seismically active.

As is developed in detail in Chapter 21, geophysicists may use still another classification scheme for identifying plates and their boundaries. Boundaries can be *active* in the sense that relative movement is occurring at the present time. Boundaries can also be *passive* in the sense that we find distinct geologic boundaries but there is no seismic activity; for example, the Appalachian Mountains are now recognized as a collision boundary between two ancient plates, once active but now passive.

Convergent Plates: The Subduction Concept. Figure 4.9 sketches a convergent boundary between plates, one of sialic land material and the other a simatic basin crust. Plate theory states that the basin crust will be *subducted* beneath the continent somewhat as shown in the sketch. We use the term subduction in the relative sense: it occurs whether the continent is being *pushed over* the adjacent basin

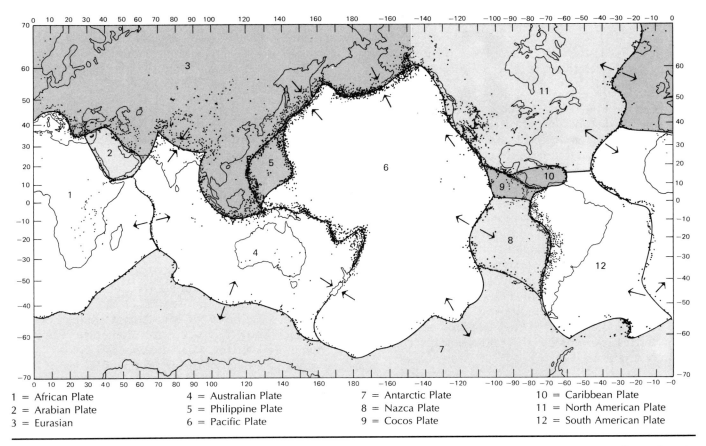

1 = African Plate   4 = Australian Plate   7 = Antarctic Plate   10 = Caribbean Plate
2 = Arabian Plate   5 = Philippine Plate   8 = Nazca Plate   11 = North American Plate
3 = Eurasian   6 = Pacific Plate   9 = Cocos Plate   12 = South American Plate

Figure 4.8 World chart of the known major plate segments and their boundaries. The earthquake epicenters for the years 1961–1967, here designated by dots, are clustered along plate boundaries. Arrows indicate the relative motion of plates at their boundaries. (After Barazangi and Dorman, 1969.)

Figure 4.9 The two types of convergence between continents and basin crusts: *(a)* slow or no convergence, *(b)* rapid convergence

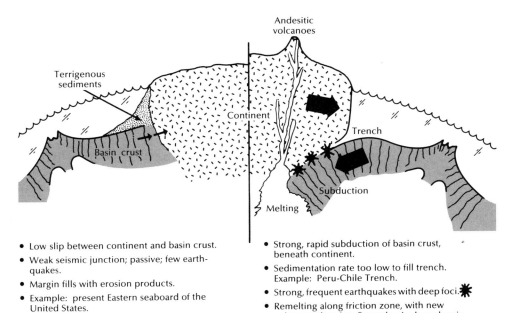

- Low slip between continent and basin crust.
- Weak seismic junction; passive; few earthquakes.
- Margin fills with erosion products.
- Example: present Eastern seaboard of the United States.

- Strong, rapid subduction of basin crust, beneath continent.
- Sedimentation rate too low to fill trench. Example: Peru-Chile Trench.
- Strong, frequent earthquakes with deep foci. ✳
- Remelting along friction zone, with new volcanoes forming. Example: Andes volcanic mountain chain.

crust or the basin crust is being *pulled under* the continent.

Figure 4.9 shows two cases of subduction, depending on the relative rate at which the two plates are converging.

1. When the rate of convergence is rapid, the basin crust is deeply underthrust beneath the continent. This is where we find the *trenches*. Sediments eroded from the continent tend to accumulate in the trench, but continuous subduction is the dominant process so the trench remains. Friction between the plunging basin crust and the overlying continental mass is the source of deep earthquakes. Friction plus the rise in temperature as the crust moves deeper are thought to cause a remelting of the crust material, with the hot new magma rising upward through fissures in the continent to produce new volcanoes at the surface. This is the origin of chains of volcanic mountains located inland from the trench boundaries and parallel to the shoreline of ocean basins. A classic example is the Andes volcanic chain of South America.

2. When the rate of convergence is slow, subduction will also occur but with less vigor. Events and processes that accompany subduction are also subdued, as shown in Figure 4.9. Erosion from the continent may deliver sediments to the trench floor faster than the tectonic motion sweeps them away, with the result that the topography of the trench vanishes. Seismic activity will also be very low. An example of such a boundary is the Pacific Northwest coast where the sediments from the Columbia River have filled the trench that was once there.

An extreme example would be where there is no relative motion between continent and basin. The Atlantic seaboard of North America is such a case (see also Figure 21.12). Here the continent and the adjacent basin move at the *same rate,* and thus it is not a plate boundary.

Connections Between Basin Tectonics and the Fluid Ocean

As oceanographers, our objectives are to learn how basin tectonics interact with the fluid ocean itself. This is best organized in terms of *time scales.* How does the present basin shape influence the behavior of today's oceans? This is basin–fluid interaction in the short term and is covered in Chapter 11 in a general description of ocean circulation. Here we examine the longer-term connections. How does the ex-

tensive system of mid-ocean ridges influence the movement of deep water? What connection is there between the trench–volcanic mountains of subduction-type plate boundaries and the distribution of water within the several oceans? What can we gain from paleooceanography studies? What happens to "sea level" in the long-term tectonic view?

Ridge Gaps and Deep Circulation

We would expect the system of mid-ocean ridges to have a direct influence on the movement of deep water, not only because ridges extend high above the basin floor but also because they exist within all the major basins. Clearly, the ridges control the way that deep waters communicate from basin to basin. More to the point, the numerous *gaps* in the ridge system are the channels through which water spills from one basin to another. One of the best-documented cases is for the Atlantic Ocean.

Figure 4.10 charts the Mid-Atlantic Ridge system. Its major known gaps are the Romanche Gap, located at the equator, a gap at about 8°N latitude, and another gap just south of the tip of Africa at 48°S latitude. Also important to South Atlantic deep circulation is the secondary ridge, the Walvis, which connects the central ridge with the African continent.

The deep water masses of the Atlantic are the coldest, densest waters of all the world oceans. The dominant water is called Antarctic Bottom Water (AABW). A large amount of this water mass forms in the surface of the Weddell Sea (Figure 4.10), from which point it sinks to the very bottom and begins to spread outward to fill the deepest parts of the ocean basins. (For details of how this water forms, see the Gigantic Antarctic Sea Ice Mixing Machine, Chapter 20.)

Some of the AABW reaches far into the Northern Hemisphere, especially in the Atlantic. In its northward progression (Figure 4.10) it first moves along the western side of the mid-ocean ridge up to the equator. Here the AABW spills through the Romanche Gap into the eastern sector of the South Atlantic basin. Part of this overflow then moves southward to cover the entire eastern abyssal plain; its progression is stopped by the Walvis Ridge. AABW also fills the basin south of the Walvis, but here the inflow occurs through a large ridge gap at 48°S.

In the Northern Hemisphere the deep flow is channeled northward along both sides of the main ridge. By this point the original AABW has been considerably diluted through mixing, however. We can trace its northward penetration to about 40°N.

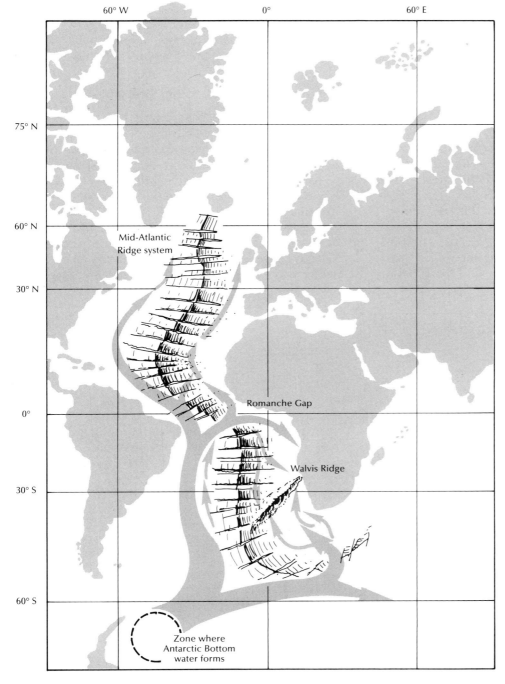

Figure 4.10 The influence of the Mid-Atlantic Ridge topography on the movement of the Antarctic Bottom Water mass as it spreads northward. Gaps in the ridge system provide the major routes through which the deep water masses are exchanged. (After Wüst, 1935.)

Basin ridge topography must also be an important element in the study of deep-ocean biological systems, especially those of the benthic community. The separation of basins by the extensive ridge system may play a key part in the evolution of such communities. So far, sampling of the deep-ocean fauna is much too sparse to yield answers to these questions.

Trench–Mountain Systems, Basin Watersheds, and Ocean Salinity

A careful look at Figure 4.1 shows that many of the major mountain ranges on the continents are the volcanic chains associated with trenches and subduction zones, as described in Figure 4.9. To the oceanographer, these volcanic chains are quite important in that they divide the continents into watersheds, which in turn drain fresh water into adjacent oceans. Figure 4.11 gives the watersheds for the Atlantic, Pacific, Indian, and Arctic oceans.

It is clear that with the major mountain ranges located on the Euro-African and American continents, the largest watershed drains into the Atlantic. Combining Atlantic and Arctic watersheds—almost all Arctic fresh water eventually mixes into the Atlantic—we find that 80% of the outflow from the major rivers feeds into the Atlantic. The Pacific and Indian oceans divide the remaining 20% about equally (see Table 18.1 for river discharge numbers).

One expected conclusion from this watershed comparison is that Atlantic seawater should be measurably less salty than that in other basins. The opposite is true. The Atlantic is in fact the saltiest major ocean (see Figure 7.4 for numbers). This is not a

trivial oceanographic result; it is imbalances like these that drive the water exchanges between basins, that is, ocean currents. Explanations of this salt–water imbalance are given in later chapters.

The Problem of Sea Level in Ocean Basins

Up to this point we have referred all vertical comparisons of heights and depths to sea level. By now, however, the reader may have acquired an uneasy skepticism about the validity of using sea surface as a reference level. Consider that the continental boundaries and ocean floors move about, and new material is injected while old material is removed from basin floors. Is the total volume of all ocean basins constant enough to maintain a usefully "steady" level of ocean water within them?

The only "level" from which to gage the height of

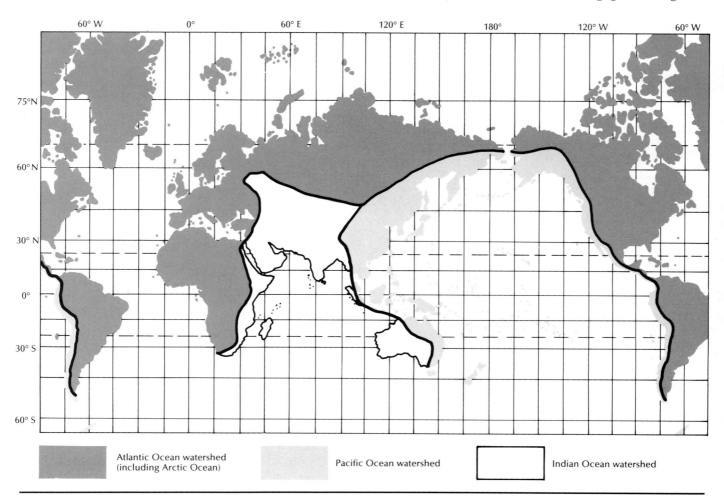

Figure 4.11 The division of lands into watersheds for the major ocean basins. The partitioning of lands is governed by the locations of the major terrestrial mountain ranges, and these in turn are the result of collision and subduction processes associated with plate motion.

SHAPES OF OCEAN BASINS

the sea surface is a fixed point on an adjacent coast. Suppose we install a suitable sensor at this point and begin to record the height of the sea surface on a continuous basis. On very short time scales, the recorded ups and downs would be surface undulation made by wind waves, tide waves, or by local storms. Such short-term variations are neglected when the

focus is on long-term sea level changes related directly or indirectly to tectonic causes. For example, what if *all* sea surface gages the world over were to record the same change in level and at the same time? Clearly, we would then conclude that sea level had changed and begin to search for the cause.

There are a number of reasons why the level of

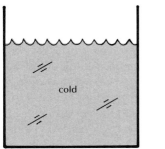 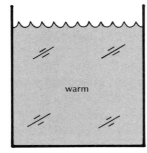

(a) Warm water is less dense and occupies more volume, so oceans of warm water stand higher.

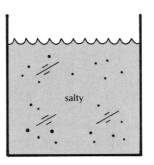 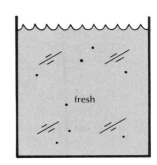

(b) The addition of freshwater to oceans dilutes the salt concentration, lessening density and causing a relative rise in sea level.

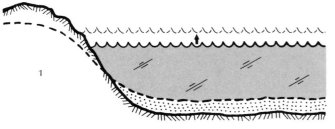 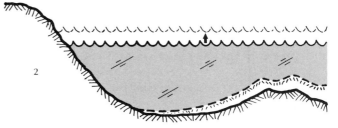

(c) The relative height of the sea can change (1) if there is an extended period of continent erosion or of mountain building, or (2) if the rate at which new magma is introduced to the ocean floor changes rapidly, hereby altering the volume of the ocean basin.

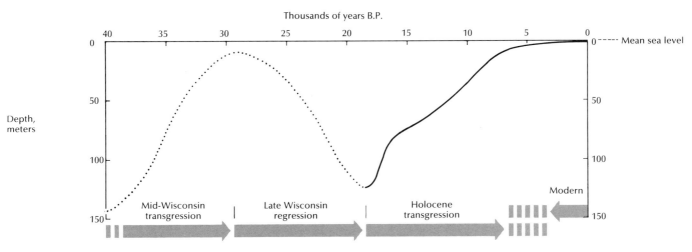

(d) Sea level rises and falls in proportion to the volume of water removed during periods of glaciation. At the height of the last glaciation about 18,000 years ago, sea level had fallen more than 100 m below its present level. The dotted curve extends estimates back to 40,000 years ago.

Figure 4.12 Causes of fluctuation in sea level. (From Komar, 1976.)

the sea can change *relative* to the average elevation of land above the sea, and each has its own time scale.

1. The average temperature of the world ocean waters may change. A 1°C cooling of all ocean waters would cause a 2-m drop in sea level everywhere. From studies of fossil sea organisms in ocean floor sediments, changes of 5 °C may have occurred in past geologic eras, causing the sea level to change as much as 10 m over long time intervals (Figure 4.9a).

2. The average salt content of seawater in one basin may change relative to that in others. The result is a change in its average density (see Chapter 9). An ocean with average density lower than that of adjacent oceans will have a higher surface (Figure 4.12 b). An example is the 30-cm drop in sea level from the less salty Pacific Ocean to the saltier Atlantic Ocean across the Panama Canal Zone. An ocean with denser waters than those of adjacent oceans will have a lower surface. Of course, such changes require long periods of time to develop. They can be caused by long-term shifts in patterns of evaporation and precipitation over global earth. Such changes can also occur whenever the shifting continents open or close various straits that interconnect ocean basins.

3. Continents erode, with the eroded material depositing in the basins. There appear to have been long geologic periods during which erosion rates exceeded the rates at which new mountain ranges were uplifted. Sea level rose as the relative volume of ocean basins decreased; these periods are associated with major transgressions of oceans over land, creating large inland seas like the one that occupied the central part of the United States during the Mississippian epoch, 350 to 300 million years ago (mega anno, Ma) (Figure 4.12c).

4. Water is removed from the oceans and deposited on land surfaces as ice. The most recent ice epoch is believed to have reduced sea level globally by about 120 m as recently as 18,000 years ago (Figure 4.12d). Such rapid changes have an important impact on studies of the shape of the margins of continents, as discussed in the following section.

5. The *relative average* depth of the basin floor can change, thus altering the volume of the basins (Figure 4.12c). New magma injected into the basin floor at the mid-ocean ridges rapidly cools to its solidification point and begins to spread away from the ridgeline. But it remains quite warm for long time periods, gradually losing both heat *and* volume. If the rate of introducing new magma is steady, we would conclude that this contraction in volume is also steady. Then the relative height of the ridge system as well as sea level oceanwide would remain steady.

If the *rate* of new magma injection is *episodic* on some large scale, however, with "pulses" of rapid sea floor spreading alternating with dormant periods, the overall volume of ocean basins could change markedly and with it the global sea level.

A major global transgression of seas over lands occurred during the period known as the Mesozoic, 225 to 65 Ma. When it reached its peak about 75 Ma, large areas of land were under shallow seas. Figure 4.13a shows the distribution of continents about 100 Ma. The Atlantic Ocean is well developed, but the drift of continents is far from complete. This, then, was a period of *vigorous* tectonic activity, and geologists believe that its transgression of seas over land was largely due to this decrease in basin volume (Van Andel, 1977).

6. The relative area of land versus seas could change. The likely areas where such changes occur are at collision and subduction boundaries. New "land" would be created if portions of the ocean floor are "accreted" to the margins of continents as material is scraped from the descending plate.

Last, there can be plate convergence in which two continents meet. For example, the land mass now geographically called India is believed to have collided with the continental lands of Eurasia. This resulted in the construction of a vast range of upthrusted and sheared mountains, the Himalayas; the total area of land was correspondingly reduced as these continents were compressed.

Plate Tectonics and Major Events in Ocean Circulation

There are two major changes in ocean circulation that accompanied the breakup of Pangea. The first was the progressive blocking of the Tethys Seaway from 65 to some 4 Ma. The second was the creation of the Antarctic Circumpolar Current from about 38 to 22 Ma. Kennett (1977) suggests that a third major event began after the Circumpolar Current was in place. A major upwelling around Antarctica caused the onset of the tremendous biological production in these waters and appearance of the baleen whales.

Blocking of the Tethys Seaway. As noted earlier, the transgression of seas over lands reached a peak about 75 Ma. At that time the Tethys Sea extended

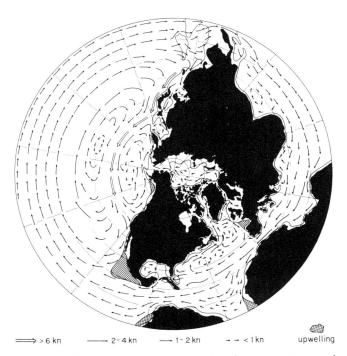

=====> > 6 kn   ——→ 2 - 4 kn   —→ 1 - 2 kn   → → < 1 kn   upwelling

(a) Results of a hydraulic experiment to chart the surface currents of the world ocean during the geologic epoch about 100 Ma. A principal feature of continent–ocean distribution at that time was the Tethys Seaway.

(b) Results of the same hydraulic experiment, except that the Tethys Seaway is blocked. Tremendous changes have occurred: the Pacific now flows into the Atlantic, which means that equatorial circulation is reversed.

Figure 4.13 Reconstructing ancient ocean current patterns. (After Luyendyk, Forsyth, and Phillips, 1972.)

across the globe in the latitude zone 10 to 25°N (Figure 4.13*a*), from Malaysia across what is now the North Indian Ocean, the Mediterranean Sea, and the Central North Atlantic and reaching into the Eastern Pacific across the then nonexistent land bridge between North and South America. The Tethys allowed for circulation of the subtropical waters completely around the globe at these latitudes. Gradually, as Africa rotated away from the Mid-Atlantic Ridge, it closed the Tethys Seaway at its eastern end, perhaps some 22 Ma. Finally, about 3.8 Ma, the Panama Isthmus rose sufficiently to close off the Atlantic–Pacific connection; from that point on circulation in the Northern Hemisphere has been much the same as today.

Luyendyk and his co-workers (1972) developed an ocean circulation model of the Northern Hemisphere that included the open Tethys Seaway. When they applied an assumed surface wind field to the model, a massive westward movement of surface water dominated the circulation (Figure 4.13*a*). In the Pacific, a current much like today's Kuroshio (see Chapter 11) provided for transport poleward of warm subtropical waters. In the Atlantic there was also poleward transport of warm waters. This circu-

lation model supports the general consensus that the polar regions were much warmer then compared to today. Europe enjoyed a subtropical climate and the Arctic had temperatures of about 14°C (Emiliani, 1961).

The Tethys Seaway vanished as the Asian and African land masses closed. The massive east–west ocean current that formerly circled the globe was now blocked. The North Atlantic began to develop a circulation scheme much like that of today (see Figure 11.1), but it kept its open connection with the Pacific across the gap between North and South America until quite recently, only about 3.8 Ma.

We now know that the Central American land bridge had a profound impact on both ocean and atmosphere circulation. For example, a comparison of the carbon chemistry in samples taken from the Caribbean and Eastern Pacific basins shows clearly that the salinity of Atlantic water began to increase markedly about this time (Keigwin, 1982). (Recall our earlier comment on comparative Atlantic–Pacific salinity.) Other evidence points to drastic changes that accompanied this tectonic event. The Gulf Stream is thought to have gradually grown in strength as the land bridge emerged, reaching its present state about 3.8 Ma. Strangely, the permanent glaciation of the North American Arctic region also began about 3.8 Ma. Oceanographers are keenly absorbed today in finding a mechanism by which closure of the Panama Isthmus could have triggered such glaciation (for example, Weyl, 1968).

Evolution of the Antarctic Circumpolar Current. From the view of oceanography, the most astounding of ocean events that resulted from the breakup of Pangea was the development of the Antarctic Circumpolar Current. This current has tremendous impact on the behavior of the world ocean today. Let us begin its history with the South Polar region as it might have looked 53 Ma (Figure 4.14a) (Kennett, 1980). During this period both Antarctica and Australia were still connected with South America, although a rift was opening between the first two. Antarctica was relatively warm during this period, and although it occupied the south geographic pole, little glaciation was occurring. Surface waters of the surrounding ocean were warm, perhaps above 15° C, and the sediments collected on the ocean floor were predominantly from calcareous-shelled marine organisms. (Temperate waters are characteristically dominated by organisms that build

calcium carbonate shells.) The arrows in Figure 4.14a are an *estimate* of how deep waters were flowing at the time.

The rift between Antarctica and Australia widened. By about 38 Ma it was wide enough to carry a major flow and in effect shortened the circuitous path that oceanic flow formerly made around the northern shores of Australia. By about 21 Ma Australia had closed against the Malaysian Archipelago, to cut off the Tethys Seaway. Figure 4.14b shows the probable Antarctic current system of that time; deep circulation was now able to carry water from the South Atlantic across the Indian Ocean and into the South Pacific.

The beginnings of deep circulation around Antarctica had considerable impact on its ecosystem. Kennett (1980) says: "One of the most dramatic changes in Southern Ocean planktonic biogeography occurred near the Eocene–Oligocene time (38 Ma). Since then, Antarctic planktonic (species) have exhibited a distinct polar aspect and reflect the beginnings in the development of the Antarctic faunal and floral provinces." Surface waters turned abruptly colder, with sea ice beginning to form; as described in Chapter 20, the Gigantic Antarctic Sea Ice Mixing Machine began to operate.

53 Ma, early Eocene

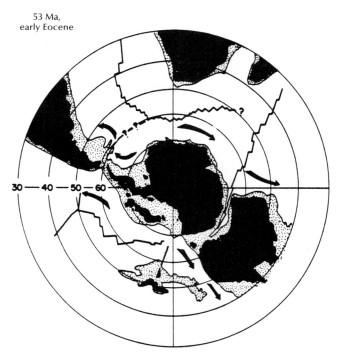

(a) Both the Drake Passage and the Antarctic–Australia passage are closed.

21 Ma, earliest Miocene

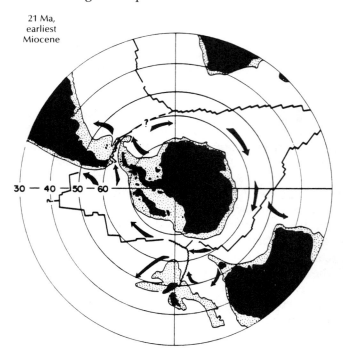

(b) The opening between Australia and Antarctica is well developed; the Antarctic Ocean begins to develop sea ice as well as a characteristic cold-water marine life; the Drake Passage begins to open.

Figure 4.14 The evolution of the Antarctic Circumpolar Current.

At about 22 Ma the Drake Passage between South America and the Antarctic Peninsula began to open (Figure 4.14b), allowing unrestricted flow to begin, of both surface and deep waters, around the entire continent of Antarctica. Because westerly winds were present at this latitude, the Antarctic Circumpolar current rapidly built up to become the largest current in all the oceans. It is both a surface and a deep current and moves a volume of water estimated at three times that of the Gulf Stream around the globe in the latitude band from 55 to 60°S.

This tremendous current had the effect of creating a huge "communication" channel between the water masses of the Atlantic, Pacific, and Indian oceans. Between the current and Antarctica, surface waters became very cold. *A unique biological environment was created*, partly by the dynamic current, but also by the Sea Ice Mixing Machine, which generates upwelling on such a massive scale that plant production here is limited by available light not by depleted nutrients. Here the ocean's shortest, most productive food chain developed. At the top are the baleen whales (Mysticeti), which evolved about 32 Ma, a date midway between the Australia–Antarctica opening and the beginning of the Drake Passage. Figure 20.10 presents the major linkages in this

Present day

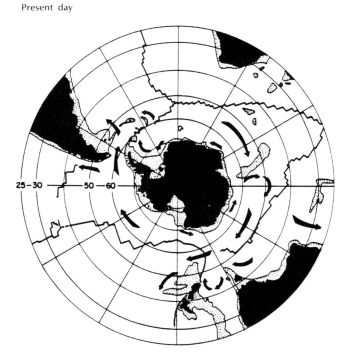

25-30 —— 50 —60

(c) The Drake Passage is now fully opened. The Circumpolar Current exists; oceans south of the current are decoupled from other oceans. (From Kennett, 1980.)

dominant food chain of the Southern Ocean. Although only three linkages, three trophic levels, exist between the microscopic plant and the macroscopic baleen whale, many other animals, birds and seals, participate. Fishes per se are a minor part of this productive food chain, which explains in part why there are no major fisheries in the Southern Ocean.

Summary

A comparison of surface area covered by lands and oceans shows a Northern Hemisphere dominated by lands and a Southern Hemisphere dominated by oceans. A comparison of elevations reveals that the earth's crust occupies two distinct modes, one for lands and the other for the crust of ocean basins, a result consistent with the concept that the rigid lithosphere rides upon a viscous mantle material in isostatic equilibrium. Figure 4.1 shows these relationships in vivid contrast. It also shows the phenomenon of the mid-ocean ridge system with its fractures and offsets. Search for an explanation for the ridge system led geophysicists into a revolution of global geology.

The new global geology has evolved during this century, beginning with the dedicated work of Wegener in proposing the concept of continental drift. Once it was accepted as a plausible hypothesis, investigators searched for a mechanism able to "move continents" and found it in the process now called sea floor spreading. This new concept explained ridges and other basin features as the result of new magma material injected into the basin floors along the ridgelines. But though sea floor spreading explained many of the *observed* phenomena such as the young age of basin floors, it did not explain what happened to old basins.

The global geology picture was finally "sutured" together through a theory of crust movement in integral segments called plates. Refinement of plate tectonics then produced explanations for the origin of trenches and chains of volcanoes as well as the cause of unique *patterns* in the distribution of earthquakes.

But the most oceanographic applications of the new global geology theories are in explaining the evolution of the largest animal, the baleen whale; in understanding why the fossils found today are where they are; in a glimpse of how the climatic events such as ice ages might be coupled to ocean circulation; in picturing how the deep ocean water masses move; and in understanding global changes in sea level.

Study Questions

1. If 68% of all land area is located in the Northern Hemisphere, does it follow that 68% of all ocean area is located in the Southern Hemisphere?

2. There is considerable information about the structure of the earth's crust in the bar chart portion of Figure 4.3. You can play games with it. For example, can you create a simplified *model* of the relative elevations of the earth's crust? One model could be created by ignoring all elevation categories *except* the two containing the largest areas—0 to 1000 m above sea level and 4000 to 5000 m below sea level. If Archimedes were alive today, what conclusion would he immediately draw from such a model?

3. *If* the ocean floor is continually spreading away from the mid-ocean rift zones (ridges), *and if* only "young" ocean floor material remains today, why isn't the entire floor covered with the rough and higher-standing stuff found near the ridges?

4. Which of the major basins is *moving* fastest, the Pacific or the Atlantic? Is it necessary that both sides of a ridge *grow away* from the ridgeline at the same rate?

5. What are fracture zones?

6. The Pacific basin is sometimes described as being encircled by a "ring of fire." Figure 4.7 shows that the basin is also virtually encircled by a series of deep trenches in the ocean floor. What might be the connection between trenches and "fire"?

7. From the view of our study of the *behavior of the seawater* contained within ocean basins, the existence of trenches is much less important than the existence of extensive continental mountains adjacent to the margins where trenches exist. Explain this.

8. The theory of sea floor spreading must also allow for the *rate* of spreading to change from time to time. How might this rate of spread affect the level of water in the basin?

9. There are several distinct contrasts to draw between the Pacific and Atlantic basin floors. Discuss at least two of these.

10. Figure 4.9 illustrates two types of plate convergence and one of plate divergence. Are there other possible types, convergence, divergence, or just slip? Name a region where divergence created an important passage for ocean currents. Name a region where convergence eliminated an ancient ocean current.

11. What is the possible connection between the evolution of the baleen whale and tectonic plate movement?

12. Here's a better one: *if* continents are continuously eroded by the hydrologic cycle, where has it all gone? It *isn't* on the ocean floor today—the ocean floor is nowhere older than about 200,000,000 years. Therefore, the ocean floor cannot have "old" sediment.

13. This is even better: Now that we know *how* ocean basins are created, we also know that they cannot be substantially *deeper* than they are today. Then what would oceans be like *if there were twice as much continental crust* than actually exists?

14. Relatively large cobbles (stone) are found on the ocean floor far removed from the continental margins. There are at least two logical *natural* processes by which these are deposited. Explain.

15. How do we arrive at the estimate that the ocean floor is *moving* away from the spreading ridges at about 4 to 8 cm/yr?

16. Write a brief essay explaining why *all* boundaries between basins and continents are not, a priori, plate boundaries.

17. It appears that the largest animal ever produced, the baleen blue whale, is a relatively recent newcomer, joining the marine biosphere only some 30+ Ma. What unique characteristics of the Southern Ocean stemming from this same point in time could explain why this beast evolved as it did?

18. It appears that until about 4 Ma a considerable flow of warm ocean surface water moved from the North Atlantic over the pole and into the North Pacific. What happened about 3.8 Ma to change this interocean connection? Write a short essay on some oceanic consequences that stemmed from this happening.

CHAPTER 5

CONTINENTAL MARGINS AND BASIN SEDIMENTS

Where does the continent end and the ocean basin begin? It is a basic question, and a useful one. The theory of plate boundaries gives us a way to classify the margins as active or passive or undergoing subduction and so on. In the field, however, we find that each margin is a bewildering patchwork of rocks and sediments. What we conclude is that the seas have transgressed over land boundaries many times in the geologic past, each time cutting new shores and setting down new layers of sediments as they rework the old margins.

Across this dynamic continental margin flows the material that settles on the floors of ocean basins. Moreover, the water within the basins often takes its chemical "imprint" from exposure to the intense processes active in the margin zones. For these reasons, the subjects *margins* and *basin sediments* are combined in one chapter.

General Views of Continental Margins

Background

Margin studies began rather more slowly than other specialty fields in oceanography. Perhaps the first application of the scientific method to margin geology was in measurements of the gravity field. The big thrust in margin geology came with World War II and the demand for petroleum as well as for accurate charts of our own continental margins for defense purposes. Oil explorers developed the technique of recording the echoes of surface explosions as these reflected from subsurface strata. Called seismic refraction, the technology became a new *probe* for marine geologists as well. Interpretation of these new data could separate the continental rock itself from the erosion products that collect so deeply along the margins. Geologists were able to "see" deep into the margins for the first time.

An Idealized Model of the Continental Margin

A standard pattern of margin structure emerged as more and more data accumulated. Figure 5.1a shows this idealized pattern in a cross-sectional view (Drake and Burk, 1974). Moving seaward from shore, we first cross the *continental shelf*, then the *continental slope*, and lastly the *continental rise*. The boundaries that separate these sectors are called the shelf "break" and the "foot" of the slope. The margin zone ends at the abyssal plain.

The characteristic that changes among these three sectors is the average *angle* at which each descends toward the abyssal floor. The shelf is very flat; it is clearly a part of the continent. The slope is the steepest part and is therefore the most likely to be the "edge" of the continent itself. We recognize the continental rise as the oceanic equivalent of the alluvial fan we often see developed at the base of a mountain, that is, it is composed entirely of material eroded from above. Later sections describe these features in more detail.

A Comparison of Cross Sections of Some Real Margin Zones

The *average* dimensions given in Figure 5.1a, although they provide a model of the typical margin, can mislead the reader. A comparison of some real margins in cross section is more useful. The selected sections in Figure 5.1b are aligned so that the continental slopes are drawn at the same point. This follows the idea that the edge of the continent is in the slope part of the margin. Three different margin sections are compared.

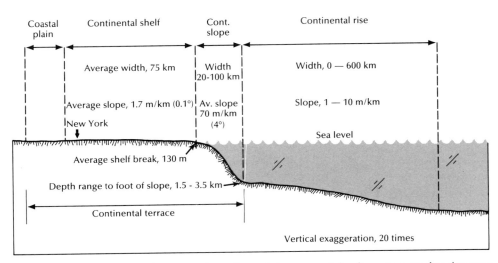

(a) The average dimensions of continental margins and the ranges of the dimensions are found to vary worldwide.

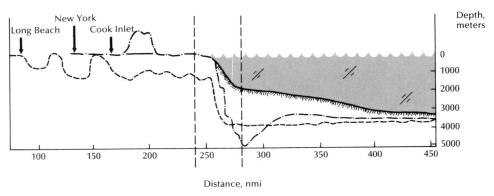

(b) Three different types of margin structures are aligned so that each shelf break occurs at the same point in the graph. The New York profile is a typical passive margin. The Cook Inlet, Alaska profile, representing a margin that is active in the tectonic sense, clearly shows the subduction zone with a well-defined trench. The Long Beach profile is an active zone without a trench, for here the plates slip past each other without convergence.

Figure 5.1 Describing the continental margins. (From Drake and Burk, 1974.)

1. The *typical* cross section is repeated here from Figure 5.1*a*; it is an actual margin section taken seaward from New York. It is selected to show that a continental margin that is *not active* in the tectonic sense will have its shape dominated by the sedimentation of materials eroded from the continent and deposited primarily in the slope and rise sections.

2. The cross section seaward from Cook Inlet, Alaska (dashed and dotted line) shows a margin that is typical of a *convergent* plate boundary. The sharp valley, representing the Aleutian Trench, is over 5000 m deep.

3. The third section, drawn with a dashed line, is the margin off Long Beach, California. The two peaks just offshore are the Catalina Islands. Notice that the entire shelf region is irregular and that very little sediment has collected at the base of the slope. The irregularity of the shelf zone indicates that faulting and folding have occurred within the margin zone itself. The shelf off Long Beach is also noted for its extensive reserves of petroleum.

A Legal View of Continental Margins

In 1982, 177 nations signed a convention called *The Law of the Sea*. The thrust of this convention was for nations to agree on two facets of the ocean domain. First, to what point seaward from a coastal nation's shore does its "jurisdiction" extend? Second, who "owns" the economic resources of the rest of the sea? The first of these facets, now called the Exclusive Economic Zone (EEZ) boundaries, is discussed in the opening part of Chapter 19. The reader should glance at Figure 19.3 to understand how the geological model described in Figure 5.1a serves also as the model for boundaries of the EEZ.

The Plate Tectonics Case for an Edge to a Continent

The global geology theories were introduced just about the time that topographic surveys of the continental margins were done. A natural application of the theory of the breakup of the supercontinent Pangea would be to see just how well the various continents could be "fitted" together. In 1965 Sir Edward Bullard and others programmed a computer to test all possible configurations of the continents joined together and to select a best fit. They found

that using the 1000-m (500 fathoms) offshore contour as the "edge of continents" gave the best fit. Figure 5.2 is a perspective view of the tectonic *edge of the continent*.

The Continental Slope

The 1000-m level falls in the zone of the continental slope (Figure 5.2) and so we begin description with this category. Most continental slopes have a well-defined upper boundary at the level where the gentle slope of the continental shelf changes sharply to that of the upper slope. Worldwide, the average depth of the break is about 130 m.

In general, the initial grade is about 4°. By grade is meant the *rate* at which bottom depth increases; a 4° grade is equivalent to a change of depth of 7 m in a horizontal distance of 100 m. It is constant down to a depth of 1000 to 2000 m, at which level the grade changes. In a trenched margin the grade often

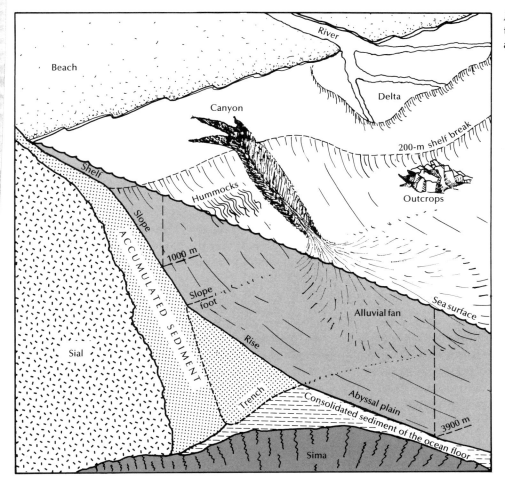

Figure 5.2 Major features of the margin between continents and ocean basins.

increases as the slope falls deeply to the bottom of the trench. In nontrench margins, such as off the East coast of the United States, the grade decreases to the more gentle grade of the continental rise.

The continental slope is a *transition* that links the two mode elevations of the earth's surface (Figure 4.3), the 840-m average level of land above the sea surface and the 3900-m average level of the ocean basin below. What is perhaps surprising is that this transition zone is so narrow, only some 35 km in general. In contrast, the continental rise can extend hundreds of kilometers away from the base of the slope toward the open abyssal plain.

We would expect the principal characteristic of the slope to be its accumulation of sediment derived from the erosion of the continents, and this is indeed the case. Nearly three-fourths of the volume of sediments in *all* the ocean basins is found in the narrow continental margins, and the major part of this is on the slopes and rises, as Figure 5.2 shows.

Two aspects of the slope regions need special discussion. The first is the relative uniformity of the grade of the slopes, and the second is the steep canyons occurring across the slope itself.

Formation of the Slope

Materials eroded off the continents are delivered to the continental margin by rivers and are usually first deposited on the continental shelf. Coastal currents then gradually move the erosion products toward the edge of a shelf, the shelf *break;* from here they spill down the face of the continental slope. Given enough time, this *lithogenous* material, meaning derived from continental rock, accumulates to form the slope itself. The accumulation continues until the vertical angle of repose of the unconsolidated sediment reaches a critical value. Adding still more sediment creates a local *unstable* condition, and a part of the slope will *slump* away. The process is continuous, so the end result is a slope sector with a *constant average grade.* Just what this grade is depends on two things. First, grain sizes of slope sediments range from sand to the finer silts; the larger grains will accumulate with a steeper grade. Second, the energy that is released into the slope sector affects its grade; for example, strong ocean currents along the slope or strong earthquake activity. One evidence of slumping is the presence of numerous "hummocks," as shown in Figure 5.2.

Geologic Faulting on Slopes.
Because there is a very large difference between average land and average basin elevations, almost 5000 m, large sections

of the continent are expected to drop away at its edge. Geologists call this a *fault* process. Faulting can impart a *terracelike* character to the continental margin, especially in the slope sector. The reader can find three such terraces along the slope of the Cook Inlet cross section (Figure 5.1*b*). Such terraces may eventually fill with sediment and disappear. An example of a very large terrace is the Blake Plateau off the coasts of Florida and the Carolinas. It extends some 200 km seaward at a uniform depth of about 800 m (Figure 5.3).

Other Slope Irregularities.
Although uniform sediments and grades are typical of slopes, their topography is often broken by outcrops of hard rocks, clearly not of sedimentary origin. This is evidence that the original margins of continents were highly irregular. The numerous sediment traps, now filled, are sometimes explored to determine whether petroleum is trapped within.

Submarine Canyons

Few marine geology features when first described so excited the curiosity of marine scientists as the large submarine canyons that cut across the continent margins. In Figure 5.2 a typical steep-walled canyon cuts across both the shelf and the slope. Figure 5.3, an expanded section from the Heezen and Tharp series of physiographic diagrams, shows several canyons. The largest, the Hudson Canyon, extends for hundreds of kilometers; it begins on the shelf in front of the mouth of the Hudson River and cuts deeply through the slope and rise to "empty" at the base of the margin onto the abyssal plain of the North Atlantic.

The Size of Submarine Canyons.
The topographic relief across submarine canyons can be astounding, fully equivalent to features on land such as the Grand Canyon. Figure 5.4 compares the relief of the Grand Canyon with that of the Monterey submarine canyon off California. Not only are terrestrial and submarine canyons comparable in size and scope of relief, but both have also developed similar branching patterns of feeder canyons.

Canyons in Relation to Rivers.
Terrestrial and submarine canyons differ in one enormously important aspect—their dependency on major rivers. Geologists know very well the forces that shape and cut river valleys, but the evolution of submarine canyons is still not well explained. All terrestrial canyons are eroded by rivers, either present or extinct.

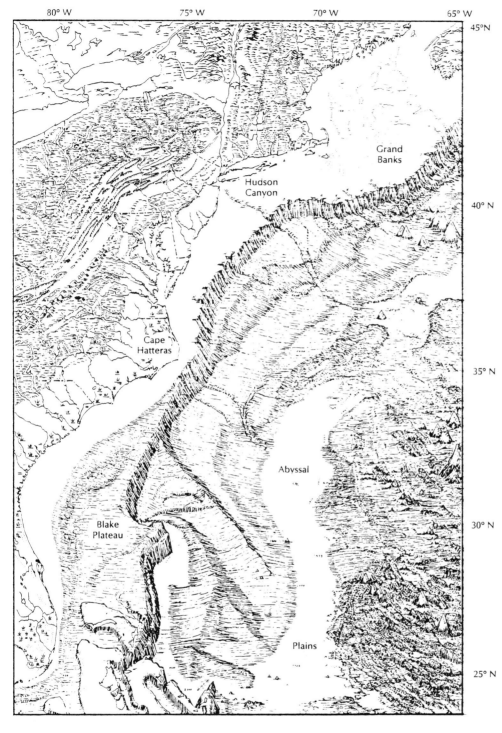

Figure 5.3 A physiographic diagram of the eastern seaboard of the United States. (From Heezen and Tharp, 1968.)

The winding path of continent erosion usually follows weaker geologic formations or the lines of structural faults. (A geologic "fault" is the name given to a zone where two large blocks of earth *slide* by each other.)

In contrast, large submarine canyons have no apparent direct connection to a major river, for example, the two large canyons off Cape Hatteras (Figure 5.3). The Monterey Canyon also falls into this category. In short, the discrepancy between known river

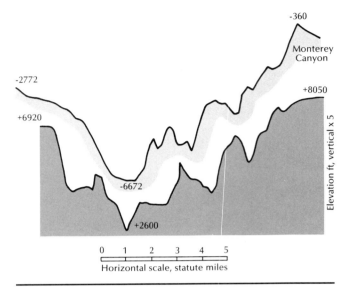

Figure 5.4 A comparison of the topographic relief for a section cut across the Grand Canyon with that of a cross section across Monterey Canyon, a submarine canyon off Monterey Bay, California. The reliefs are very similar in both size and shape. (From Shepard, 1963.)

sites and major submarine canyon sites is one of the vexing marine geology problems today.

Canyons and Changes in Sea Level. Could the canyons have formed during past epochs when sea level was much lower, when river channels themselves were extended seaward at least as far as the shelf breaks? Sea level was some 100 to 150 m lower during the Wisconsin glacial epoch (10,000 to 120,000 years ago), and during the Illinoian (240,000 to 360,000 years ago) it retreated about 200 m.

Most of the major submarine canyons extend much deeper than the shelf break or even 200 m. Many extend far down the face of the slope itself before disappearing, and others, for example, the Hudson channel, extend all the way across the rise to the abyssal plain. For river flow itself to cause the canyon erosion, it would have to plunge to great depths on entering the ocean. This does not happen. Even sediment-laden river water is still less dense than seawater and simply floats away from the river mouth along the surface of the sea.

An alternative explanation for canyon erosion is that the coastal margin has been uplifted. Over geologic time, existing rivers would slowly erode the canyons as the uplift occurred; when the margin

If suddenly sea level were to drop by about 1000 m, we could likely "see" submarine canyons with topography much like this view of the upper Grand Canyon of the Colorado River. The bluffs in the background would be the former shoreline, and the plateau in the foreground would be the continental shelf into which a submarine canyon has been eroded, complete with dendritic feeder channels. (Courtesy of John S. Shelton.)

later subsided, the canyons would remain as part of its permanent topography. The difficulty with this argument is that so many of the world's continental margins would have had to undergo the massive uplift to explain the widespread distribution of canyons today. Major submarine canyons are found in all ocean margins, including the Arctic basin.

The Turbidity Current and Canyon Erosion. During the 1950s, a group of marine geologists at the Scripps Institution of Oceanography demonstrated that a mixture of sand, mud, and water could travel for substantial distances *underwater* without the flow mixture losing its sediment load. Called *turbidity flows,* or more appropriately *density currents,* these flows are believed to reach very high speeds. Theoretical calculations show that speeds of up to 80 km/h are possible, depending on the type of sediment, the load, and the slope of the ocean floor down which the flow travels.

What is the evidence that turbidity flows occur in the ocean? The classic case described by most oceanography texts is an event that broke a series of transatlantic telegraph cables in the North Atlantic. The cables were laid along known routes on the slope off Newfoundland. Because the cable companies knew exactly when each cable broke, and where each cable lay, they were able to show that a massive submarine slide could have ruptured all cables if it moved at speeds of about 60 km/h! More recently, we have mapped the Congo Canyon and discovered that it penetrates up the riverbed itself to perhaps 20 km inland. The Congo River carried enormous sediment loads. Measurements in the Congo Canyon show that some 50 flows per year could be classified as turbidity currents.

What physical factors can initiate turbidity currents? Prime candidates are earthquakes. These are able to shake an unconsolidated sediment bed into a semiliquid state. Such a liquid would be very dense because of its great load of suspended particles. Even if the particles are freely suspended only for a short period of time, the fluid can begin to move. As it does so, it becomes more turbulent, entrains even more sediment, and grows unstably into a powerful density flow.

Even so, experts disagree on the erosion powers of density currents. There is agreement that turbidity flows are responsible for the channel–levee structure seen on the surfaces of alluvial fans of the continental rises (see Figure 5.2). But whether they have the power to erode the hard granite or quartz rock materials making up the walls of many deep canyons remains a question.

Turbidity Flow in Canyons Not River Connected. There is another source of turbidity flows that might explain how canyons were formed away from river mouths. The reader will learn in Chapter 18 that considerable amounts of sand are continuously moving along shorelines everywhere. Called littoral transport, the process is powered by the energy of waves that break across beachfronts; it is a major process by which the oceans distribute sands along their shores. Eventually, such sediments encounter the head channels of a canyon and are captured. The canyon then becomes the route by which sediments proceed to the outer limits of the continental margin, generally in episodic turbidity flows.

The Continental Shelf

The continental shelf is the shallow, submerged borderland of the continent. It is an integral part of the continent itself, not of the ocean basin. The fact that all continents have these platformlike features is evidence of the ocean's power to erode into and over the surface of its continental borders. Erosion and sedimentation, then, are the dominant processes that shape the shelf. "It is a basic principle of geology that erosion and deposition processes seek to bring land levels either down to or up to sea level" (Shepard, 1963).

In fact, the continental shelf can be viewed as a single dynamic system in which there are three controlling factors: (1) the flux of sediment material that moves *through* the continental boundary from land to ocean basin, (2) the *rate* at which energy is available to move the material into, around, and out of the shelf zone, and (3) the construction–destruction history of the shelf.

World Shelf Distribution

Figure 5.1a shows that the average width of all continental shelves is 75 km, and that the average depth of water at the point where the shelf breaks into the slope is 130 m. The unique aspect of the shelf regions is that there is only a small range of depth values for the break average, but the *range of shelf widths* is the greatest of all the margin zones, from zero to over 1000 km for the shelf off Siberia in the Arctic Ocean. Figure 5.5 charts the distribution of continental shelf margins around the world.

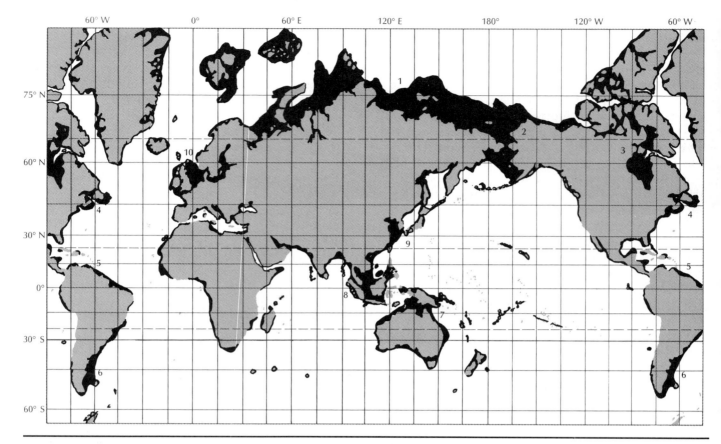

Figure 5.5 The worldwide distribution of continental shelves. The major shelf systems are: (1) the extensive but relatively unexplored shelf areas off Siberia; (2) the rich Bering Sea and Chukchee Sea shelves off Alaska, (3) the Canadian Archipelago and Hudson Bay, also relatively unexplored; (4) the eastern seaboard of the United States and Nova Scotia; (5) the Gulf of Mexico with the Caribbean and the extensive shallows of the Orinoco and Amazon rivers of South America; (6) the Argentina-Falkland Island Shelf, also not well explored; (7) the Great Barrier Reef with the Australia-New Guinea Shelf system; (8) the Indonesian Archipelago system, (9) the East China Sea Shelf, and (10) the North Sea-Baltic Sea complex of Europe. (The distortion inherent in a Mercator projection makes the high-latitude areas appear larger than true area.)

Looking first at the American continents, notice that the Atlantic seaboard has a much wider shelf area than does the Pacific side. The reason is clear. The western shores are by and large collision boundaries between the American and Pacific plates. In contrast, the Atlantic seaboard is a passive margin. Its extensive coastal plains and estuaries and shelf areas are a rich resource of fisheries. The Gulf of Mexico is also rich in resources, both fishery and mineral. Continuing down the eastern seaboard, notice the large shelf zones along the northeast coast of South America, coincident with the deltas of the mighty Amazon and Orinoco rivers. Far to the south, the shelf zone off the south of Argentina is massive, includes the Falkland Islands, and is relatively unexplored.

Along the Pacific coast shelf zones are narrow, in many regions but a few kilometers. Only in the Bering Sea do we find extensive shallow margins. The Bering Sea Shelf is enormous, extending some 500 km in width and over 700 km long.

The Bering Sea Shelf is extremely productive in fisheries. Far to the north the Canadian Archipelago is now being explored for its petroleum resources. On the eastern seaboard of North America, Hudson Bay and Baffin Bay are large, shallow inland seas. These bays have long played an important role in the culture of the Inuit peoples of the north.

Perhaps the most extensive and least explored continental shelf is off the cold arctic shores of Siberia. Off eastern Siberia the shelf extends over 1000 km into the Arctic Ocean; it is covered most of the time with perennial ice pack, and its resources are poorly known. In contrast, the shelf of northern Europe is not only extensive, ranging from the British Isles to the inland shallows of the Baltic Sea, but is also both rich in resources and well studied.

Asia also enjoys extensive shelf zones, from the ice-covered Sea of Okhotsk in Siberia to the archipelago of Indonesia. Australia too has extensive, wide shelf areas, particularly to the north with the island of New Guinea.

Some of these are probably well known to the reader in different contexts. For example, the vast Bering Sea Shelf between Alaska and Siberia, including the even vaster shelf that lies against the poleward shores of these lands, is the "land bridge" across which early humans migrated to North America. During the Pleistocene epoch, a period covering the last 1,000,000 years of geology, several cycles of glaciation caused sea level to retreat below the average depth of this vast shelf area. In another sense, the large shelf regions like the Bering Sea and North Sea off Norway, western Europe, and Britain are major fishing zones. Some also contain major petroleum reserves. Others, principally the shallow shelf that interconnects the thousands of islands of the archipelago Indonesia, the fifth most populous nation in the world, are considered by nations as sovereign territory. Indeed, wars are now waged over ownership of the continental shelf. The brief 1982 war between Britain and Argentina, ostensibly over the Falkland Islands, was in reality a conflict over the potentially rich shelf zones around the islands. Perhaps the best known of the shelf areas in the Far East is the one extending eastward from northeast Australia and terminating at its break in the famous Great Barrier Reef.

World Shelf Descriptions

Given that the depth of the shelf break is surprisingly uniform the world over, does it follow that shelves everywhere have a characteristic type of structure? The answer is yes and no. Yes, if we look only at the *veneer* of surface sediment; no, if we examine the shelf as but a part of the margin in cross section. The central point is that the continental margins have seen the oceans transgress and regress many time during geologic time. The most recent series of advances–retreats caused by the glacial–interglacial oscillations have worked and reworked the shelf regions; these are the causes of the uniformity in topography of shelf and shelf break as we see them today.

Deep Structure. We look first at the deep structure of continental shelves (Figure 5.6). Beneath the veneer of recent sediments we find a confusing array of sediment structure, ranging from consolidated muds to carbonate sands to even coarser material. The margins of continents are the first and most effective repository for material that erodes from the interior of continents. Over geologic time, the added weight of layer upon layer of sediments causes a subsidence; the deeper layers are forced downward into the mantle as the coastal region seeks its own isostatic equilibrium. Where these layers are exposed, they are called "relict" structure.

In a margin zone that is passive (in the plate boundary sense) we find a layered structure of a type called *trailing edge* (Figure 5.6a). Successive layers are wedge-shaped; they may lie above a faulted deeper continental rock or may themselves be faulted. Knowing as we do that seas alternately *transgressed* over the lands and then *regressed*, we view these layers as the historical record of ocean–continent interaction along a shifting, always dynamic boundary. Cores taken from petroleum boreholes are our prime evidence for these structures, especially the locations of fault zones (which often trap oil and gas).

In the active margin zone the layering is different (Figure 5.6b). The ocean crust being subducted may also be losing its own veneer of sediment. Successive layers can be sheared off the descending crust and accumulate as a prism of new "land" joined to the continent (geologists use the term "accretion prism"; see also Figure 21.13). As expected, the shelf itself will be very narrow. Inland, we may find relict terraces that are old beaches and shorefaces uplifted during earlier accretion events. One of my favorite mementos is a fossil shark tooth I recovered from one such terrace inland from the Chilean coast near the city of Coquimbo; it is special to me because the same location was once visited by Charles Darwin (*Voyage of the Beagle,* 1823), who also recovered fossils and recognized the feature as a relict shoreface.

Present Veneer of Shelf Sediments. In general, we find that the texture of surface sediments grades from coarse-grained sand along the shoreface and out to depths of perhaps 50 m to the siltlike fractions and even finer muds along the mid- and outer-shelf zones. To a large measure, this distribution conforms with the *power* available from the physical processes that move sediment.

The sketches of Figure 5.6 may give the impression that the shelf floor is smooth and featureless.

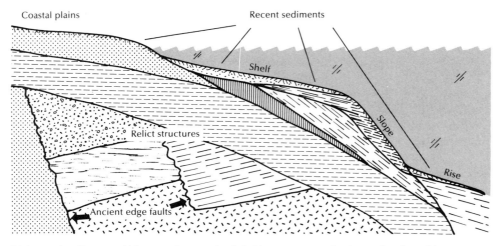

(a) Layered sediments within a passive margin. Subsidence accounts for the wedge-shaped layers.

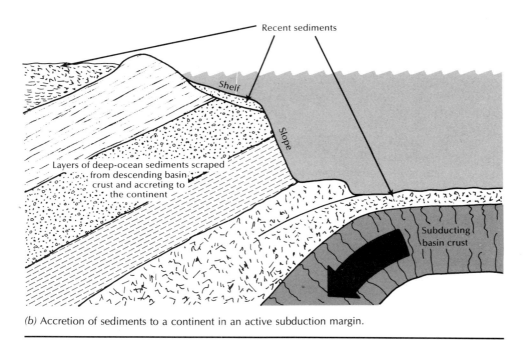

(b) Accretion of sediments to a continent in an active subduction margin.

Figure 5.6. Two types of continental margins shown in cross-section.

This is definitely not the case, particularly where the shelf is wide and the hydraulic forces available to move sediment are dominated by waves and storm-generated shelf currents. Our Atlantic coast shelves fit this category well. Figure 5.7 illustrates the morphology charted for the shelf of the Mid-Atlantic Bight, a well-known coast structure that extends from Long Island to Virginia. It has numerous topographic features: shoals, defined as shallows constructed of unconsolidated floor material; channels, which are important to ship traffic; and many others. It is the task of the National Ocean Survey

(formerly the U.S. Coast and Geodetic Survey) to keep up-to-date charts of the variable shelf topography.

Energy Sources on the Shelf. Here we take a brief look at the sources of energy operating on the continental shelf and the power available to move sediments and to shape the coastlines (Figure 5.8). This is a generic model, useful for any shelf region. The main energy sources are wind waves, other long waves, coastal currents, and river discharge.

It has long been known that water flowing above

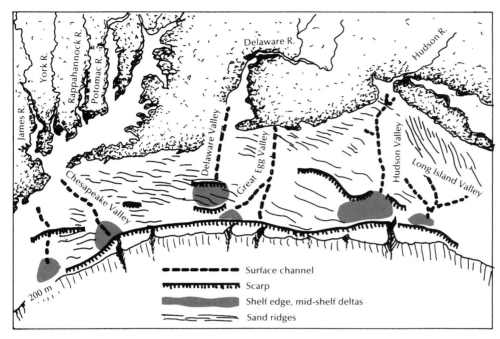

Figure 5.7 Topographic features typical of extended continental shelves, formed by waves, river and coastal currents, and storm-generated currents. The area shown is the Mid-Atlantic Bight. (From Swift, 1974.)

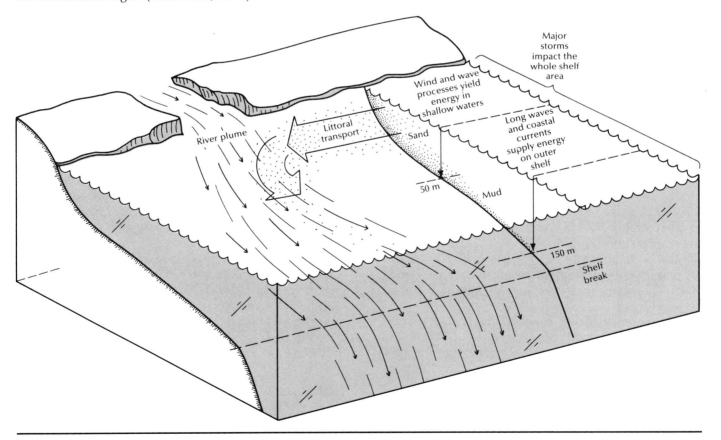

Figure 5.8 Physical processes effective in moving, eroding, or depositing coastal sediments.

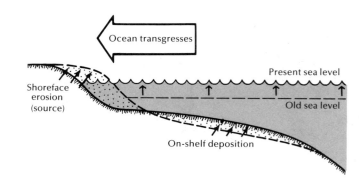

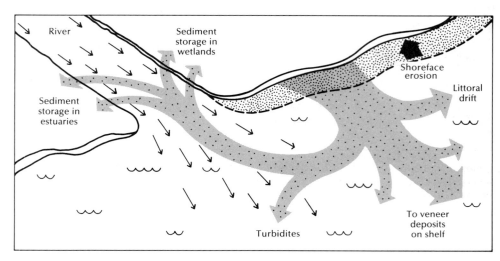

a sandy bottom can form the sand into patterns of ripples. To discover if such ripple evidence exists over a shelf, oceanographers made a photographic reconnaissance of the shelf off Oregon out to water depths of over 200 m (Komar et al., 1972). They found ripples throughout the shelf and demonstrated that wind waves could cause these and, by inference, move sediments. As the reader will learn in Chapter 17, wind waves release most of their energy when breaking in the surf zone, in water depths of 15 m or less. This is why the nearshore sediments are coarse-grained sands; the fine-grained fractions remain in suspension in this high-energy zone.

Tide currents do not produce much *net* transport of sediments along an open coast. In channels and estuaries, however, these are prime movers. The steady, unidirectional coastal currents are the principal movers of the fine-grained silts and muds found on the middle to outer shelf zones. Many examples show this. In the Gulf of Mexico, we find muds from the Mississippi River discharge as far west as the tip of Texas at Brownsville, 800 km

away. Muds and silt from the Orinoco River mouth cover the shelf for over 1000 km along the northeast coast of South America.

It appears that the most important energy source is the local winter storm. Experiments off the coast of Washington (Hopkins, 1971) show that bottom currents strong enough to move the silty bottom sediment occur about 4% of the time, and these times correlate directly with the occurrence of local winter storms. Off the Atlantic coast the storms are the traditional "nor'easters." These rise as cells of atmospheric low pressure (storm cells) pass northward off the Atlantic coast; the winds and waves that reach the coast are from the northeast. Storm currents generated during these events are responsible for the massive shifting of shelf features such as the shoals shown in Figure 5.7.

Continental Shelf and Sea Level Change

Of all the segments that comprise the continental margin, shelf, slope, and rise, the shelf is most directly affected by changes in sea level. The coastal

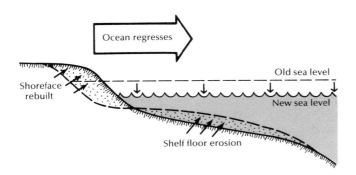

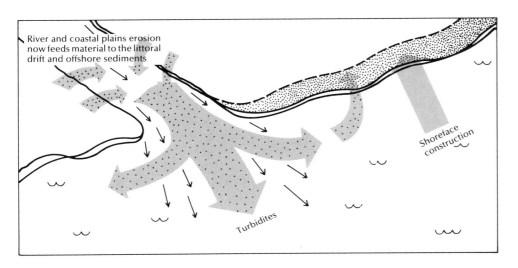

Figure 5.9 Sources and sinks of coastal sediment during (*a, b*) transgressing seas and (*c, d*) regressing seas.

plains with tide flats and marshlands, together with the estuaries, are all a part of this dynamic regime (see also Chapter 18). All are impacted by the rise and fall of sea level, as sketched in Figure 5.9.

The Transgressing Sea. Sea level rose rapidly from its last minimum 18,000 years ago (Figure 4.12*d*), reaching almost its present level about 2000 B.C. Since then it has continued to rise but at a much slower rate. Since 1930 sea level along the Atlantic coast has risen at the rate of 3 mm/yr, or about 15 cm in all (about 1/2 foot in English units). We are in a period of a *transgressing* sea, albeit a relatively slow one.

What happens to a shoreface during a slow transgression of the sea? Figure 5.9 sketches this situation in two ways. First, in cross section (Figure 5.9*a*), the rising sea slowly erodes the shoreface. This becomes a major *source* of new sediment to the shelf and is distributed across the shelf as a veneer of recent deposits (Figure 5.9*b*). A large part of this new material is moved *along* the shoreline (see Figure 18.10 for an explanation of littoral transport).

Some of this alongshore drift finds a submarine canyon and thus makes its way to the abyssal plains; some is moved *into* estuaries and deposited there.

The river, of course, delivers its sediment load to the coastal zone regardless of the sea condition. But as sea level rises slowly, the wetlands and marshes of a coastal plain also receive more water during tide cycles, and with this new flooding, additional sediment settles out (Figure 5.9*b*). Studies of the Atlantic coastal plain show that the vast acreage of wetlands are "soaking up" virtually all the fine muds that its rivers are delivering (Meade, 1972).

This is a surprising result, with some important consequences for us. We know, for example, that many pollutant chemicals reach the coastal zone already attached to other particles, usually the finer fractions such as clay. That these are accumulating *within* the wetlands, and not being exported across the continental shelf toward the abyssal depths, is important to coastal ecology.

The Regressing Sea. The opposite situation, that of a regressing sea, is depicted in Figures 5.9*c* and *d.*

As sea level falls, waves erode downward into the existing sediment on the shelf floor. The broad shelf itself becomes a source of materials. Some material is deposited in the shoreface as this advances in step with the receding water.

As sea level recedes, the rivers acquire a higher "head" and consequently more eroding power. Instead of a floodplain that accepts new sediment, erosion removes material from the coastal plains and delivers this to the shelf and to the open ocean.

Deep-Ocean Sediments

We look next at the materials covering the vast expanse of basin floor and ask the following questions. How thick is the sediment covering the basin floors? How rapidly do they accumulate? Are the sediments of the same type everywhere, that is, are differences in types of materials *in the sediment* useful clues to biological and geochemical processes in the ocean itself? And what connections do we find between the basin material and coastal processes?

Thickness of Ocean Sediment

Figure 5.10 shows how the thickness of the sediment that overlies the basin crust changes from place to place. Shown in solid lines are the mid-ocean ridges, the spreading centers of new basin crust. When viewed against the background of spreading centers, the contours of sediment thickness show some immediate, interesting patterns. Before we discuss these, the reader should note the categories of thickness used: shaded areas are *less than 100 m thick*, black areas are *more than 1000 m thick,* and the white region between shade and black should be interpreted as covered with sediments at thicknesses between these two extremes.

Sediment Thickness related to Ridges. Without question, ocean ridges everywhere have the thinnest layers of accumulated sediment. This is expected, of course, because these are the youngest parts of ocean basins. The *type* of sediment that correlates with the ridges (Figure 5.11) is primarily *calcareous ooze*. And as shown earlier, in Figures 4.1 and 4.6, the ridge zones are relatively shallow parts of ocean basins, rising up to 2000 m above the level of adjoining abyssal plains. The composite picture is that the shallow parts of ocean basins are dominated by calcareous ooze sediment; we examine this correlation in more detail shortly.

Locations of Thickest Basin Sediments. The black areas in Figure 5.10 carry sediments in excess of 1000

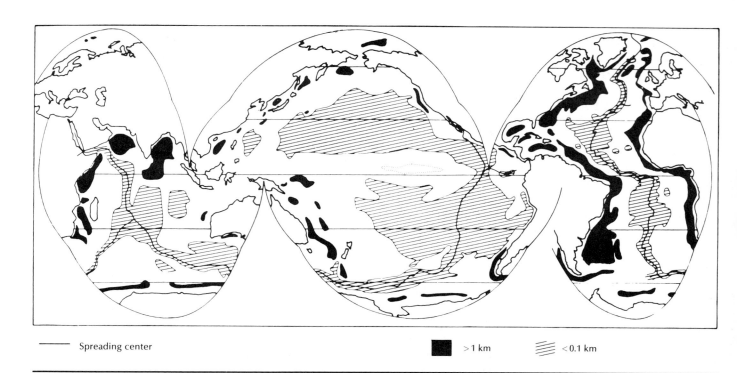

——— Spreading center   ■ >1 km   ▨ <0.1 km

Figure 5.10 The thickness of bottom sediments in the world oceans. (From Berger, 1974.)

m thick. We are struck with an immediate conclusion—the thickest sediments occur adjacent to continents. Specifically, they occur just seaward of the shelf break. To verify this, the reader may try a "fit" of the black zones in Figure 5.10 against the worldwide distribution of continental shelves in Figure 5.5.

There are at least four points to bring out regarding this correlation. First, we know from the plate tectonic arguments that the basin crust cools and deepens with distance from the spreading ridge; a natural result is that sediment trends toward the deepest part of a basin. Second, these black areas include most of the continental rise zones; reference to Figure 5.2 shows that these are the final depository of terrigenous materials on the abyssal plains. Third, the basin with the largest area of thick sedimentation is the Atlantic, the basin that is surrounded almost entirely with passive margins and with these well-developed continental rises. In contrast, the Pacific basin, which is surrounded almost entirely by subduction margins, has not acquired such extensive, thick deposits. Fourth, there is a direct correlation between the black zones of thick sediment deposits and the major world rivers. The Atlantic gets sediment from a tremendous watershed via the Amazon, Orinoco, Congo, Nile, Mississippi, and all the major rivers of the Arctic. Further evidence is seen in the direct correspondence of black zones on the west side of India with its Indus River and on the east side with the greatest sediment-laden river in the world, the Ganges–Brahmaputra (see Table 18.1 for river data).

Types of Ocean Sediment

Figure 5.12 summarizes the principal sources and pathways by which sediments are derived and deposited. Principal sources are the mineral products eroded from continental or ancient sedimentary rocks, the ash debris ejected from volcanoes, and the hot solutions emanating from the hydrothermal vents along submarine ridges. The routes through the seawater phase are either as suspended particulates or as solutions.

Sediments Derived from Suspended Particulates. All the lithogenous material deposited on the continental margins is clearly transported in particulate phase. But we also find lithogenous material far out on abyssal plains, away from any possible deposition by turbidity currents. These are the "red" clay deposits first discovered in deep samples taken during the HMS *Challenger* expedition. Scientists still are not agreed on how this material collects in the

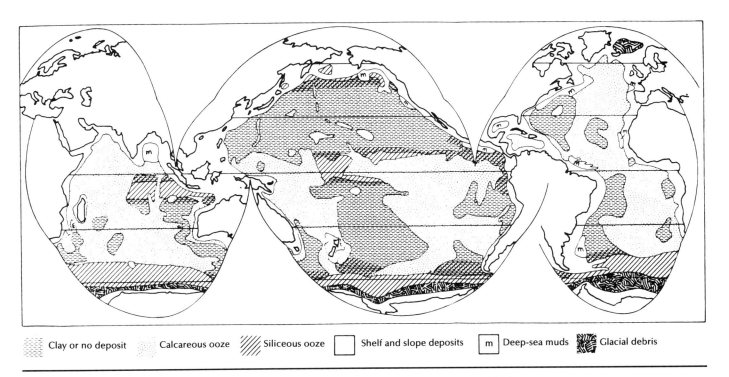

Clay or no deposit   Calcareous ooze   /// Siliceous ooze   ☐ Shelf and slope deposits   m Deep-sea muds   Glacial debris

Figure 5.11   The distribution of recent sediments on the ocean floor. (From Berger, 1974.)

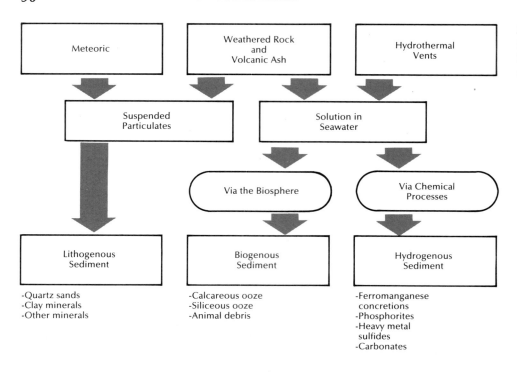

Figure 5.12 Types and pathways of sediment deposition. (Based on Shepard, 1963, and various other sources.)

deep basin, but a major contribution comes from continental dusts that are airborne far out to sea.

The distribution pattern of the red clay (Figure 5.11) has two features; (1) the clay occurs in mid-latitudes, perhaps in association with the westerly wind belts, and (2) in the Pacific it covers the central parts of the main gyres where sediments are not thick. The deposition *rate* of the red clay is the lowest of all sediment types, as Table 5.1 clearly shows.

One other process that moves terrigenous material out to sea is the rafting of rocks and other debris aboard icebergs. Icebergs are pieces of land glaciers

Table 5.1 Rates of accumulation of recent sediments on the ocean floor. (From Berger, 1974).

Facies	Area	mm/ty*	Reference
Terrigenous mud	California Borderland	50 – 2000	Bandy, 1960
	Ceara Abyssal Plain	200	Hayes et al., 1972
Calcareous ooze	North Atlantic (40 – 50°N)	35 – 60	McIntyre et al., 1972
	North Atlantic (5 – 20°N)	40 – 14	Schott, 1935
	Equatorial Atlantic	20 – 40	Schott, 1935; Ericson et al., 1956
	Caribbean	~28	Emiliani, 1966
	Equatorial Pacific	5 – 18	Hays et al., 1969 (last 1 my)
	Eastern Equatorial Pacific	~30	Blackman, 1966
	East Pacific Rise (0 – 20°S)	20 – 40	Blackman, 1966
	East Pacific Rise (~30°S)	3 – 10	Blackman, 1966
	East Pacific Rise (40 – 50°S)	10 – 60	Blackman, 1966
Siliceous ooze	Equatorial Pacific	2 – 5	DSDP (in Berger, 1973)
	Antarctic (Indian Ocean)	2 – 10	Lisitzin, 1972
	North and Equatorial Atlantic	2 – 7	Turekian, 1965
	South Atlantic	2 – 3	Maxwell et al., 1970
Red clay	Northern North Pacific (muddy)	10 – 15	Opdyke and Foster, 1970
	Central North Pacific	1 – 2	Opdyke and Foster, 1970
	Tropical North Pacific	0 – 1	DSDP (in Berger, 1973)

*Millimeters per thousand years.

that break off after a land glacier has pushed its way offshore. There is on my desk a rock that I picked off Ice Island T-3 when it was floating some 300 km from the North Pole, well into the central Arctic Ocean.

Sediments Derived from Solution Via the Biosphere. That the bulk of deep-sea deposits were biogenic in origin was recognized soon after the *Challenger* samples were analyzed. Three solid materials in marine sediments originate in the skeletal products of marine plants and animals: calcium carbonate, in the form of calcite or aragonite; silica, in amorphous forms, opaline silica; and calcium phosphate, as apatite.

1. *Calcite.* Organisms that contribute to calcite deposits are mollusks, some algae, sponges, and echinoids such as starfish, but by far the most important contributors in the open ocean are members of the order Foraminifera. These are small planktonic animals of the pelagic province. They are dominant grazers (actually omnivores) in the warmer waters; their distribution pattern in Figure 5.11 clearly shows a tendency toward the middle and tropical latitudes. (See Plate 15.4 for a sketch of one species of Foraminifera.) Some plant families also fix calcite; the coccolithophores (see Plate 15.3) in past epochs laid down enough calcite to form the famous White Cliffs of Dover, England.

There is another marked feature in the calcareous sediment pattern. These sediments occupy the shallower parts of the basins, in particular the areas on the ridges. Some geologists refer to the calcareous oozes as the "snow that covers the mountains of the deep ocean" (Berger, 1974). The rate at which it accumulates (for example, 35 to 60 mm per thousand years, Table 5.1) is much greater than that for any other type of sediment in the open ocean; therefore it dominates.

But if calcareous material is generated so rapidly, why does it not dominate the entire ocean floor? A first answer is simply that the calcite is dissolved at great depth and therefore does not accumulate in the sediment. Questions about the dissolution of calcite at depths have occupied geochemists for a full century; answers are still being sought and new data still add complexity. The ultimate reason why the abyssal waters dissolve so much of the shells that fall is that the surface life removes calcite from the upper waters *much faster* than river runoff can resupply it. Upper waters therefore become highly undersaturated; then, when the oceans turn over, the undersaturated water eventually contacts the calcite accumulating in deep waters and dissolves it.

2. *Silica.* The silica cycle differs from that of calcite. In the photic zone silicate is readily used by plants, in particular the diatoms, and the protozoan animals called radiolarians (see Plate 15.4, for example). At depth, dissolution does not occur as rapidly or easily as for calcite. The net result is that we find siliceous ooze deposits even to great depths. There is a strong correlation between zones of very high plant production and zones of siliceous sediment. Bottom deposits in areas of upwelling will contain many skeletons of diatoms, as for example in the Southern Ocean (Figure 5.11).

Where ancient deposits of siliceous material have been uplifted above sea level, they are mined for industrial use. Called "diatomaceous" earth (after diatoms), the material has many uses. Because the shells are of very small and uniform size, diatomaceous earth is, for example, an excellent filter for municipal water systems. Perhaps the most notable application was invented by Alfred Nobel when he combined it with nitroglycerin to make a safe-to-handle explosive called *dynamite.*

3. *Calcium phosphate.* Calcium phosphate deposits consist of things like fish teeth and the earbones of whales. Although these contribute only a minor portion of the volume of marine sediments, they seem to serve a useful role as the nuclei around which other materials are caused to precipitate from seawater solution.

Sediments Derived from Solution by Chemical Process. A number of deposits found on the ocean floor occur by precipitation out of solution with seawater; we call these *hydrogenous* sediments. A variety of elements are involved. One type of deposit that has

Figure 5.13 A polished cross section of a typical manganese nodule. The size of nodules varies, but a typical diameter is on the order of 6 cm. Nodules often accrete in thin layers about a nucleus such as a shark tooth or the earbone of a whale. Accretion rates are very slow, on the order of a few tens of atomic thicknesses per day. (From W. E. Dean USGS.)

Table 5.2 Chemical analysis of a typical manganese nodule collected at *Challenger* Station 248, latitude 37°41′N, longitude 177°04′W, at a depth of 5304 m.* (From Shepard, 1963.)

Oxide	Weight Percent	Oxide	Weight Percent
Al_2O_3	6.04	NiO	0.53
BaO	0.58	P_2O_5	0.30
B_2O_3	0.010	K_2O	1.45
CdO	0.0011	Sc_2O_3	0.0015
CaO	2.10	Ag_2O	0.0016
CoO	0.17	Na_2O	3.37
CuO	0.80	SrO	0.18
Ga_2O_3	0.0029	Tl_2O	0.009
GeO_2	0.0007	SnO_2	0.03
Fe_2O_3	21.4	TiO_2	0.56
La_2O_3	0.02	WO_3	0.006
PbO	0.15	V_2O_5	0.071
MgO	0.90	Y_2O_3	0.003
MnO	31.62	ZnO	0.70
MoO_3	0.06	ZrO_2	0.015

Source: Data from J. P. Riley and P. Sinhaseni, *J. Marine Res.* 17: 466–482, 1985.

*Data are given in weight percent of the oxide in the soluble fraction of the nodular material.

been studied for over a century is the so-called *manganese nodule.* The nodule is a concretion of many elements; it occurs as a free, irregular-shaped solid body of some 6 cm in size (Figure 5.13). Table 5.2 lists the many compounds, mostly metallic, that precipitate into the typical nodule. Manganese and iron are the main components–they are now called ferromanganese nodules—and aluminum, sodium, and calcium are others. Nodules have a layered structure, metal oxides alternating with clay minerals.

The origin of nodules is still being studied, but it is generally agreed that their growth rate is extremely slow, only a few millimeters per million years! Stated another way, the daily growth rate is but a few atoms of thickness. These deposits are found in all basins but are especially dense in the Pacific; estimates are that 50% of the Pacific basin carries visible nodules. Reconnaissance has been photographic (see Figure 19.14 for an example).

One question continues to plague researchers. Given the slow growth rate, how is it that the *nodules remain on the surface?* Although the densest concentrations of nodules occur more or less in the same (hatched) area of Figure 5.10 where sediment thickness is less than 100 m, even these clay sediments accumulate much faster, 1 to 2 mm/ty. At this rate a typical nodule would be buried in about 60,000 years! Perhaps the worms and other burrowing animals keep the nodules alive. It is useful to

note that the *Challenger* recovered the first nodules over 100 years ago!

Recently, we have found encrustations of ferromanganese oxides over vast areas of undersea volcanoes and on the Blake Plateau off our Atlantic seaboard. A necessary condition for this type of accretion is that currents be strong enough to keep the rock surfaces swept clean of other sediments, for example, the calcareous oozes.

Phosphorites (also called apatites) also precipitate out of solution as irregular-shaped nodules or as a coating over the seabed. In general, we find these deposits in shallow water where dissolved oxygen content is low. Upwelling areas fit this category because their high organic production also results in high organic content in the sediment. Thus two requisites for phosphorite formation are met: a high demand for oxygen in decomposing organic compounds and a large supply of organic phosphates liberated during such decomposition.

The most recent discoveries of hydrothermal vents along the active mid-ocean ridges have added still another major category of hydrogenous sediment, the heavy metal sulfides that collect on the surface near the vents. The superhot seawater ejected from the fissures in the new magmatic crust is rich in sulfur compounds. A number of metal sulfides precipitate out of solution as the vent water is rapidly cooled by mixing with cold ambient seawater. These collect as coatings over the surrounding area. Researchers now use submersibles to visit these sites,

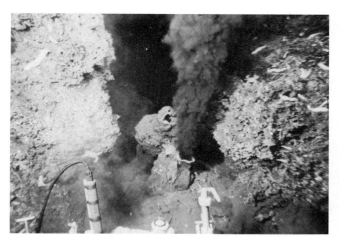

A "black smoker" photographed from the research submersible ALVIN on the East Pacific Rise, a spreading zone. Such vents of superheated seawater bring dissolved chemicals from fissures in the extruding magma into the overlying cold seawater; many chemical components, such as sulfides of iron and manganaese precipitate around such events.

take samples, and attempt to estimate their economic potential.

Last, there is oolite. Every reader is likely some day to spend an afternoon lying about in oolite. Oolite sands are formed in shallow, warm tropical surface waters by direct precipitation of calcium carbonate. The process occurs when water saturated with CO_2 is rapidly warmed so that the normal extraction of carbonate by organisms forming shells does not keep pace with the supersaturation. Calcium carbonate then forms around some nuclei such as fecal pellets or quartz grains to form the rounded pellets that make up the "white sand" beaches of tropical islands.

An Ocean Application of Deep-Sea Sedimentology

Oceanographers know that the oceans and the atmosphere are closely coupled, and that oceans must play an important part in the glacial–interglacial cycles. Recently, with the great success of the Deep-Sea Drilling Project in taking core samples of deep-sea sediments, we have gained new knowledge about how the fluid ocean behaves during a glacial period. Figure 5.14*a* shows the locations of cores taken from the North Atlantic.

One of the first studies of these cores was to catalog the species of fossil shells found in the core layers deposited 18,000 years ago. This was the time of the last global glaciation, the end of the Wisconsin regression (see Figure 4.12*d*). Researchers isolated the fossils belonging to an *assemblage of Foraminifera* animals that today are known to be *polar ocean spe-*

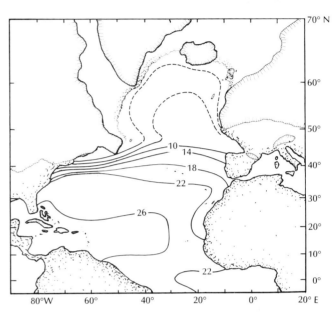

(b) A reconstruction of sea surface temperature for the month of August, 18,000 B.P., based on the distribution of foraminiferal assemblages.

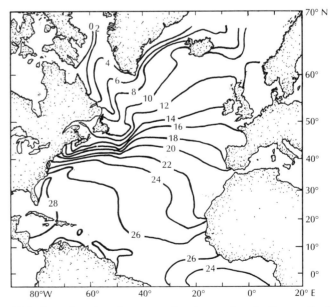

(c) A chart of modern-time sea surface temperature for the month of August. (From McIntyre et. al., 1976.)

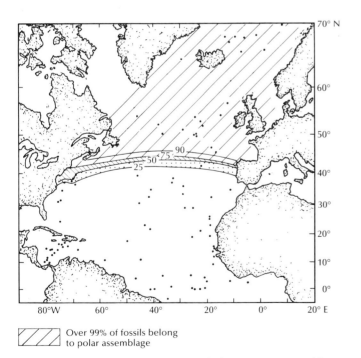

Over 99% of fossils belong to polar assemblage

(a) Contours of the percentage of fossils belonging to an assemblage of high-latitude foraminiferans.

Figure 5.14 A study in paleoceanography. (Photo by Dudley Foster, Woods Hole Oceanographic Institute.)

cies. These were counted and map contours were drawn to show regions where these animals dominated the fossil community, as in figure 5.14a. What emerged was a clear picture that 18,000 years ago the edge of the *polar ocean* stretched along a latitude band from Spain across to Nova Scotia, far, far south of where the edge of the polar ocean is today.

Since we also know the surface water temperature preferred by each species of Foraminifera, we could construct the probable pattern of sea surface temperatures for the month of August, 18,000 years ago, in Figure 5.14b! Compare the 10°C contour then against what it is today (Figure 5.14c); then it reached no farther north than the tip of Spain, but today it reaches to Iceland.

The fossil data can be studied another way. If we choose one certain species of fossil, then count the number found in each successive layer (read time) in the core, and then plot the data against time, we find a strong cyclic pattern of abundance. Major cycles occur at intervals of 20,000, 40,000, and 100,000 years. Interestingly, the Yugoslavian mathematician M. Milankovitch wrote years ago that ice ages might be triggered by changes in the orbit of the earth around the sun. For example, the perihelion of the earth's orbit moves across the seasons of the year on a cycle of 21,000 years; in modern time, earth is closest to the sun during the northern winter. In the 40,000-year cycle the angle between the polar (spin) axis and the ecliptic plane changes from 22 to 24.5°; today the angle is about 23.5°. The latitudes of the Tropics of Cancer and Capricorn are 23.5°. In the 100,000-year cycle the eccentricity of the earth's orbit about the sun changes from an ellipse to a near-circle and back again. Although there is reason to correlate ice ages with these orbit parameters and with sediment history, we still do not know the precise order of ice age events. The next time for a peak in one of these orbit parameters is 3000 years in the future.

Summary

An ideal model of the *shape* of continental margins is compared to several real shapes. We find that the topography of the slope sectors is the most uniform part among the different margins: width ranges from 20 to 100 km worldwide, and *grades* average 4°. The slope contains the true edge of the continent, located between 1000 and 3000 m in depth.

One of the more fascinating features of the edges of continents is the *submarine canyon*. Some have a topographic relief greater than that of the Grand Canyon. One argument to explain how these have formed is the *turbidity current*. The fact that many canyons do not align with present-day rivers is evidence that they were not formed during past glacial epochs when sea level was lower. The discovery that surface waves can move large amounts of beach sands parallel to the coast, and eventually into the feeder network of a submarine canyon, offers one explanation of how these canyons form without proximity to river mouths. Coastal sediment also collects on slopes by direct transport over the shelf edge, the *break*. The rather uniform grades among world slopes is evidence that these are maintained by periodic *slumping* of slope sediments.

Continental shelf sectors are of great importance to humankind. They are the communications links between continents and open oceans, in the sense that all material eroded from lands must pass across the shelf. Topography of the shelf is controlled by wave action near shore, but the rather uniform world average depth of *shelf break* indicates a strong connection to past changes in sea level. During geologic epochs when the seas were transgressing over land (as is occurring at the present time), the shoreface was eroded and its materials deposited over the entire shelf as well as in estuaries and wetlands of the coastal plains. During regression periods, shorefaces are reconstructed and estuaries are severely eroded, as are the marshes and wetlands.

Deep-sea sediments give us fascinating insights into ocean behavior. They are a storehouse of historical data on how the global geology works. They provide information on past volcanic activity and on the ancient climate of the earth. More recently, we have discovered vast potential mineral resources on the ocean floor ranging from manganese to phosphorite.

Study Questions

1. What logic do you suggest is the most appropriate to develop an answer to the question "Where does the continent end and the ocean basin begin?"

2. Stated another way, which feature of the entire continental margin appears to be the most consistent, from continent to continent the world over?

3. Which aspect of the continental slope is the most

consistent, its width or its grade or both? Why should there exist a *typical* grade of about 4° in the slope? (that is, why not 8° or 2°?)

4. Each of us has seen the alluvial fans along the foothills of mountain ranges on land. To which continental margin features are those analogous? Some of you have climbed mountains and probably seen (and traversed) features called *talus*. To which continental margin feature are these analogous?

5. Samples taken from a continental rise are likely to be a *mix* of different sizes of particles, from cobbles to sand to silt. What deposition mechanism can produce such a wide range of particle sizes in one place?

6. Submarine canyons can be deep or shallow, long or short. What is the one feature of canyons in general that is strong evidence *against* their having been formed during geological epochs (glacial epochs) when sea level was lower?

7. Two *ranges* of elevation dominate the continental margins by far, the 0 to 135-m zone for shelves and the 2000 to 4000-m zone for the rises. How then can we explain the formation of the Blake Plateau, whose elevation is about 800 m below sea level?

8. What is the argument that explains why (a) worldwide all shelf zones are very *flat* and (b) the break is so consistently found at about 135 m average?

9. What geologic phenomenon makes the sedimentation structures of shelf zones so complicated to interpret?

10. During a transgressing sea epoch, what is the main *source* of sediments to the shelf sedimentation process? What are the *sinks* (areas where sediment is deposited)?

11. At first thought, it would appear that during a transgressing sea the sediments carried out by rivers would be rapidly mixed in the sea. This does not seem to be the case. Explain.

12. Basing your answer on Figure 5.11, describe accurately where we find the *thickest* deep-ocean sediments and why.

13. In the Atlantic and Indian Ocean basins, we find calcareous ooze-type sediments in association with the mid-ocean ridge features. Explain how this comes about.

14. What does "glacial debris" mean in terms of sediments on the ocean floor?

15. Which type of ocean floor sediment accumulates at the most rapid rate? In what latitude zone do we find its accumulation most rapid? Which type of sediment is the slowest to accumulate?

16. If manganese nodules form at such an extremely slow rate, how is it that they remain "on top of" the other sediment on the ocean floor?

17. Write a brief essay on *how* paleontologists were able to reconstruct the probable sea surface temperature during one month of August in the year 18,001 B.C.?

PART III

The Southern Ocean near Antarctica. Here we find water in all three of its phases, liquid, ice and vapor. Vapor plumes rise from the liquid surface in the left central part of the scene; these are visible because the evaporated moisture has condensed and then frozen into tiny particles that reflect light. On the horizon are numerous huge, tabular icebergs reflecting the low summer sunlight, a scene much like that reported by Captain James Cook from his first circumpolar voyage during the years 1772 to 1775.

SEAWATER: PHYSICAL PROPERTIES, CHARACTERISTICS, AND BEHAVIOR

In which we learn how the basic properties of substance water force ecosystem earth to function in the way it does.

CHAPTER 6   PHYSICAL PROPERTIES OF WATER

We study the behavior of the water molecule in its different states of solid ice, liquid fluid, and gas, and the direct consequences of these states on how the earth's system works.

CHAPTER 7   SEA SALTS

What are the origins, residence time, and ultimate fates of salts dissolved in seawater, the additional consequences to ocean behavior stemming from its salinity?

CHAPTER 8   TWO OCEANOGRAPHIC CRUISES

We go out to sea to observe the nature of the ocean at first hand and formulate new concepts on how the water column is structured.

PHYSICAL PROPERTIES OF WATER

It is essential in oceanography that the reader learn the basic physical properties of ordinary *water*. By these are meant its specific heat, its latent heats of vaporization and fusion, its viscosity, and several other properties. Water may be an ordinary substance, but it is very uncommon in its characteristics.

In oceans water exists in both liquid and solid phases, and in the atmosphere as a vapor. The earth ecosystem is powered by energy from the sun, but it is the substance water that distributes this energy over the surface of the earth and through the interior of oceans. To learn how water accomplishes this key role is to learn *oceanology*, the science of large-scale bodies of water on earth. To master this science we must come to full grip with these basic properties of H_2O—properties that derive from its molecular structure.

What Is Water?

Water exists in the states of fluid, solid, or gas. In each of these states it has certain unique properties that it derives from the structure of its molecule (Figure 6.1).

It Is a Fluid

Clearly, water is a fluid. But so is the gaseous atmosphere in which we live. Water is different from air in that the behavior of a *single* molecule is governed mostly by the nearby presence and influence of all *adjacent* molecules. In contrast, a molecule of a gas such as air has a random motion in which the effect of a neighboring molecule is felt only on collision. A water molecule "feels" the presence of adjacent molecules at all times, and two important properties of liquid water can be described as immediate consequences.

1. *Water is virtually incompressible.* The unit of pressure used in ocean study is the *bar;* one bar is defined as the weight per unit area on sea surface of a column of overlying atmosphere (for a standard air, this is 1.013 bars, or 1013 millibars). Below the sea surface pressure increases rapidly with depth because of the high density of seawater, about a thousand times higher than that of air. Pressure increases to 2 bars at a depth of about 10 m (one bar is due to the overlying atmosphere, the other to the weight of the water column overlying the 10 level), pressure is 3 bars at 20-m depth, and so on. At the bottom of a typical 4000-m-high column of seawater above the abyssal plain the pressure is around 401 atmospheres. But in spite of very high pressures in its interior, the compression of ocean water is quite low. Stated another way, the volume occupied by a given number of water molecules changes very little even under high pressures. This *resistance* to volume change gives it a low compressibility, a "property" inherent in the way nature constructs H_2O. For a pressure change of 1 millibar at the sea surface, the corresponding change in height of the typical 4000-m column is only 0.01 m (one centimeter, or about 0.00025%); if the earth were to lose completely its atmosphere, sea level would rise by only 10 m or so. From the preceding we infer that the *pressure-caused change in seawater density* has negligible impact on whether water rises or descends within the liquid ocean column. For example, suppose we lower a sampling bottle to the 4000-m level, close the valves on the bottle to entrap the water sample, and then raise the bottle to the deck of the ship. The increase in volume of the enclosed water would be negligible. (Oceanographers are most grateful! To take samples of a highly compressible fluid at great pressures

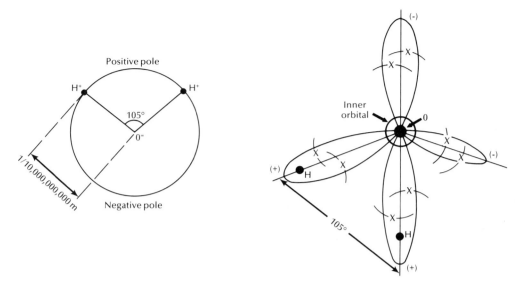

● The atom's nucleus, 0 for oxygen, H for hydrogen.

X An electron occupying orbits in both the oxygen and the hydrogen atoms.

(+) and (-) Regions where the molecule exhibits net electric charge.

If we could look deeply into a single molecule of the substance water, we would see its shape to be much like that in this sketch: two hydrogen atoms (H) closely coupled with one oxygen atom (O). Its molecular weight is 18; 16 units are gained from the oxygen atom and 2 from the hydrogen atoms. Its size is described as the average distance between the O and H nuclei, about 0.1 millimicrometer (1/10,000,000,000th part of a meter).

Why does nature choose this particular arrangement of one O with two H? A simple answer is that neither oxygen nor hydrogen atoms behave as single entities. Each prefers to join with other atoms. Like all other atoms, O and H each have a nucleus of tightly packed neutrons and protons surrounded by a network of orbiting electrons. It is the specific arrangement of electrons in the *outermost* orbits of an atom that determines just how one atom interacts with other atoms. An oxygen atom has eight electrons, the hydrogen atom only one.

A three-dimensional sketch of how the oxygen and hydrogen atoms combine might look like this. The heavy oxygen nucleus is the central part: two of its electrons are paired off in a tight inner orbit, leaving the remaining six in an outer orbit. According to electron network rules, this outer orbit is the most stable when filled with eight; therefore, the oxygen atom is two electrons shy of a *preferred* state. Each hydrogen atom also needs one electron to pair off with

the single electron in its orbit. By combining, two hydrogen atoms and one oxygen atom share two electron spaces, forming a molecule that is highly stable; stated another way, the combined molecule with shared electrons closely fits the network rules.

But, although electron sharing has created a stable-physical molecule, the electric field around the molecule is unbalanced. Whereas the hydrogen's own electron is separately able to neutralize the proton charge in its own atom's nucleus, once sharing begins the electron spends a part of its time within the network field of the oxygen atom. This leaves the vicinity of each hydrogen nucleus with a small, net average positive electric charge. By a similar argument we find a small net negative charge near the oxygen side of the molecule. These opposing electric charges together form what is called an "electric dipole." This dipole characteristic is responsible for much of the unique behavior of water. For example, when an external electric field is applied to water, its molecules twist to align their own dipoles with the external field, thereby giving water a high *dielectric* value, about 80; for contrast the dielectric value of a vacuum is one. This means that two foreign electric charges placed into the water itself would attract each other only one-eightieth (1/80) as strongly as when placed in a vacuum. Water thus becomes an extremely effective solvent, one reason that life on earth is so critically dependent on it.

Figure 6.1 The water molecule.

would be tricky, costly, and not a little dangerous.) For contrast, consider the expansion of the helium gas in a balloon released from sea level to rise upward into the atmosphere. At a height of 5 km the air pressure is *one-half* of its value at sea level; all other factors being the same, the volume of gas inside the balloon would double. The conclusion is immediate—oceanographers may neglect the expansion–contraction of seawater that undergoes large

vertical movement, as for example in regions of upwelling.

When air is compressed its temperature rises. In our atmosphere air temperature at sea level is about 10°C higher than at 1-km altitude above sea level. The almost incompressible ocean waters behave differently. Even though pressure increases rapidly with depth, the resultant rise in water temperature with pressure is low, only about 0.1°C/km of depth change. If water behaved like the gas atmosphere in this respect, high temperatures at the bottom of the ocean would "cook" all life as we know it.

Being almost incompressible also allows water to propagate sound waves at high speeds and without absorbing much of their wave energy. Pressure waves of sound move through seawater at about 1500 m/s, roughly five times their speed in air. Moreover, sound travels great distances in the oceans; it is no accident that much animal life in the sea has evolved to communicate via sound. We are all familiar with the recent studies of the "speech" of whales and porpoises. Still another facet of marine animal behavior that appears closely tied to the ease and speed of pressure waves in water is the schooling phenomenon of many fish species. Each individual fish senses changes in the local pressure field created by the turning of an adjacent fish and makes a similar turn. Propagated rapidly through the group, these turnings are seen as the apparent simultaneous change in the direction of the whole school of fish.

2. *Water is a viscous fluid, although its coefficient of viscosity is quite low* (see Table 6.1). The motion of individual water molecules depends on the presence and influence of adjacent molecules also in motion. It involves a degree of *friction drag* between them—this is the meaning of "viscosity." Some fluids such as motor oil have high viscosity.

But viscosity often changes for motions at scales much larger than that of molecules. Large ocean gyres and currents move with very little internal "viscous drag."

As orientation to the length scale at which viscous drag becomes important, let us consider the copepod whose body length averages about 2 mm. At this size scale water viscosity is low; the copepod is therefore able to swim without great difficulty. But the phytoplankton it feeds on are so much smaller that water viscosity become a major factor in the copepod's feeding behavior. (See Figure 2.3 for a true scale comparison between copepod and phytoplankton.) The separation between the hairlike structures of the copepod's feeding appendages is only a few micrometers—within a few thousand times the size of a water molecule itself. Therefore, the copepod "feeds" from a viscous fluid! Stated another way, the microscopic phytoplankton "feel" the ocean as an extremely viscous fluid in which they "sink" very slowly. In fact, phytoplankton often construct elaborate hairlike appendages to increase their "suspension" against the pull of gravity(see figure 9.9).

It Is Also a Solid or a Gas

Water is an uncommon liquid in that its solid and gas phases *also* readily occur within the climate conditions on earth.

A dense school of bait fish off Cape San Lucas, Baja California, photographed by a submerged diver. (Ron Church/Photo Researchers.)

Table 6.1 Properties of substance water that are important in oceanographic studies

Property	Characteristics	Relevance to Oceanograhic Studies
GROUP I: *Properties Affecting Earth Climate and the Marine Biosphere*		
Specific heat (also called heat capacity)	One calorie of heat energy changes the temperature of 1 gram of water 1°C; highest value among all other common substances.	With its great ability to absorb or release heat energy, water moderates the climatic extremes of planet earth. Equally important, because temperature extremes within the ocean are kept within mild limits, most marine life forms are cold-blooded.
Heat energy of vaporization (also called latent heat)	At its boiling point, pure liquid water needs to absorb 540 additional calories to convert 1 gram to the vapor state.	Because this large quantity of heat energy is a property of water in its vapor form, its transport through the atmosphere is a key process by which heat is transported poleward from equatorial zones of surplus radiation heating by the sun. It is also the process that energizes cyclones, typhoons, and hurricanes.
Heat energy of crystalization (also latent heat of fusion)	At its freezing temperature, pure liquid water must release 80 calories of heat energy for each gram that crystallizes into ice.	When ocean currents move ice out of polar zones, the heat of fusion remains behind, making ice an efficient way of packaging heat energy for transport around the earth. In the biosphere, solid water floating on the surface gives cover and support for a wide diversity of forms, from birds to walrus.
Heat conduction	Highest of all common liquids.	At the molecular level, heat energy conducts through liquid water about a hundred times faster than salt is diffused through the liquid. This difference supports a unique mixing process, called double diffusion, by which water masses in the ocean interior exchange heat and salt.
GROUP II: *Properties Important in Chemical and Biological Processes*		
Dissolving power and dielectric strength	The electric dipole of a water molecule allows it to dissociate many compounds. Its high dielectric strength sustains and isolates individual elements.	Seawater becomes a matrix of water molecules with numerous solvated ions of elements or inorganic complexes. In biological terms, seawater with all its dissolved substances is a nourishing soup to the microscopic plants and animals living in it.
Surface tension	Highest of all common liquids.	Because most marine plants, and many marine animals (protozoans), are single-cell forms suspended in seawater, they absorb needed chemicals across the membranes separating body tissue from the surrounding soup; the surface tension of water exerts an important control over these processes.
Transparency and refractive index	Water, because of its high dielectric strength, readily absorbs electromagnetic radiation at all wavelengths except those of visible light.	The transparency window of water precisely coincides with the wavelength seen as blue–green, the wavelength at which the sun emits radiation at peak strength. Thus, plants can grow at depths up to about 100 m. Moreover, some marine life forms have evolved with the ability to bioluminesce in this same transparency window of blue–green light and with their eyes the most sensitive to blue–green.

The Phases of Water. The phase behavior of pure water is demonstrated using a so-called "phase diagram" such as Figure 6.2. At a sea level atmospheric pressure of 1 bar, water exists in all three states depending on its temperature. In fact, the Celsius scale of temperature now in common use is defined at its end points by the freezing point (0°C) and boiling point (100°C) of pure water at standard sea level atmospheric pressure. The decision to use 100 divisions (degrees) for the Celsius scale was arbitrary—the original fixers of this standard could have chosen any other number. But *the selection of fluid water as the medium whose internal properties would be the standard for temperature measurement the world over* was a wise one; water is universally available, and sea level is known throughout the world.

Water does not change its state freely. Each phase boundary requires a "toll" of heat energy; *calories* are the standard units used to define a *quantity* of heat energy. A specific number of calories must be added to a water molecule to change its state from liquid to vapor, or extracted from it to change it from vapor

Table 6.1 (continued)

Property	Characteristics	Relevance to Oceanographic Studies
GROUP III: Properties Important in Physical, Geological, and Biological Processes		
Density	The density of seawater is inversely proportional to temperature and directly proportional to its load of dissolved salts.	Because of the fluid properties of water, very small differences in density throughout the interior of oceans can generate and sustain very large currents. This imposes a severe constraint on oceanographers who must measure temperatures and salinities very accurately in order to compute ocean circulation. From the view of biology, water density is so high that organisms float in the sea; marine plants invest little energy in stalks, and marine animals little energy in bones. From the view of geology, water density is high enough to suspend sand and silt and to transport these for long distances.
Compressibility (the rate at which the volume of a fluid is decreased as external pressure is applied)	Water exhibits a low compressibility; the ocean column of average depth will change its total height by 1 cm for each 1000 units of change in atmospheric pressure at the sea surface.	A direct consequence to oceanographers is that a sample of water drawn up from the deepest ocean level will not expand enough to rupture its container or injure the ship's crew. A second effect is that water readily transmits sound (pressure) waves for long distances; sound thus becomes an important communication mechanism for a wide diversity of marine life forms.
Viscosity (the ability of a fluid to resist change in shape)	With low viscosity, water is readily placed into turbulent motion	Waves can transport energy for thousands of miles across the ocean with only small losses of friction effects. Marine life forms can locomote at relatively low cost in energy; only at the microscopic size scales is viscosity high enough to be an important constraint. Marine plant cells are almost completely suspended because water at this scale is very viscous.

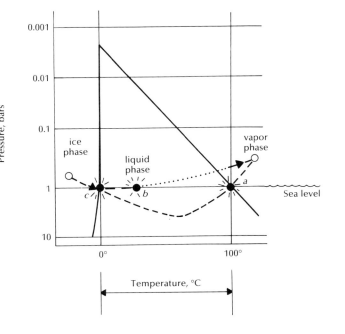

Figure 6.2 A phase diagram for pure water.

At sea level water may exist in any of its three states, ice, liquid, or gas, depending on its temperature. For a molecule of H_2O to make the transitions from one phase to another, it must undergo a substantial change in its *internal* energy.

For example: (1) To cross the phase line from ice to liquid (point c), each gram of H_2O must *gain* 80 cal of heat energy. (2) To cross the phase line from boiling liquid to vapor (point a), each gram must gain 540 cal of heat energy. (3) For a gram of liquid at ordinary sea surface temperature (point b) to vaporize, it must acquire the same 540 cal plus an additional amount. Called the total heat of vaporization, it is calculated by formula:

vaporization heat (in calories) = 596 − 0.52 × Temperature (in °C).

to liquid. What is more, the precise number depends on the pressure and temperature at which the change of state occurs; this is why a *phase diagram* is used to explain this basic property of water.

From Liquid to Vapor—The Heat of Vaporization. For example, at sea level pressure and a water temperature of 20°C (point *b* in Figure 6.2), 585.6 cal of heat energy are needed to "evaporate" each gram of H$_2$O. This is a substantial quantity of energy—the same energy would raise the temperature of a cup of tea by about 6°C. The world ocean annually evaporates about 1.2 m of water from its surface layer; we will study this aspect in detail in later sections.

From Liquid to Solid—The Heat of Fusion. About 80 cal of heat energy must be "lost" by each gram of water at freezing temperature to convert liquid into crystalline solid water (point *c* in Figure 6.2). Water in its solid phase also plays an important role in the earth's ecosystem. First, an ice cover over the sea tends to *insulate* the surface waters from intereaction with the atmosphere. Less solar energy is then gained by the ocean because much of the incident light is reflected off the icy surfaces, and less heat is lost by the ocean because it is isolated from direct contact with the atmosphere. Second, when sea ice forms, most of the salts dissolved in seawater are excluded from the crystalline solid water. These rejected salts then drop into the liquid water below and increase its density; the process may continue to the point that plumes of very cold, very salty seawater "fall" toward the bottom. This is a common occurrence in the Southern Hemisphere, where the surface ice pack around Antarctica expands to cover 8% of the hemisphere's oceans during the austral winter and, in summer, meltoff reduces the area to about 1%. Over the world oceans, the seasonal change in ice cover affects some 18,000,000 km^2 with an annual freezing–melting volume of about 18,000 km^3 of ice. The quantity of heat energy alternately released and absorbed by the melting of ice in the Southern Ocean is a major factor in the global ocean heat budget.

Heat–Temperature Properties of Pure Water

The energetics of phase changes in water is shown more clearly in a chart of the *temperature* of a single gram of pure water as *heat* is added (Figure 6.3). Remember, *temperature* in degrees is just one way of describing how much *heat* energy in calories a ma-

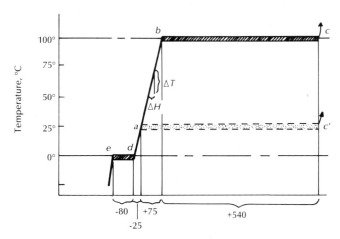

Calories of heat added or removed from 1 g of H$_2$O

Section a–b. Here the temperature of water increases linearly as heat is added. The slope of the graph, $\Delta H/\Delta T$, defines the basic property called *specific heat*. This is a measure of how much heat is needed to raise the temperature by a given amount. Water has the highest heat capacity of all common fluids or solids, 1 cal/(g)(°C).

Section b–c: Water that has been brought to its boiling point requires additional heat energy to make the phase transition from liquid to vapor. This energy is called the latent *heat of vaporization;* at sea level (pressure of one atmosphere) the amount of required heat is 540 cal/g. The needed energy is gained by individual molecules, either directly from *external* heat sources such as solar radiation or *indirectly* by absorbing internal energy from collisions with other water molecules.

Section a-d. The removal of heat from the water sample reduces its temperature. Pure water begins to freeze at 0°C, but seawater temperature must be reduced to about −1.6°C (because the freezing point of a salt solution is lower.)

Section d-e: The phase transition from liquid to crystalline water involves the *latent heat of fusion.* Individual molecules of water at 0°C must lose additional heat at the rate of about 80 cal/g in order to crystallize.

Section a-c'. Seawater evaporates directly from the liquid state to vapor without being "boiled." The phase transition here requires more heat energy than the 540 cal/g (see the equation in the extract of Figure 6.2).

Figure 6.3 The thermal state of a gram of pure water as heat is either added or removed; the initial temperature (point *a*) is 25°C.

terial contains within itself, which is a definition of *internal energy*. For this discussion, enter the chart at a typical laboratory temperature of 25°C. Now, as heat is added in section *a–b*, 75 added calories takes the gram of water to its boiling point, 100°C. Each calorie of heat added to the single gram of H$_2$O raises its temperature 1°C. What we have just defined is a basic physical property of water—its *specific heat*, sometimes referred to as heat capacity. Table 6.1 shows that water has the highest specific heat of all commonly encountered solids and fluids. This means that all other materials undergo a greater rise in temperature when the same amount of heat is added. In a following section we will explore some direct oceanic consequences of this property, but for now a single statement is made: *the capacity of water to absorb and release large quantities of heat while at the same time undergoing small changes in temperature accounts for its key role in moderating climatic extremes over the entire earth.*

With the addition of 75 cal to the 25°C gram of water, it is heated to its boiling point of 100°C. Additional heat does not further increase its temperature, instead, individual molecules of water begin to break their bonds with adjacent molecules and escape the liquid surface as molecules of water vapor. The process continues until all liquid is vaporized; it requires a total of 540 cal after the liquid is initially brought to 100°C temperature. This is called the *latent heat of vaporization.* It is important to note that the vaporized gram of H_2O "carries" this heat energy away from the liquid surface. Water vapor injected into the atmosphere carries with it a substantial "load" of *latent* energy, energy that will be released only when the vapor is condensed back into liquid form (as rain, snow, or sleet). If in the meantime the vapor is carried by prevailing winds to a point distant from its origin, the ocean–atmosphere system has succeeded in transporting a substantial quantity of heat energy across the globe.

Returning to Figure 6.3, notice that the removal of 25 cal from water at the initial laboratory temperature of 25°C brings its temperature down to the freezing point for pure water, 0°C. From this point about 80 additional calories must be removed in order to "freeze" the gram of liquid H_2O into its crystalline form, ice. Throughout the crystallizing process the temperature of both liquid and ice phases remains at 0°C. Once all liquid crosses the phase boundary into ice, the removal of additional heat will begin to lower the temperature of the ice itself. But note that the rate of drop of ice temperature per calorie of heat removed is about twice that for liquid water; this means that the specific heat of ice H_2O is about 0.5 cal/(g)(°C).

In the real ocean, the extracted heat that causes phase transition to ice is given off to the atmosphere directly, tending to warm the cold air, or to the surrounding yet-unfrozen water, tending to warm it. The process of phase transition to ice tends to inhibit the formation of more ice. Other oceanic consequences are examined later. But one idea can be posed here: suppose you need to desalinate some

CASE IN POINT

"SEA SMOKE" OVER AN OPEN LEAD IN THE ARTIC PACK ICE

Water evaporated from the liquid ocean is quickly frozen into ice crystals in the frigid polar air; we can see the evaporite. Here a surface wind is blowing from the lower right toward the upper left of the photograph, causing the sea smoke to be organized into row streams along the downwind direction. Water vapor over temperate and tropical oceans is similarly organized but is not visible to the eye. On occasion clouds of crystallized vapor rise hundreds of meters into the sky; many early explorers mistook such clouds for distant mountains or land masses. (Courtesy Victor T. Neal.)

seawater to get drinking water, and you must pay for the energy required for the process. Would you choose an evaporation process of desalinization over a freezing process? Think about it.

Some Oceanic Consequences of the High Specific Heat of Water

Large bodies of water strongly moderate the climate variations of adjacent lands, a fact known to most of us who have enjoyed or dreamed of living on a coast. The principle of physics involved is straightforward: water is capable of storing *or* releasing large quantities of heat while keeping its temperature change small.

We have learned (Figure 3.1) that the patterns of surface winds and currents are the response of the atmosphere and oceans to unequal global heating by the sun. They are the response mechanisms that act to remove inequality. In the steady state, the rate of unequal solar heating is offset by the same rate of transport of excess heat from low latitudes to regions of deficit solar heating in polar regions. The whole system is sketched in Figure 6.4. What is important here is to understand the role of water as the "working fluid" within this global "heat engine."

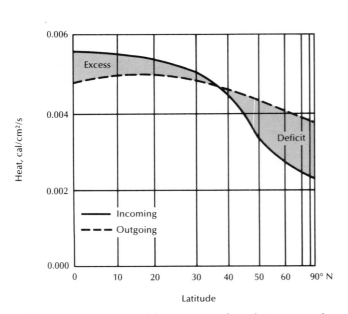

(a) Variation with latitude of the incoming solar radiation versus the outgoing back radiation from earth. Values are the mean annual radiation averaged over latitude bands in the Northern Hemisphere. At about 40°N the incoming and outgoing radiation fluxes are balanced. Poleward of this point the earth loses more heat than it gains; the deficit must be made up by poleward transport of heat via the oceans and atmosphere. (From Weyl, 1970.)

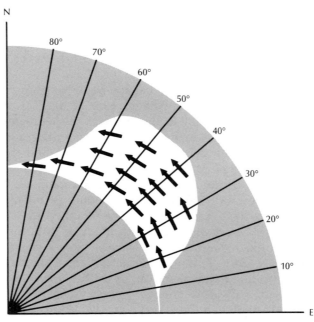

(b) A quadrant cut from a cross section of earth to show that the maximum poleward flux of heat takes place *across* the 40°N latitude. The diagram combines the poleward heat transport of both oceans and atmosphere, which share about equally in the overall process. Oceans carry warm water poleward, and the atmosphere transports water vapor with its large amount of latent heat. (From Weyl, 1970.)

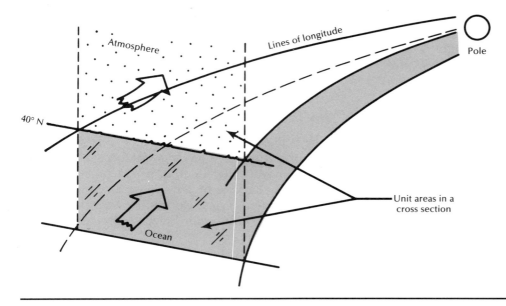

(c) The term *flux* is applied to the poleward transport of heat in the oceans and atmosphere. Each arrow represents a quantity of heat transported through a unit area during a fixed interval of time.

Figure 6.4 The concept of poleward transport of heat by oceans and atmosphere.

Poleward Heat Flux

Here is how the picture develops. Figure 6.4a compares the mean annual incoming solar radiation (short-wavelength, ultraviolet, visible, and infrared light waves) with the mean annual outgoing radiation (long-wavelength, "black-body" electromagnetic radiation, similar to the "heat" waves radiated by a potbellied stove). Clearly the tropical regions receive a surplus of incoming over outgoing radiation, and vice versa in polar latitudes. The two radiation curves cross over at about 40°N latitude in the Northern Hemisphere; it follows that *all* the excess heat of lower latitudes must be transported across the 40° parallel to satisfy the deficit at higher latitudes.

We need an appropriate way to express the *rate* at which heat is transported poleward. Oceanographers use the term "heat flux," of which a diagram is given in Figure 6.4c. It is defined as the amount of heat transported across a unit area during a given interval of time, for example, 1000 cal/(m²)(s). In the diagram, arrows of equal size imply that the oceans and the atmosphere share about equally in the task

of transporting heat poleward; this is in accord with our best estimates of winds and currents today. The ocean arrow indicates that heat flux is poleward because poleward currents carry warm water and equatorward currents carry colder water. Poleward heat flux in the atmosphere means that poleward winds carry more moisture (water vapor) than do equatorward winds. The requirement that maximum poleward heat flux occurs across the 40°N parallel is diagramed in Figure 6.4b.

Just how the ocean performs poleward heat transport is neatly illustrated by a chart of circulation in the North Atlantic. Figure 6.5 shows that the principal current crossing the 40°N parallel in a poleward direction is the North Atlantic Current, an extension of the powerful stream of warm water that flows northward along the Atlantic seaboard of the United States, the Gulf Stream. The Gulf Stream collects 26 million m³/s, almost half of its maximum flow, from oceanic regions equatorward of the 20°N parallel; these are tropical and subtropical surface waters with temperatures in the range of 25°C or higher. A total of *48 million* m³/s of warm water are transported poleward across the 40°N latitude, and branches of

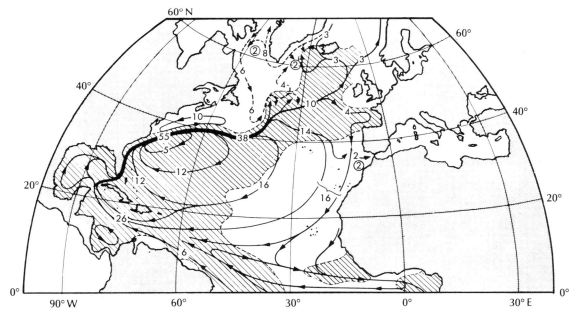

The shaded area represents surface waters that are warmer than the average temperature of all ocean waters, (computed in latitude bands). Numbers give the transport volumes of seawater in millions of cubic meters per second, and the arrows show the flow direction. The intense dark line along the Atlantic seaboard of the United States is the Gulf Stream. It is the prime mover of warm water, (and therefore heat), from the tropical to polar latitudes; some of the warm water is transported into the basin of the Artic Ocean off Norway.

Compare the information in this sketch with that of latitudinal heat flux shown in Figure 6.4. If we select 40°N as the crossover line, a comparison of the northward warm-water transport with that of the currents moving southward shows a near balance: 48 million m³/s poleward and 44 million m³/s equatorward. (From Von Arx, 1962.)

Figure 6.5 The poleward transport of warm surface waters in the Atlantic.

this North Atlantic Current reach far into the polar seas of Norway and Greenland. Adding the numbers in Figure 6.5 for waters flowing southward across the 40°N parallel, we find only *44 million units* and these returning waters are colder. (Apparently 4 million m³/s are evaporated or otherwise "lost" in the northern regions?)

How does the high heat capacity of water enter into this discussion? If the average temperature of water in the poleward transport is 10°C *higher* than the temperature of water returning equatorward, *net transport of heat is 10 cal poleward for each gram of exchanged water, or 10 million cal for each cubic meter of water, or a grand total of over 40 trillion cals for the entire North Atlantic per second!*

An Alcohol Ocean Analogy. We can play the numbers game in another way. Suppose the ocean basins were filled with a liquid different from water. How might the transport numbers just given change? For example, ethyl alcohol has a specific heat of only 0.5 cal/(g)(°C), one-half that of water. *An ocean made of ethyl alcohol would have to flow twice as fast as the real ocean to produce the same poleward heat transport!* But ocean currents are driven by winds, and it is not im-

mediately clear that wind speeds over an alcohol ocean would also double. If this fictitious ocean could not double its flow speed, the present balance between surplus and deficit radiation patterns could not be maintained. The result would be an increase in earth temperature at low latitudes and a further cooling of polar regions. Such changes would continue until a new equilibrium condition was reached, one in which the temperature difference, pole to equator, would be much higher than now, and this would force stronger winds than now. We are indeed fortunate that the oceans are water, not whiskey. Virtually every aspect of the behavior of the present world ocean is a *direct consequence of the high heat capacity of water.*

Monsoons and Sea Breezes

The fact that the specific heat of liquid water is substantially greater than that of rocks and land masses explains other phenomena. Figure 6.6 concerns phenomena of the Arabian Sea and the adjacent coasts of India and the inland mountain ranges of the Himalayas. During the summer the Himalayas absorb large amounts of solar heat, as does the adjacent

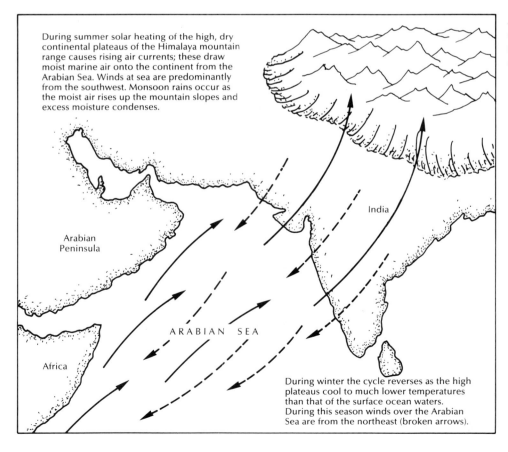

During summer solar heating of the high, dry continental plateaus of the Himalaya mountain range causes rising air currents; these draw moist marine air onto the continent from the Arabian Sea. Winds at sea are predominantly from the southwest. Monsoon rains occur as the moist air rises up the mountain slopes and excess moisture condenses.

Arabian Peninsula

India

Africa

ARABIAN SEA

During winter the cycle reverses as the high plateaus cool to much lower temperatures than that of the surface ocean waters. During this season winds over the Arabian Sea are from the northeast (broken arrows).

Figure 6.6 The monsoons of India, a seasonal phenomenon caused by differences in the specific heats of rocks and water.

ocean, but the temperature of the rock masses increases much more than that of the ocean because of its lower specific heat. The consequence is that the hot rock masses heat the overlying air (by direct conduction of heat), causing it to rise in plumes above the mountain ridges. This in turn creates a suction pattern by which moist air is drawn inland across

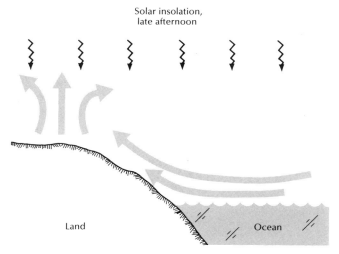

Under uniform heating by the sun, land temperature increases more than that of adjacent seawater because its specific heat is lower. Rising plumes of air warmed by the land surface create a suction that draws cool marine air onto the beach, creating the afternoon sea breeze.

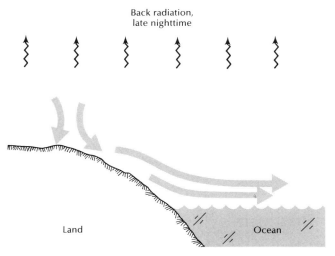

During nighttime the land masses cool more rapidly than the ocean as heat is lost by back radiation. By the late evening hours air that has been cooled over the land begins to spill out across the shore toward the sea, creating the evening sea breeze.

Figure 6.7 The sea breeze phenomenon, a daily cycle of onshore and offshore coastal breezes caused by the difference in the specific heats of rocks and water.

the surface of the Arabian Sea. As these air masses rise up the slopes of the Himalayas, much of the contained moisture condenses as rain. The process continues during most of the summer, or monsoon season. In contrast, winter periods of reduced solar radiation cause the circulation to reverse, with cold, dry air sweeping downward from the mountain slopes to spill out over the Arabian Sea.

James Michener, in his epic narrative *The Covenant* (1982), indicates how this semiannual reversal of surface winds determined the sailing patterns of the early Indian and Arabian traders. They followed the winter winds southward along the coasts of Africa and returned northward during the following season's monsoon winds.[1]

The common sea breeze effect known by all coastal residents is also caused by the difference in the specific heats of land and of water. Figure 6.7 illustrates why the sea breeze reverses direction each day. During the day, given that both land and sea receive the same amount of heat from the sun, land temperatures rise faster than ocean temperatures. The result is that warmed air over land rises, suctioning air inshore from the ocean along the surface—the afternoon sea breeze. During nighttime the temperature of the land falls more rapidly than that of the nearshore water. Cooled air flows down the shore and out onto the ocean—the evening breeze.

Role of Water Properties in Ocean–Atmosphere Coupling

We have so far shed light on two aspects of the statement that the oceans tend to moderate extreme variations in global climate. Figure 6.4 demonstrated that oceans share with the atmosphere the task of maintaining an equilibrium in the distribution of heat over global earth. Warm surface water moves into poleward regions and cooler water returns equatorward. But what are the processes by which heat is *transferred* from the warm waters to the lands and weather systems in polar regions? Recall that the budget of ocean waters crossing the 40°N parallel

[1]An interesting point: Early sailing ships were unable to sail in any direction except downwind. It required the invention of the "keel" before ships under sail power could actually move *against* the wind as well. Therefore, early mariners were by natural selection "good meteorologists"—their lives depended on the direction of the winds and on their ability to predict wind patterns. Until the invention of satellite sensing of wind patterns, our best estimates of winds over the oceans came from detailed analyses of thousands of observations logged by sailing pilots.

of Figure 6.5 was not balanced; the apparent loss by evaporation of 4 out of 48 million m³/s transported northward must play a key role in the transfer. Further, explanations of these roles must include both oceans and atmosphere and the interaction between them.

Air–Sea Interaction I—The Transfer of Heat by Conduction

Whether the oceans warm the overlying air, or vice versa, depends on the size and direction of the difference in temperature between them. Figure 6.8 shows the two cases: cold air over warm water and warm air over colder water. In both cases heat is transferred through physical contact, so that a first explanation of the *rate* at which heat transfers between them involves the *thermal conductivity* property of each, properties that are intimately tied to their molecular structure (see Table 6.1 for water). But the actual air–sea heat transfer involves much more than simple molecular conductivity of heat, as we shall see in the following descriptions.

Case 1—Cold Air over Warm Ocean. There is a direct upward flux of heat as the warm water conducts heat to the layer of cooler air in direct contact (Figure 6.8*a*). Now a new factor comes into play: as this air is warmed it becomes less dense than the still-unwarmed air at higher altitudes above the sea surface. This creates an *unstable* layering in which the upper and lower parcels of air move to exchange vertical positions, that is, the air column will "convect" vertically in an attempt to regain equilibrium.

When warmed from below, marine air rapidly organizes itself into a series of vertically convecting *cells*, one of which is sketched in figure 6.8*a*. How high the convecting plumes rise depends on many factors, but in general the larger the temperature difference between water and air the more vigorous is the heat flux and the height to which the warm plumes will rise. It is not uncommon to encounter cells hundreds of meters high.

The ocean responds in a similar but inverted manner. Whereas the air column is forced into convection by *heating from below*, the ocean column can convect vertically by being *cooled from above*, that is,

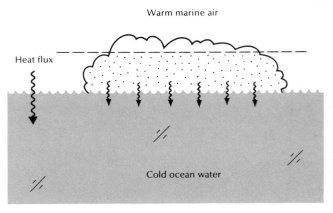

(a) Cold air near the sea surface is warmed by direct conduction of heat. Columns of air become unstable when heated from below and eventually break into vertical convection cells in which heated air rises while cooler air from aloft descends. Cells may reach hundreds of meters in height. Vertical convection can also develop in the near-surface layer of the ocean, provided it is sufficiently cooled by heat loss to the air; cells in water will be much smaller, only 10–20 m in height.

(b) Heat is conducted from air into the sea. Upon losing heat, the near-surface air gains density, so that vertical convection tends to lessen rather than grow. If the air cools vigorously, its temperature may drop below dew point; moisture may condense, usually as small droplets and fog. This situation often occurs in upwelling areas where cooler waters are forced to the surface.

Figure 6.8 Air–sea interaction I: consequences of the transfer of heat by direct conduction between oceans and atmosphere.

at the sea surface. Cooling increases the density of the surface layer. If cooling is vigorous, the water column can become unstable in density and break into vertical convection cells similar to those described for air. The ocean cells are usually much smaller, however, reaching at most to a few tens of meters depth (Figure 6.8*a*). This heat transfer process continues until enough air is warmed and water cooled that the original temperature difference vanishes.

Case 2—Warm Air over Cool Ocean. Case 2 is sketched in Figure 6.8*b*. Here the direction of heat flux is downward, from warm air to cooler water. On losing heat, the near-surface air is cooled and becomes denser. This tends to *stabilize* the air column and vertical convection cells will not develop. In fact, given sufficient cooling down in temperature, the near-surface air may reach its dew point, at which excess moisture vapor condenses out to form fog.

Fog banks tend to occur near shore wherever upwelled, cool ocean water meets the air. I once drove the Pan-American Highway north from central South America; enroute the road parallels the Peru coast, where strong offshore upwelling is a persistent feature. I could spot the zones of upwelled cold water by the telltale banks of fog lying about 1 km offshore. Along the Oregon–California coast, where upwelling is also persistent, some coastal plants have evolved to take moisture directly from droplets of fog condensed on leaves. An example is the redwood tree, which grows to heights unattainable by other tree species that rely only on root pressure to pump water up the tree stem. Tree-ring experts contribute to marine science by providing a historical record of "strong" and "weak" upwelling seasons; in places, the record can be traced backward thousands of years.

Air–Sea Interaction II—The Transfer of Heat by Vaporization

Direct conduction of heat from ocean surface water to the overlying air is *not* the most important process in air–sea exchange or in the study of global heat budgets. The process was used in the previous section to introduce the concept of convection as an important air–sea interaction process.

In fact, the transfer of heat via the evaporation of water is *about ten times greater*, as demonstrated in the ocean heat budget of Table 6.2. How and why

this is so is a direct consequence of the exceptionally large quantity of heat energy that water vapor retains on leaving the liquid ocean, the latent heat of vaporization.

Dry Air. Air is a gas composed mainly of nitrogen molecules, N_2 (78% by volume), oxygen molecules, O_2 (21%), and argon atoms, Ar (1%); minor constituents exist in small amounts, of which carbon dioxide (0.03%) and ozone are the most important because of the absorption of long-wave radiation by CO_2 and ultraviolet short waves by O_3.

Air is highly compressible. At any given altitude above sea level, the pressure of air is simply the weight of a unit column containing the remaining portion of the atmosphere above that level. Pressure decreases exponentially with altitude, and thus an air parcel that is convected upward tends to expand in volume. In accordance with the universal gas laws, the expansion reduces the absolute temperature of the rising parcel. For the standard atmosphere, temperature drops at the rate of about 10°C per kilometer of altitude.

Wet Air. When a molecule of water vapor enters the atmosphere from the liquid ocean, it displaces one of the standard molecules that make up air. This displacement concept is based on early work in gases by Amedeo Avogadro (1776–1856) and John Dalton (1766–1844). Avogadro found that at the same temperature and pressure, equal volumes of different gases will contain the same number of molecules. Dalton established that the total pressure exerted by a gas that is a mixture of several gases is the sum of the partial pressures of the gas components. We apply these ideas to our next study of the air–sea interaction. Figure 6.9 shows two column of air, of *equal cross-sectional area*, resting on the sea surface. The *dry* air column contains only the standard atmosphere gases, nitrogen with molecular weight 28 and oxygen with molecular weight 32. The *wet* air column, however, has received some water vapor molecules and, by the gas laws, loses an equal number of molecules of combined N_2 and O_2. But because the molecular weight of the vapor is only 18, that is, less than that of the N_2 and O_2 molecules displaced, the net result is that the *wet column must weigh less*. Stated a different way, the pressure exerted by the wet column on the sea surface is *lower* than that under the dry air column.

This is the origin of the terms meteorologists use to describe a dry air mass—an atmospheric "high"—

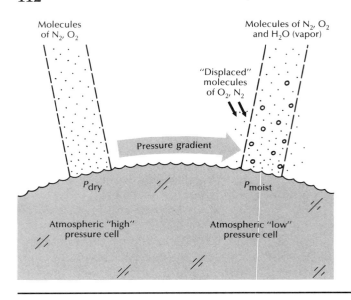

Cold air is denser than *warm* air—because the volume occupied by a gas increases with its temperature (when pressure is constant).

Dry air is denser than *moist* air—because a mixture of nitrogen, oxygen, *and* water vapor has lower density than a mixture of nitrogen and oxygen alone.

molecular weight of nitrogen (N_2), 28
molecular weight of oxygen (O_2), 32
molecular weight of water (H_2O), 18

Therefore, as N_2 and O_2 molecules are displaced by H_2O vapor, the density of the gas mixture with water vapor is less than that of air containing no water vapor.

The consequence is that the pressure at sea level under a dry air column is greater than that for a moist air column. An atmospheric dry column is therefore called a *high,* and a moist column is called a *low.* When moist and dry masses of air are adjacent, as in the sketch, a *pressure difference* exists between two points, directed along the sea surface. Winds are generated along the sea surface by these pressure differences—they are the response of the atmosphere as it attempts to move air from the high cell toward the low.

Figure 6.9 An explanation of how an air column becomes "dry" or "wet."

or a wet air mass—an atmospheric "low." In a later chapter we study how wind patterns are created over the oceans as a direct result of pressure differences beneath air masses that are either dry or moist, that is, how evaporation is responsible for organizing patterns of winds. For now, our focus stays on the heat exchange between oceans and the atmosphere.

Thundercloud over the open ocean. While the ocean surface loses about 120 cm of height each year through evaporation, much of that loss is regained through precipitation at sea. (Official U.S. Navy photo.)

Table 6.2   Heat budget of the ocean. (From Harvey, 1976.)

Net Gain of Heat	Units	Net Loss of Heat	Units
By short-wave radiation	100	By long-wave radiation to atmosphere and space	41
		By conduction of sensible heat to atmosphere	5
		By evaporation	54
			100

Role of Latent Heat of Vaporization in the Heat Engine

We now understand clearly a previous statement that the oceans and atmosphere share about equally in the work of transporting heat energy poleward (as diagramed in Figure 6.4c). The atmosphere extracts heat energy from the oceans in the form of water vapor and transports the vapor molecules, together with their latent heat energy, poleward. The question how this energy is eventually "released" in polar regions is also straightforward. It is released to the surrounding air, and eventually to the land masses, when the vapor condenses into liquid rain or crystalline snow or sleet. We conclude that water is truly the working fluid of the earth's heat engine, whether in liquid form in which its high heat transport is due to the property of high specific heat, or in vapor form in which its high heat transport is due to high latent heat of vaporization.

Heat Budget of the Ocean. In the end it is of course the oceans that supply the water for the heat engine. Table 6.2 compares the total heat received by the ocean with that given off by it. Note that the portion of total heat lost by the oceans by direct conduction to the atmosphere is minor compared to that lost by back radiation or by evaporation. The evaporation loss is the largest and most important process; it exists because water has such a high latent heat of vaporization. The oceans annually evaporate about 1.2 m of water, on the average.

Consider again the hypothetical ocean filled with alcohol. Ethyl alcohol has a latent heat of vaporization of only 200 cal/(g)(°C) (less than one-half that of water). If this alcohol ocean lost the same 1.2 m of liquid, its heat budget would be greatly unbalanced, for heat loss would not compensate for the solar radiation absorbed. Temperature would rise toward a new equilibrium state. For such an ocean to evaporate twice as much liquid (to satisfy its heat budget), the atmosphere would need to transport twice as much vapor alcohol, and it is not at all clear that it could do so. The failure of the atmosphere to do its share of poleward heat transport (now made more difficult because the latent heat of vapor alcohol is low) would force a decline in polar temperatures. As a result of an overall *increase in temperature difference between equatorial and polar regions*, the speeds of both winds and currents would increase. In summary, the earth is precisely the way it is today because of the specific physical properties of the substance water.

Role of Vaporization in Atmospheric Convection

In Figure 6.8 the phenomenon of vertical convection cells in the layers of cool air over warm ocean surface water is related directly to the warming (and destabilization) of the air column from below. *What if, in addition to sensible heat flux, there occurs a flux of water vapor into the layers of air just above the sea surface?* The answer is now clear: the injection of water vapor into the lower levels of the air column also decreases air density and drives the column toward greater instability. Vertical convection becomes much more vigorous than when heating is by conduction heat transfer alone. The rising columns of warmed and moisture-laden air penetrate higher into the overlying atmosphere, often to levels at which the moisture begins to condense to form clouds.

We may carry this discussion one step further. In fact, vertical convection can occur, and vigorously, even though air and water temperatures are the same; *vaporization alone can drive convection* motion. This condition is often found in tropical oceans and is responsible for the fluffy cloud formations and general haze of tropical skies. The converse statement also applies: there can be sensible heat flux even though no evaporative flux is occurring. This condition is found whenever the near-surface air is warm and at its moisture-saturation point. The fog banks that I described earlier in areas of upwelling off Peru are a case in point.

On the basis of the arguments given, we conclude that the conditions for *optimum heat flux from ocean to air occur wherever a warm ocean is overlaid by cold and*

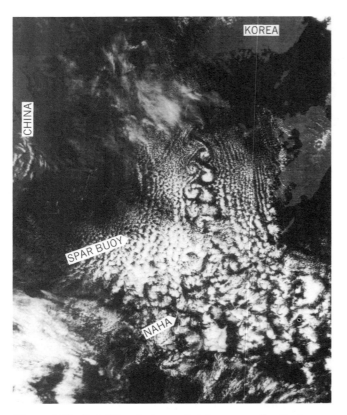

Figure 6.10 Satellite photograph taken over the East China Sea at 1136Z on February 17, 1975, during a large-scale ship (NAHA) and buoy (SPAR) research experiment; the separation between ship and buoy is 315 km. The photo shows the remarkable development of convection cells during an outbreak of cold, dry air off the Manchurian-Korean plateau and over the East China Sea (from the top toward the bottom of the photograph). We see the cells by virtue of the cloud pattern that accompanies vigorous upward air convection. The stream of vortexes in the center are the result of air moving across a mountainous island off the southern tip of Korea. (From Trump, Neshyba and Burt, 1982.)

dry air. These conditions exist in several ocean regions. Off the northeast coasts of North America, cold, dry air sweeps off the Canadian Shield and Greenland to spill out over the warm surface waters of the North Atlantic Current. This is a region of intense evaporation and cooling of surface water. Similar phenomena occur over the Sea of Japan and the East China Sea, which are swept by cold, dry air masses breaking out from the Siberian Shield of eastern Asia. Figure 6.10 is a satellite photograph of the Asian region taken in February 1979 during an outbreak of cold and dry continental air masses over the East China Sea. Cellular cloud patterns developed as the convection cells carried moisture-laden air to high altitudes. There the air cooled through release from pressure, and much of its moisture condensed to form clouds.

Role of Evaporation in Oceanic Convection

Figure 6.8 shows that the loss of sensible heat from the ocean surface causes higher density in the cooled seawater, a destabilization of the water column, and the subsequent development of vertical convection cells. We now show that the evaporation of water from the surface can also set up vertical convection within the upper ocean. In this case, however, there are two contributing effects, and the end result is a convection much more vigorous than that caused by cooling alone.

1. Heat loss by vaporization is ten times greater than sensible heat loss (see Table 6.2). Water molecules acquire the energy needed to escape the liquid surface from two main sources, solar radiation and a transfer of energy from adjacent water molecules, thus leaving *cooled* fluid behind. (This second effect is well known in arid lands of the world where evaporative water coolers are used to air-condition buildings.) As in sensible heat loss, cooling the ocean from the top tends to destabilize the water column and generate vertical convection.

2. Seawater is a binary fluid composed of H_2O and dissolved salts. These salts do not participate in the evaporation. The liquid that remains after some surface water evaporates must accept the salt fraction left behind, and the result is again an increase in density of surface liquid relative to that at deeper levels and a forcing of vertical convection.

Taken together, the heat loss plus salt loading in a zone of high evaporation rapidly brings about overturn of the surface layers of the ocean. The reader now understands the significance of a statement made earlier: "Each mass of water in the interior of the oceans, which we can identify today by its unique temperature and salinity values, acquired those values at an earlier time when exposed to and interacting with the atmosphere." Now the reader may ask additional questions. To what depth can the convective overturn penetrate? Where in the world oceans is this surface process of greatest importance? Before answering these inquiries, we need to examine in some detail another physical property of seawater—its density, how density changes with heat and salt, and just what are the magnitudes of such changes. We will return to more discussion of the air–sea interaction afterward.

Summary

The point of Chapter 6 is a crucial one—we cannot lay claim to a study of oceanography without first getting to know water itself. Central to the *knowing* of water is to accept that the earth behaves just the way we observe it to behave precisely because of the *basic physical properties of water*. Water is almost incompressible, which is important for two reasons: we can sample it easily and it does not boil at high pressure. It has the highest heat capacity of any common liquid or solid, a property that allows the oceans to move large amounts of heat poleward to offset unequal solar heating of the earth. Water's heat of vaporization is the highest of common liquids, enabling the atmosphere to do its share of poleward heat flux. Water is really the "working fluid" of the heat engine of the earth.

Nowhere is this role of water more clearly seen than in air–sea interactions. The fact that vapor injected into the air makes it more "buoyant" than dry air explains how the atmosphere gets organized into the wind and weather patterns that we observe (recall, for example, Figure 3.1). The interaction goes both ways, for it is the wind that drives the surface circulation of the ocean itself. As we shall see in the next chapter, the global "shuffling" of water by the atmosphere, through evaporation and rain, creates the important differences between the major oceans, differences that show up in all the marine sciences from water chemistry to sedimentology to the specific adaptations seen in marine life.

Study Questions

1. From Table 6.1, list the physical properties of matter for which the substance water possesses higher values than any other commonly encountered liquid.

2. Because of its high density, seawater in the ocean depths exists at high pressures. What is the pressure (measured in bars) at about 100-m depth? At 4000-m depth?

3. A gaseous substance when compressed undergoes a rise in its temperature. In our atmosphere, even though air at sea level is compressed by the weight of overlying air, its temperature is relatively low because the heat generated during compression can be radiated away to outer space. Not so in the internal depths of ocean fluid. There radiation is rapidly absorbed within the fluid. Why then is not the temperature of deep water much higher than it actually is?

4. In this chapter we defined the quantity *specific heat;* for water it is 1 cal/(g)(°C). We often refer to this property as the *heat capacity* of water. In one sentence, write the meaning associated with the statement "Water has a high heat capacity."

5. The heat energy associated with the vaporization of water is often referred to as "latent" heat. Why do you think the word *latent* is used here?

6. In the study of the importance of the processes of freezing and evaporation of seawater to the behavior of oceans on planet earth, it is useful to describe the heat energy changes that accompany the phase changes liquid to ice and liquid to vapor, or ice to liquid and vapor to liquid:

   (a) What happens to the 80 cal given up by a gram of liquid water when it turns to ice? (Where does the heat come from? Where does it go?)

   (b) What happens to the 540 + cal of heat energy involved in vaporization? (Where do the calories come from? Where do they go?)

   (c) How do you describe the reverse processes (ice to liquid, etc.).

7. An oceanic consequence of the high heat capacity of water is that poleward currents of warm surface water are highly important, moderately important, or of minor importance in moderating the temperatures in polar regions.

8. A consequence of the high heat capacity of seawater is manifested in the daily occurrence of sea breezes across shorelines. Explain.

9. The heat from a warm ocean surface is vertically transferred to a cooler air mass over the ocean through two processes. Name the processes and indicate which of the two is the most important in overall global sea–air heat transfer.

10. From what you have learned, it appears that the atmosphere is heated from below and the ocean from above. Do you agree or disagree? State your argument.

11. *If* in question 10 you chose to defend the *agree* position, how would you then defend against the argument that both evaporation and conduction represent heat *losses* from the ocean surface, and therefore the ocean, over the long term, is *cooled* from above? Would the numbers in Table 6.2 be of help in resolving the argument?

12. *Suppose* that in Figure 6.4 I had deliberately left out sketch *b*. Could you deduce from sketch *a* that a *maximum* in poleward heat flux *has to exist* at some latitude between 0° and 90°N?

13. Simply stated, what is the difference between *dry* air and *moist* air?

14. Is it true that an atmospheric "low" can be a low-pressure cell *only* if it contains substantial quantities of water vapor? Stated another way, must *all* highs be dry?

15. Winds are the atmosphere's response to differences in pressure along the earth surface. Could we make the same statement about ocean currents along the ocean floor?

CHAPTER 7

SEA SALTS

The saltiness of the sea is its best-known characteristic as well as one of its most interesting. Without dissolved salts and gases the ocean fluid would not support life as we know it today. Most of the oldest dated fossils of life forms are of marine species, and it leads us directly to the question of *when* the early oceans acquired sufficient dissolved salts for life to evolve, if indeed it did in the early oceans. A second question is whether the salinity of seawater is continuously increasing. We now believe that it is not.

The study of sea salts impacts all marine science disciplines. The diffusion rates of salts in water are an integral part of the mixing phenomena studied by physicists. Biologists studied the limits to plant growth on the basis of the cycling of salts. Chemists are engrossed in measuring the rates at which gases are exchanged with the atmosphere, the lifetimes of salts, and how they are assimilated in the sediments. Geologists measure the distribution and ages of such sediments, and geochemists worry about how the entire picture of sea salts fits together. It seems fitting that salts be given a chapter of their own.

Origin of Salts

At just what point in geologic time the oceans first became salty is not known. What we do know comes to us from three sources of evidence that suggest that the waters became salty shortly after they began to fill the earliest basins. The first of these is the fossil record. Fossils of the late Precambrian period (over 600 Ma) reveal the existence of burrowing worms similar to the types of worms that still bur-

Salt flats near Wendover, Utah. When ancient seas or brackish lakes evaporated, the salts that had been dissolved were left behind as mineral salts of various types. (Grant Heilman)

Table 7.1 Elements found in seawater: the concentrations of 81 of the known 92 elements in seawater as measured in the surface layer, in the deep layer, or both.

Atomic Number	Element	Species	Behavior	Predicted Mean Water Concentration	Residence Time Years Based on Sediment Accumulation	Based on River Input
1	Hydrogen	H_2	Biogenic or hydrothermal origin	108 g/kg		
2	Helium		Nonnutrient gas	1.9 nmol/kg		
3	Lithium		Conservative	178 μg/kg	2.0×10^7	
4	Beryllium		Nutrientlike, but increases with depth	0.2 ng/kg	150	
5	*Boron	Inorganic boron	Conservative	4.4 mg/kg		
6	†Carbon	ΣCO_2	Nutrient	2200 μmol/kg		
7	†Nitrogen	N_2	Nonnutrient gas	590 μmol/kg		
		Nitrate	Nutrient	30 μmol/kg		
8	Oxygen	Dissolved O_2	Biological dependence	857 g/kg		
9	Fluorine		Conservative	1.3 mg/kg		
10	Neon		Nonnutrient gas	8 nmol/kg		
11	*Sodium		Conservative	10.781 g/kg	2.6×10^8	2.1×10^8
12	*Magnesium		Conservative	1.28 g/kg	4.5×10^7	2.2×10^7
13	Aluminum			1 μg/kg		
14	†Silicon	Silicate	Nutrient	110 μmol/kg	8×10^3	3.5×10^4
15	†Phosphorus	Phosphate	Nutrient	2 μmol/kg		
16	*Sulfur	Sulfate	Conservative	2.712 g/kg		
17	*Chlorine	Chloride	Conservative	19.353 g/kg		
18	Argon		Nonnutrient gas	15.6 μmol/kg		
19	*Potassium		Conservative	399 mg/kg	1.1×10^7	1×10^7
20	*Calcium		Correlates with carbonate alkalinity	415 mg/kg	8.0×10^6	1×10^6
21	Scandium			<1 ng/kg	5.6×10^3	
22	Titanium			<1 ng/kg	160	
23	Vanadium		Conservative	<1 μg/kg	1.0×10^4	
24	Chromium		Nutrient-correlated; silicate and phosphate or nitrate	330 ng/kg	350	
25	Manganese		Surface maximum; at depth, correlated with the labile nutrients and negatively correlated with dissolved oxygen	10 ng/kg	1400	
26	Iron		Correlated with the nutrients; negatively correlated with dissolved oxygen	40 ng/kg	140	
27	Cobalt		Similar to manganese	2 ng/kg	1.8×10^4	
28	Nickel		Nutrient-correlated; phosphate and silicate	480 ng/kg	1.8×10^4	
29	Copper		Resembles nutrients with sedimentary release; scavenging at intermediate depths	120 ng/kg	5.0×10^4	
30	Zinc		Nutrient-correlated; silicate	390 ng/kg	1.8×10^5	
31	Gallium			10–20 ng/kg	1.4×10^3	
32	Germanium		Nutrient-correlated; silicate	5 ng/kg	7×10^3	
33	Arsenic	As(V)	Nutrient-correlated; phosphate	2 μg/kg		
34	Selenium	Se(IV)	Nutrient-correlated; silicate and phosphate	170 ng/kg		
35	*Bromine	Bromide	Conservative	67 mg/kg		
36	Krypton		Nonnutrient gas	3.7 nmol/kg		
37	Rubidium		Conservative	124 μg/kg	2.7×10^5	

row today in the soft sediment of the sea floor. In the Cambrian (over 500 Ma) skeletonized fauna existed, examples of which are the trilobites, crablike animals that had segmented exoskeletons and apparently fed on detrital material upon the ocean floor; similar marine animals exist today. These data suggest that life forms had at that early period already adapted to life in a saline fluid. A second source of evidence comes from analyses of the evaporated salts left behind when the seas of ancient basins dried up; these ancient salts are similar to those we find in modern evaporites. The third clue

Table 7.1 (continued)

Atomic Number	Element	Species	Behavior	Predicted Mean Water Concentration	Residence Time Years — Based on Sediment Accumulation	Based on River Input
38	Strontium		Nutrient-correlated; phosphate	7.8 mg/kg	1.9×10^7	
39	Yttrium		First approximation: conservative	13 ng/kg	7.5×10^3	
40	Zirconium			<1 µg/kg		
41	Niobium			1 ng/kg	300	
42	Molybdenum		Conservative	11 µg/kg	5.0×10^5	
44	Ruthenium			0.5 ng/kg		
45	Rhodium					
46	Palladium					
47	Silver			3 ng/kg	2.1×10^6	
48	Cadmium		Nutrient-correlated; phosphate	70 ng/kg	5.0×10^5	
49	Indium			0.2 ng/kg		
50	Tin		Nonconservative; anthropogenic	0.5 ng/kg	5.0×10^5	
51	Antimony		Conservative	0.2 µg/kg	3.5×10^5	
52	Tellurium					
53	Iodine	Iodate	Nutrient-correlated; nitrate and phosphate	59 µg/kg		
54	Xenon		Nonnutrient gas	0.5 nmol/kg		
55	Cesium		Conservative	0.3 ng/kg	4×10^4	
56	Barium		Nutrient-correlated; silicate, alkalinity	11.7 µg/kg	8.4×10^4	
57–71	Lanthanum and the Lanthanides		Nutrient- or depth- correlated		1.1×10^4	
72	Hafnium			<8 ng/kg		
73	Tantalum			<2.5 ng/kg		
74	Tungsten			<1 ng/kg	1.0×10^3	
75	Rhenium			4 ng/kg		
76	Osmium					
77	Iridium					
78	Platinum					
79	Gold			11 ng/kg	5.6×10^5	
80	Mercury		Nutrient-correlated; silicate	6 ng/kg	4.2×10^4	
81	Thallium		Conservative	12 ng/kg		
82	Lead		Nonconservative; anthropogenic	1 ng/kg	2×10^3	
83	Bismuth			10 ng/kg	4.5×10^4	
84	Polonium					
85	(Astatine)					
86	Radon					
87	(Francium)					
88	Radium					
89	Actinium					
90	Thorium			<0.7 ng/kg	350	
91	Protactinium					
92	Uranium		Conservative	3.2 µg/kg	5×10^5	

*These eight elements are called *conservative*; they exist in uniform relative concentrations throughout all oceans.
†These are the commonly labeled plant *nutrient* elements.
After Quinby-Hunt and Turekian, 1983; Shephard, 1963; Goldberg and Arrhenius, 1958; Barth, 1952.

came with the recent discovery that each salt constituent has a characteristic residence time, (or cycling period) between the time of first entry into seawater and that of final exit when it is fixed permanently in ocean sediment. Further, for the elements whose residence times have been measured (Table 7.1), we find average cycle times much shorter than the estimated age of ocean water itself. This eliminates any hypothesis that present-day salt concentrations are simply accumulations over the geologic history of the earth.

Average seawater has a concentration of about

34 g of dissolved salts per 1000 g of mixture. This ratio is referred to as the *salinity* and is expressed as 34 parts per thousand (ppt). Although dissolved salts are well mixed throughout the oceans, the salinity values vary over wide limits that reflect areas of localized high rates of precipitation or river runoff or high rates of evaporation. Stated differently, salinity is really a measure of how rapidly H_2O moves into and out of the salty "soup" we call oceans.

Elements in Seawater

The dissolving power of water is the highest of all commonly occurring liquids on earth (see Table 6.1). Its ability to separate the atomic parts of chemical compounds stems from the bipolar form of the water molecule itself. Figure 7.1 diagrams the dissociation of ordinary table salt, sodium chloride (NaCl), into its separate elements called *ions*: the element sodium is called a *cation* because it carries an excess positive electric charge in its ionized state, Na^+; the element chlorine is called an *anion* because it carries an excess negative electric charge in its ionized state, Cl^-. In nature, Na^+ and Cl^- fit together in solid form because their excess electric charges are mutually neutralized. In seawater, because the H_2O molecule is itself an electric dipole (see Figure 6.1), free ions tend to be charge-neutralized by water molecules, which arrange themselves around the ion in specific ways, as shown in Figure 7.1.

The composite structure of ion plus water, called a "solvated" ion, is large in size and seldom com-

pletely neutral in charge.[1] Individual ions can still combine with each other in water. Table 7.1, which presents the most recent (1983) tabulation of dissolved elements, also lists the "preferred" chemical state that each element takes in seawater: for example, carbon is found in the compound CO_2 (dissolved carbon dioxide) and sulfur is present as SO_4 (the sulfate ion), and so on. In short, the high dissolving power of H_2O explains why most of the known elements are found in seawater salts.

Concentrations

Table 7.1 lists the mean concentrations of the many elements found in seawater. The reader is cautioned that these are not precise numbers and will surely change in the future as better chemistry is developed. As expected, concentrations vary widely, from the high (chlorine alone accounts for more than 55% of total salts, at 19.353 ppt), to the minor (fluorine is present at approximately one part per million), to

---

[1] One way to desalinate salty water is to pass it between a pair of electrodes connected to the positive and negative terminals of an electric battery. The solvated Na^+ ion is attracted to the negative (−) electrode, whereas the Cl^- composite ion moves toward the positive (+) electrode. The fluid midway between the electrodes is depleted of salt and can be pumped off as "fresher" water. In practice, the process works only for brackish waters, of less than 5-ppt salinity. The fact that water is such a "good" dissolver (read neutralizer) means that strong electric fields are needed to separate the nearly neutral, solvated ions, and this translates into high costs.

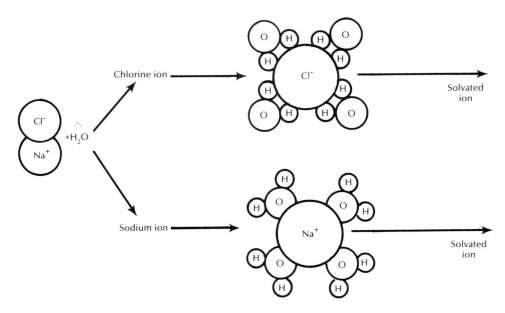

Figure 7.1 Sodium chloride, NaCl, is dissociated by water into ions of elements sodium (with one positive charge), and chlorine, (with one negative charge.) Because the water molecule is itself polarized, it can aggregate to the charged ions to neutralize them; the result is a larger particle, called the solvated ion.

Vapors issuing from fumaroles in volcano Mt. Aso, Kyushu, Japan. Superheated steam coming from active volcanoes consists mostly of ground water that has seeped into the fissures of magma, but some part of the fumarole steam is "juvenile" water, H_2O that is coming to the earth's surface for the first time. (Leo de Wys, Inc.)

extremely low values (gold is present at 0.00001 parts per million.)

Distributions

Not all elements are distributed in the oceans in the same way. The various patterns of concentrations that we measure allow us to separate the dissolved elements into four groups.

1. *Conservative elements.* The eight most abundant elements together account for 99.5% of all dissolved material. Their concentrations are everywhere much larger than the rate at which new quantities are introduced either through continent erosion or through vulcanism. Stated another way, their residence times are much longer than the few thousands of years required for the oceans to become well mixed. Their relative concentrations occur in fixed ratios, everywhere; this rule of constancy of proportions for the conservative elements is discussed again later in this chapter. In Table 7.1 conservative elements are marked with an asterisk.(*).

2. *Nutrients.* Concentrations of elements that are vital nutrients to plant life are unequally distributed, vertically as well as horizontally. Concentrations of nutrients tend to be low in the surface waters where

plants absorb them and higher at deep levels in the water column where decomposition of organic matter returns the elements to solution. In Table 7.1 primary nutrients are marked with a dagger.

3. *Trace elements.* Third in order of abundance, trace elements are found at concentrations less than one part per million (ppm). In Table 7.1 various elements in negligible concentrations are described as *nutrientlike.* Although not considered nutrients, their similar distribution results from the tendency of many marine species to assimilate them just like nutrients. Trace elements are known to play vital roles in body chemistry.

4. *Dissolved gases.* The most commonly occurring gases are those that also comprise the atmosphere. Therefore, surface waters are generally saturated with dissolved nitrogen and oxygen. The distributions in deeper waters, however, are not alike. Carbon dioxide, for example, is the prime source of carbon for building organic protoplasm and is classed as a nutrient in Table 7.1; its concentration increases with depth, a behavior that is nutrientlike.

Gross (1977) presents these groups in a different way (Figure 7.2) by using the format of the classical periodic table. The trace elements that are essential

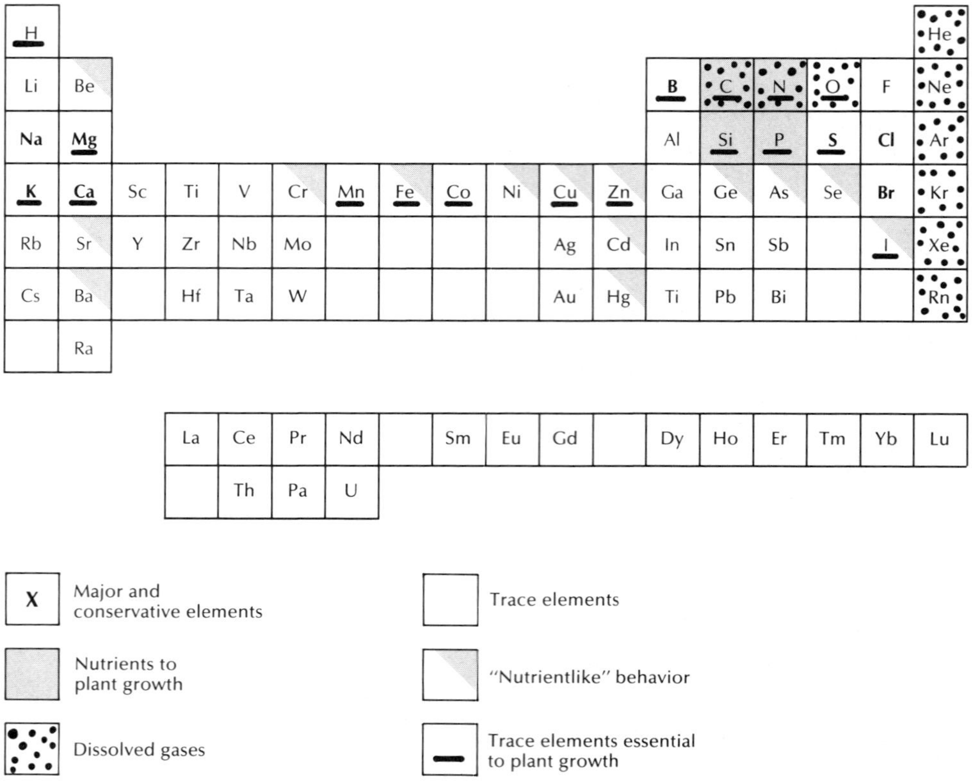

Figure 7.2 The dissolved elements in seawater displayed on a periodic table. (Based on Gross, 1977.)

for plant growth are clearly identified as the *transition* elements.

Sources, Sinks, and Residence Time

Sources and Sinks

Table 7.2 groups the eight species of conservative ions into categories of cations and anions.

Cations. Cations are defined as positively charged element particles. They are introduced to oceans through the weathering of continental rocks during the hydrologic cycle. Sodium is by far the dominant cation-forming element. Although the erosion of rock is a continuous process within the hydrologic cycle, this does not mean that cation concentrations are cumulative in seawater. All the cations listed are also eventually removed from seawater by incorporation into sediments.

The point is that rates of introduction and removal are controlled by different processes. Erosion rates are controlled mainly by dissolution in water during the hydrologic cycling of water, a continuous process in the heat engine of the earth. In contrast, removal rates depend mainly on the actual concentrations of the elements in seawater, for many chem-

ical reaction *rates* are functions of the numbers of atoms or molecules available to join the reactions. Therefore, the concentration of a salt always increases until its removal rate just balances its rate of introduction; this is its "steady-state" concentration. For example, geochemists estimate that 600 million years were required for the concentration of sodium to reach its present steady state; for contrast, the estimate for potassium is only 30 million years.

Anions. Anions are defined as negatively charged ions. The principal known sources of the anions dissolved in seawater are the fumaroles and hot springs associated with volcanic activity. The "outgassing" of mantle material deep in the earth's lithosphere is done by fumaroles and hot springs. Indeed, water itself is a component of the outgassing process, so the ocean fluid itself represents an accumulation.

Table 7.2 Principal constituents in sea salts

Cations	Anions
Sodium	Chloride
Magnesium	Sulfate
Calcium	Bromide
Potassium	Borate

With new material being continuously ejected, it is reasonable to believe that the volume of ocean waters is slowly increasing, but at an extremely low rate. One estimate places the volume of new water at 0.4 km^3/yr, a rate that would require 4 billion years to fill the present ocean.

In contrast to the cations, concentrations of anion-forming elements do not appear to have reached steady-state equilibria. The anion-forming elements not incorporated in sediments are called "excess volatiles"; chlorine is the prime example. The composition of excess volatiles probably changed during the phases in which the crust of the earth cooled, as the upper mantle selectively outgassed different constituents. In contrast, the composition of materials weathered from continental rock is more constant; the *rate* of weathering, however, has probably varied over geologic times. The uncertainties implied here mean that the ratios of concentrations of cations to anions may have varied in the past, but the time scales of such variations are large. A great amount of research is still needed, particularly since the recent discovery of hydrothermal vents on the ocean floor. For our studies here, we accept the principle that the ratios of the eight major ion concentrations are constant.

Residence Time

Each ion species has its own characteristic *cycling* process as it passes through the ocean reservoir and hence a specific cycle time scale, called the "residence" time. Some, like sodium, require millions of years to cycle. Iron and aluminum, on the other hand, pass through the ocean fluid in only hundreds of years. Because both ions play useful or important roles in biological processes, they are likely to be incorporated into sediments quickly.

Generally, the low-reactivity ions have the longest residence times; sodium, magnesium, potassium, and lithium are examples. These tend to occur as free ions, not compounded with other elements, and so do not participate in organic cycles to the extent characteristic of aluminum and iron. There are many reasons why it is useful to know the average cycling time or *residence time* of each element, and two methods are used to get such numbers.

Residence Time Based on River Input Rate. The earliest method (Barth, 1952) for determining residence time requires knowing both the *total amount* of a specific ion in the world ocean and the *rate* at which the element is injected into the ocean. For example, let X = 150 million tons be the total amount

of dissolved gold in the world oceans today. Next, add up all the gold that is transported to the oceans by the rivers of the world in, say, one year; let this number be x = 600 tons. Then *600 tons per year* is the rate that gold is injected to the sea. Dividing the total X by rate x yields 250,000 years as the *average residence time* that one atom of gold will spend dissolved in seawater before being incorporated permanently into the bottom sediments.

Not all the salt carried by a river is being introduced to the ocean *for the first time*, however. For most salt species, the greater fraction of the river load is salt that is being *recycled*. For our gold example, suppose that half is "new" gold eroded out of continental rock and the other half is recycled gold. Then the *rate* we need to use for residence time calculation is 300 tons/yr. Taking the ratio X/x again, we compute the residence time of a gold ion to be about 0.5 million years. Table 7.1 lists residence times for a number of elements calculated by the Barth method.

The reader may ask how can salt be recycled? What is the recycling mechanism? As far as we know, the principal mechanism is through the ocean spray that is generated both at sea and along the shore when waves break. If a droplet of seawater thrown into the air is small enough so that the wind can keep it airborne, the water will evaporate leaving a still smaller crystal of sea salts, which can in turn be lofted to high altitudes by air currents. In fact, such salt particles are the main type of nuclei around which raindrops condense; they become an essential part of the total hydrologic cycle. Still a different source of droplets at sea is the "unfoaming" of the brine. When a wave breaks at sea, countless tiny bubbles of air are entrapped—hence the poetic "foamy brine." As each bubble bursts after arriving at the surface and the surface tension of water collapses the cavity left by the bubble, one or more minute droplets are ejected forcibly into the air. These can become airborne and evaporate to form salt nuclei for moisture condensation.

Resident Time Based on Rate of Inclusion in Sediments. The second method (Goldberg and Arrhenius, 1958) for determining residence time is based on measuring the rate at which a specific salt is incorporated into sediment. Using core samples of floor sediments, Goldberg and Arrhenius measured the annual volume of deposition and the percentage of a specific salt in the sediment layer. Knowing the areal distribution of the selected layer, they estimated the annual *rate* at which the salt was deposited. By dividing the *known* total ocean volume of the dissolved salt by its rate of deposition, they found the residence times of a number of elements. Their results correlate quite well with the estimates published by Barth, so that we gain reasonable confidence in the residence times of major ion species.

Nutrients, Trace Elements, and Dissolved Gases

An interesting way to display the four groups of dissolved elements is by superimposing them over the periodic table (Figure 7.2). In this way we can visualize the overlaps by which some elements appear in more than one group. The chart also shows which elements are essential to plant growth and which have distributions similar to that of nutrients but are not usually classified as such. Carbon is the most conspicuous: it is a major element, occurring for the most part in the form of bicarbonate ions (CO_3, HCO_3); it is a dissolved gas, in the form of CO_2; and it is an essential element in the formation of organic molecules, and hence could be classed also with nutrients.

The Nutrient Elements

The principal nutrients are nitrogen (nitrate NO_3), phosphorous (phosphate PO_4), and silicon (silicate $SiOH_4$). Because of their essential role in plant growth, their distributions differ from those of the major constituents. Nutrient concentrations are almost always low in surface waters, because nutrients are taken up by active plants, and increase with depth, because organic material decomposes within the deeper layer. In Chapter 8 we will study actual nutrient profiles from a *real* oceanographic station in the Northeast Pacific.

Trace Elements

Figure 7.2 includes elements whose presence in seawater has been measured at concentrations less than 1 ppm. Some of these trace elements have vertical distributions similar to those of nutrients.

1. Biologists have discovered that many marine organisms concentrate ions of some trace elements. The concentration can be far out of proportion; for example, the ratio of iron to potassium *within* a plant cell is far larger than that in the ambient seawater. Most of the transition elements, such as iron, manganese, zinc, copper, and cobalt (Figure 7.2), follow this pattern. These metals are important in forming

sites on the protein enzyme structures that enable the enzymes to carry out their special functions; their atoms are usually found near the sites where chemical transfer between enzyme and substrate occurs.

We could also classify these trace elements as "nutrients." Early attempts at manufacturing seawater for use in aquaria experiments by mixing the known seawater concentrations of major salts into pure water generally failed. In fact, this was the impetus for analyzing seawater for those ingredients missing in the initial experiments.

2. Trace elements can constitute particulate matter suspended in seawater. Fine particles of terrigenous dust exist in trace quantities throughout the oceans. Industrial activity has contributed substantial quantities, as have natural events such as volcanic explosions.

Dissolved Gases

The gases dissolved in seawater are the series of inert gases (see Figure 7.2) as well as the principal gases of the atmosphere, nitrogen and oxygen. Carbon is shown as a gas through its presence in the atmosphere as carbon dioxide. Nitrogen in elemental form is not generally considered a nutrient, for open-ocean phytoplankton prefer their nitrogen in the ion form NO_3. An important exception to this rule occurs in the coral reef community where the principal plant production is by blue-green algae, which take their nitrogen supply as dissolved gas, N_2. An isotope of radon, ^{222}Rn, is sometimes used as a tracer element in the study of fluid motion near the ocean floor; it is a gas component of the mantle exhalations occurring in the zones of sea floor spreading.

Determining Salinity—The Rule of Constant Proportions

William Dittmar (1884) analyzed 77 samples of seawater taken during the HMS *Challenger* expedition and reported that although the concentration of total salts varied over wide limits, the *relative proportions of the major salts were the same in all samples*. We now know the reason: the major salts have long residence times and are so well mixed through the oceans that their concentration ratios are referred to as conservative properties of seawater. This is the basis for the "rule of constant proportions" by which we infer the *total* salt concentration, that is, salinity, by measuring the concentration of only one of the major species. The chloride ion is usually chosen as the key, partly because it is the dominant ion, but also because the chemical titration technique needed to define its concentration is straightforward. (But the technique is delicate; technicians require substantial training to obtain precise and repeatable results). So that chlorinity titration the world over be referred to a common base, the ocean community has set up a standards laboratory from which all oceanographic institutions obtain water standardized for chlorinity. In this way, nations and scientists can exchange data with confidence that all data were obtained in the same, standard way. Once chlorinity is measured, salinity is calculated using the empirical formula

$$\text{salinity (ppt)} = 1.8065 \times \text{chlorinity (ppt)}$$

The resultant values of salinity are then used to calculate density (see Chapter 9) or to study patterns of ocean motion, or the behavior patterns of marine life, by mapping the field of salinity directly.

Modern oceanographic cruises are outfitted with sophisticated electronic gear that measures the electrical conductivity of seawater—conductivity is a function of ion concentration and temperature—and converts this datum directly to equivalent salinity in parts per thousand. There are distinct advantages to electronic sensing. Chemical titration requires the taking of a physical sample, but conductivity is measured *in situ*. We obtain a much more *continuous* vertical profile of salt distribution, that is, many more point values within the water column. In addition, the conductivity method has an inherently higher resolution, 0.003 ppt versus about 0.01 ppt for the titration method.

Distribution of Salinity in World Oceans

By combining measured salinity values from all oceans, together with estimates of the volume of water at each value, Montgomery (1958) produced the first total distribution graph of salinity for the world oceans (Figure 7.3). By describing his technique, we can gain a solid understanding of this graph. Montgomery first divided the total ocean volume into a number of equal-size cubes. From the data bank of salinity values available at that time, he found the measured salinity for each cube and then sorted the cubes into categories of 0.1-ppt salinity increments. Next he calculated the percentage of cubes that appeared in each category. For example, about 6% of all cubes from all oceans had a salinity value be-

tween 34.4 and 34.5 ppt (dashed line in Figure 7.3). The category with the largest number of cubes, having salinity values of 34.6 to 34.7 ppt, accounted for over 40% of all ocean volume, located almost entirely in the Pacific Ocean.

To those who have compared the size of the Pacific Ocean with the others, it is no surprise that it should dominate the distribution charts of *any* ocean characteristic. What is surprising, as the reader may recall from Chapter 4, is that the Pacific is *less saline* on the average than the Atlantic, 34.62 versus 34.90 ppt. Yet the Atlantic basin receives some 80% of the *total river* runoff. So why is the Atlantic saltier? We form the explanation using the knowledge gained about evaporation in Chapter 6, about wind patterns in Chapter 3, and about the concentrations of salts in seawater.

First, study the map of sea surface salinity for the world oceans (Figure 7.4). The water with the *highest* salinity, over 37.00 ppt, occurs in the North Atlantic between 20 and 30°N latitude. Three thousand kilometers southwest from this central point, in the Gulf of Panama, is the surface water with the *lowest* salinity, less than 28.00 ppt. Separating these two regions is the narrow land bridge connecting the two American continents. A first hypothesis to explain this difference might be that the central North Atlantic is evaporating excessive amounts of surface water, leaving salts behind, whereas the Eastern Pacific is receiving excessive precipitation that dilutes the surface layer to very low salinity. But how can water move from the Atlantic to the Pacific? The answer is that a system of tradewinds that blows generally westward (easterlies) carries moisture across the land bridge and dumps it as rain in the Pacific. In fact, closer study shows that the 34.0-ppt contour in

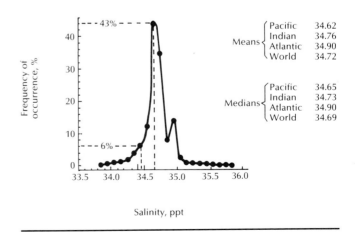

Figure 7.3 The distribution of salinity in the world oceans. (Based on Montgomery, 1958.)

CASE IN POINT

FLUX OF AIRBORNE WATER BETWEEN MAJOR OCEAN BASINS

Why should the Atlantic be saltier than the Pacific? In view of the fact that the Atlantic receives about 80% of all of earth's freshwater river runoff, it should be the most diluted of all oceans!

An explanation to this paradox must identify a freshwater "leak" out of the Atlantic. In fact, there are two such leaks, both of them large in scope. From the east, Mediterranean Sea water, made very salty by excess evaporation (over 36 ppt; see Figures 9.3 and 9.4), pours into the Atlantic at intermediate depth. In the west, water vapor evaporated from the Central North Atlantic and the Caribbean Sea is transported out of the Atlantic basin by the northeast trade winds. Eventually, this moisture is precipitated into the Pacific along a zone

meteorologists call the Inter-Tropical Convergence Zone (ITCZ).

The evidence for interbasin freshwater transfer is seen on both sides of the Central America land bridge. Very high surface salinity is found in the North Atlantic, over 37 ppt, and very low salinity, less than 34 ppt, is found in the Pacific, here seen in the tongue-shaped contours of low-salinity surface water trending westward away from the Central American land bridge.

From the scarce data available, I estimate that about 0.5 million m³ of fresh water are moved by these trade winds every second. In part, this transfer forces the Pacific Ocean to stand about 0.3 m higher than the Atlantic across the Panama Canal. Of course, this water must return to the Atlantic some way, and the logical mechanism is the Antarctic Circumpolar Current via the Drake Passage, around the Horn of South America.

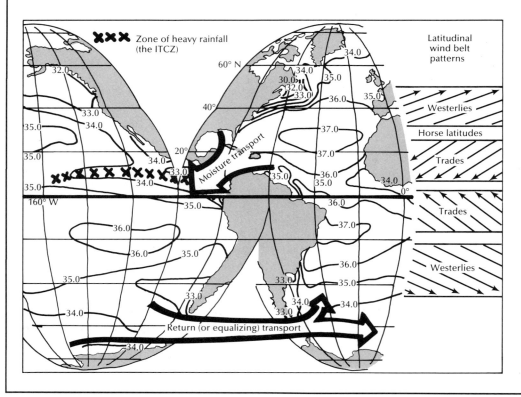

Figure 7.4 Base chart from Sverdrup, Johnson, and Fleming, 1942.

the Pacific coincides very well with a meteorological phenomenon of heavy tropical rainfall, called the Intertropical Convergence Zone. For continuity, water must return to the Atlantic. It does so via the energetic Antarctic Circumpolar Current that pours through the Drake Passage around Cape Horn.

Physical Properties of Water That Change with Salinity

Most of the physical properties of water change with changes in salinity (Table 7.3). Some of these de-

pendencies are best described in this chapter on sea salts and the consequences thereof; others, particularly density and the coefficient of thermal expansion, fit better in a special treatment of seawater density in Chapter 9.

Properties That Increase with Salinity

Electric Conductivity. The number of solvated ions per unit volume of seawater governs its electric conductivity. In addition to its use as a way of measuring salinity, conductivity and its change with salt content have other uses.

The binary mixture of water and dissolved salts contains large numbers of electrically charged particles. As ocean currents move these across the earth's geomagnetic field, electric currents are induced in the fluid. A possible consequence (Figure 7.5a) was foreseen by Michael Faraday in a lecture before the Royal Society (1832):

> Theoretically, it seems a necessary consequence that where water is flowing, there electric currents should be formed: thus, if a line be imagined passing from Dover to Calais through the *sea,* and returning through the *land* beneath the water to Dover, it traces out a circuit of conducting matter, one part of which, when the water moves up or down the channel, is cutting the magnetic curves of the earth. . . .

In the ocean, seawater flowing at a speed of 1 cm/s generates an electric field; we can measure about 10 millivolts of potential between two electrodes spaced 100 m apart *across* the direction of flow. In practice, we trail such a set of electrodes behind the ship, to get them away from the ship's electrical noise, and convert the recorded voltage into its equivalent ocean current flowing perpendicular to the ship's course (Figure 7.5b).

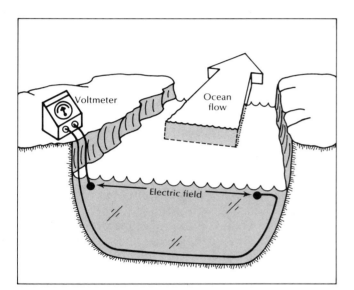

Figure 7.5 An ocean current that moves salty water through the earth's magnetic field actually generates an electric potential.

(a) The Faraday effect across the English channel. Insulated cables within the seawater are terminated in exposed electrodes, one on each side of the channel. The flow through the channel generates a measureable voltage whose size is proportional to the water transported, and whose sign depends on the flow direction. A typical tide current of 1 knot (0.5 m/s) produces a signal of several volts.

(b) Towing electrodes at sea to measure ocean currents. Two cables that terminate in electrodes about 100 m apart will sense an electric potential if the ship track is perpendicular to the moving ocean. In the example, with a strong ocean current flowing from left to right, the device senses a positive voltage as the ship crosses the current from *A* to *A'*; during the turn, with the ship's path *A'* to *B* parallel to the current, no signal is measured; on the return leg *B* to *B'*, the device records the same signal strength as in leg *A* to *A'*, but the signal is opposite in polarity.

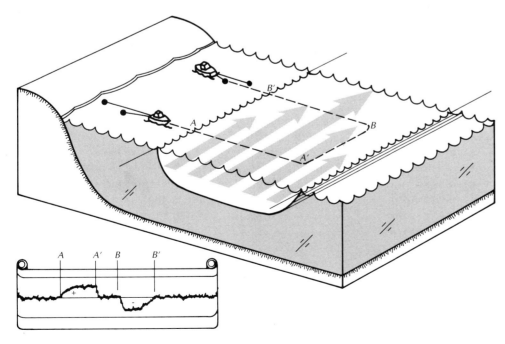

Table 7.3 Physical properties of seawater that change as salinity changes.

Properties That Increase with Increasing Salinity	Properties That Decrease with Increasing Salinity
Density	Specific heat
Molecular viscosity	Freezing point temperature
Surface tension	Temperature of maximum density
Refractive index	Vapor pressure
Electric conductivity	Thermal conductivity (molecular)
Coefficient of thermal expansion	
Speed of sound	
Osmotic pressure	

It is entirely reasonable to assume that marine life also makes use of this phenomenon, perhaps for navigation during extended migrations. If the reader will look again at Figure 3.4, it should be clear that marine animals have some way to "know" that a current is carrying them in a desired direction. Most aquatic animals that have evolved with the ability to generate their own electric currents are freshwater species; this follows because the presence of conducting charges, as in seawater, tends to short-circuit the gland electrodes in these animals. (It is said that the electric eel found in the Amazon River will electrocute itself if a wire is tied between tail and head; I haven't tried this.) Some Amazon fish species generate an electric field and sense the perturbations to their own field caused by currents or turbulence in the water. Here the electric field may act as a navigation sensor for use in waters so murky that the fish cannot rely on visual navigation.

Index of Refraction. The index of refraction is a measure of the speed of light in a medium. Using the index for air as a reference (set at 1.00), the seawater index is about 1.33; light travels more slowly in water.

The refractive index of seawater changes slightly with salinity, about 0.02% ppt change in salinity. It is not known whether marine life uses this phenomenon in any way. Oceanographers have attempted to construct salinity meters based on the principle, but no such device is commercially used. The phenomenon has had one extraordinarily useful application, however, in the sensing of convection motions that are driven by salinity changes. We know, for example, that because salt is excluded from crystals of ice, the freezing of sea ice creates a density change in the fluid just below the ice–water interface. The density change then powers a vertical convection within the fluid. The behavior of such verti-

cal convection in laboratory experiments has been photographed using a polarized light shining through a transparent tank into a camera. As the salt plumes leave the ice–water interface and travel downward through the water tank, they change the refractive index and thus alter the passing light that is recorded on film. The plumes are very thin, on the order of 0.1 cm thick, but can extend vertically for meters. No one has yet carried out the ice experiment in the real ocean, but photographs of oceanic salt plumes generated by other mechanisms have been made.

Speed of Sound. Pressure waves are an efficient method of transmitting information through ocean waters. It is no accident that many marine animals communicate using this method. Sound waves are used in ocean research in many ways: for depth measurements, for echo sounding in search of fish schools or for assessing the population of fishes (even for identifying fish species), and in the transmission of information.

The speed of sound varies with salinity, temperature, and pressure. For water of 35 ppt, 30°C, and at sea level pressure, the speed is 1546.16 m/s, or 5072.6 ft/s. Sound speed *increases* with salinity at the rate of about 1.3m/s(ppt). Salinity changes, in general, are not the dominant influence on sound speed. Temperature effects dominate in the upper ocean layers, because sound speed increases by about 3 m/s(°C), and pressure dominates in the deeper layers, where the speed increases by about 1.8 m/s for every 100-m change in depth.

The way the speed of sound changes with depth in the real ocean is a very special phenomenon. In Chapter 8 we will study the vertical changes in temperature and salinity that occur with depth. We here preview that study to demonstrate that there is a *minimum* in the vertical profile of sound speed (see the shaded curve in Figure 8.3); it occurs at about 400-m depth in the water column off the coast of Oregon, but this type of minimum is found in virtually all the world oceans. It now has a name—it is called the SOFAR zone. If one places a sound transmitter in the water at this depth, its sound waves travel *so far!* The reason is that as sound waves try to travel either up or down from the 400-m level, the waves are *warped back toward* the SOFAR zone, much as a guide channel. Scientists at the Woods Hole Oceanographic Institute have deployed such transmitters within the SOFAR channel in the Gulf Stream and tracked their drift for thousands of kilometers.

OSMOTIC PRESSURE EFFECTS
ON MARINE LIFE FORMS

Hypotonicity. Teleost (bony) fishes in the marine environment are *hypotonic.* To keep their body fluid down to the required osmotic pressure for the species, they secrete concentrated brine through special cells located in the gill structure. Some marine species have a kidney whose function is opposite in sense, that is, the kidney acts to keep body fluid at a higher osmotic pressure by extracting and secreting water from the animal's body fluid. Some biologists believe that the kidney is an organ that evolved originally as fish moved from salt to fresher water. The subsequent return to the oceans by some species left these with a vestigial organ at cross-purpose to the animals' new needs, hence the development of the chloride cell method.

Isotonicity. Marine invertebrates are generally *isotonic* with their environment. This means that the relative "salinity" of their body fluid is similar in value to that of the seawater. Many species are able to adjust themselves to a wide range of salinities. For example, the lugworm *Arenicola marina* found in water of freezing-point depression −1.72°C has body fluids of −1.7°C depression; the same species in the less saline waters of the Baltic Sea (−0.77°C) has fluid of only −0.75°C depression. When a marine invertebrate encounters seawater of salinity outside the range to which it can adjust, it will perish.

Hypertonicity. Brackish-water and freshwater fishes are hypertonic with their environment. They must avoid the inevitable result of reverse osmotic pressure, which would increase the water content of cells to the point of *turgidity.* In fact, cell membranes can rupture if their internal volume is overpressurized with water. The kidney evolved as the necessary defensive mechanisms.

But other fish features may also have evolved to the same purpose. The scales of fishes are bony plates that serve to isolate the underlying skin tissue from contact with water and hence from the undesired diffusion of water into body fluid.

Osmotic Pressure. Osmotic pressure increases as salt is added to water. A fundamental use of this property is illustrated in a study of osmotic pressure effects on marine organisms (see Case in Point, page 129). *Osmosis* is defined as the diffusion of solvent through a semipermeable membrane separating two different solutions, in a direction that tends to equalize the concentrations. In the ocean it is the solvent water that diffuses across the membrane *into* the solution of higher salt concentration.

In unicellular marine plants the membrane that contains the cell is also an osmotic membrane; the plant takes in desired chemical ions and compounds (for example, PO_4, NO_3) and discharges its metabolic waste products (for example, NH_3) directly through its cell membrane. In a sense the entire ocean is a "vascular" system that carries nutrient-laden fluid past the plant cell. The plant cell regulates the diffusion process by adjusting the osmotic pressure difference across its membrane; it does this by controlling the "salinity" of its tissue fluids.

We find a fundamental distinction in how vertebrate and invertebrate marine animals handle their osmotic pressure problems. The invertebrate in general adjusts its own body fluids to a "salinity" nearly equal to that of the surrounding seawater. Such animals are said to be "isotonic" with their environment.

Most marine fishes must work continuously against the difference in the osmotic pressures of seawater and of their own body fluids. Osmotic pressure in biological systems is often quantified in terms of the freezing-point depression caused by the addition of salt to a fluid. Ordinary seawater of salinity 35 ppt has a freezing point of −1.9°C; Figure 7.6 shows how the freezing point is depressed with salinity. In contrast, the body fluid of a marine bony fish has a depression of about −0.8°C; the difference, or osmotic pressure, is then of the order of 1.1. The consequence is that the marine fish must work continuously against the loss of water to the ocean by osmosis across its skin and other exposed tissues. It performs this feat by ingesting salty seawater, concentrating the excess salt ions into a brine, and then secreting the brine through a specialized organ called "chloride cells" located in the gill structure.

On the other hand, freshwater fish with body fluids having a salt content similar to that of marine fish (actually slightly lower, around −0.6°C depression) must also fight the osmosis battle but in an opposite direction—they must avoid the tendency for their body fluid to be overly diluted. These fish have developed kidneys to extract excess water from body fluid.

A discussion of osmosis effects would be incomplete without mention of species such as salmon, which migrate between fresh and marine waters. The salmon has this problem licked because its vestigial kidney serves its needs in fresh water and its chloride cells in salt water. Crabs that live in waters of higher osmotic pressure deposit some excess salts in the carapace exoskeleton; periodic shedding of the carapace thus rids the animal of excess salts. Marine mammal females undergo high osmosis stress during the time of nursing young, for the production of milk requires extra water beyond the parent's own needs. In some predator–prey interactions, for example, starfish predation on oysters, a difference in tolerance to salinity provides the prey with a natural escape; oyster beds are found higher upstream in estuaries and thus in water of lower salinity than the starfish can tolerate.

Properties That Decrease with Increase in Salinity

Specific Heat. We have learned that the specific heat of pure water is 1 cal/(g)(°C). The addition of salt decreases the specific heat of the mixture; water at 35 ppt salinity and at 20.0°C has a specific heat of 0.932 cal/(g)(°C). Temperature changes also affect this property: decreasing the temperature of 35-ppt water from 20 to 0°C increases the specific heat from 0.932 to 0.942. The impact of this lower specific heat of salt water on the climate-moderating capacity of the oceans is slight.

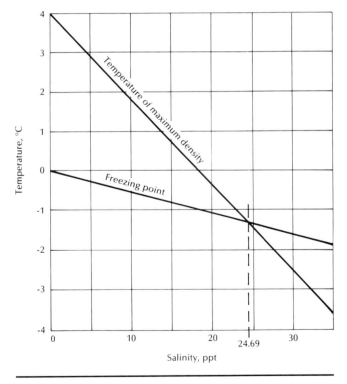

Figure 7.6 The dependence of freezing temperature and the temperature of maximum density of water on the salinity of seawater. The value of salinity at which the two temperature lines cross, 24.69 ppt, separates bodies of water into those that behave like lakes and those that behave like oceans.

Vapor Pressure. Seawater at normal values of salinity has a vapor pressure about 98% of that of pure water at the same temperature. In most ocean research it is not necessary to consider the effect of salinity because variations in temperature of surface waters have a much greater impact, which is why warmer waters evaporate faster than colder waters, all other factors being equal.

Freezing–Point Temperature. The addition of salt to pure water markedly decreases the freezing point of the saline mixture (which is why we add salt to the ice when making ice cream at home). Figure 7.6 diagrams the dependency for water from 0 to 35 ppt salinity; at 32 ppt, a salinity typical of polar oceans, surface water freezes at a temperature of −1.7°C.

Ocean surface waters in prolonged contact with an ice pack will have such lowered temperatures. From the discussion of the freezing-point depression of the body fluids of fishes, which is about −0.6 to −0.8°C, it is clear that ordinary fishes cannot survive for prolonged periods in ice pack waters without additional adaptations that protect against the freezing of body fluid. They have particular difficulty in "breathing," for their gill membranes, which absorb oxygen and excrete waste products, are extremely thin and therefore susceptible to freezing in ambient seawater at −1.7°C. They solve this problem by the internal manufacture of an "antifreeze" substance that, when mixed with normal blood and body fluid, reduces their freezing point enough to protect the animals, in this case to −1.7°C or lower. In temperate climates, fishes living in shallow bays during winter, where exposure to subzero temperatures is likely, adjust freezing-point depression on a seasonal basis.

Temperature of Maximum Density. Ocean waters differ greatly from lake or brackish waters in their temperature of maximum density. Figure 7.6 shows this property superimposed over the same salinity scale as the graph of freezing-point temperatures, and for a reason. Notice that the two lines cross over at the salinity value of 24.69 ppt: waters of lesser salinity behave like brackish lake water. Figure 7.7 shows the unique feature of pure water that its maximum density occurs at +4°C, not at its freezing point of 0°C. Let us now compare the behavior of lakelike versus sealike fluids as each approaches freezing conditions.

The water of Crater Lake in Oregon is almost pure, dissolved salts are only about 0.08 ppt, and water at this salinity has its maximum density at

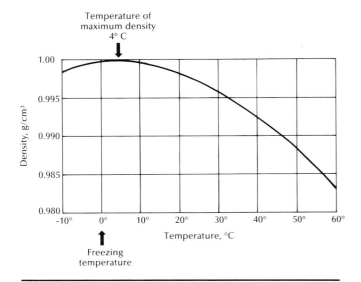

Figure 7.7   Pure water reaches its maximum density of 1.000 g/cm³ at a temperature of 4°C. Its density at the freezing temperature of 0°C is slightly lower, about 0.9995 g/cm³.

+3.8°C (Figure 7.7). The temperature of deep water in Crater Lake is +3.8°C; it is a straightforward exercise to explain why. Surface water cools as the freezing season sets it. When the temperature of surface water is lowered to +3.8°C, it is as dense or denser than all other water in the lake. If at this time the bottom water in the lake is either warmer or cooler than +3.8°C it is less dense than surface water; hence the column becomes unstable and the lake "turns over," depositing +3.8°C water on the bottom. Further surface cooling continues until the entire column is at this temperature of maximum density. But still further cooling results in the surface water becoming less dense as the freezing point is approached. Thus prevented from sinking by the denser water below, the surface water rapidly cools to its freezing point.

In contrast, ocean surface waters freeze *before* reaching their temperature of maximum density. The cooling of surface water as the freezing season begins will increase water density and cause vertical convection to occur. Since the ocean is already density stratified, however, the depth to which such convection reaches is limited at most to a few hundreds of meters. The Antarctic Ocean is a notable exception.

Summary

Eight elements comprise more than 99.5% of all dissolved salts in the oceans, but over 80 different elements are found in concentrations ranging from 19.353 ppt for chlorine to traces that are at the limits of chemical detection, such as indium at less than one part per million million. The eight are called *conservative* elements because their concentrations are always the same *relative to one another* regardless of ocean or of depth, or of the total amount of dissolved salts that we call *salinity*. Of these eight, four are cations believed to originate for the most part from the erosion of continental rock, and four are anions thought to originate in the volcanic emanations from the earth's mantle of molten material.

Gradually, oceanographers came to appreciate the importance of elements that occur in trace quantities but that are essential to living processes. Living protoplasm is often found to contain an element in much greater concentration than its natural concentration in the ambient water, for example, iodine. Living cells may horde these elements because they are enzyme-specific, acting as an essential *communication link* between enzymes and the organic molecules they control. Trace element chemistry is the key to some pollution problems.

Every element has a *residence time* in seawater, an average length of time between its first introduction and its eventual capture by a sedimentation process. It is important to know residence times because they serve as controls for many geochemical theories.

The chemistry of nutrient reactions in seawater controls the rate of cycling of supplies of nitrates and phosphates, which are essential to plant growth. Oceanographers continue to measure and analyze the distributions of these elements throughout the oceans.

The concentration of salts also controls many of the basic physical properties of seawater. Of these, the most important are the partial pressure of salt in water, called osmotic pressure, the freezing-point depression of water, and the temperature of maximum density. The first of these determines much of the *physiological* evolution and behavior of marine life, and the others control the physical behavior of ocean surface waters in the polar regions, for example, the amount of sea ice formed and melted in response to cooling and heating, a part of the great heat engine of the earth.

Study Questions

1. Once oceanographers learned that the salts in seawater were the same as those derived from weathering rock and from volcanic hot springs, the following hypothesis had to be tested: "Today's ocean salt concentrations are the accumulation of such over geologic time." Defend or refute.

2. Eight ion species (of the 81 elements found in seawater) comprise 99.5% of the dissolved salt mass. Name the eight elements.

3. Which of the eight are important nutrients to plant growth? Which of the eight has the longest residence time in the ocean?

4. Contrast the vertical distribution of concentrations of *conservative* versus *nutrient* ion species.

5. Some trace elements are found to have the same pattern of distribution in the vertical water column as the nutrients, but they are not called nutrients. Why not?

6. Chapter 7 attempts to explain the origin of all the dissolved salts. But what explains the origin of the water itself?

7. Now that sources of salts are identified, what are the possible sinks?

8. Which plant and animal species contribute in a major way to the formation of calcium carbonate sediments?

9. Which plant and animal species contribute to the siliceous sediments?

10. Commercial mining of some deposits on the ocean floor may soon become a reality. There are two major candidates for such mining. Discuss.

11. Most of the mass of salts being carried to the oceans by rivers is said to be a recycling of salts. How can salts recycle *if* salts are left behind when seawater evaporates?

12. Describe the two different methods that have been used to establish the *residence time* of ions in seawater.

13. Which are the principal nutrient ions?

14. Iron, manganese, zinc, copper, and cobalt are called trace elements. What is the concentration level that distinguishes *trace* quantities from those that are more abundant?

15. If we examine the concentrations of trace elements within living tissue, we find concentrations that are much higher. Why is this?

16. Briefly, define the rule of constant proportions. Does the rule apply to all ion species?

17. Which ocean is the saltiest? If you answer the Atlantic, explain why this is so in spite of the fact that the Atlantic receives the lion's share of runoff from the continental watersheds (refer back to Figure 4.11).

18. Discuss the impact of osmotic pressure on marine life. For example, a marine fish placed in fresh water will die of *turgidity* of tissue cells. A freshwater fish placed in seawater will die of dehydration. These are extreme statements, but they serve to make the point. You explain the point.

19. The temperature of water in the deepest part of Crater Lake, Oregon is +3.8°C. Why?

20. As briefly as you can, explain *why* the Navy Arctic Research Lab lost a bulldozer at Ice Island T-3 when the driver drove the machine over a freshwater pond that had frozen over. The machine broke through the ice and sank to the bottom of the pond. This occurred in December, four months after the first freeze in that region of the Arctic! Your answer should address the question of *how* there could still be *liquid* freshwater under a thin ice cover only 300 miles from the North Pole in December!

CHAPTER 8

TWO OCEANOGRAPHIC CRUISES

Probably few readers will have a chance to share in the work of a real oceanographic cruise. Nonetheless, mastery of the previous chapters of this book qualified each of you to do so. You are prepared to probe the real ocean, to gather real data, and to develop reasoned explanations of what is found.

Begin by studying the brief cruise plans for two hydrographic cruises off the coast of Oregon, as outlined in Figure 8.1. Continue with the discussion of the routine expected during the cruises in (Figure 8.2.) Then "Bon voyage" (and don't forget the Dramamine); with luck you will return with about 80% of the data you set out to take!

This objective of this chapter is to analyze these cruise data. The Northeastern Pacific Ocean is typical of most oceans, so that techniques of analyses and general conclusions about the structure of water columns based on these cruise data are applicable elsewhere.

The Basic Hydrographic Data

Beginning in 1966, oceanographers at Oregon State University began a series of bimonthly cruises from Newport, Oregon west to a point 165 nautical miles (nmi) offshore. The objective was to gather physical, chemical, and biological data in a systematic pattern that could yield a picture of the seasonal and interannual variations in that sector of the Northeastern Pacific. Table 8.1 lists the observed data from two selected cruises taken during winter (January 4, 1968) and summer (July 9, 1968).

The Research Vessel *Yaquina* operated during the 1960s and 1970s by the School of Oceanography, Oregon State University. (Courtesy Oregon State University News Bureau.)

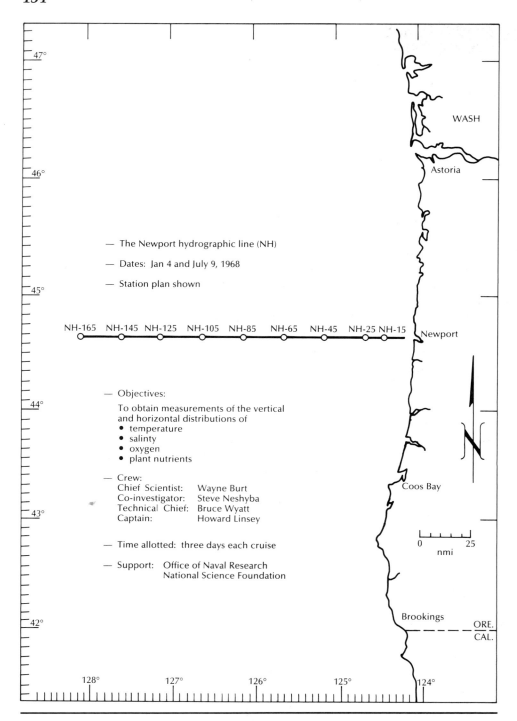

Figure 8.1 Cruises of the Research Vessel *Yaquina* during January and July 1968.

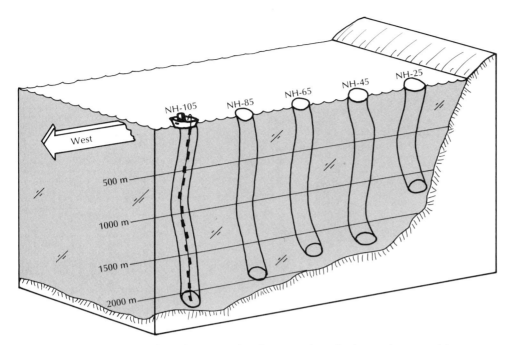

The only practical way to bring home samples of seawater from depths is to lower a cable strung with a series of special sampling bottles, usually 20 per cast, which are spaced on the cable to match the desired sampling depths. Once the string is fully deployed and a pause time allotted for thermometers to come to equilibrium temperature, a weight is clipped to the cable and released. As the weight falls, successive bottles are tripped shut, and the thermometers are inverted in position to fix the mercury column at its *in situ* volume. The technician measures the angle that the cable takes on as a result of either ship drift or deep currents, both of which can cause the cable to hang at an angle other than vertical. The vertical is a goal seldom achieved on the dynamic oceans.

Hydrocast stations such as these are at positions predetermined during cruise planning, as in Figure 8.1. The task of the ship's captain is to locate these positions and maintain the ship on station during the cast, not an easy task considering that very deep casts require more than 4 hours from start to finish. In the event that bad weather forces changes in the cruise plan, the chief scientist designs a new schedule, deciding which stations may be omitted or altered to maximize data return. At each station it must also be decided how close to the bottom the final bottle should be set on the string. There is the ever-present danger that the line may snag on a seamount and break. Equipment costs range from a few hundred dollars for plankton nets, to $40,000 for a full string of bottles and thermometers, to hundreds of thousands of dollars for specialized electronic sensors. The alertness of both the scientific crew and the ship's complement is a major factor in safety as well as in protecting equipment. Work is carried out in 4-hour shifts with 8-hour rest periods, round the clock, day after day.

Once ashore, equipment is immediately serviced to avoid saltwater corrosion. The all-important data including net samples of plankton and other marine life, are immediately processed or properly stored for later analysis. Only then does the scientific crew retire for a steak (rare, of course), a bottle of wine, and a well-earned rest.

As soon as possible after the cruise, all data are analyzed, data reports are prepared and published, and participating scientists exchange information from which succeeding cruises are planned. Most United States ships are funded by the National Science Foundation; for the most efficient use of costly ship facilities, scientists are obliged to plan several years in advance.

Figure 8.2   Cruise of the Research Vessel *Yaquina* along the Newport hydroline.

Table 8.1 Hydrographic data taken off Oregon during both winter and summer of 1968: Depth at which the sample was taken, *D*; water temperature *T* and (derived salinity *S*; the amount of dissolved oxygen, O_2, and the concentrations of nutrients PO_4, NO_3, and $Si(OH)_4$ measured in the samples*

NH-105, 44°39.1'N 126°31.1'W, Date 04 Jan 68 0434 GCT, Wire 2, Dry 49.8 Wet 46.1, Wind direction 33, VEL 5 kts, BAR 32, Swell direction 31 H 4 T 7, Cloud 6 AMT

			Observed						Interpolated		
D, (m)	*T,* (°C)	*S,* (ppt)	$O_2,$ (ml/liter)	$PO_4,$ (μmol/kg)	$NO_3,$ (μmol/kg)	$Si(OH)_4,$ (μmol/kg)		*Z,* (m)	*T,* (°C)	*S,* (ppt)	σ_t
0	9.13	32.399	6.54	0.74	3.1	4		0	9.13	32.40	25.09
10	9.16	32.396	6.66	0.69	2.8	3		10	9.16	32.40	25.09
30	9.16	32.395	6.55	0.76	3.1	4		20	9.16	32.40	25.09
50	9.13	32.399	6.51					30	9.16	32.40	25.09
75	7.92	33.231	4.49	1.73	17.0	22		50	9.13	32.40	25.09
100	8.16	33.596	3.62	1.99	24.0	28		75	7.93	33.24	25.93
125	7.94	33.834	3.02	2.19	27.5	33		100	8.16	33.60	26.18
149	7.78	33.851	2.96					150	7.77	33.88	26.46
199	7.11	33.944	2.58	2.23	31.2	35		200	7.10	33.94	26.60
300	6.14	34.008	1.79	2.73	36.7	53		250	6.57	33.98	26.70
401	5.48	34.072	1.11	2.95	39.1	66		300	6.15	34.01	26.78
601	4.65	34.201	0.42	3.28	43.1	82		400	5.49	34.07	26.91
801	4.10	34.308	0.26	3.39	44.7	99		500	5.01	34.14	27.02
1001	3.50	34.414	0.32	3.38	44.0	125		600	4.65	34.20	27.11
1005	3.47	34.421	0.39	3.39	44.6			700	4.36	34.26	27.18
1202	3.05	34.473	0.60	3.32	45.2	137		800	4.10	34.31	27.25
1207	3.06	34.480	0.56	3.36	44.2			1000	3.51	34.41	27.40
1408	2.65	34.512	0.67	3.30	44.4			1200	3.05	34.47	27.49
1609	2.33	34.549	0.94	3.23	43.3			1500	2.49	34.53	27.58
1810	2.08	34.574	1.24	3.17	42.5			2000	1.91	34.61	27.69
2012	1.90	34.608	1.56	3.05	41.8			2500	1.76	34.64	27.73
2213	1.78	34.616	1.77	3.02	41.5						
2414	1.78	34.635	1.89	3.01	41.4	169					
2615		34.645	2.49	3.00	42.4	175					

*The data at actual observed depths were interpolated so that they could be reported according to a set of international standard depths.

Explanation of the Data

Decoding the Header Notes. Like all scientists, oceanographers are very careful to annotate each station data sheet with the information that is needed in later analyses. Many of these data are recorded in shorthand notation, as explained below for the January 1968 cruise.

NH-105 — Each station is assigned an identifying number.

44°39.1'N; 126°31.1'W — The station is located at this latitude and longitude.

Date 04 Jan 68 0434 GCT — The date is January 4, 1968; time is always reported in Greenwich, England time, using the 24-hour clock system.

Wire 2 — The angle that the dangling cable makes with the local vertical is measured as 2°.

Dry 49.8 Wet 46.1 — These are air temperatures measured first with a dry-bulb thermometer and then with a wet-bulb one; the difference is a measure of humidity.

Wind direction 33 — The wind is blowing from a true heading direction of 330°.

VEL 5 kts — The measured speed of the wind is 5 knots.

BAR 32 — Barometric (air) pressure is 1032 millibars.

Swell direction 31 H 4 T 7 — These data specify waves conditions during the station sampling; the waves have an average height of 4 feet, a period of 7 seconds, and are moving in the direction of 310°.

Cloud 6 AMT — Cloud cover six-tenths (6/10).

Table 8.1 continued

NH-105, 44°39.1'N 126°31.1'W, Date 04 Jan 68 0434 GCT, Wire 2, Dry 49.8 Wet 46.1, Wind direction 33, VEL 5 kts, BAR 32, Swell direction 31 H 4 T 7, Cloud 6 AMT

Observed							Interpolated			
D, (m)	*T,* (°C)	*S,* (ppt)	O_2, (ml/liter)	PO_4, (μmol/kg)	NO_3, (μmol/kg)	$Si(OH)_4$, (μmol/kg)	*Z,* (m)	*T,* (°C)	*S,* (ppt)	σ_t
0	17.68	29.751	5.80	0.30		8	0	17.69	29.76	21.37
10	16.11	31.428	5.90	0.30		2	10	16.11	31.43	23.01
30	14.84	32.468	6.05	0.43		1	20	15.50	32.20	23.74
50	10.64	32.505	7.12	0.48		1	30	14.84	32.47	24.09
75	9.17	32.573	6.65	0.65	0.9	3	50	10.64	32.51	24.93
100	8.18	32.949	5.42	1.28	10.8	13	75	9.17	32.58	25.22
125	7.59	33.294	4.73	1.36	10.7	13	100	8.18	32.95	25.67
150	7.71	33.656	3.95	2.00	22.3	28	150	7.71	33.66	26.29
201	7.21	33.868	3.45	2.10	27.9	35	200	7.22	33.86	26.52
251	6.66	33.913	3.09	2.38	30.6	39	250	6.67	33.91	26.63
301	6.09	33.962	2.25	2.36	28.3	47	300	6.10	33.96	26.75
402	5.24	34.000	1.58	2.65	32.0	57	400	5.25	34.00	26.88
603	4.66	34.202	0.42	2.87		79	500	4.87	34.09	27.00
805	4.03	34.306	0.23				600	4.66	34.20	27.11
996	3.53	34.398	0.23				700	4.35	34.26	27.19
1005	3.50	34.402	0.27	2.93			800	4.05	34.30	27.26
1195	3.13	34.471	0.42	3.04		121	1000	3.52	34.40	27.39
1206	3.03	34.459	0.43	2.97			1200	3.08	34.47	27.48
1394	2.70	34.512	0.93	3.11	39.6	133	1500	2.50	34.53	27.58
1594	2.34	34.548	1.20	2.97			2000	1.93	34.60	27.69
1793	2.11	34.577	1.46	3.06	38.8		2500	1.71	34.64	27.73
1992	1.93	34.603	1.81	2.86	35.4	143				
2191	1.85	34.623		3.03						
2390	1.77	34.633	1.84	2.93	37.9	158				
2590	1.65	34.641	2.04	2.73	34.0	161				

Nonuniform Depth Intervals. The samples are spaced quite closely together near the surface; 10, 20, and 25 m separate adjacent bottles in the first 100 m of depth range. A 200-m spacing is used in the deep range, however. The reason is that we know from experience that water characteristics change much more rapidly within shallow depths than at deeper ones. We need data points much closer together near the surface. A quick look at the vertical profiles plotted from these data confirms this (Figure 8.3).

Interpolated Data. All hydrographic data are published by international agreement and are reported at *standardized depth intervals*. We construct new data sets by interpolation from the original data, either by computer or by first plotting the vertical profiles and then extracting new point values from the curves at the desired standard depths.

Units. Units are always shown on lists. Oxygen, the only dissolved gas measured on these cruises, is given in milliliters of dissolved gas per liter of seawater. The concentrations of the nutrients PO_4, NO_3, and $Si(OH)_4$ are in micromoles per kilogram, (a chemistry notation), and are not interpolated.

Precision of Measurements. Temperature is reported to the nearest 0.01°C; experienced oceanographers with reliable thermometers can actually derive *in situ* temperatures to this precision. Today, electronic temperature sensors yield a resolution 10 times better, 0.001°C, but thermometers are still used for periodic checks of the electronics. Salinity is reported here to the nearest 0.001 ppt, which tells us that the actual measurement was taken with electronic sensors instead of less precise chemical techniques. (The inherent accuracy of shipboard salinometry, however, is only about 0.003 ppt.) Why do

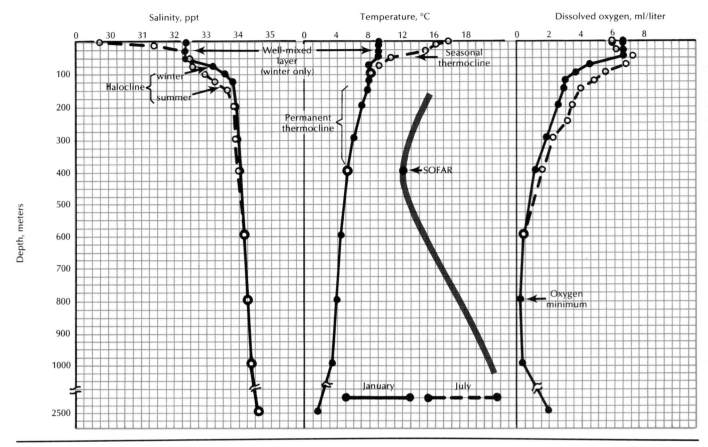

Figure 8.3 Vertical profiles of temperature, *T*, salinity, *S* and oxygen, O_2 at NH-105 (44° 39.1'N; 126° 31.1'W), determined by cruises of the RV *Yaquina* on January 4, 1968 and July 9, 1968. (From Barstow, Gilbert, and Wyett, 1968.)

we require such precision in temperature and salinity? Because oceanographers base their *estimates* of ocean currents on the way that sea water density changes from station to station, and density can be calculated only from precise values of water temperature and concentration of dissolved salts (see Appendix 2).

Every effort is made to perform the nutrient concentration chemistry as soon as the sample is brought aboard. The arduous tasks of shipboard oceanography do not always permit this, however. In such situations a fixative is added to the water sample to prevent things from growing inside the sample bottle. In spite of such precautions, results from postponed analyses are often unreliable.

Bad Data. Some bad data points are the rule rather than the exception. One of the prime reasons for preparing vertical profiles of measured quantities (as

in Figure 8.3) is to reveal data points that are so different from the general trend of the curves as to be *suspect*. Sources of errors are almost legion in number: premature tripping of the sample bottle before the entire string is deployed; seasickness on the part of the scientific crew; errors in winch operation; temporary disruption of power during critical phases of chemical analyses, which may cause a change in temperature in cold-storage areas where samples are stored—the list goes on and on.

Data Analyses I—Vertical Distributions

Once the cruise data are processed, initial study begins with plotting graphs to show the vertical distribution of each of the items measured. Which item is graphed first depends on the investigator. Here we

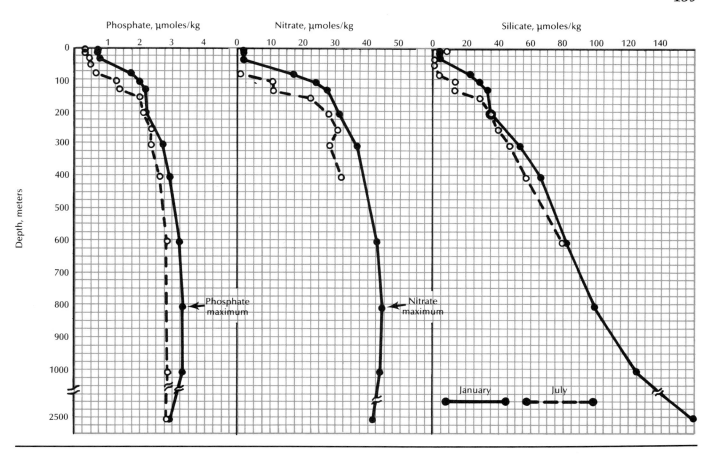

Figure 8.4 Vertical profiles of the nutrients phosphate, PO_4; nitrate, NO_3; and silicate, $Si(OH)_4$, at NH-105 (44° 39.1'N; 126° 31.1'W), determined by cruises of the RV *Ya-quina* on January 4, 1968 and July 9, 1968 (From Barstow, Gilbert, and Wyett, 1968.)

begin with the temperature and salinity data (Figure 8.3) in order to study some basic structures of oceans using the distributions of *conserved constituents.* Oxygen is not a conserved quantity (because animals breathe it and plants produce it), but a minimum oxygen value yields important interpretations.

Temperature Profiles

Thermoclines. A characteristic feature of almost all water columns except those in high-latitude oceans is the *permanent thermocline;* "-cline" refers to the *rate* at which the temperature decreases with depth. The permanent thermocline at NH-105 separates the warmer upper ocean layer from the relatively cooler, more uniform temperature waters of the deep layer. Figure 8.3 also shows a *seasonal* thermocline; the July water temperature changes by more than 4°C between 30 and 50-m depth.

Well-Mixed Layer. By January the surface water at NH-105 has cooled to 9.13°C. In place of the seasonal thermocline we find that the first 50 m of the water column are at uniform temperature. This vertical uniformity is seen in the salinity and oxygen profiles as well. The causes of this *well-mixed* layer are well known—it results from vertical convection cells driven by surface cooling (described in Chapter 6; see Figure 6.8) and from vertical mixing caused by winter storm waves. The depth of the well-mixed layer can be interpreted as a summation of effects from successive storms and mixing during a period prior to the cruise. Information about the well-mixed layer is useful to biologists who study the time history of primary production by plants (see Chapter 14 and the definition of "critical depth"). It is also useful when doing heat budget studies as a measure of how much heat has been transferred from ocean to atmosphere.

Deep-Water Temperature. Deep waters are always the coldest, but the temperature drop with depth becomes very small below 1000 m in all oceans. In drawing graphs of vertical distribution, we customarily change the vertical scale of the graph for the deep sections and present *all* water column data on one sheet. This is why the depth scales of Figure 8.3 are broken, with a tenfold compression of the deep-water depth scale. Otherwise the graph would extend to impractical lengths.

Salinity Profiles

The depth zone over which salinity changes most rapidly with depth is called the *halocline*. Because salinity and temperature are independent characteristics of seawater, there is no fundamental reason why the halocline should occur in the same depth zone as the thermocline.

One reason I chose NH-105 data for this exercise is that the abnormally low surface salinity value for summer, 29.75 ppt, is wholly *uncharacteristic* of normal ocean surface waters. There must be a source of fresh water to cause such an extreme dilution—in this area it is the plume of Columbia River water. But why is the river influence absent in January?

The answer to the Columbia River question came from a series of oceanographic cruises that crisscrossed the coastal waters from Northern California to British Columbia. We found that the near-coastal currents change with the seasons. During summer the California Current is close to shore and carries the river outflow southward, hence its presence in the July station at NH-105. In winter the north-flowing, near-coastal Davidson Current develops and carries the river plume toward the Gulf of Alaska.

Oxygen Profiles

Near-Surface Saturation. Surface seawater in contact with the atmosphere is normally saturated in oxygen. This is what happens at NH-105.

The Oxygen Minimum at 800-Meters Depth. The process of photosynthesis generates free oxygen as plant cells divide CO_2, keep the carbon, and discard the O_2. But photosynthetic activity is limited to the sunlit photic zone, which is at most 100 m deep. Deeper waters can only *lose* oxygen to the respiration needs of living organisms or to the oxidation requirements of decomposing organic compounds. Because much of the dead plant and animal material

in the productive photic zone sinks downward through the water column, organic decomposition extracts free oxygen from the interior water.

In this sense, water that has a minimum of oxygen is thought to be older, to have been subjected to oxygen extraction over the longest period of time. The oxygen minimum found at the 800-m level in Figure 8.3 can be interpreted this way—water at this level was last exposed to the atmosphere (hence to oxygen recharging) at an earlier date *than either waters above or below this level.* By locating depths of oxygen minima in many water columns sampled over a broad ocean basin, oceanographers have been able to trace the movements of water masses from their earliest formation at the surface to their eventual disappearance in the ocean interior. A classic example is the pioneer work of George Wüst, given in Chapter 9 (see Figure 9.6).

Seasonality—To What Depth Do the Oceans "Feel" the Seasons?

Vertical profiles of some conservative components are useful in the study of seasonal variations and their causes. On the basis of temperature profiles alone, we would select 100 m as the depth below which the water column does not "feel" the change of seasons; below 100 m the summer and winter profiles are identical. On the basis of salinity profiles alone, we would select 400 m as the level below which there is no seasonal change. From oxygen profiles we would select 600 m. The point to make here is an important one for it has plagued oceanographers from the very beginning of the science. How "far" can we push the interpretation of ocean behavior on the basis of *one station alone?*

The problem is that the ocean is in constant movement. It is impossible to tell the difference between changes that occur because of vertical mixing and those that occur because horizontal movement carried *different* waters past the measuring station. For example, currents moving different water masses across the NH-105 station, could cause the difference in oxygen concentrations at any level in the range 100- to 600-m depth, summer and winter. Stated another way, this troublesome problem is one of separating a given measurement into two parts : the "average" condition and the "variable" part that must be explained *either* as a local phenomenon *or* as an "imported" one.

Generally, we accept the depth of the *base* of the permanent thermocline as the level below which the impact of seasonal change vanishes. At NH-105 the base is at 400-m depth.

Nutrient Profiles

Nutrient profiles in (Figure 8.4) are much more variable than those of the conserved properties described earlier. Note also that some data points are missing. Even so, we can gain three useful results from their study.

Low Surface Concentrations. Using the 100-m level as the base of the photic zone, we conclude that nutrients within the 0- to 100-m plant production zone are *always low*, summer *and* winter. This is a good example of the nutrient depletion effect that is commonly encountered in open-ocean surface waters. Further, summer concentrations are naturally lower than those of winter; because primary production is greatest in summer sunlight, nutrient depletion is also greater.

Nitrogen Depletion: The Limiting Factor. Nitrogen is usually depleted to critical levels *before* the other nutrients are. The July readings at NH-105 show the upper 75 m to be virtually exhausted of nitrate while some phosphate remains. Almost all the available nitrate has been taken up in the living plants themselves, so that no "free" nitrate reserves are left for new growth and reproduction. We conclude that plant production probably peaked at an earlier date than July. (Confirmation of this behavior is discussed in Chapter 12.)

Nitrate and Phosphate Maxima. Both the nitrate and phosphate profiles have maximum values in the same depth zone in which dissolved oxygen is a minimum. Is this a coincidence or is there a valid, explainable correlation? As the decomposition of organic material consumes available oxygen from the water, the products of decomposition are simultaneously released into the same water. Therefore, we expect a correlation between high concentrations of NO_3 and PO_4 and low concentrations of dissolved oxygen. If we recall that organic detritus is continuously settling downward through the water column, the joint processes of O_2 depletion with NO_3 and PO_4 enrichment are also continuous. Further, the longer that these processes are allowed to work, the lower is the level of O_2 and the higher are the concentrations of nutrients. This also supports the conclusion that an oxygen minimum is an indicator of relatively *older* water.

Data Analyses II—Cross Sections

Much insight can be derived about the structure of the ocean by plotting the vertical profiles of each station's data. The next analytical step is to combine into one composite picture the data from all water columns (stations) sampled during the cruise. For an example let us use the station data from NH-5 (5 nmi offshore) to NH-165 (165 nmi offshore) and from the sea surface down to a depth of 200 m. The resulting charts are called *cross sections;* they show the distributions of things within a vertical slice through the ocean. We first construct a *grid* of sample points that can be used for all cross sections, for example, for temperature or salinity or oxygen; Figure 8.5 shows the grid. Each grid point corresponds to a set of data taken from a single sample bottle. We transfer the data for a chosen variable from Table 8.1 onto the grid (for example, phosphate concentrations) and then contour the grid with lines representing specific values (for example, $PO_4 = 2.00$ μmol/kg) as in Figure 8.6a. Again, we will compare summer and winter conditions by plotting both sets on the same graph.[1]

Nutrient Distribution

There are dramatic differences between the summer and winter distributions of both phosphate and nitrate concentrations, as demonstrated in the cross sections of Figure 8.6.

The winter contour of phosphate at the 2-μmol/kg concentration is relatively flat over the 165-nmi section for the January 1968 cruise; its depth varies be-

[1]The process of contouring cross sections *always* generates an internal conflict in the mind of the oceanographer. On the one hand there is the natural reaction to interpolate *precisely* between the measured values. On the other hand there is the inescapable reality that no two water columns were sampled at the same time (we would call such simultaneous measurements "synoptic" data). Stated bluntly, there exists an uncertainty whether contours are a legitimate tool of analysis of data extracted from a fluid, moving ocean using a moving ship platform. We get around this by tacit agreement: do not "push" the conclusions too far.

The renowned oceanographer Joe Reid (from the Scripps Institute of Oceanography) put it this way: "There are *lumpers* and there are *splitters.*" *Lumpers smooth out the data* to erase the smaller variations (station to station, or depth to depth, or time to time) because they wish to describe the "big" picture. *Splitters keep all data points,* preferring to tear hair over why the variation exists. They both seek explanations of the physics that governs ocean behavior.

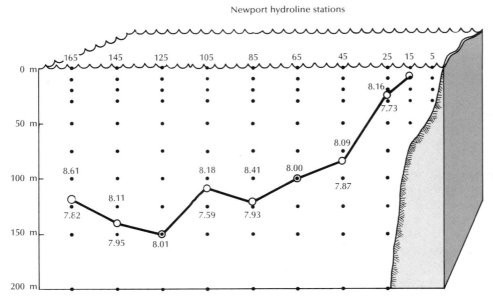

Figure 8.5 Constructing a grid network over which to graph the distributions of ocean variables measured during the Newport hydroline cruise of the *Yaquina* in January and July 1968. Station locations are plotted accurately; dots represent the depths at which the data were taken. The normal procedure for contouring the cross-section is to print the measured values alongside each point and then draw the contours for selected integer values. In this example the contour along which temperature is a constant, 8°C, is interpolated between the data points that bracket the 8.0°C value. (From Table 8.1.)

tween 115 and 135 m. In marked contrast, the July contour for the same phosphate concentration varies from more than 200 m depth at NH-165 to an intersection with the sea surface itself at NH-5! Other July contours also intersect the sea surface; summer surface waters inshore of 15 nmi are as rich in phosphate as are waters at 100 m or more depth at 165 nmi offshore.

The nitrate contours exhibit a similar pattern change from summer to winter. July contours for 10 and 20 μmol/kg contours also intersect the surface inshore of the 15-nmi station. The one major difference between the phosphate and nitrate cross sections is the virtually complete depletion of nitrates within the upper 60 m of water, beginning at NH-45 at the 30-m level; we will return to this "event" shortly. Notice that the winter contour of the 20-μmol/kg nitrate concentration is as "flat" as that of the phosphate 2-μmol/kg contour.

It appears that the change from winter to summer patterns, for both phosphate and nitrate, occurs inshore of the 85-nmi station, with the most drastic changes happening within 35 nmi of the coast. Offshore from NH-85 the contours are relatively flat out to NH-165 for both summer and winter. To demonstrate that the NH-165 data are typical of open-ocean conditions, Figure 8.6 has two data points from NH-565. Being 565 nmi west of Newport, this station is clearly open ocean. The 1-μmol/kg phosphate point is only 12 m deeper than at NH-165, although 400 nmi away; the 10-μmol/kg nitrate depth is precisely the same at NH-565 as at NH-165.

At this point the reader should ask how the summer nutrient contours can rise from depths in excess of 100 m to intersect the surface if (1) surface layers are *always* depleted in nutrients and (2) the stratified ocean is *always stable*, with lower-density water on top and higher-density water on the bottom? Should not the nutrient-rich deeper water remain "captive" beneath the less dense upper strata? The answer is that the density stratification barrier can be *broken* if enough force is applied. Along the ocean shore, coastal winds can be strong enough to "warp" the density field. What we have graphed in Figure 8.6 is a snapshot of an ocean undergoing active *upwelling*.

Seasonal changes in nutrient distributions are of intense interest to biologists and fishery specialists. Summer primary production is very high in the nearshore high-nutrient zones, and it follows that the rest of the food chain must also produce heavily (recall the fish production figures for upwelling versus open-ocean regions in Chapter 1). Physical oceanographers are closely involved with the upwelling phenomenon and describe the physical processes involved, for example, the forces, the currents, the time scales during which the wind field forces the upwelling, and so on.

A major project, the Coastal Upwelling Ecosystems Analysis (CUEA), was carried out during the 1970s to investigate the upwelling processes. Scientists from all disciplines joined in field experiments of coastal upwelling regions off Oregon and California, off the northwest coast of Africa, and off the northwest coast of Peru. (For a justification of why

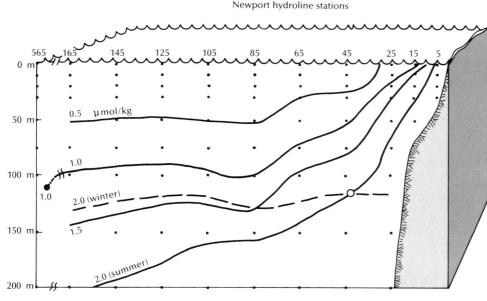

(a) Phosphate cross section

Figure 8.6 Cross sections of nutrient distributions along the Newport hydroline. Solid-line contours indicate the summer (July 9 1968) condition; dashed-line contours indicate winter (January 4, 1968) conditions selected to contrast with those of summer. Crossover points are circled. Station NH-565 has been appended to demonstrate that the distribution fields of both phosphate and nitrate are relatively flat across the open ocean. Therefore, the upward warping of these contours near the coast is a direct result of coastal upwelling. (From Barstow, Gilbert, and Wyatt, 1968.)

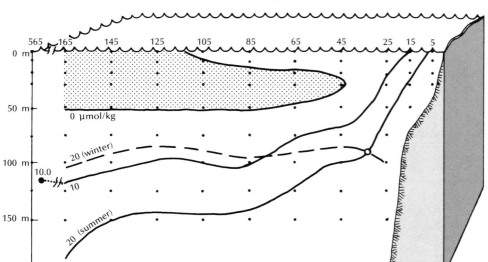

(b) Nitrate cross section

The Distribution of Density

What is the physical process that creates upwelling? We gain some answers by plotting the field of seawater density in the same format (Figure 8.7a).

Density data are extracted from Table 8.1 under the column σ_t, called sigma-t. This is a "convenience unit" adopted by oceanographers to avoid writing out the full decimal for density. Observe the first row of measured data for January 1968: seawater at a temperature of 9.13°C and salinity 32.399 ppt has a density of 1.02509 g/cm^3 (see Chapter 9 for a method of getting density from temperature and salinity data). Under the interpolated heading, find the σ_t column and the value 25.09. In Figure 8.7a the density contour marked 25 is a line of constant density of 1.02500 g/cm^3 or, simply, a constant sigma-t of 25.

(these different geographic regions were studied under one project envelope, see Figure 15.4.)

Newport hydroline stations

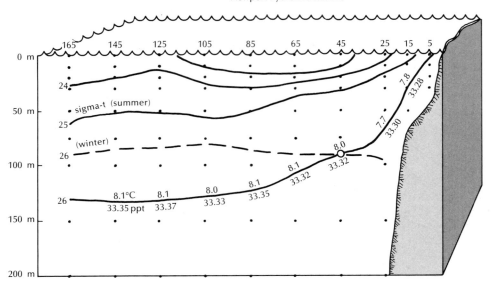

(a) Density contours in the Newport cross section during summer. Contours are labeled in sigma-*t* units of density (e.g., 1.025g/cm³ = 25 sigma-*t*). Along the 26-sigma-*t* contour, the values of temperature and salinity measured at the depth of this density are given above and below the curve, respectively. The relatively constant *T,S* values along the curve suggest that water moves *along* the 26-sigma-*t* contour from the 120-m depth at NH-165 to the surface itself near NH-5. This is upwelling. For comparison, the dashed line shows the 26-sigma-*t* contour during winter, nonupwelling conditions.

Newport hydroline stations

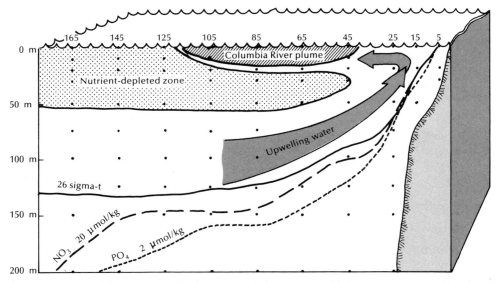

(b) A comparison of the 26-sigma-*t* density contour during summer with the summer 20-μmol/kg nitrate and 2.0-μmol/kg phosphate contours. The interpretation of an onshore flow of waters *along* the contours of constant density, derived from part (a), now explains that the upwelling phosphate and nitrate are carried with the upwelling water, from levels of about 120-m depth offshore up to the surface itself. (From Barstow, Gilbert, and Wyatt, 1968.)

Figure 8.7 The mechanism of wind-driven coastal upwelling.

Features of the Density Field in Cross Section. In general, the density contours of Figure 8.7a are remarkably similar to those of nutrient distributions. Their seasonal changes also occur inside of the 85-nmi station and the winter contour of 26 sigma-t is similarly flat. The summer sigma-t contour of 26, which offshore is in the same depth range as the 1.5-μmol/kg phosphate and 10-μmol/kg nitrate contours, intersects the sea surface inside the 15-nmi station just like those corresponding nutrient contours.

There is one major difference, however. A lens of low-density water occurs between stations NH-45 and NH-105, its maximum depth at only 15 to 20 m. This, in cross section, is the plume of the Columbia River, the same diluted seawater we found in the vertical profile analyses.

What we learn from study of the density cross sections is that seawater appears to move horizontally *along* surfaces of constant density. Stated another way, seawater does not seem to mix *across* a level that separates water layers of different density. In the following section, this idea is developed into a theorem.

A Theorem for Ocean Mixing: Seawater Moves Along Constant-Density Surfaces

In Chapter 6 we learned that vertical convection cells develop when the density stability of a water column is destroyed by surface cooling. What is the ocean's response when a force is applied that does not destroy the stratification? The answer is that the fluid moves *along constant-density surfaces;* if the forcing is strong enough, a fluid layer will change depth as necessary to keep within the same "density" stratum.

This theorem for horizontal motion, although difficult to prove mathematically, is readily demonstrated in Figure 8.7a by examining the values of *conserved properties along the contour of density.* If the temperature and salinity values are the same along a selected density contour, we must rule out vertical mixing as the process by which the density contours of Figure 8.7a rise to intersect the surface. The only remaining and acceptable explanation is that the surface water that formerly occupied the nearshore zone has moved offshore, and deeper waters have risen to fill the *void* left by departing surface water. In Figure 8.7a the temperature and salinity values of the rising contour of density 26 sigma-t during summer (July) are almost the same all along the length of the contour, from its deepest level of 115 m at NH-165 to the surface inside the 15-nmi station. This deeper water that rises also carries to the surface its high concentrations of nitrates and phosphates—*this is the upwelling process.*

To complete the theorem development it is necessary to show that the upwelling process is a continuous one. The mechanism that drives upwelling is the friction force between a surface wind and the ocean surface layer (described in Chapter 10). *Any water at the surface will be forced away from shore so long as the correct winds continue to blow.* Thus, newly upwelled water from deeper depths is *also* moved offshore, this requires more replacement water from below, and so on in a continuous process. Because the winds that favor upwelling begin off Oregon in early April, by July the process is well established, as our cross-sectional study confirms.

I would like to complete this scenario by describing just how far offshore the upwelling regime continues. Unfortunately, the experts on upwelling are not yet in agreement on this aspect, and more work needs to be done before a final description is made. There is one clue, however. The mouth of the Columbia River is 150 nmi north of Newport, yet by the time the river plume crosses the Newport hydroline, its westward edge is at NH-105 (Figure 8.7b). Clearly, the winds have pushed the plume offshore. Can we take the plume position as the offshore limit of the upwelling region? Yes, and the reason is shown in Figure 8.7b, on which is superimposed the zone of almost total nitrate depletion that we discovered in Figure 8.6. Notice how nearly the nitrate-depleted waters dovetail with plume water. Plume water obtains nutrients from two sources—its own river supply plus the upwelled waters with which it mixes as it is pushed offshore—and is not yet depleted. Surface waters to the west of the plume (and beneath it), have no resupply process, so production is not sustained once the available nitrates are consumed.

Summary

We have taken a real oceanographic cruise and returned with real data. These are processed and made available to the oceanographic community at large in the form of data points listed at standard depth intervals. Great care is taken in the chemical analyses of water samples; wherever feasible we carry International Standard Seawater along on the cruise for periodic checks on technique and calibration of our work.

A tanker vessel plowing through high seas. Wind waves generated by winter storms are a major source of energy for creating the "well-mixed layer" at the top of the ocean. (Courtesy Exxon Corporation.)

In general, we first plot the vertical profiles of each property measured, searching for the clues these offer in explaining ocean and marine life behavior. Among them are the depth of the well-mixed layer; measures of many things, from primary production to geochemical phenomena; and evidence for deep-water behavior, such as oxygen minima and nutrient maxima.

It is standard procedure that cruise tracks are laid out perpendicular to shorelines. The objective is to gather "section" data in the offshore direction. Much of what we study involves the movement of *things* from land to sea, or vice versa, and section lines are best for showing their distributions. The almost classic example in our cruise data is the cross section showing the lens of Columbia River water as it exists in July off the Pacific Northwest. We infer much from cross sections: the rate of development of the overall upwelling season, the potential impact on fish migration, and so on. In our sections special attention is given to the analyses of coastal upwelling, an ocean behavior characteristic of the Pacific Northwest and important to the regional economy. The July cruise produced a good sample of a well-developed upwelling system. We are able to see how the local winds have warped the fields of density and other parameters so that the area very near shore gets nutrient-rich waters upwelled from levels of 100–200 m.

By comparing the cross sections of nutrients with that of density, we discover a theorem for ocean mixing: water in a density-stratified ocean tends to move along surfaces of constant density.

Study Questions

1. Figure 8.2 shows that the water columns being sampled on the Newport hydroline cruise are anything but straight in the vertical. Why is this, and what does the ship's captain do to offset the vertical crookedness as much as possible?

2. Why during the sampling operation does the Nansen sampling bottle have to be reversed, in the sense of end-up and end-down?

3. Before the invention of the reversing thermometer, sampling the interior temperatures of the ocean was indeed a formidable challenge. One device used an ordinary glass thermometer with a large ball of beeswax cast over the mercury bulb end. What was the idea behind this device? Another early attempt was a thermometer gage with a special second "hand" that recorded the *maximum* swing reached by the primary pointer during the cast to some depth. Can you describe the disadvantages of this device?

4. As our studies of ocean water advanced, it became more and more imperative to *get* samples as close to the ocean floor as possible. Is this a difficult task? Why? Describe one device that enables the scientist on board to measure the distance above bottom at which a sample is taken.

5. What is the *precision* with which we can measure temperature *in situ*? What is the precision obtainable in measuring salinity? Why do oceanographers strive for such high precision in these measurements?

6. Why are the sample depths, (as reflected in the vertical spacings of bottles on the lowering cable), generally *not* at uniform intervals?

7. Once the values for temperature, salinity, concentrations of chemical ions, and so on have been obtained from each sample point, analyses begin. Generally, the first of these is to plot the vertical profile of each variable that has been sampled, and for each station. From the view of data verification, why? From the view of a physical oceanographer, why? From the view of a biological oceanographer, why? From the view of a chemical oceanographer, why?

8. Near the surface, the concentration of oxygen is at its maximum. The amount of oxygen in surface water is higher in winter than in summer. Saturation levels for most gases dissolved in seawater change with water temperature, colder water "holds" more oxygen.

   (a) Why then do deeper ocean levels generally contain less oxygen than surface levels?

   (b) If biological oxygen demand (BOD) reduces oxygen concentration in deeper waters, why do the bottom waters often contain a higher concentration of oxygen than water at middle depths? Can the answer have any connection with the "age" of water at the different levels? Would you then conclude that the *deepest* water may actually be "younger" than water at middle levels? If so, how can this hypothesis make *physical* sense?

9. Explain how we might decide the *depth* below which ocean waters do *not* feel the effects of seasonal climate changes at the surface.

10. Can you give a plausible explanation why concentrations of the nutrients NO_3 and PO_4 (Figure 8.4) show a *maximum* at about the same depths at which an oxygen *minimum* occurs?

11. The vertical profile of nitrate shows that, at 400-m depth, the July content is about 7 μmol/kg less than the January value. Is this because phytoplankton *consume* the nitrate during summer, even at this level? If not, what other explanation is there?

12. After gleaning the vertical profiles of their clues to ocean fluid behavior, we then plot cross sections of each variable. This is done by transcribing data values from tables of each cast (as from Table 8.1) onto a grid (as in Figure 8.5) and then *drawing* contours of equal value of the variable (as in Figure 8.6 for No_3 and PO_4).

   (a) The "huge" difference between summer and winter cross sections of nutrients is typical of an upwelling area. True or false?

   (b) Now that you have taken a position on the subject, tell me what *is* the huge difference?

   (c) The physical oceanographer claims that the relatively constant values of temperature and salinity along the constant-density contour 26 in Figure 8.7a are evidence that the ocean fluid moves *along* constant-density surfaces rather than *crossing* (moving across) surfaces of different density. Do you agree? Is this characteristic behavior found *only* in an upwelling area, or does it occur everywhere in the liquid ocean?

PART IV

Sun, Sea and Sky. (H. Edel/Leo de Wys.)

THE MOVING FLUIDS: OCEANS AND ATMOSPHERE

In which we develop a physical picture, first of how the ocean's water masses are structured and second of how prevailing forces control their motion.

CHAPTER 9   SEAWATER DENSITY: PHYSICAL AND BIOLOGICAL EFFECTS

Seawater density is the key to understanding fluid motion as well as the behavior of marine organisms.

CHAPTER 10   FORCES THAT MOVE FLUID OCEANS AND ATMOSPHERE

Both the oceans and the atmosphere are fluids whose motions are driven by several types of forces, one of them the mystical Coriolis force.

CHAPTER 11   CIRCULATION OF THE OCEANS

How is the ocean's motion organized into patterns, both for the wind-driven surface layer as well as for the density-driven deep layer?

CHAPTER 9

SEAWATER DENSITY: PHYSICAL AND BIOLOGICAL EFFECTS

The total of all research cruises since that of the HMS *Challenger* (1873–1876) numbers in the thousands, with hundreds of thousands of stations reported. As the numbers grew, oceanographers became aware of differences between the major oceans, for example, the varying distribution of salinity (Chapter 7), and began searching for a systematic way to "map" the oceans. They chose seawater density, a basic physical property, as the best one to chart. And they devised a technique of graphing density in terms of the two independent factors that control it, temperature, T, and salinity S. Thus began the now universal method of mapping oceans, *the T–S nomogram.*

The plotting of T and S values from station data onto a T–S nomogram (for example, from NH-105 in Table 8.1) indicates how the density of seawater changes from top to bottom of the water column. The *shape* of the graph shows *how* the changes occur. *Each ocean region possesses its own unique T–S graph shape.* The graph then is a logical tool for mapping the oceans.

The T–S Nomogram

We construct a nomogram using graph paper with a grid of horizontal and vertical lines (Figure 9.1). The vertical lines are scaled to show salinity in parts per thousand, covering the range of salinity values we expect to find in typical oceans, 30 to 35 ppt; the horizontal lines are scaled with the expected range of temperatures, -2 to 24°C. Because the density of water is controlled for the most part by its temperature and salinity, each point inside the grid space has a specific density value associated with it.

The basis of the T–S nomogram is the computing of seawater density from pairs of data, T and S, using an empirical equation. (Readers who are mathematically minded and wish to examine the equation are urged to consult Appendix 2.) The equation is programed to cover all combinations of T and S found in the oceans. From the set of computed densities is constructed a family of curves over the T–S space. In Figure 9.1 these curves are drawn for constant values of seawater density and are labeled in sigma-t units (the same as those used in analyzing the cruise data in Chapter 8), but the full unit for density is labeled ρ; for example, in Figure 9.1 the sigma-t contour 22 is really the density contour $\rho = 1.02200/cm^3$.

The process of constructing a T–S nomogram provides some insight into why the ocean behaves the way it does.

1. The contours of constant density are systematically *curved*, which indicates that the density of seawater *changes nonlinearly* with changes in T,S. This behavior of water relates to two of the basic physical properties of water listed in Table 6.1, its thermal coefficient of expansion and its dissolving power. Most of the curvature in the density contours results from the nonlinear behavior of this expansion coefficient.

Because the density contours are curved, changes in temperature have only minor effect on the density of *very cold* water. For example, point A is typical of T,S conditions in the Arctic Ocean surface water; because the density contours are oriented almost vertically in the vicinity of A, large changes in temperature cause only small changes in water density. An important concept emerges: salinity changes control ocean dynamics in polar latitudes.

Point B, however, is located in the ranges of T,S found in tropical ocean surface water. Here both temperature *and* salinity changes alter the water density, so both factors control ocean dynamics in these low latitudes.

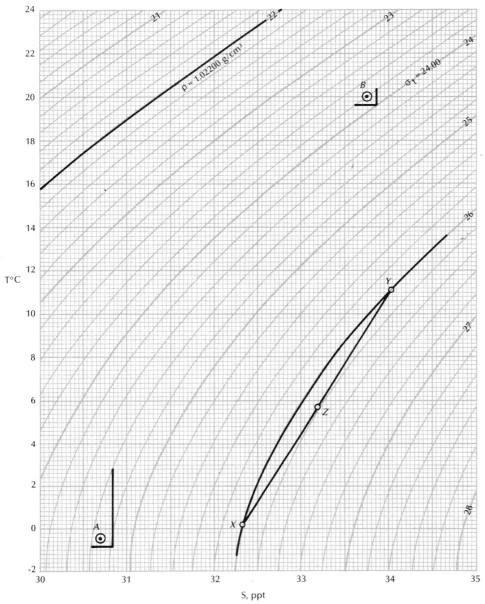

Figure 9.1 The temperature-salinity (*T-S*) nomogram. Simply stated, this nomogram solves for seawater density if we enter the scales of *T,S* with any pair of these data. For example, at 19.0°C and 31 ppt, seawater has a density of 1.02200 g/cm³ (or 22 sigma-*t*). The curved lines are the family of density contours that fit over this domain of temperatures and salinities normally found in seawater.

2. The straight line *XY* in Figure 9.1 demonstrates another unique, *nonlinear* fluid ocean behavior. Suppose that we take two identical buckets of seawater, one each at points *X* and *Y*. The density of water is the same at both points, even though both temperatures and salinities are different. Now if we mix the two buckets in a larger container, we expect to obtain a mixture that has the characteristics of point *Z*, that is, temperature and salinity values midway between those of *X* and *Y*. This is the case, but the density of the mixture *Z* is a *full 0.2 sigma-t units greater than that of either component alone!* To what use can we put this "esoteric" result? Simply stated, it means that two different pieces of ocean surface wa-

ter might come together, mix, and *promptly sink.* More will be given on this later.

How To Use the T–S Nomogram— Mapping and Mixing

Graphing a Real Water Column on *T–S* Space

How can the *T–S* graph technique be used to study a real ocean? Figure 9.2 is a graph of the same Newport hydroline station NH-105 that was analyzed in Chapter 8, but this time plotted in *T–S* space and only for the July 1968 cruise. The reader should ver-

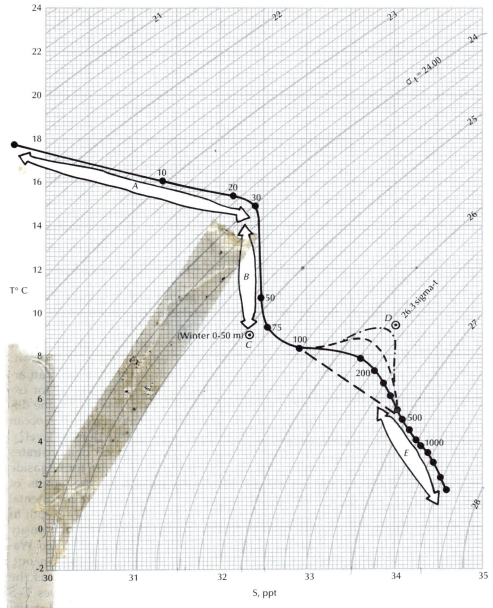

Figure 9.2 The *T-S* structure for the water column sampled at NH-105, July 9, 1968 (44°39.1'N, 126°31.1'W). Temperature and salinity were sampled at the surface and at 20 depths. Each pair of data points yields a point on the nomogram (●); the curve connecting the points has a *shape* that is *unique* to this water column. For this reason *T-S* nomograms are sometimes called water mass *structure* diagrams; they reveal the complex structure that results when layers of different masses of water are superimposed.

ify the accuracy of the plot. From Table 8.1 take the July 1968 pair of *T,S* data measured at 50-m depth (*T* = 10.64°C, *S* = 32.505 ppt) and locate the corresponding point on the *T–S* space of Figure 9.2. It should check out as the data point labeled 50 on this graph. Oceanographers usually print the depths alongside the plotted points for a variety of reasons. For one, notice that fully half the length of the graph for NH-105 covers only the uppermost 75 m of the water column.

When all the plotted points are connected, the resulting curve takes a definite "shape," one that is *unique to this part of the world oceans*. No other water column anywhere will duplicate the precise shape of

NH-105. The graph crosses many of the curved density lines on the *T–S* nomogram, from 21.2 sigma-*t* at the surface to 27.7 sigma-*t* at 2500 m depth. We can use the graph, then, to study how the density of the NH-105 water column changes with depth. In a real sense, the reader is beginning to study an "ocean map."

Density Structure within a Water Column

In Figure 9.2 the NH-105 density structure is marked off in four separate sections, *A* through *E*. Each of these is now examined for its meaning:

A. There is not much question what *A* means. The graph moves from 0 to 30 m by spanning a *very large change in salinity,* from less than 30 ppt to 32.5 ppt. This has to be the *Columbia River plume,* which is slowly being mixed into the upper ocean.

B. Part *B* also has a simple explanation. It is the section below the river plume, from 30 to 75 m, in which only temperature changes, and it changes a great deal. This has to be the effect of *summer heating,* a process that does not alter the salt content itself (except at the surface where evaporation occurs). Can the solar rays heat the surface water to depths of 75 m? The answer is yes; we see here the evidence that light penetrates at least to 75 m depth in this water column. (For a quantitative look at how light penetrates into oceans, see Figure 14.1.)

C. Part *C* is not a piece of the July water column, rather it is a plot of all the January data between the surface and 50-m depth. Its use here emphasizes how the upper-layer conditions vary over the seasons. All the water down to about 100 m can be lumped together and treated as *coastal waters* that are highly variable and subject to coastal runoff and summer heating and winter cooling.

D. Part *D* is a point. It is not on the graph of NH-105, but its density is about midway between the 100- and 500-m levels of NH-105, a section of the graph that has a decided *curvature.* More important, this density is the same as that of the NH-105 graph at the apex of the "hump" in its curvature, 26.3 sigma-*t*.

We can now use the theorem of ocean mixing developed in Chapter 8. Is it possible that a mass of water of this specific density has been formed elsewhere and moved into the NH-105 water column at this density (depth) level? If the *D*-type water originated far away, it may have lost its original *D* values of temperature and salinity by mixing with waters above and below the level at which it moved. Stated another way, if the NH-105 station is close to the place in the ocean where *D*-type water is being produced, its graph should pass almost exactly through point *D.* Conversely, if NH-105 is far removed from the *D* source, mixing will have so diluted the *D* water that its presence can no longer be detected, and the *T–S* graph will then have *no* curvature, as for example the straight, dashed section added to Figure 9.2.

E. The deep water from 500 to 2500 m appears to be very "regular." Viewed in terms of its density, it is a layer of weak stratification. Temperature changes slowly but consistently over the entire 2000-

m-thick layer, and so does salinity. The entire layer can be treated as though it were a single water mass.

The Concept of Water Mass Structure: A New Theorem. Analyzing *pieces* of the NH-105 *T–S* graph was actually an exercise in identifying the *different types of water* in the column at this station from top to bottom. If we generalize about the ocean on the basis of this one station, it follows that the *oceans are made up of layered masses of water, piled one above the other.* This is precisely the case, and an additional generalization leads to a new theorem on ocean behavior. Each ocean station reveals a layering of water masses that is unique to its ocean location; the shape of its *T–S* graph is similarly unique and is called the *water mass structure of the ocean region* in which the station was taken.

Using the *T–S* Nomogram To Trace Water Masses. The theorem just stated is a powerful analytical tool for oceanographers. The *T–S* graph for one station, as analyzed for NH-105, reveals how different water masses are arranged at one point in the ocean. Could a series of *T–S* graphs, prepared from an array of stations within a large ocean sector, for example, the entire North Atlantic, reveal how the different water masses are moving across the ocean basins, each at its own characteristic density level?

The answer is yes, and Figure 9.3 demonstrates how a mass of water called *Med Water* forms inside the Mediterranean basin, overflows the Straits of Gibraltar, and sinks down the shelf and continental slope on the Atlantic side until it reaches a depth of about 1500 m. This is the level at which the density in the Atlantic water column equals that of Med Water. At this level the Med Water begins to fan outward into the interior of the North Atlantic, as the sketches in Figure 9.3b show. Figure 9.3c gives *T–S* graphs for each of the hydrographic stations located west from the Straits. It is clear that the curvature in the graphs can be used to "trace" the mass of Med Water back to its source point.

Mapping the Oceans Using Water Mass Structures. A "water mass" is just what the term implies—a mass of water. The term further denotes a mass of water throughout which prevails a more or less uniform temperature and salinity, (sometimes called the *characteristic T,S*). The existence of large volumes of seawater of characteristic *T,S* suggests that all the seawater within the mass originated in the same locale and under identical circumstances. Recall that seawater can only "change" its character

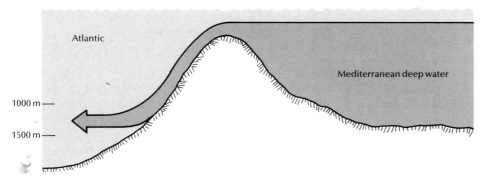

(a) Cross section showing the overflow of Mediterranean Deep Water into the North Atlantic basin.

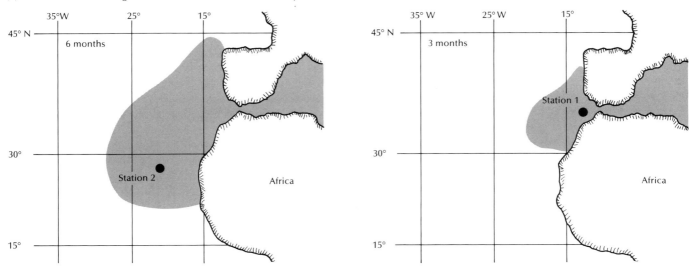

(b) Med Water spreads laterally as a lenslike water mass into the interior of the North Atlantic Central Water, at depths ranging between 1000 and 1500 m.

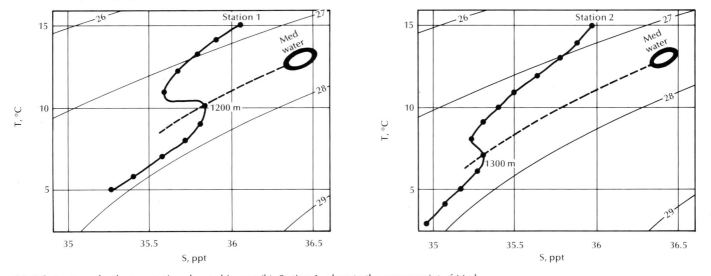

(c) T-S structures for the two stations located in part *(b)*. Station 1, close to the source point of Med Water, has the characteristic sharp curvature in its T–S graph; at station 2, far from the source point, Med Water has been mixed with adjoining layers of water, so that the curvature signal is also reduced.

Figure 9.3 Spreading of Mediterranean Water in the North Atlantic.

when exposed to the atmosphere and the exchange fluxes that occur across the air–sea interface.

Combining all these ideas, we deduce that (1) water masses are unique in characteristic T,S because they form at different global points; (2) once removed from the surface, water masses retain their characteristics as (3) they move through the oceans along constant-density levels; (4) their presence within a water column being sampled is detected by their influence on the "shape" of the density structure at that station. This concept of mapping the oceans through water mass structures is basic to classifying the world ocean.

Comparison of Water Mass Structures from Different Oceans

The differences in water mass structures among the several major oceans are best shown by plotting on a single $T–S$ nomogram some hydrographic stations selected as representative of these ocean sectors. A comparison of five representative water columns is given in Figure 9.4 for stations in the North Atlantic, South Atlantic, North Pacific at mid-latitude, North Pacific at high latitude, and Subartic; also shown as the dashed line is the NH-105 station off Oregon.

Clearly, each of these $T–S$ graphs is different.

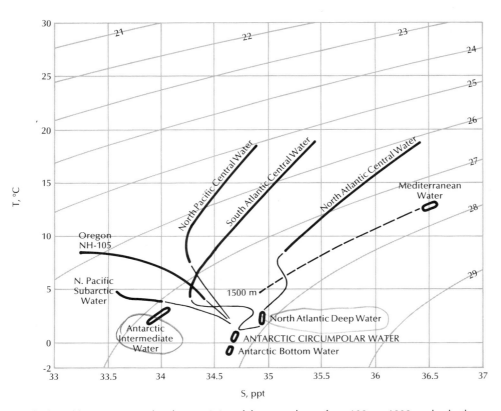

The broad lines represent the characteristics of the upper layer, from 100- to 1000-m depth; the thinner sections of the curves represent the deep-water density structure. The Oregon NH-105 structure is transferred here from the $T–S$ plot of Figure 9.2, (but only the part from 100- to 2500-m depth).

The isolated elliptical zones mark the characteristic temperature and salinity of the major water mass types: Antarctic Bottom Water, Antarctic Circumpolar Water, North Atlantic Deep Water, Antarctic Intermediate Water; and Mediterranean Water.

The structure of the South Atlantic water column is the most unique. For the first 1000m it is dominated by Central Water. Just below this the structure responds to the presence of Antarctic Intermediate Water, whose core forces the sharp bend. Still deeper, the column is filled with North Atlantic Deep Water. The graph then passes into the domain of Antarctic Circumpolar Water and finally into Antarctic Bottom Water

Figure 9.4 $T–S$ graphs showing the different types of water mass structures found in different ocean basins.

Most noticeably, they differ in the sectors that graph the upper 1000 m at each station (shown in heavy black lines on the composite nomogram). The two stations most nearly alike for these surface layers are the North and South Atlantic, with the North Atlantic considerably more saline. This observation fits our earlier discussion of high surface evaporation in the North Atlantic and hence more saline water. The most saline water of all is Med Water, shown on the nomogram at salinities around 36.5 ppt. The North Pacific, on the other hand, is quite a bit less saline, including NH-105, as mentioned earlier.

In addition to the station graphs, Figure 9.4 shows five different water types superposed over the T–S space. Much like point D examined in the analysis of NH-105, these water types have characteristic T,S values. Med Water was described earlier. North Atlantic Deep Water is unique; the reader previewed the formation of this water in Figure 3.8 in a discussion of how tritium is injected into deep water. Antarctic Bottom Water was discussed in Chapter 4 in connection with deep-water movement across mid-ocean ridge gaps. It is the densest of all

water types, as mentioned earlier. Antarctic Intermediate Water is a very special water, formed at the conjunction of the South Atlantic and Southern oceans. We will return to it later; for now, we note only that its depth is about 1000 m.

The *shape* of the South Atlantic T–S graph is the most contorted of this set. This is because its water column intersects layers of Antarctic Intermediate Water at about 1000 m and North Atlantic Deep Water at about 3000 m. The curvatures in the graph are quite extreme, indicating that both of these water layers have survived extinction by mixing.

Last, note that *all* the graphs converge on *one specific water type* at very deep levels. This is Antarctic Circumpolar Current Water, the dynamic channel through which all oceans mix with each other.

Classifying Pacific Ocean Sectors Using Water Mass Structure

Figure 9.4 demonstrates that different oceans are structured differently in terms of their layered water

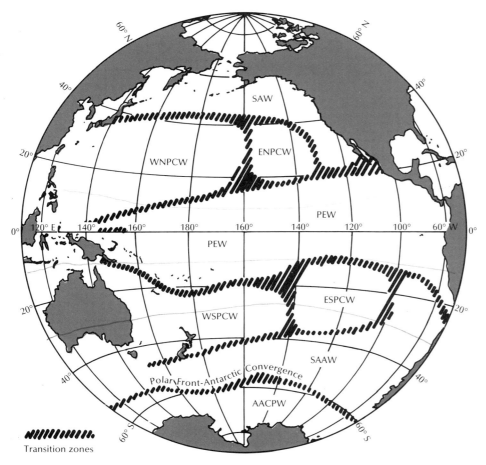

Figure 9.5 Classification of Pacific Ocean waters according to their water mass structures. (From Johnson, 1972.)

SAW	Subarctic Water
WNPCW	Western North Pacific Central Water
ENPCW	Eastern North Pacific Central Water
PEW,	Pacific Equatorial Water
WSPCW	Western South Pacific Central Water
ESPCW	Eastern South Pacific Central Water
SAAW	Subantarctic Water
AACPW	Antarctic Circumpolar Water

Transition zones

masses. But thus far we have examined only a very few of the hundreds of thousands of stations now archived. The pattern that emerges as we plot and compare many T–S graphs is that (very large) ocean sectors have almost identical graph shapes, hence layered water mass structures. This then is a strategy for *mapping the oceans;* we combine all ocean sectors with similar T–S structures and give each such region a name.

The work of earlier oceanographers was a true pioneering effort. They demonstrated that the oceans can indeed be classified and, surprisingly, that the *number of ocean regions thus identified is quite small!* Figure 9.5 shows the *eight sectors* that comprise the entire Pacific Ocean, from Subarctic to Antarctic latitudes. One of the striking features of this chart is the mirror-image nature of sectors in the Northern and Southern hemispheres of this vast ocean.

Oceanographers also assign names to the transition zones between sectors. Examples are the Subarctic Convergence, the Subtropical Convergence, and the Polar Front. (The Polar Front is often called the Antarctic Convergence; it is the zone where Antarctic Intermediate Water originates.).

Located with a star is our now-familiar NH-105. This station samples the water mass structure called Subarctic Water (SAW), which occupies the coastal zone along the entire length of the western United States.

Density Structure in the Atlantic Ocean

The Atlantic Ocean, being smaller than the Pacific and much closer to the European centers of early oceanographic research, was the first to be mapped for density structure. The data base came from the famous German Atlantic expedition of the Research Vessel *Meteor* (1925–1927), from which base Georg Wüst published his comprehensive treatise on the Atlantic Ocean in 1935. In his translation of Wüst's work, Gordon (1978) writes:

> Wust made his most lasting imprint on Oceanography—as scientific supervisor of the German Atlantic Expedition. It was a time when countries launched large-scale ocean expeditions, equipped with the newly developed precision tools and scientific methods of oceanography. The objective was simple: to determine the most basic fundamental nature of ocean stratification and circulation. It must have been a particularly exciting and challenging time for the creative mind, perhaps not unlike our own pioneering effort into outer space.

It was the first time that a complete cross section of a major ocean was analyzed for its density stratification. I have selected one diagram from Wüst's work, not only because of its historical nature but also because the Atlantic, North and South, is a quite special ocean from the view of stratification. Figure 9.6, taken directly from Wüst, summarizes the water masses and inferred circulation in the Atlantic Ocean. Wüst's chart divides the Atlantic cross section into *troposphere,* (the uppermost layer), and *stratosphere,* (the deeper layer).

Troposphere Water Masses

In both the Northern and Southern hemispheres the troposphere contains a central, lenslike mass of warm saline water, the Central Water; these masses are the Atlantic equivalent of the Central Pacific Waters (see Figure 9.5). The thick, lenslike Central Water of the North Atlantic has a special name, the *Sargasso Sea.* Notice also the hatched area labeled M just below the Sargasso water. This is Med Water.

Deep Water Masses

There are four principal masses of water within the stratosphere of the Atlantic.

North Atlantic Deep Water. Water mass *I*, originating at the Polar Front of the North Atlantic, sinks to fill virtually the entire northern basin and continues its slow movement toward the Southern Ocean off Antarctica. Near Antarctica the North Atlantic Deep Water rises to the surface, taking about 700 years to make the transit from the Arctic.

Antarctic Intermediate Water. In the vicinity of the Polar Front around Antarctica, at about 50°S (see this feature in Figure 9.5) two waters arriving from quite different directions meet and mix. Moving northward from the continent itself is a thin surface layer of very cold water (about 0°C), heavily diluted by melting sea ice. Moving southward toward the Polar Front comes the South Atlantic Central Water, quite warm and saline. When these two waters are mixed, the mixture sinks and begins to spread northward beneath the Central Waters as the Antarctic Intermediate Water, so-called because it occupies an intermediate position between the upper troposphere and the deeper masses. It flows across the equator to enter the northern basin before losing its characteristic identity as Z_s in Figure 9.6.

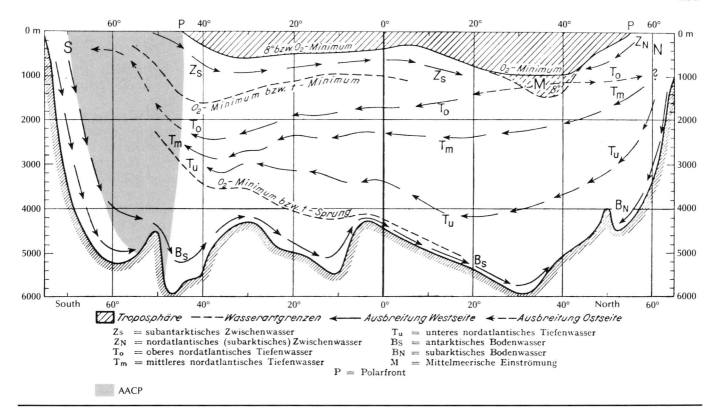

Figure 9.6 Meridional cross section of the Atlantic showing water masses spreading north and south. This diagram, published by Wüst in 1935, is a summation from the pioneer work of the RV *METEOR* in the Atlantic Ocean. The cross section stretches from Iceland in the north to the Antarctic in the south. The shaded area if an approximate cross section of the Antarctic Circumpolar Current.

Antarctic Bottom Water. The coldest, densest water in the world oceans, B_s, forms near the Antarctic continent during the winter freezing period. Its northward progression was described in Chapter 4 in connection with the topography of basin floors.

Antarctic Circumpolar Water. Superimposed on Wüst's chart is Antarctic Circumpolar Water, so-called because it is the water carried around the continent in the massive Circumpolar Current. This feature serves as the grand and common meeting ground of water masses from all major oceans, and as the means by which all major oceans are interconnected. (AACP in Figure 9.6).

The O₂ Minimum: More Evidence of Water Mass Structure. Wüst used the oxygen minimum zones as evidence of separation between water masses. For example, the uppermost O_2 minimum in Figure 9.6 lies between the north-flowing Antarctic Intermediate Water and the south-flowing North Atlantic Deep Water; the lower O_2 minimum separates the latter from the Antarctic Bottom Water. The explanation here is like that used in explaining the O_2 minimum in the NH-105 profile. If two water masses are flowing in opposing directions, between them must be a thin layer of water that does not move at all and so is slowly depleted of oxygen.

Biological Consequences of Density Structure—Whales to Plankton

In this section we examine how some forms of marine life cope with a fluid environment when relative density differences, body versus seawater, have an immediate impact on life-style and behavior. We begin with a statement: the *average* density of living matter, including its internal or external skeleton, is greater than that of average surface seawater, and so marine animals tend to sink. All forms of marine life therefore employ specific adaptations to achieve the desired end result of remaining at selected depths for lengthy periods of time. Species must maintain position within a specified depth range because they require proximity to prey species and need to safeguard the activity of body enzymes that are somewhat pressure-range specific.

Buoyancy Control in the Sperm Whale

For reasons that we do not understand, the preferred prey of the sperm whale is the deep-water squid found at depths of 300 to 2000 m. As an air-breathing mammal, this whale has its *resting* level at the air–sea interface. To gain the flotation required to offset its skeleton weight, and at the same time avoid excessive heat loss to the ambient water, it surrounds itself with a layer of fatty tissue called blubber. Blubber lipids (fat) are less dense than seawater; all whale species use them for insulation. The sperm whale has evolved with an additional buoyance mechanism, however, a huge snout filled with oil-saturated tissue, the spermaceti organ (Figure 9.7). Why should this whale employ two different mechanisms of flotation and other species only one? Or is it possible that the prime purpose of the spermaceti organ is not for flotation? Whalers have known for centuries that the oil from the spermaceti organ is "different." A very recent study suggests that the spermaceti organ is an adaptation that provides the animal with *negative* buoyancy, rather than positive or (flotation) buoyancy, and that without

this organ the whale would be unable to dive to depths of up to 2 km to seek out its preferred prey.

To see how this mechanism might work, let us examine the nature of the sperm oil itself. Like most fluids, it can be "frozen," except that the solid phase is really a "gel" (like gelatin), not a solid like ice. Just as important, it *contracts in volume* as it gels. Further, the phase change from liquid to gel takes place at the temperatures and pressures normal to the oceans at about 100-m depth.

The implications are intriguing. First, assuming that the animal finds some way to cool down its oil supply, the volume of the animal could *shrink* as it dives down deep into the water column, and substantially so seeing how large the nose is. This means that its average density would increase, just as the density of the ambient water increases with depth in every water column. It follows that the animal could adjust to neutral buoyancy and even to negative buoyancy during its dive, and thereby save the energy it would otherwise have to use in *swimming* down (remember, its normal resting buoyancy is at the surface). The potential advantage is clear: with less energy used in reaching its "hunting" level, more energy and time is reserved for the hunting exercise itself.

The next question to ask is whether the whale *needs* assistance to get down? Figure 9.8 shows that at the equator the seawater density at 1500-m depth is 0.5% greater than surface water density. A 50-ton whale that started its dive at neutral buoyancy

An unusual view of a male sperm whale. The belly of the whale is facing us. Notice the small lower jaw with teeth that fit into the sockets of the upper jaw; the huge snout containing the spermaceti oil organ, occupying a third of the body length; the small flippers and the penis farther back. (Photograph by Professor H. Boschma of the University of Leiden. From A. Hardy, *The Open Sea*, Part 2, Houghton Mifflin Company, Boston, 1965.)

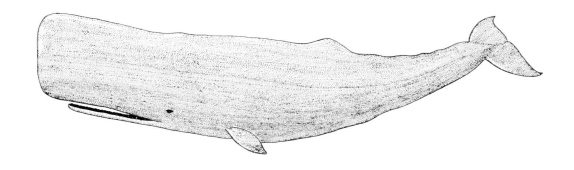

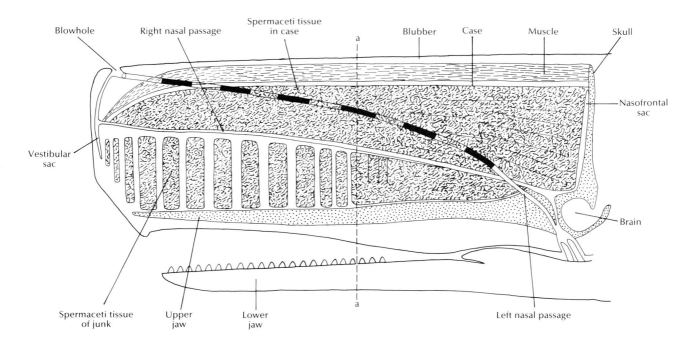

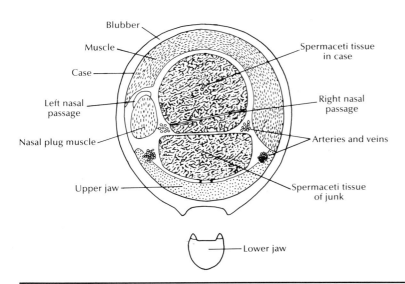

The head of the sperm whale takes up about one-third of its body length. A 30-ton animal may carry as much as 2.5 tons of spermaceti oil in a huge organ cradled within its upper jawbone. The oil is a clear fluid at the internal temperature of an animal resting at the surface, about 33°C (or 90°F). But if removed from the animal and cooled to air temperature, about 65°F, the oil solidifies to a soft, pliant gel.

A longitudinal section through the head shows the main features of the spermaceti organ, along with the right and left nasal passages which connect the blowhole to the animal's lungs during surface breathing. The nasal passages can also circulate seawater through the spermaceti organ during the animal's dive.

One way that the animal loses heat from the spermaceti oil is by conduction of heat through the muscle and outer blubber layer directly to the cold, ambient seawater. Another way is to take cold seawater in through the blowhole, into the vestibular sac and then through the wide, thin right nasal passage that traverses the length of the spermaceti organ; heat transfers from the ingested seawater to the oil of the surrounding organ, causing gelling. Muscle fiber surrounding the organ provides the power to ingest and expel seawater.

Figure 9.7 The sperm whale *Physeter catedon*. (From Clarke, 1978.)

would experience an upward force of 250 kg at the 1500-m level. It could use a boost, so to speak.

Buoyancy Control during Dive. Apparently, the whale begins its dive with the major part of its sperm oil reservoir in liquid form. As it dives, it ingests seawater through the blowhole normally used for breathing air and forces the cold water through a wide, flat passage that threads through the interior of the spermaceti organ, as shown in Figure 9.7. The result is a rapid exchange of heat from the oil into the ingested water, which is then "exhaled." Repeated operations remove additional increments of heat until enough oil has been jelled so that the whale has achieved the overall volume contraction it needs to bring its mass density *up* to the value corresponding to that of ambient seawater. In a sense, this animal could literally "breathe" its way down to deep levels! But because its dive time is inherently limited, the buoyancy adjustment is probably augmented with active downward swimming.

Buoyancy Control during Resurfacing. The power needed for the resurfacing swim could be reduced if in some way the whale could reliquefy the now-jelled spermaceti oil. This requires a source of heat, of which there is precious little in the ambient water at these depths. Moreover, with ambient water at a lower temperature than that of the gel, such a heat exchange is impossible. The whale does have a source of internal heat, however, the heat built up within muscle tissue. But because the whale is so well "insulated" with blubber, it cannot lose enough of this excess heat through the single process of conduction to surrounding seawater.

Instead, the animal pumps blood through its muscle tissue to "pick up" excess heat; the heated blood is then pumped to the arterial system of the spermaceti organ where its excess heat is given off to reliquefy the oil. The whale gains a distinct additional advantage through this process—it saves time. Normally, with the heavy insulation of blubber, the whale can only exchange body heat with the air that it breaths, a relatively slow process because of the low specific heat of air. By using up body heat to reliquefy its spermaceti gel, the whale arrives back at the surface with its internal (muscle) temperature cooled down *and* its density readjusted to neutral buoyancy with surface water! Now its only requirement before the next dive is to recharge its oxygen supply and discharge gaseous metabolic by-products such as CO_2, a job for which its lungs are well suited.

The whale has been clocked at about 4 knots during a typical dive, (by analogy, a fast walking rate for a human). This gives it a diving time of about 8 minutes to 1000 m depth; it returns at a faster rate, about 5 knots. Total swimming time down and up is about 15 minutes, and some 35 minutes more are used at the selected depth for the search and predation effort. When the whale returns to the surface, it normally rests for about 10 minutes, taking between 40 and 50 breaths to recharge the oxygen levels in muscular tissue. The whale is then ready for another dive.

Flotation Mechanisms of Plankton

No less remarkable are the many adaptations by which the *microscopic forms* of marine life sustain themselves within the sunlit near-surface layers of the ocean. *Plankton* is a collective term for very small marine plants and animals, but there are divisions in the terminology. *Phytoplankton* are the plants, and *zooplankton* refer to the larger elements of plankton and are generally animals. If a large adult species reproduces via eggs and larval stages, the larvae are termed *meroplankton* to denote that only a portion of the life cycle is in the world of plankton. Many species remain small even as adults and are called *holoplankton* because the whole life cycle is planktonic.

For organisms smaller than about 0.1 mm (100 micrometers), the fluid ocean is indeed a very viscous medium in which to live. "Swimming" as we know it is not possible for them. The very small forms, including most marine plants, are completely at the mercy of vertical movements within the fluid itself. To these individuals, the fact that the water column is density stable, and therefore not subject to violent vertical motion, is a major life-sustaining factor.

Phytoplankton. Marine plants reproduce by asexual budding; there is no egg stage. In a nutrient-rich surface layer with adequate sunlight, the reproduction time of a diatom can be as short as 24 hours. Their strategy is simply to stay in the photic zone long enough to reproduce. To do so is a problem for even microscopic plants because their average density is about 1.04 g/cm^3 in seawater of about 1.026 g/cm^3. Like it or not, they tend to sink. Their principal defense against sinking is through adaptations that take advantage of the high viscosity of water at very small scales.

Zooplankton and Pelagic Larvae. Most marine animals employ an egg stage in their life cycles. A first

COMPARISON OF AMBIENT CONDITIONS SEEN BY THE SPERM WHALE

In its annual migrations along the surface, the sperm whale passes from equatorial waters to subpolar waters. In doing so, it encounters changes in surface water temperatures of the order of 20°C. The difference in density of tropical and surface polar waters is on the order of 0.4% (if salinity is constant), with polar waters denser.

How does the animal adjust its buoyancy so that it is neutrally buoyant along the entire migration route? One way is to adjust the volume of air in its lungs. Another way is to control the temperature of the spermaceti oil in its snout. Because the density of the oil *increases* as its temperature *drops*, the animal has a built-in device to compensate for the higher density of polar surface waters. Because migration is slow, the rate of change of ambient temperature is also slow. The animal has ample time to adjust oil temperature.

The sperm whale encounters a similar range of temperature change in the equatorial zones as it *dives* from the surface to depths of 1500 m or more. (The Pacific equatorial zone is a known feeding area for sperm whales. The dynamic upwelling there brings about high primary productivity and plentiful numbers in the biomass food chains, including large numbers of the sperm whale's favorite prey species.)

In polar waters, a dive to 1500-m depth exposes the whale to a *density* change of about 0.2%: if the animal is neutrally buoyant at the surface, it has a 0.2% positive buoyancy at this depth. In equatorial waters, because the density stratification of ocean layers is more definite, the animal encounters a density change of about 0.5% for a dive of similar depth. Clearly, a *single* method of buoyancy control would leave the animal at a disadvantage in one area or another, hence the need for multiple buoyancy control methods.

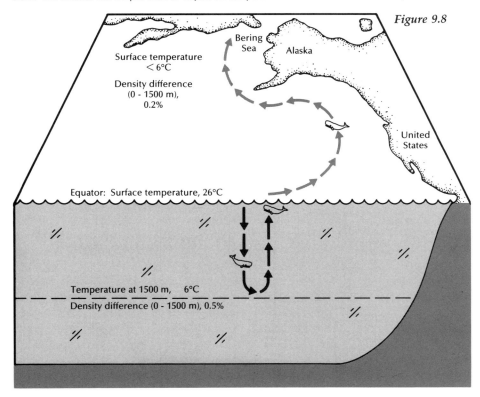

Figure 9.8

Bering Sea

Alaska

Surface temperature
< 6°C

Density difference
(0 - 1500 m),
0.2%

United States

Equator: Surface temperature, 26°C

Temperature at 1500 m,   6°C
Density difference (0 - 1500 m), 0.5%

---

requirement is that the larva on hatching is within feeding reach of food of correct "size," usually very small. This means that the parent must concentrate enough fats in the egg that it will float. Another requirement is placed on the egg, that it be osmotically protected against loss of fluid to the salt water outside, or against the infusion of undesired ions.

On hatching, a larva must sustain itself within the photic zone. Many larvae are capable of swimming. Even so, the advantage lies with the larva that can

grow spines, plumes, or other devices to provide frictional reaction against sinking in the viscous water.

Flotation of Plants and Animals

In general, the individual will not rely on a single mechanism for flotation. The following are different techniques used by plants and animal larvae to stay afloat. Some are illustrated in Figure 9.9.

STRUCTURES OF PLANKTONIC FORMS THAT REDUCE THE TENDENCY TO SINK

(a) Small phytoplankton and zooplankton organisms frequently have body shape adaptations that slow their rate of sinking by maximizing the ratio of friction-producing cross-sectional area to weight. One way to accomplish this is by being extremely small. For a simple spherical shape the cross-sectional area increases as the square of the radius, but the volume of the sphere, and therefore the organism's weight, increases as the cube of the radius. Another way to maximize the ratio is to avoid a spherical shape. Many phytoplankton cells are thin and flat, others long and thin. Both shapes slow sinking by causing the cell to sink with a "falling leaf" motion.

(b) Another way of reducing the tendency to sink is to maximize the ratio of surface area to volume enclosed. Long thin spines accom-

plish this. Such spinosity is found in many diatoms, such as the *Chaetoceros* spp.

(c) Dinoflagellates of the species *Ceratium macroceros*, individuals native to high latitudes where cold water is relatively dense and viscous have less of a sinking problem than their tropical cousins. Their spines are therefore shorter than those of dinoflagellates living in low latitudes where the warm water is less dense and has lower viscosity.

(d) Plumose (featherlike) appendages can be even more effective in maximizing the area-to-volume ratio as well as in taking full advantage of the viscosity effect. The individual hairs on a plume are close enough together that water, which is highly viscous at this small scale, is unable to flow between them. So a plume generates the same resistance to sinking as a solid flat plate, but weighs only a fraction as much. In tropical or mid-ocean copepods such as *Calocalanus pavo*, plumosity can be extensive and elaborate.

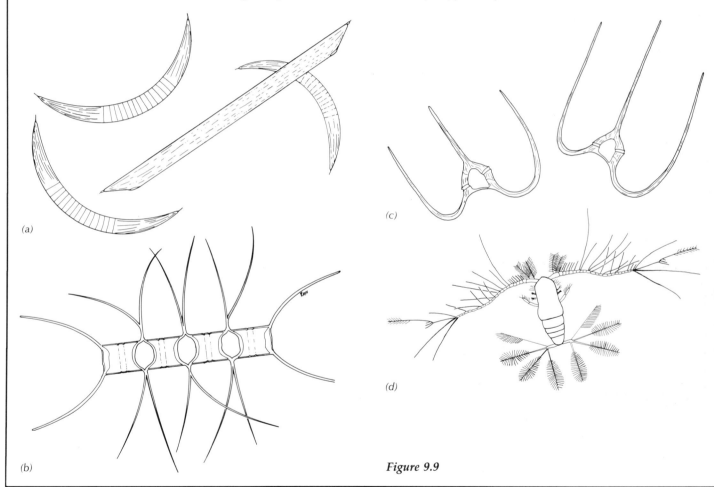

(a)

(b)

(c)

(d)

Figure 9.9

1. *Changes in solutes in body fluid.* One way to change specific gravity is to substitute lighter ions for heavier ones, as long as the physiological function of the ion in the body fluid is not compromised. For example, many of the gelatinous plankton animals such as jellyfish substitute sodium or potassium cations, Na^+ or K^+, in place of calcium or magnesium cations, Ca^{2+} or Mg^{2+}, and thereby achieve lower specific gravity. Or the substitution may be of the lighter chlorine anion, Cl^- for the heavier sulfate, SO_4^{2-}. In general, the substitution trick is to rid the body fluid of divalent ions. The monovalent ammonium ion NH_4^- is also often stored, perhaps because it is readily available within body tissue as a by-product of catabolism (the breaking down of organic tissues.

2. *Reduce shell materials.* Many organisms build exoskeletons. For example, members of the order Foraminifera build shells of calcium carbonate; other protozoans such as Radiolaria construct an exoskeleton of silica, SiO_2, as do the diatom plants. Studies of species of Radiolaria that are pelagic reveal that their shells are thinner than those of their benthic cousins, or they have larger holes in the shells, or both. Members of the squid group, (Cephalopoda), contain an internal skeleton of cartilaginous material: members of the genus *Loligo* are pelagic and have a thinner skeleton than members of the genus *Sepia*, which are found in the benthos. (Squid also store relatively large amounts of ammonium ions.)

3. *Change body form.* A favorite trick is to change body form, and there are number of ways to accomplish the same end. Let us look first at diatoms (Figure 9.9).

Some diatoms are found to have bilateral symmetry rather than radial and are elongated into a "pennate" shape. The *Nitschia* are a good example. The dynamics of such a body shape as it falls through a fluid under the force of gravity is easily envisioned—we have all seen the convolutions of a tree leaf as it falls. The net result is that the pennate diatom actually "flies" or "glides" downward rather than falls per se.

Some diatoms such as *Rhizosolenia styliformis* have radial symmetry except for a nonsymmetrically shaped tip. It accomplishes the same slowdown of fall rate as does the pennate-shaped plant.

Still other diatoms grow appendages in the form of long, thin spines, for example, *Chaetoceros decipiens* (see also Plate 15.1). These plants are often seen linked together into chains of individuals. Spinosity, the use of spines, increases the frictional drag of the body as it falls in a viscous fluid, and hence fall rate is lowered.

Viscosity enters this discussion in a very real way. In polar oceans, where surface waters are at very low temperatures, the viscosity of water is higher than that of surface waters in tropical zones. Consequently, plants and animals at low latitudes tend to greater spinosity. Members of the dinoflagellate species *Ceratium macroceros* taken from both high- and low-latitude oceans are a good example; those taken in tropical waters have longer spines.

Tropical species are also more "plumose" than those in polar seas. This tendency to grow feathery appendages is an adaptation that produces high friction drag to offset the low viscosity of warm water. The warm-water copepod *Calocalanus* has plumose appendages, but *Acartia clausi*, a species common to cooler coastal waters does not.

Still another trick allows an animal to change form not by adding additional cells but simply by incorporating seawater within existing tissue. Mesoglea is a tissue that readily stores water. Bodies of some pelagic coelenterates such as the medusae (jellyfish) consist mostly of mesoglea. By increasing size without increasing body weight, the animals achieve greater drag resistance against motion through the water. This technique is common among the invertebrates that have no osmotic pressure problems because they are able to adjust body fluid to near isotonicity.

4. *Use of fat.* Fats and lipids incorporated in body tissue or structure are a common method of buoyancy adjustment. Radiolarians are found with droplets of fat attached to their spiny appendages, and copepods incorporate fat globules within their body tissues. Vertebrates, for example the sperm whale previously discussed, can also incorporate large quantities of fat. The basking shark stores so much fat that it can float at the surface. Codfish store fats in their livers.

5. *Use of gas.* The low density of gas makes it a very efficient material for acquiring buoyancy. Fish employ gas bladders into which gas is discharged out of a network of blood vessels. The process of charging or removing gases is relatively slow, so fish must guard against rapid changes in depth. (In fact, a fish may literally "fall up;" if for some reason it is carried or swims to a lesser depth, the gas within its bladder expands according to the gas laws. If the positive buoyancy gained is greater than the fish's ability to swim down, or to discharge excess gas, or both, the fish loses control of its fate and is pressure-thrusted upward at ever-increasing rates to a probable doom—it literally explodes inside.)

The cuttlefish (actually a species of squid) stores gas inside pores within a rigid skeletal bone—the cuttlebone. Some mollusks such as the chambered nautilus seal off one or more chambers of their spiral-shaped shell and store gas inside for buoyancy.

6. *Swimming.* If all else fails, swim for life! The variety of techniques used are many. The copepod, because of its small (2 mm) size and its resulting problems with viscosity, uses a snapping action of two large antenna appendages. These species, which participate in the diurnal migration associated

with the deep scattering layer phenomenon (see Figure 2.2), may swim vertical distances of the order of 100 m twice daily.

Members of the order Dinoflagellata employ two flagella whose whiplike action gives them slow motion through the water. One flagellum is attached to the bottom of the body and its motion offsets sinking. A second flagellum resides within a lateral groove that circumscribes the midsection (see Plate 15.2); it is believed to serve as part of a feeding mechanism. The Dinoflagellata are classed somewhere between plant and animal, for some species are carnivorous, but all carry chloroplasts for photosynthesis.

The bell-shaped mesoglea body of the medusa jellyfish has a contractile fiber ring around the bottom rim of its bell. By spasmodic contractions the animal succeeds in jetting water outward and thus gains a type of vertical motion. Balance organs around the bell's rim tell the animal which way is up.

Summary

The density of seawater is a fundamental physical property. Its value is a function of temperature, T, and salinity, S, and the function is very much non-linear. By constructing a grid with dimensions T and S, and superposing contours of constant density, oceanographers developed a useful analytical tool called the T–S nomogram. The data taken at a hydrostation, when plotted into this T–S space, reveal a *shape* to the graph that is unique to the water columns in this sector of the ocean. Oceanographers have sorted through hundreds of thousands of station data sets to classify all the world oceans into a discrete number of such sectors, each of which has its own characteristic T–S graph shape. The Pacific has eight major sectors, for example. Transition zones between sectors become important physical boundaries in that they separate different currents or mark zones of convergence of surface water masses. They become important biological boundaries in the sense of separating different marine communities.

The biological world is intimately connected to the physics of seawater density. This is especially true of the world of microorganisms; many of these have adapted to the density field and slow down their rate of sinking by having thin shells, by growing appendages and changing body shapes in other ways, by incorporating fats, and by other methods. But the problem of buoyancy control is not confined to the world of the tiny. One of the world's largest creatures, the sperm whale, has evolved a unique method of controlling its buoyancy in either a negative (sinking) or a positive (floating) direction, so that it *can* dive to deep depths and return to the surface with a minimum of muscular work.

Study Questions

1. Write short, one-sentence distinctions by which we classify plankton into *(a)* phytoplankton, *(b)* zooplankton, *(c)* holoplankton, and *(d)* meroplankton.

2. List four different buoyancy-adjustment techniques used by plankton to aid in flotation.

3. Describe two different ways by which plankton with shells slow down their sinking rate.

4. The shark does not have a gas bladder to achieve buoyancy as do many species of finfish. It must therefore stay *in motion* to avoid gradual sinking; its forward fins are used as hydroplanes and its tail fin is nonsymmetric. The shark has another technique for flotation that results in fresh shark meat having an odor of ammonia? What is it?

5. The sperm whale has evolved a very unique system for buoyancy control. Describe the mechanism?

6. Whales that migrate between polar and tropical waters, for example, gray and sperm whales, encounter changes in the density of surface waters. The sperm whale can adjust for this easily by altering the average temperature of the oil in its spermaceti organ. How do other whales adjust buoyancy?

7. The central theme of Chapter 9 is that oceanic regions differ in their water mass structures. Can you write a short, clear essay on precisely what is meant by the term "water mass structure?"

8. What explains the *curvature* of the density structure in the vicinity of point D on the T–S nomogram of Figure 9.2?

9. From the way the T–S curves of different water mass structures converge in Figure 9.4, we are tempted to say that all of them converge on the type called Antarctic Bottom Water. Do you agree with such a statement or would you argue

that all structures converge on the type called Antarctic Circumpolar Water?

10. I sometimes refer to the AACP water as the water involved in the "Great Communicator Current." What is the basis for describing the AACP in this way?

11. The "structure of" North Atlantic Central Water passes through the water type called North Atlantic Deep Water. But why does the graph of South Atlantic Central Water also appear to be influenced by this same water (Figure 9.4)?

12. How many distinct water mass structures can be assigned to the oceans around Antarctica? What is the name of the distinct ocean feature that separates the so-called Southern Ocean of Antarctica from the rest of the oceans? If you were to cross

this feature in a ship, what water characteristic would you sample to learn whether you had crossed into the Southern Ocean?

13. How many water mass structures (Figure 9.5) are used to classify the Pacific Ocean (do not include the Southern Ocean)? In a few words, describe the symmetry between North and South Pacific water mass structures.

14. Explain how the existence of an oxygen minimum layer can be used to infer the existence of a "boundary" between superimposed water masses.

15. Why does Mediterranean Deep Water sink to about 1500 m depth in the Atlantic before beginning its spread into the interior of the Atlantic? (Hint: follow its dashed curve in Figure 9.4.)

CHAPTER 10

FORCES THAT MOVE FLUID OCEANS AND ATMOSPHERE

This chapter begins with an explanation of the *forces* involved in moving air and water. The discussion is generalized so that the reader will not "run aground" on shoals of details *before* acquiring the basic tools of the meteorology and oceanology trades. The "forces" themselves are simple to explain. In the process, we prove to ourselves that indeed water does not run downhill but rather around the hill.

Types of Forces

Surface Forces

We are inherently familiar with the surface force—it goes by the name of friction, or surface stress. Defined briefly, a surface force is the result of one body exerting a "drag" on another body. When the two bodies are solid, for example, a hockey puck moving over an ice surface, our terrestrial experience gives us an inherent "feel" for the coefficient of friction drag between the two bodies. But when one or both of the bodies are fluids, the process of drag becomes complex. We observe that the fluid is actually deformed, often beyond recognition of its original form, for example, in the process of stirring cream into a cup of black coffee using a rigid spoon. In such instances it is more appropriate to think of "stress" as the *rate of deformation* of the fluid elements.

In fact, two fluids can interact; when air moves over water, each will exert drag on the other. We know quite well that winds drive the ocean surface water, and this must be via a transfer of momentum of wind into the water. This is a "stress" imparted to the ocean surface. Similarly, one ocean water mass moving alongside another mass, or over it or below it in the case of layered water masses, exerts a stress on the other.

This chapter focuses on the surface force of wind acting on water. It is this *surface stress* that generates the waves we see and feel. It also is responsible for setting the upper ocean layer into the motion patterns we call *surface circulation*. And it *mixes* the surface layer down to some depth, a process that is intricately involved with the photosynthetic behavior of plants.

Oregon State University TOTEM spar buoy, used in air-sea interaction research. (Courtesy S. Neshyba).

Body Forces

A fluid can also respond to a force interior to its mass. The classic example is the force of *gravity*, which acts uniformly on each particle of a fluid regardless of where that particle is, on the surface or interior. There are other body forces that also act on fluids. In the oceans, the additional forces are two, the *pressure force* and the *Coriolis force*.

Gravity Force. Gravity force is clear to us. If we attempt to float on the sea surface, the gravitational attraction on our bodies is a direct downward force—it is counteracted, we hope, by the buoyancy force of the water we displace. The same statement holds for a parcel of water in the interior of the ocean. If it is less dense than the surrounding fluid, the parcel is forced upward because its buoyancy exceeds gravity, and vice versa.

In some ocean motions the simple unbalance between gravity and buoyancy forces is the dominant driving force. Take the example of the iceberg melting in salt water (Figure 10.1a). Icebergs are derived from snow compressed into glacier ice; thus they are almost pure solid water.

Meltwater from an iceberg, at any depth, will have a density close to that of pure water at its freezing temperature, about 0.99800 g/cm^3, and so the meltwater will be buoyed upward by a body force and will rise toward the surface along the side of the iceberg. Whether it actually arrives at the surface depends largely on how rapidly the rising plume of fresh water mixes turbulently with the ambient seawater, but it is clear that the meltwater mixture is "pumped" upward.[1] It is possible that this "pump" will move nutrient-rich deeper water upward into the photic zone. Thus the iceberg can be a source of local "upwelling" (Neshyba, 1977).

The opposite of ice melt is ice freezing. Sea ice fresh-frozen on an ocean of 35 ppt retains only about

---

[1]Some years ago it was popular to debate the pros and cons of towing icebergs from Antarctica to water-thirsty regions like South Australia and San Diego. Some reasoned that an iceberg at rest inside a saltwater harbor would melt mostly from the top down, from solar heating, and that the meltwater would collect on the surface of the seawater without mixing with it. Users would simply pump away the fresh water as fast as it collected. Actually, icebergs melt much faster along their sidewalls than from solar heating of the relatively small portion exposed above the sea surface. Stated another way, heat for melting is much more readily available from surrounding seawater than from the sun directly. Buoyant meltwater making its way up the side of the iceberg becomes so thoroughly mixed with seawater that it loses all value as "potable" water.

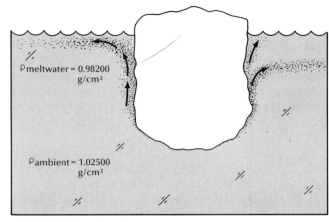

(a) What happens to the meltwater from a melting iceberg? The answer is that meltwater drifts upward toward the surface because it is less dense than the surrounding seawater at the same depth. As it moves upward, it entrains some of the ambient seawater to form a mixture that becomes saltier as the entrainment progresses. Whether the mixture reaches the surface level itself, as shown on the left side of this berg, or comes to equilibrium with the ambient seawater at an intermediate level, as shown on the right, depends on (1) how strong is the density stratification of the ocean in which the berg is melting, and (2) the degree of turbulence in the waters around the berg. The greater the turbulence, the more rapidly meltwater is mixed with seawater, and the more likely it is to move away from the berg along an intermediate depth level rather than rise all the way to the surface. In any case, the melting berg is a *magnificent pump*, raising nutrient-rich water closer to the surface. Over a million million tons of icebergs are discharged from Antarctica each year.

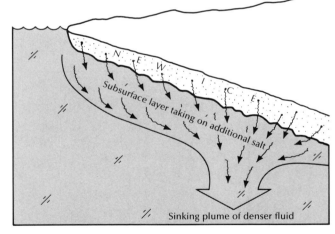

(b) What happens to the salt that is excluded when seawater freezes?

The answer is that the salt is initially concentrated into pockets of brine, with salinity so high, (more than 100–200 ppt), that it does not freeze. Rather, the brine pockets "melt" their way downward through the new ice and enter into the fluid water below. Eventually, subsurface water becomes so dense that it sinks, carrying salt down into deep ocean layers.

Figure 10.1 Vertical mixing in polar oceans, driven by buoyancy forces.

5 ppt of salts within its crystallized structure. The remainder, 30 ppt, is excluded as a concentrated brine. In Figure 10.1b this excluded brine is shown to mix downward into the fluid under the ice. When that fluid has absorbed a critical load of additional salts, it can become dense enough to force its way downward to much deeper levels. This is how Antarctic Bottom Water is formed.

Pressure Force. Figure 10.2a shows a cross section of an ocean inside of which is drawn a "level" plane. By definition, a level plane is everywhere perpendicular to the local gravity force; a marble placed upon a solid and level plane will not roll.

There are, of course, no solid surfaces in the interior of oceans. If a "parcel" of seawater is substituted in place of the marble, it may tend to rise, as

does iceberg meltwater, or sink, as does Med Water entering the North Atlantic. The pressure force is vertical in direction, and vanishes only when the parcel is surrounded by ambient fluid of the same density as that of the parcel itself. This is called "hydrostatic" balance. It implies a balance of forces in the vertical direction, and we assume that the water parcel in Figure 10.2a is in this equilibrium hydrostatic condition.

The parcel, however, may independently experience a net *horizontal* force when subjected to unequal pressures acting *along* the level plane. This force is brought about by one or both of the following conditions.

1. *A tilt in the sea surface (Figure 10.2b).* The height of water columns above the level plane

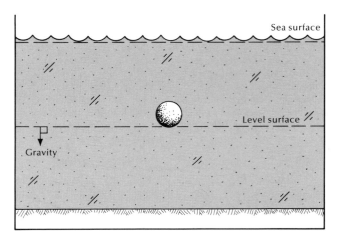

(a) A marble or a parcel of water placed on a *level* surface will not move except in response to *horizontal* pressure forces.

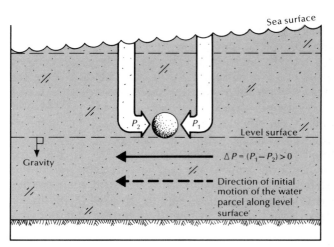

(b) Under a sea surface that is itself *tilted*, pressure on a level surface varies simply because the height of the water column overlying the level surface varies from place to place. Column 1 is taller than column 2, so P_1 is greater than P_2. The difference is a *net* pressure force that will accelerate the parcel from right to left.

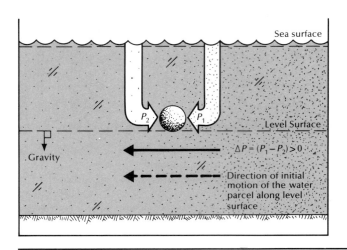

(c) Another way that a pressure difference can exist is for column 1 to contain water that is *denser* than that in column 2. Then P_1 will again be greater than P_2, with results just like those described in part (b).

Bear in mind that the accelerations described in parts (b) and (c) apply only to the *initial* direction in which the parcel will be moved. As soon as the parcel actually moves, it must also respond to the other body force, the Coriolis. See Figure 10.9 for an explanation and more diagrams.

Figure 10.2 Origin of pressure forces in the interior of water masses.

changes from point to point when the sea surface tilts. The pressure P_1 exerted on one side of the parcel is not balanced by pressure P_2 on the opposite side, so a "net" horizontal pressure force ΔP begins to accelerate the fluid parcel along the level plane. A tilt in sea surface can cause currents to form in the interior of an ocean.

2. *A non-uniform distribution of seawater density (Figure 10.2c).* Even with a level sea surface there can be differences in pressure along an interior level plane because the average densities of adjacent water columns differ. The pressure developed under a column of dense fluid, P_1, exceeds that under the same-size column of less dense fluid, P_2. Again, the water parcel experiences a *net* pressure difference, ΔP, and will move accordingly—along the horizontal.

Coriolis Force. The motions of fluids over our spherical and *spinning* earth are influenced by still another body force that we call the "turning effect due to earth rotation." The French mathematician Gaspard Coriolis (1792–1843) was the first to define the effect in rigorous terms and so it carries his name. We all experience the effect but may not be aware of its presence because we are so strongly coupled to solid earth by friction, between our feet and the ground, or between the automobile tires and highways. But in the world of fluids that have low viscosity, as do fluid oceans and the atmosphere, the effect must be taken into account to describe the *path taken by moving particles* over the spherical, spinning earth. The Coriolis effect is a body force.

Figure 10.3 explains the principles that govern the Coriolis effect, but the cannonball in the example is solid, not fluid, and it is small, not vast as are the oceans. A more appropriate example is needed. Consider a large rectangular piece of ocean; in thickness let it be 2 m, and in size the same dimensions as a football field. To simulate the physical property of low viscosity for water, think of the football field as being covered entirely by moving cheerleaders, but not one of them touching another, therefore having no friction between them. Each cheerleader, then, represents an individual parcel of water. Begin with all cheerleaders facing the same direction. For the initial movement, have all step forward one step. Then considering that this motion would be "deflected" by the Coriolis, have all step sideways one step to the right. Again considering that the sideways motion was nevertheless true motion, and must be similarly deflected to the right, have all step backward one step; then, to account for the deflec-

tion on this last increment of motion, have all take one step to the left. Notice that each successive step was at right angles to the "motion" that required deflection. Notice further that each "parcel" has returned to its starting point, and that the *trajectory* of each parcel is a squared "circle." Last, the facing direction of each person did not itself change, that is, no cheerleader turned although each traversed a circular path. Now, to make the transition to the ocean case, you should go up in the blimp high above the stadium. The impression you would have is that of an ocean in motion, the entire fluid body executing a translatory motion as if continuously undergoing deflection, but not rotation.

From this we derive a series of statements describing how motion is affected by the Coriolis force.

1. The deflection of moving objects is to the *right* in the Northern Hemisphere and to the *left* in the Southern Hemisphere.

2. The rate of deflection is greatest at the poles and zero at the equator.

3. Although the turning effect is described as a result of a force acting on a moving object, it is only an "apparent" force since it does no work on the object. The force always points at *right angles* to the direction in which the object is moving; with no displacement in the direction of force, no "work" is done.

4. The deflection is greater and the circular trajectory smaller in diameter as the speed of the moving object decreases. For example, both the moving atmosphere and the oceans are influenced by the turning effect, but the oceans move more slowly so the resultant deflection of ocean currents is more readily apparent than for winds.

For an additional explanation of the turning effect, work through the discussion of the Coriolis force in Figure 10.3. For now, it suffices that you accept its reality. James Michener, in his epic narrative *Centennial* (1974), described the trail drive of a herd of cattle from Texas north to the Midwest. He pointed out that at the head of the column of moving cattle, flank riders were posted on either side for the purpose of maintaining the drive in the desired direction. But if the cattle stampeded, one of the two flank riders was in much more imminent danger of being trampled. Which one? The answer is the right flank rider. Why? We cannot say for sure, but Michener's account of the history of cattle drives was correct in that stampeding cattle columns do tend to turn to the right, at least in our hemisphere.

CASE IN POINT

A SHORT COURSE ON CORIOLIS ACCELERATION

An ice skater begins a "spin movement" with arms and legs extended horizontally as far as possible (see illustration a). Then the leg muscles are used to twist the body into an initial spin, and the skater immediately rises onto the tip of one skate to reduce friction against the ice. Finally, arms and legs are pulled inward as tightly as possible (see illustration b). The result is a *spin-up*, ever faster as arms and legs contract. To stop rotation, the skater extends arms and legs back to the horizontal.

The physical principle involved is the "law of conservation of angular momentum." The equation is

$$\frac{\text{angular}}{\text{momentum}} = \frac{\text{mass of}}{\text{skater}} \times \frac{\text{spin rate}}{\text{rad/s}} \times \frac{\text{distance of mass}}{\text{from spin axis}}$$

Excluding the effect of friction, the angular momentum must stay *constant*: the skater's mass is constant, but because the distance of arms and legs decreased during the pull-in, the spin rate had to increase to satisfy the equation.

Now extend the concept to objects moving along the earth's surface. Consider a cannon on the equator firing a cannonball toward England along the Greenwich meridian (see illustration c). Assuming that the cannonball lands at the latitude of Greenwich (52°N) will it land to the east of town, in town, or west of town?

Answer. When the ball touches earth at 52°N, it will in fact be closer to the spin axis of the earth than it was at the time of launch. Its spin rate must have increased as it moved closer to the spin axis of the earth; therefore the ball must land to the *east* of Greenwich, as shown. Solving another way, we find that at the time of firing the ball had an eastward velocity corresponding to its position at the equator, about 1000 mph; it does not lose its eastward motion during its flight northward. Greenwich has an eastward velocity of only 700 mph. Therefore the ball traveled farther eastward than did Greenwich itself during the time of flight. For this reason the cannonball had to fall to the east of the city.

From these examples we make a generalization. *Any* object in motion over the Northern Hemisphere of spinning, spherical earth will appear to be deflected *to the right* of its initial motion, in the Southern Hemisphere *to the left*. An acceleration caused by the earth's rotation, the Coriolis force, is responsible for the deflection. The force is always directed at *right angles* to the motion of the object being accelerated; it is a "turning" force. The *strength* of the force is given by the equation

$$\frac{\text{acceleration}}{\text{(turning force)}} = 2 \times \frac{\text{spin rate}}{\text{of earth}} \times \frac{\text{sine of the}}{\text{latitude}} \times \frac{\text{velocity of}}{\text{the object}}$$

Note that the turning force is stronger as the object moves faster; also, it increases in strength from *zero* at the equator (sin latitude = 0) to a maximum at either pole.

We can compute the actual turning force from the original path of motion as

$$\text{deflection} = \times \frac{\text{spin rate}}{\text{of earth}} \times \frac{\text{sine of the}}{\text{latitude}} \times \frac{\text{object's}}{\text{velocity}} \times \text{(time of flight)}^2$$

Clearly, then, an object moving slowly between two fixed points on earth will be deflected *more* than will an object moving rapidly between the same points, simply because the *time of traverse* enters into the equation *twice*. Some examples will help. With the spin rate of earth at 7.3×10^{-5} rad/s and at a latitude of 45° where the sine of latitude is 0.707, then

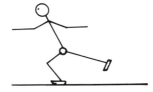

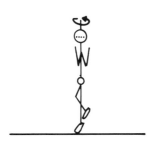

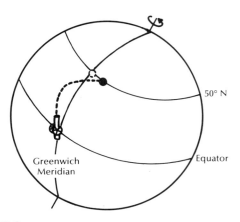

Figure 10.3

50° N

Equator

Greenwich
Meridian

Example 1. A rifle bullet fired at a velocity of 300 ft/s takes 0.3 s to travel 900 ft; its deflection is about 0.17 in., a minor amount.

Example 2. An automobile moving at 60 mph (88 ft/s) takes 60 s to travel 1 mi; its deflection, ignoring the friction of its tires with the pavement, is about 16 ft.

Example 3. An ocean current moving at 1 knot (1 nmi/h) requires 1 h to move 1 mi. Its deflection would be about 0.186 mi, or about 20% of the distance it traveled in its original direction.

Conclusion. The Coriolis force is always with us, and ocean currents are strongly deflected.

Steady-State Motion

The reader has now acquired a complete set of physical concepts by which to explain *how* the fluid oceans and atmosphere move in response to forces. Before proceeding to the final picture, however, it is important to point out that what we actually observe in these moving fluids is the so-called "steady state." This means that we must accept as fact the idea that the patterns of ocean and atmosphere motion are the same today as a year ago, or ten, or a thousand years ago. This does *not* mean that there is no variation in currents or winds today versus yesterday or tomorrow. Rather, it is a statement that there are no observable long-term mean trends—the Gulf Stream will *not* double in velocity during the next ten years, for example. We shall continue to see variations in atmosphere motion, such as periodic cyclones in the Caribbean Sea, and in ocean motion, such as the spin-off of large eddies from the Gulf Stream, but we view these variations as a *steady state* in the continuum of complex motions of these fluids—we can "see" them any time.

Wind Patterns over Oceans

With comprehension of the Coriolis force comes the ability to explain why the winds over the earth are organized into a pattern of latitudinal belts. As a gaseous fluid, the atmosphere must also respond to the turning effect as well as internal pressure and gravity forces.

Following the Route of an Air Parcel

Figure 10.4 shows the route taken by an air parcel as it responds to these physical processes. This tour begins near the equator in the zone of easterly winds with an air parcel just evaporated from the sea surface. As the parcel rises in the air column, by convection, it also moves westward; this is the cause of the equatorial easterlies. As it begins westward movement, it is deflected northward, by the Coriolis force, even while steadily rising because of its high vapor content. High in the atmosphere and by now traveling more north than west, it rains out excess moisture and cools by expansion and by radiating

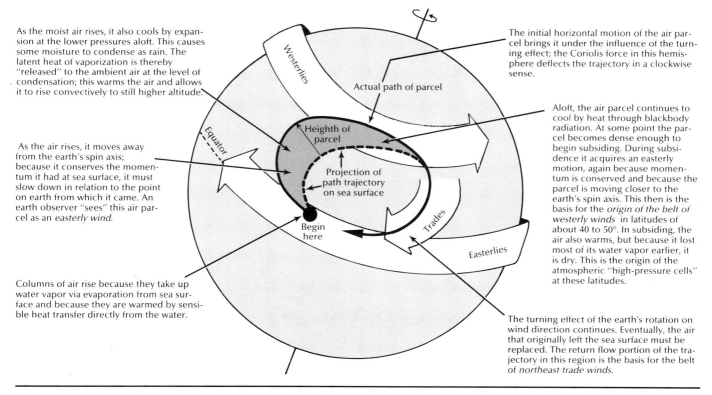

As the moist air rises, it also cools by expansion at the lower pressures aloft. This causes some moisture to condense as rain. The latent heat of vaporization is thereby "released" to the ambient air at the level of condensation; this warms the air and allows it to rise convectively to still higher altitude.

As the air rises, it moves away from the earth's spin axis; because it conserves the momentum it had at sea surface, it must slow down in relation to the point on earth from which it came. An earth observer "sees" this air parcel as an *easterly wind*.

Columns of air rise because they take up water vapor via evaporation from sea surface and because they are warmed by sensible heat transfer directly from the water.

The initial horizontal motion of the air parcel brings it under the influence of the turning effect; the Coriolis force in this hemisphere deflects the trajectory in a clockwise sense.

Aloft, the air parcel continues to cool by heat through blackbody radiation. At some point the parcel becomes dense enough to begin subsiding. During subsidence it acquires an easterly motion, again because momentum is conserved and because the parcel is moving closer to the earth's spin axis. This then is the basis for the *origin of the belt of westerly winds* in latitudes of about 40 to 50°. In subsiding, the air also warms, but because it lost most of its water vapor earlier, it is dry. This is the origin of the atmospheric "high-pressure cells" at these latitudes.

The turning effect of the earth's rotation on wind direction continues. Eventually, the air that originally left the sea surface must be replaced. The return flow portion of the trajectory in this region is the basis for the belt of *northeast trade winds*.

Figure 10.4 The trajectory of an air parcel, the forces that propel it, and the resultant pattern of wind belts around the earth. (As you read, keep in mind that the meteorologist refers to winds in terms of the direction *from* which they blow.)

heat waves away from earth. By the time the parcel reaches latitudes of 30°N, it has lost moisture and heat and shrunk in volume to the point that it begins to subside, toward the sea surface. Subsiding forces the parcel to take an eastward direction, hence the beginnings of the mid-latitude westerlies. Its horizontal motion continues to force right-hand deflection, so that the ground track of the parcel's trajectory is somewhat circular by this time. As the parcel arrives near sea level, it is traveling rapidly and is deflected back toward the southwest, forming the set of winds called the trades. In the Northern Hemisphere they are called the northeast trades;

vice versa south of the equator. The tour finishes with the parcel back at the surface, near the equator. The end result of all these processes is the belted pattern of winds over the ocean, much like that shown in Figure 3.1 for the physicist's view of earth.

High- and Low-Pressure Cells and Winds

The presence of large land masses lying across the various wind belts modifies the actual picture. The belts tend to interconnect along the margins of major continents. The net result is that each major ocean basin has a characteristic *cell* of high atmospheric

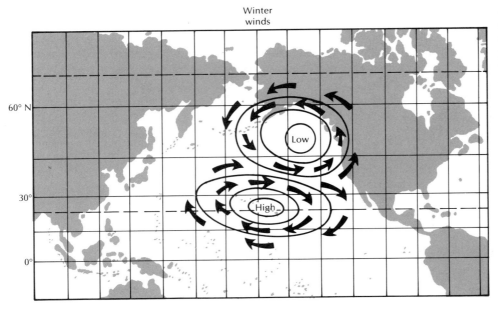

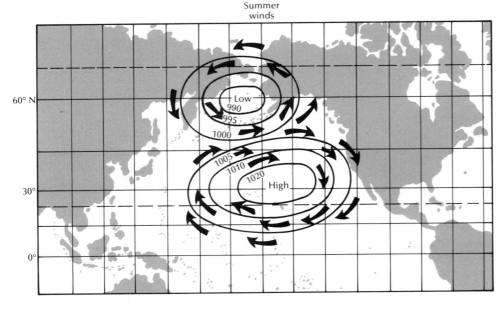

Figure 10.5 Seasonal shifts in the average locations of the North Pacific High and Low atmospheric pressure cells. Each closed contour is a curve of constant atmospheric pressure at sea level: pressure at the center of the high is over 1020 millibars; that in the low is less than 990 mbar in summer. The summer winds along the coast of the Pacific Northwest are favorable to upwelling (see also Figure 10.8).

pressure over its central region. Figure 10.5 shows this *high* over the North Pacific; its center is located to the northeast of Hawaii. During the summer a *low*, a cell of low atmospheric pressure, hangs over the Bering Sea. Between these two pressure cells is the belt of westerlies; arrows show this flow. To the north of the Bering Low are the counterflowing easterlies, and south of the Hawaiian High the trades of the north converge with the trades of the south to form the equatorial easterlies.

In Figure 10.5 the surface streamlines of winds *circle counterclockwise* around the Bering Low. The cause of this lies in the effect of the Coriolis force on wind direction as surface winds *initially* attempt to move air directly from the center of the high-pressure cells to the center of the low-pressure cells. In other words, air "tends" to move down the pressure gradient (see also Figure 6.9). Again, as soon as air parcels begin horizontal movement, their trajectories are deflected to the right; the final "steady-state" balance of forces requires that the *winds circle around the centers of both the high and the low cells.*

One additional point: Does the excess air under a high-pressure cell *ever* find its way into the low and thereby equalize the sea level pressures of these two cells? The answer is yes. In the real world, the streamlines of flow slowly spiral outward from the centers of the cells as friction takes its inevitable toll. A corollary question: If spiraling wind patterns around high-and low-pressure cells eventually "relax" the pressure differences between highs and lows, why are the cells quasi-permanent features over ocean basins? The answer must be that the heat engine of earth is continuously "reinforcing" both the high- and low-pressure cells, and that surface winds over the ocean are a continuous effort by nature to bring about a relaxation of the pressure difference. It means that *observed* patterns of winds are a steady-state phenomenon.

A final question: Does it then follow that there is little variability in the day-to-day and season-to-season wind patterns? Not at all. The steady-state picture just described is an "average" result. For example, we observe that the atmospheric high off Hawaii migrates toward the northeast in the spring and retreats toward the southwest in the fall; at the same time, the Bering Low migrates toward the Gulf of Alaska, that is, to the southeast, during the fall season and returns to its Bering Sea niche in spring. This two-cell "system" oscillates on a seasonal cycle, but the average is still a steady-state pattern. The impact on the ocean of such seasonal changes is great. In the following sections we shall see how the

Fridtjof Nansen's ship *Fram,* departing from Oslo in June of 1893. (Courtesy Fram Museum, Oslo.)

The 355-ft research platform FLIP floating vertically has over 300 ft of its tubular structure below water level, making it extremely stable against roll and heave motions. Developed by Scripps Institute of Oceanography primarily for research in underwater sound, it has no propulsion engines and must be tended by mother ships. Its stability makes it ideal for measuring the rate of evaporation and heat flux from the sea surface. (Courtesy Scripps Institute of Oceanography).

CASE IN POINT

DEVELOPMENT OF THE SOLUTION TO THE "WIND DRIFT" PROBLEM

1. 1893: Fridtjof Nansen set his ship *Fram* into the Arctic Ocean near the New Siberian Islands. The ship became frozen into the pack ice and thereafter drifted *with* the pack ice. Repeated observations of the drift direction of ship and pack ice relative to the *downwind* direction of prevailing winds indicated that the ship and ice usually drifted off at an angle of about 20° to the right of the wind (see illustration a).

2. 1896: Nansen returned to Norway after three years and resumed university duties. He assigned the problem of deriving an explanation for the behavior of his ship and ice drift relative to wind direction to a graduate student, Vagn W. Ekman.

3. 1905: Ekman published a theory of wind drift currents. For the Northern Hemisphere, the theory states that a surface wind will cause the topmost, thin layer of surface water to move in a direction 45° to the *right* of the wind. Successively deeper layers will, in turn, be moved in directions to the right of the water layer above, whence comes the frictional driving force. The speed of each layer is less with depth, however. Eventually, the water motion is directed *opposite* to that of the surface wind. (see illustration b). Oceanogaphers call this layer the *depth of frictional resistance* of ocean surface water to the driving force of the surface wind.

For our studies, the most important aspect of the theory is that, when all motion above the friction depth is summed, the *net transport* of surface layer water is directed 90° to the right of downwind direction in the Northern Hemisphere, to the left in the Southern Hemisphere.

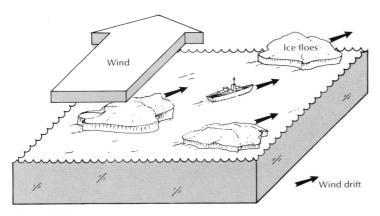

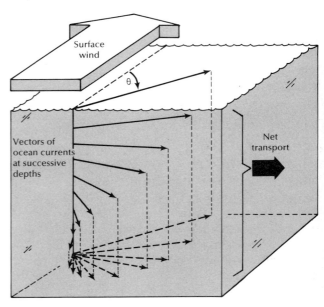

Figure 10.6

---

seasonal upwelling phenomenon off the coasts of Washington, Oregon, and northern California is directly related to the oscillation in the atmospheric cell structure. To do so we need to describe and understand one final facet of the air–sea interaction, the Ekman effect.

Air–Sea Interaction III— Response of Surface Water to Wind Stress

In 1893 Fridtjof Nansen began an epic three-year drift across the Arctic Ocean in the now legendary ship *Fram*. Nansen was a keen observer of natural phenomena, and after months of drifting with the ice pack, he concluded that the surface ice pack seemed to drift at an angle of about 20° to the right

of the prevailing wind (Figure 10.6a). He hypothesized that the cause was the turning effect of the earth's rotation, and on completion of his voyage he assigned the problem of explaining the process to a graduate student named Vagen W. Ekman. The event was destined to make oceanographic history.

After making several simplifying assumptions, Ekman derived a theoretical solution that showed the following. If a steady wind exerts friction drag over an ocean of uniform density (Figure 10.6b), then:

1. A thin layer of surface water moves in a direction 45° to the right of the wind in the (Northern Hemisphere) or to the left of the wind in the (Southern Hemisphere). The cause is the turning effect of the earth's rotation, that is the Coriolis force.

2. Successively deeper layers of water react in the same way to the above flow of the shallower layers, that is, a sublayer moves off to the right of the motion of the layer immediately above.

3. Therefore, there is a depth at which the water moves in a direction *opposite* to that of the surface wind.

4. The summation of transport of water from all the layers affected is in the Northern Hemisphere a net transport directed 90° to the right of the surface wind.

The theory of Ekman was an "oceanmark" event. For the first time oceanographers could explain many observed phenomena in ocean motion. Among these are upwelling along coastal boundaries as well as in oceanic zones of wind divergence, and the fact that sea level stands higher in the central portions of major ocean basins, forming hills of seawater. Let us examine some of these in detail.

Hills of Surface Water in the Central Oceans

In the North Atlantic basin, the surface of the sea stands about 1 m higher in the center than along the periphery; further, the surface layer water in the central portion is extraordinarily thick, that is, hundreds of meters, and of quite uniform temperature. Oceanographers refer to surface water here as "18° water"; navigators have for centuries called this the Sargasso Sea. In the Wüst diagram (Figure 9.5) it is the lens of central troposphere water. How do we account for this heap of uniform surface water? The answer is shown in Figure 10.7: the surface

wind pattern under the North Atlantic high-pressure cell causes the surface water to form a standing hill. Eventually, the hill becomes high enough so that the *rate* of wind-forced piling of central surface water is just *balanced* with the *rate* of runoff from the hill. This, then, is the *steady-state equilibrium* for the wind–water system. What is important to realize is that the hill remains as a *permanent* feature; in a later section we learn that currents flow around the hill.

Upwelling

In a sense, upwelling is the other side of the "pileup of hills" coin. If the wind system is such that the resultant Ekman transport is divergent, a topographic "valley" is created. In the ocean the divergence of surface water out of a given area is offset by an inflow to replace the water that has been moved away. We illustrate with some examples.

1. In the Northern Hemisphere the ocean surface beneath a cyclone, hurricane, or typhoon center is exposed to counterclockwise winds. Using the Ekman idea, it follows that *surface waters are transported radially away* from the storm center. This is a clear divergence of surface waters. The ocean will replace the lost surface water in the easiest way it can, and during the most intense phases of the storms the replacement water comes upward to the surface from deeper levels—hence *upwelling* by the cyclones.

2. If a surface wind blows parallel to a coast, the situation is similar to that of the cyclone, except that the coast effectively eliminates one-half of the cyclonic area. Upwelling will still occur.

Let us look at some examples of winds along a

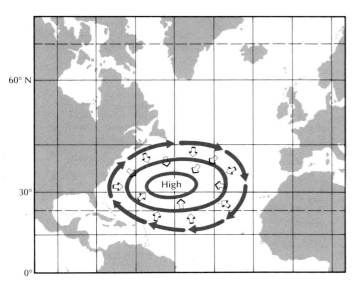

Figure 10.7 How hills of water are created in the Central ocean basins.

Each major ocean basin carries a centralized atmospheric high-pressure cell with anticyclonic winds. (Cyclonic winds are associated with low-pressure cells, as in cyclones, typhoons or hurricanes.) In the Northern Hemisphere anticyclonic winds are clockwise winds, as indicated here for the North Atlantic. Because of the Ekman effect, the net transport of surface water in response to anticyclonic winds is toward the center of the cell; stated another way, surface waters are caused to *converge* beneath an atmospheric high. This creates hills of water. In the North Atlantic the region of elevated sea surface is called the Sargasso Sea. The double-line arrows indicate the convergence of surface water being transported 90° to the right of the prevailing wind vectors.

Launching a surface toroid buoy carrying instruments to measure the transfer of momentum from storm winds to the ocean surface layer, a part of the STREX exercise in the Gulf of Alaska, 1980. The vessel is the USGC *Oceanographer*. (Courtesy Jay Simpkins, Oregon State University.)

coast. Figure 10.8 shows a cross section through a coastal ocean and shoreline. Assume the shore is a part of the Oregon coast and that the wind is northerly (Figure 10.8*a*). Because of the Ekman effect, surface water is forced to leave the shoreline toward the open sea. The replacement fluid comes from a deeper level, one that does not feel the wind stress, to occupy the volume emptied by the forced removal of surface water. (See also the analysis of NH-105 in Figure 8.7*b*.)

A reversal of the prevailing winds off Oregon eliminates upwelling and substitutes downwelling along the coast (Figure 10.8*b*). Historically, in mid-September the wind regime off the Oregon coast changes from *northwesterly*, (favorable to upwelling,) to *southwesterly* (favorable to downwelling). Upwelling conditions return during the last week of March, like the swallows return to Capistrano on the winter equinox.

How Forces Are Balanced in Steady Motion

Throughout the text we have separated the initial *tendency* of motion by ocean and atmosphere from the *steady-state* observed patterns of flow. The reason for this approach was to establish that the initial response of fluids to a forcing is different from the final equilibrium condition. What is actually observed in the way of motion of air and water at any instant is the "steady-state" condition. The mechanics of the earth have long since gone through evolving states and now are in a "constant" mode.

But it is very important to distinguish clearly between "uniformity in motion" and "constancy in the mechanics of motion." Uniformity implies *no change*, and this is not what we observe—we find all manner of variations in hour-to-hour, or day-to-day, or season-to-season, and even year-to-year observations of ocean motion. Still, there must be a way of describing accurately the true "steady-state" condition. Is it that the steady state is one of continuous change? If so, is it necessary to specify the time and space scales over which the changes occur? The answer is both yes and no.

Steady Motion

After decades of collecting data on the motion of ocean surface water, oceanographers have concluded that about 75% of the observed flow is explained by a balance of forces in which the measured "hills" and "valleys" in the level of the sea provide the *pressure force* and the observed velocity of current provides the magnitude and direction of the *Coriolis force*, and *the two forces are in balance*. This statement is illustrated in Figure 10.9.

The sequence of diagrams in Figure 10.9 progresses through the development of ocean currents. Consider that a previously level, quiet surface of the sea is disturbed, most likely by wind forces, and that the disturbed topography of the sea surface is a hill of excess water (Figure 10.9*a*). Regardless of how the disturbance occurred, the ocean, because it is a fluid of low viscosity, attempts immediately to restore itself to its original level condition. Parcels of water *begin to move downhill* in direct response to the downhill pressure. But immediately on initial flow the parcel begins to be deflected from its direct downhill path by the Coriolis force. Recall from Figure 10.3 that the *strength of the turning force*, which is zero when there is no motion, now increases as the water picks up speed downhill. So too does the deflection

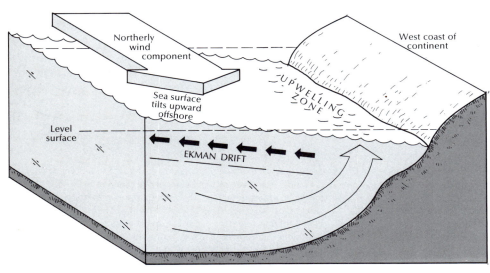

(a) The phenomenon called the Ekman transport (or Ekman drift) also explains why upwelling occurs along some coastlines and not along others. Here the Ekman transport, driven by a surface wind that has a northerly component, is in the offshore direction. These are the necessary conditions for upwelling. Notice the upward tilt of sea level in the offshore direction.

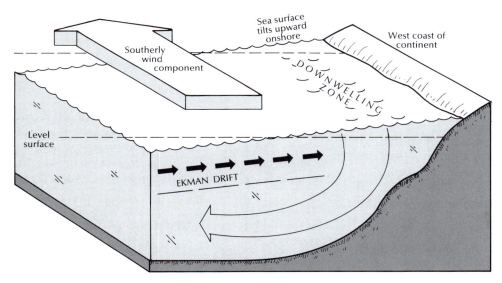

(b) The same coastline sees a wind with a southerly component. The Ekman transport is onshore, moving warm water against the coastline and creating a tilt in the sea surface near the shore. Strong winds will actually force enough water against the shore that the warm water sinks somewhat.

Figure 10.8 Upwelling and downwelling Ekman effects produced by surface winds along coastlines.

increase in strength as the speed increases (Figure 10.9*b*). But the two forces, pressure and Coriolis, do not balance each other perfectly until (1) the direction of the current is parallel to the hill, so that the Coriolis force *points* uphill directly opposite to the pressure force pointing downhill, and (2) the speed of flow has increased until the *strength* of the Coriolis force just matches that of the opposed pressure force. At this point the system arrives at equilibrium: the hill remains in place (for the most part), and the ocean current goes around the hill (Figure 10.9*c*).[2]

[2]To learn that ocean fluid does *not* "run downhill" can be a traumatic experience to the dedicated landlubber.

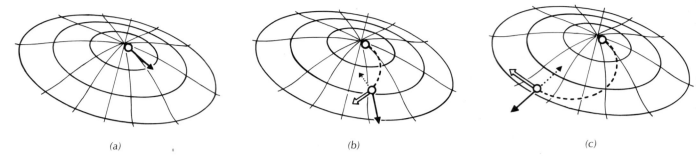

(a)   (b)   (c)

This series of sketches demonstrates (1) how transient currents eventually develop to a steady-state condition, and (2) the ocean currents flow *around* hills and valleys on the sea surface.

Sketch *a* depicts a hill of water. A parcel of water is shown near the top of the hill, just at the point at which its downhill motion begins. The solid arrows denote the direction of the *pressure force* that initially accelerates it.

Sketch *b* shows the position of the parcel after a short time. It has begun downhill motion, but because of the Coriolis effect, its trajectory is deflected to the right, as indicated by the dashed curve. The double arrows now represent its velocity vector, and the dotted arrow the magnitude of the Coriolis force. The Coriolis force does not yet *balance* the downhill pressure force, and the parcel continues spiraling outward.

In sketch c the parcel has been accelerated to a point at which its Coriolis force vector just balances the pressure force, both in size and in opposing direction. At this point the parcel has reached *steady-state* balance *and* is in motion *around the hill*. In a completely friction-free ocean, these currents would continue revolving around the hill forever. Clearly, however, some small friction is present, and the currents slowly spiral away from the center of the hill, eventually causing the hill to disappear.

But if the atmospheric high-pressure cell that created the hill in the first place continues to "build" the hill anew, we conclude that the final steady state is one of balance between the rate at which winds cause waters to converge upon the hill and the rate at which the spiraling currents remove water from the hill.

Figure 10.9 How transient and steady-state currents are created.

Unsteady Motion—The General Circulation Model

But what of the remaining 25% of ocean motion that is *not* explained by this balance? The answer is that there is still a balance of forces, but additional forces must now be included in the force diagram. Such forces are (1) the wind stress surface force; (2) the inertial force related to the momentum of fluid mass in motion, (a body force); (3) the internal friction drag related to the viscosity of the fluid itself, (also a body force); (4) the friction drag of the ocean floor, (a surface force analogous to wind stress); and (5) the mass attraction forces that cause tides.

Physical scientists are continuously devising new models to explain observed ocean motion. With so many possible forces to include in the *instantaneous* balance, such models are often programed through a wide, dynamic range of magnitudes of force variables in an effort to duplicate mathematically what is actually observed. They are called General Circulation Models (GCMs), and each requires a large, dedicated computer to process the mathematical model through countless steps, each step a tiny increment of change in one or more of the forces that move the oceans. The atmosphere is a part of the model. At present (1986) oceanographers are racing to see who designs the "right" model first. How will we *know* the right model? The answer is that *when the model can predict the unsteady motion,* it is right. The following section discusses two such "test cases."

The Pacific Equatorial Undercurrent (Cromwell Current)

In 1952, the young fishery scientist Townsend Cromwell accompanied an exploratory Japanese fishing expedition in the equatorial Pacific. He observed that when the longline gear was set at depths of about 100 m, the entire fishing gear moved rapidly toward the east. After taking accurate measurements of the cross-equator distribution of temperature and salinity, Cromwell described for the first time a major ocean current previously unknown. It was initially named after him but is now referred to as the Equatorial Undercurrent.

The current is very unusual (Figure 10.10). It is found along most of the length of the equator across the Pacific Ocean. It exists as a very thin layer of eastward-moving water in the zone between 100 and

THE PACIFIC EQUATORIAL UNDERCURRENT

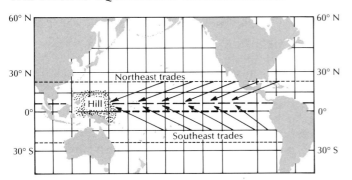

(a) The zone of convergence of the Northeast Pacific and Southeast Pacific trade winds (the "meteorological" equator) is displaced north of the geographical equator by about 5° of latitude. Together, the trade winds cause a pileup of surface waters against the Asian continent, to a height of about 1m in relation to sea level on the American side of the Pacific. The return flow of these waters is directly along the equator, but below the surface. Shown here as a heavy dashed line, this return flow is called the *Equatorial Undercurrent.*

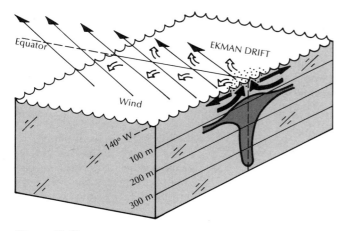

Figure 10.10

(b) The Equatorial Undercurrent is shown here in cross section along the 140°W meridian. It extends horizontally some 200 km to the north and to the south of the equator, but it is confined to a thin depth zone between 100 and 300 m. More aptly called a "sheet" current, it reaches speeds of up to 2 knots and transports about as much volume as the Gulf Stream off Florida.

The symmetry of the Undercurrent in cross section, plus the location of its core directly along the equator, suggests that it is a "trapped" current. An eastward flow is maintained at the equator simply because if it attempts to leave, the Coriolis effect, both to the north and to the south, deflects the excursion back toward the equator.

(c) Because the southeast trades in the Pacific *straddle* the equator, as shown in illustrations a and c and because the direction of Coriolis deflection changes from right turning on the north side of the equator to left turning on the south side, the net result is that equatorial surface waters are forced to *diverge*, sucking deeper water to the surface and causing *upwelling.*

300 m, and its breadth extends some 200 km both north and south of the geographical equator. It transports roughly the same volume of water as does the Gulf Stream, at speeds ranging up to about 2 knots.

Cause and Control. Once discovered and mapped, an explanation of the cause of the Equatorial Undercurrent was rapidly formulated. Recall from our earlier discussion of the Intertropical Convergence Zone (Figure 7.4) that the Northern and Southern Hemi-

sphere trade wind systems converged not at the equator but at 5 to 10° *north* latitude. The winds that *straddle* the geographical equator are therefore the southeast trade winds of the South Pacific. A direct consequence is the "piling up" of equatorial surface waters against the Asian continent; thus a hill of water is created on the western margin of the equatorial Pacific. This is possible because the Coriolis force is *zero* in magnitude at the precise geographical equator; the surface water is therefore driven directly downwind and gathers against the Asian shore to a height of about 1 m above that of sea level on the American side.

Eventually, water from the hill begins to slide away, in this case toward the east as a thin lens of water between the depths of 100 to 300 m. It is this return flow that moved the fishing gear for Cromwell. It is a flow directly downhill. Because the Coriolis force vanishes at the equator, the Equatorial Undercurrent responds to only one driving force—the downhill pressure force. It is the only major current in the world ocean to do so.

But what keeps this current trapped along the equator over the thousands of kilometers it traverses from the Asian to the American shores? The answer is that as the current tries to deviate from its eastward progress along the equator, it outflows into either the Northern or the Southern Hemisphere to higher latitudes at which the Coriolis force *does* have an effect. A northward excursion brings the current under deflection to the right, which forces the flow back toward the equator. If the flow then overshoots into the Southern Hemisphere, a deflection to the left forces its trajectory back toward the equator, and so on. The Cromwell is therefore a "trapped" current.

Upwelling along the Pacific Equator. Is westward-directed flow along the Pacific equator similarly trapped? No. In response to the southeast trade winds surface water here does move toward the west, as just described. Surface water just to the north of the equator is deflected to the right, however, away from the geographical 0° latitude line. Similarly, water moving westward but located south of the equator is deflected to the left, again away from the equator. The net result is a zone of *divergence* of surface waters, (in the water layer between the Equatorial Undercurrent and the surface), and a consequent upwelling of deeper water to the surface (figure 10.10c). The equatorial Pacific is thus a relatively rich biological zone; among other animals such as the tuna, we also find the sperm whale feeding here.

El Niño—Special Effects in the Equatorial Pacific. In 1983 Peru suffered drenching rains across its usually dry uplands, and floods caused vast damage and loss of life and homes; in its history since about 1500 A.D., no flood has been more severe. The same year, violent waves lashed the California coast, eroding the shoreface and wrecking many houses. In the Caribbean the coral reef sea urchin *Diadema*, normally present on the reefs at a density of 14,000 per hectare, declined in numbers to less than 1 per hectare by 1985! What caused those violent disturbances in the normal climate? The answer offered by almost everyone is—El Niño.

What is El Niño? Traditionally, it is the name

The drama of unusually high waves tearing out houses in California was but part of the worldwide phenomena associated with the 1982–1983 El Niño climactic event. (Mark Leet/Sygma.)

given to the appearance along the coast of Peru of unusually warm surface water, generally beginning during December or the Christmas season, hence the term El Niño, or Christ Child. The normal coastal waters off Peru are quite cold because of the strong upwelling that occurs there. (Recall the discussion of fog banks offshore in Chapter 6.) It is a zone of rich fishery for the anchoveta, a schooling fish that thrives in productive upwelling waters. But with the intrusion of warm surface water, plankton productivity drops sharply and the anchoveta disperse. In modern times, the dispersion has had severe impact on the Peruvian fish meal industry, with concomitant repercussions across the world.[3]

Oceanographers have searched for the cause of El Niño for a long time. In 1979 William H. Quinn showed that differences in the atmospheric pressures over Darwin, Australia and over Easter Island in the Southeast Pacific changed with a pattern that correlated closely with the irregular appearances of El Niño recorded over the past century. The pattern itself is controlled by a very large scale atmospheric phenomenon called the Southern Oscillation. Quinn proposed an index by which the behavior of the atmospheric cycle could be quantified and an El Niño might be predicted. Since then, great strides have been made, with the attention of the scientific community even more sharply focused by the 1983 event. So what is the connection with the Equatorial Undercurrent?

Our current understanding is this. Under control of the Southern Oscillation, sharp reductions occur in the average strengths of the trade winds along the Pacific equator. This means that the large hill of water normally pressed against the Asian continent cannot be sustained by the weaker winds, so a return flow begins. Waves of warm water progress rapidly across the Pacific equatorial zone, with the Equatorial Undercurrent a main participant. Masses of warm surface waters return to the eastern shores of the Pacific, impinging the coast of Ecuador and dividing into northern and southern branches. In short, the southern branch moving south along the coast becomes El Niño. To the north the flow is more complex, but the movement of warm surface waters over vast areas of the eastern tropical Pacific spawned the storms that wreaked so much havoc on Malibu Beach.

Gulf Stream Meanders and Eddies. Figure 6.5 depicts circulation in the North Atlantic, providing a simplified picture of heat transport in this ocean basin. Since 1960, our observations of the Gulf Stream have been vastly expanded by charts of sea surface temperature taken by satellite. Such charts are synoptic in time and cover a large segment of the North Atlantic. We find that the core of the Gulf Stream is not at all "regular" and smooth as implied in Figure 6.5; rather, the core appears to meander with ever-widening excursions after the current leaves the Atlantic seaboard and begins its migration eastward across the North Atlantic. This is shown in Figure 10.11.

In fact, time sequences of satellite charts reveal that the large meanders tend eventually to "pinch off" from the core and become separate "eddies" up to 1000 km in diameter. The curvature in the flow streamlines is substantial, so that a simple balance of pressure and Coriolis forces does not apply. The inertial force associated with accelerating motion along curved paths is a significant element in the force balances applied to these meanders and eddies. What is more, once pinched off, the eddies appear to migrate slowly toward the west, indicating that friction forces must also play a part.

Following the discovery of eddies in the Gulf Stream, oceanographers studied other major current systems to discover whether the phenomenon occurred elsewhere. Growing evidence indicates that such eddies exist virtually everywhere in the ocean, although the dynamics that are responsible for their genesis may differ. They are found in the Polar Front separating the Antarctic Surface Water from Central Waters of the Southern Hemisphere and have been reported throughout the North Pacific. In fact, some oceanographers believe that at any one instant in time there is more kinetic energy in all the eddies together than exists in the main circulation gyres. We have much to learn.

---

[3]In 1973, following the 1972 El Niño, which was also severe, I chanced to share a beer with two Dutchmen while they shared with me their lamentations over fish meal. Holland imports large quantities of fish meal to support its chicken industry (see Table 19.2). Because of the 1972 El Niño, Peru could not fill its fish meal contracts. The Dutchmen were forced to New York to bid for soybean meal as a substitute, and the price of soybeans had skyrocketed to over $10 a bushel. The profit potential in soybeans enticed many Texas cattle ranchers to plow up rangeland and convert to rowcrop. Mink ranchers in the Pacific Northwest organized to buy and stockpile protein to protect their markets, which forced companies that made food pellets for the aquaculture market to purchase high-protein whey from whiskey distillers because the scrap fish market was tied up by mink ranchers. All because of *one* El Niño!

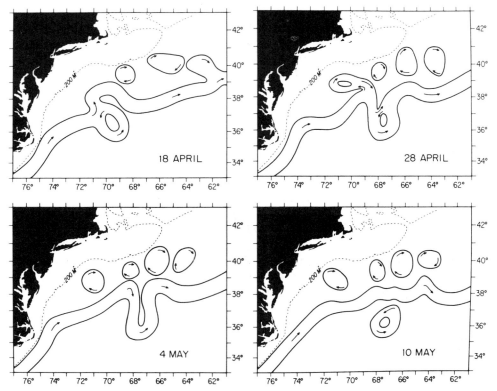

This intense current, which is quite well behaved during its course along the United States coast from Florida to Cape Hatteras, begins to meander after veering eastward away from the coast. Notice the excursion in the upper right on April 18; by April 28 it had broken away from the core of the Stream to form an eddy, which is filled with *warm* water from the Sargasso Sea bordering the Stream on the south. This warm-core eddy is now imbedded in the colder waters of Labrador origin to the north of the Stream. The large meander southward, seen in early stages on April 18, breaks off by May 10 to form an eddy of *cold* water, now isolated inside the Sargasso Sea. The warm and cold eddies "rotate" in opposite senses; compare these data with the currents rotating around the hill of surface water shown in Figure 10.9. (From Richardson, 1980.)

Figure 10.11 Meanders within the Gulf Stream.

Summary

Fluid masses can respond to both surface and body forces. The surface force that concerns most oceanographers is the stress exerted by the winds on the ocean surface; it is this stress that makes waves and drives the surface currents. There are three body forces acting on large fluid masses: the earth's gravity force; the pressure force, (which we identify with the terrestrial "downhill acceleration," but which in the oceans can rise either from a tilt in the level of the sea surface or from an uneven distribution of density); and the Coriolis force. Most of the ocean movement is steady, by which is meant that it is continuous over time.

It is in the study of the stress imposed on the oceans by surface winds that much insight is gained on the large-scale motions in both fluids. Equatorial easterly winds occur because rising parcels of air must slow down relative to the earth's spin. Westerly winds in mid-latitudes are generated as dry, cold air subsides toward the ocean and is forced to gain speed to the east. Both occur because angular momentum is conserved. The continents alter the wind belt pattern into cells that cover the entire ocean basin. The zone of subsidence becomes an atmospheric high over the central part of the basin, and the high is flanked both poleward and equatorward by low cells; it is this pattern that establishes the wind system.

Because of the turning effect of the earth's rotation, surface winds do not immediately force the surface water downwind. Rather, the initial reaction is a net transport of surface water at right angles to the downwind direction. This is the mechanism responsible for so much coastal upwelling, as well as that found along the Pacific equator. It is also the mechanism responsible for building the standing hill of water that is found in each major ocean basin, in direct relation to its characteristic atmospheric high. Once the hill is built, however, the ocean rapidly adjusts to an equilibrium of forces in which its surface currents move *around* the hills, not down them.

One exception to the "around" hill current is the Pacific Equatorial Undercurrent. At the equator the Coriolis force has zero value. The Equatorial Undercurrent is found to be an integral part of the aperiodic phenomenon known as El Niño.

Study Questions

1. "Big *whirls* have smaller ones that feed on their velocity, and these in turn have smaller ones, and so on to viscosity." This is a statement about a turbulent fluid. Of the four types of forces studied here, which are most likely involved?

2. Although *not* discussed in the text, can you explain the *difference* between *mixing* and *stirring* cream into a cup of coffee? Which process requires a spoon?

3. (a) Friction is the mechanism by which the atmosphere couples energy into ocean fluid. There are two principal *size scales*, enormously different, on which the coupling occurs. One of these (the larger) is covered in this chapter. State the scale and name the corresponding circulation features in the atmosphere and the ocean.

   (b) The second of these two scales manifests itself in such things as the depth of the *well-mixed layer* mentioned in connection with analyses of profiles in Chapter 8. What ocean features are generated by the atmosphere, the dissipation of which *creates* the well-mixed layer?

4. What has to happen to the meltwater created by ice melting from a deep-draft iceberg?

5. What has to happen to the seawater into which meltwater is mixed?

6. Explain two ways by which a *horizontally directed* pressure force can be brought to bear on a parcel of fluid in the ocean's interior.

7. Is the interior parcel also subjected to the pressure exerted by the overlying water column? If so, does the parcel necessarily have to move in a vertical sense?

8. All the preceding questions are tough. Here is an easy one. What are two different names we give to the force acting on a parcel *only* when the parcel is in motion?

9. If someone asked you why the Coriolis force does not *perform* work, how would you answer?

10. If the Coriolis force does *not* perform work, why call it a "force" at all?

11. Do all objects moving across the earth's surface "feel" the Coriolis force? If so, why is it you never complain about it while driving an automobile?

12. What is the connection between the "displaced meteorological equator" and the fact that the equatorial Pacific zone is a place where sperm whales are caught in large numbers?

13. Does the Coriolis force also impact on objects (or parcels of fluid) that initially are placed into motion along an east–west direction?

14. Explain why the coastal winds off Oregon display an annual, *monsoonlike* behavior.

15. Now we come to the mind-boggling hill of water standing in the central part (more or less) of each of the major ocean basins. What creates the hill?

16. The Wüst diagram, Figure 9.6, provides evidence that these hills do exist. What is the evidence?

17. Now here is the tough one—answer this and you understand it all: If the surface layer of the ocean responds to the friction stress of winds by moving (being transported) at right angles to the wind direction, how do you explain the observed fact that the currents of major ocean gyres are in the downwind direction?

18. Write a short essay on the subject of water column modification by icebergs melting and sea ice forming.

CHAPTER 11

CIRCULATION IN THE OCEANS

The first mariners must have begun the practice of giving names to ocean currents and oceanographers continue the practice today. You have encountered many such names in the course of your education, including those used in prior chapters without definition because they are considered a part of a general educational background. But no course in general oceanography is complete without the exercise of naming all major *-ography* parts of the oceans. That is the task set forth in this chapter.

But unlike the practice of mariners who see only the sea surface, here we must also describe and name the elements of three-dimensional oceans. The process is not without learning value. Up to this point the reader has been exposed to bits and pieces of ocean motion. In the process of collecting all names, we shall also achieve a final goal of describing how the pieces fit together.

Gyre Structure in the Central Surface Waters

When viewed on a large scale, the pattern of surface circulation is that of single gyres in each of the major basins (Figure 11.1), each gyre bearing the basin name. But early mariners were unable to see the large scale. Public knowledge of surface circulation developed piecemeal, with the pieces of the gyres named as individual currents. This practice survives today.

In the listings for Figure 11.1, the names of surface currents are grouped in gyres. Notice that each gyre has four components. The reason for this is not to provide the reader a memory guide, but because each gyre has a distinct *western boundary flow*, an *eastern boundary current*, and the *interconnecting east–west flows* that complete the gyre.

Salient Features of Central Gyres

Hemisphere Symmetry. The gyres in the Northern and Southern hemispheres have a symmetry about the equator. This is expected because the principal features of surface circulation result from wind forcing, and wind patterns in the two hemispheres are mirror images. The cause of the symmetry is the reversal at the equator of the direction in which the Coriolis force deflects the moving currents.

Intense Western Boundary Currents. The poleward-flowing currents located along the western boundaries of the major ocean basins are much more *intense* than the counterflowing, (equatorward) currents along eastern boundaries. By *intense* is meant that these currents are narrower in width, flow much faster, and extend to deeper depths than currents along the eastern boundaries. The five major western boundary currents are the Gulf Stream (North Atlantic), the Kuroshio (North Pacific), the East Australian (South Pacific), the Brazil (South Atlantic), and the Agulhas (South Indian).

This difference between eastern and western boundary currents has been known for centuries, but not until 1948 was an explanation developed. Henry Stommel (1960) of the Woods Hole Oceanographic Institute has shown that the difference is due to the fact that the Coriolis force increases in magnitude from the equator toward the poles (see Figure 10.3). Overall, the latitude variation in the turning effect of the earth's rotation has a twofold impact: gyre flow is concentrated along western boundaries and the *hill* of water that occupies the central part of each gyre is also displaced toward the west.

Western boundary currents flow much more rapidly than the eastern counterflows. Poleward flow in the core of the Gulf Stream has been measured at over 4 knots, or about 2 m/s; in contrast, equator-

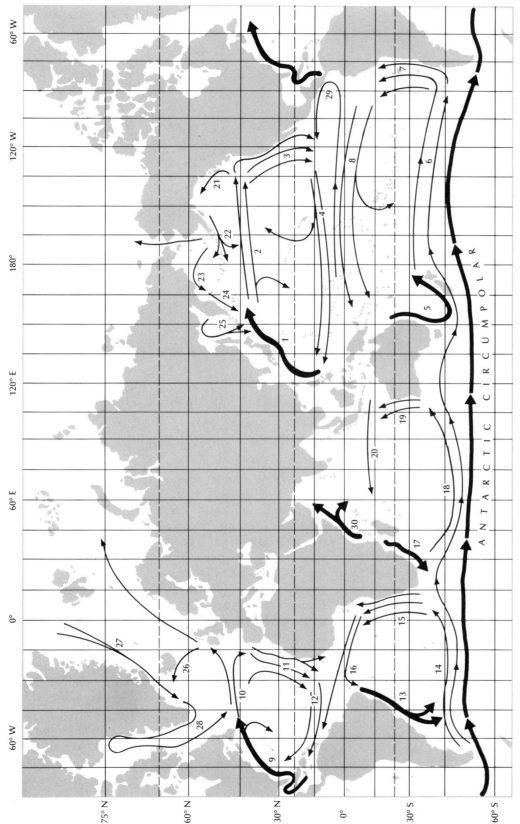

Figure 11.1 Surface currents of the world's oceans.

North Pacific Central Gyre:
1, Kuroshio Current.
2, North Pacific Current.
3, California Current.
4, North Equatorial Current.

South Pacific Central Gyre:
5, East Australian Current.
6, West Wind Drift (this leg is a part of the Antarctic Circumpolar Current).
7, Humboldt Current.
8, South Equatorial Current.

South Indian Central Gyre
17, Agulhas Current.
18, West Wind Drift (this leg is a part of the Antarctic Circumpolar Current).
19, West Australian Current;
20, South Equatorial Current.

North Atlantic Subarctic Gyre:
26, Irminger Current.
27, East Greenland Current.
28, Labrador Current.
10, North Atlantic Current (repeated).

North Atlantic Central Gyre:
9, Gulf Stream.
10, North Atlantic Current.
11, Canary Current.
12, North Equatorial Current.

South Atlantic Central Gyre:
13, Brazil Current.
14, West Wind Drift (this leg is a part of the Antarctic Circumpolar Current).
15, Benguela Current.
16, South Equatorial Current.

North Pacific Subarctic Gyre:
21, Alaska Current.
22, Alaska Stream.
23, Bering Sea Slope.
24, Kamchatka Current.
25, Oyashio Current.
2 North Pacific Current (repeated.

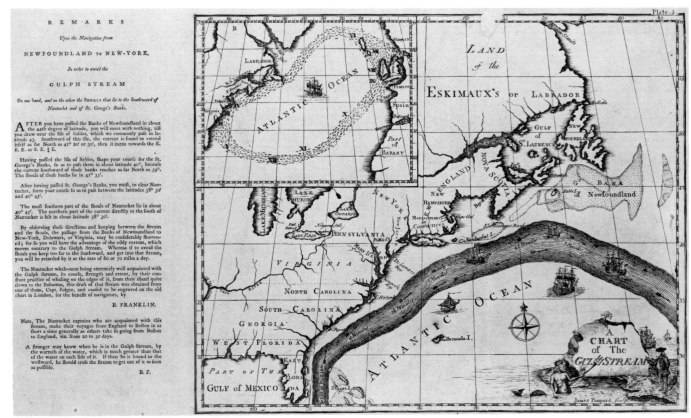

Benjamin Franklin's chart of the Gulf Stream, published in London in 1786, was the world's first chart of an intense western boundary current. (From the Library of Congress.)

ward flow in the Canary Current is slow, 0.1 to 0.2 m/s, roughly tenfold more laggardly. Similar analogies can be drawn for the other western–eastern gyre current pairs.

Western boundary currents also extend to great depths. The Gulf Stream reaches to depths of at least 2000 m as it flows northward along our eastern seaboard. Today almost every book on ocean currents shows a cross section of the western Atlantic on which the field of Gulf Stream speed is graphed, thereby dramatizing the extreme nature of such flow. The data for Figure 11.2 are taken from a cruise track that began at Cape Hatteras and sampled the Gulf Stream waters directly eastward across the main stream flow. The Gulf Stream begins just at the break between continental shelf and slope and increases in speed to a maximum value of 180 cm/s at the surface some 75 km offshore; virtually all of its poleward transport is contained inshore of the 160-km mark. Deeper in the water column the core shifts offshore a bit, eventually touching bottom along the continental rise at about 3000-m depth, but its speed is by then reduced almost to zero.

Countercurrents exist below and to either side of

the Gulf Stream core. The offshore portion of this counterflow marks the principal route taken by North Atlantic Deep Water in its cross-ocean journey from its origin at the surface off Greenland to its final surfacing off the Antarctic continent (see also the Wüst diagram, Figure 9.6). Thus deep-layer circulation is also concentrated along western boundaries, as shown in Figure 11.3.

Eastern Boundary Currents. In contrast, eastern boundary currents are slow in speed, and their flow is distributed over wide zones and only to relatively shallow depths. Figure 11.1 illustrates these characteristics by using multiple light lines for currents such as the California, Humboldt, Benguela, and Canary, all eastern boundary currents. Most of the transport in these currents occurs within 500 m of the sea surface, and speeds rarely exceed about 20 cm/s; widths may extend up to 1000 km offshore (see also the North Atlantic flow chart in Figure 6.5).

The Coincidence of West Wind Drifts with the Westerly Wind Belts. Transport in the western boundary currents extends poleward only to about

45° latitude, north or south. At these latitudes, and in coincidence with the belts of westerly winds, streamlines of gyre flow turn eastward across the major ocean basins. In the Northern Hemisphere these easterly flows are the North Pacific and North Atlantic currents. (The North Indian Ocean does not have a corresponding flow because there is not enough ocean north of the equator to allow a steady-state gyre to develop).

In the Southern Hemisphere, because there is no land barrier to eastward flow at latitudes between 53 and 65°S, the westerly winds drive surface waters completely around the globe. This is the Antarctic Circumpolar Current (AACP). Therefore, the Southern Hemisphere counterparts of the North Pacific and North Atlantic currents are actually incorporated into the AACP. Many oceanographers refer to the AACP as the West Wind Drift of the Southern Ocean. Early navigators knew all too well of this drift, for they faced a terrible risk in sailing their ships around Cape Horn to get from Europe to the lands around the Pacific. Not only did they have to tack against the westerly winds, but the strong currents and high waves made the crossing around the Horn one of the most infamous and dangerous of all ocean sailing routes.

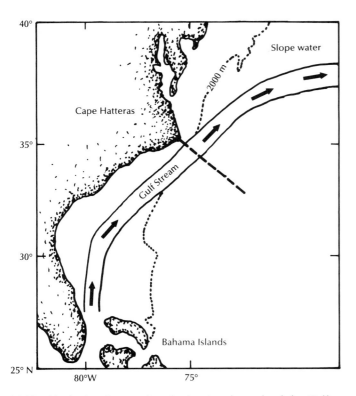

(a) The North American seaboard, showing the path of the Gulf Stream. The Stream closely follows the break between shelf and slope from Florida to Cape Hatteras. Beyond Hatteras the Stream breaks away from the continent to begin its eastward route across the North Atlantic. The dashed line is the track of the oceanographic cruise that gathered the data from which the velocity contours of illustration (b) were calculated.

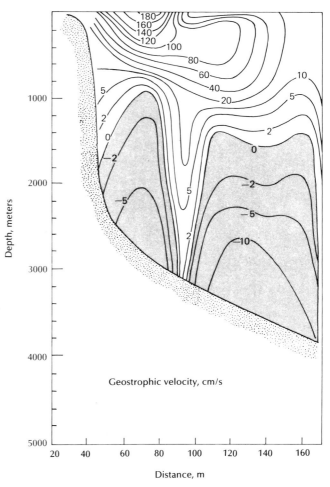

(b) Contours of the Gulf Stream velocity off Cape Hatteras. The surface core of the Stream is located about 75 km offshore and has a speed of about 4 knots. The Stream is detected to depths of over 2000 m. Shaded areas indicate the equatorward counterflow of North Atlantic Deep Waters formed off Greenland, enroute to Antarctica. (From Knauss, 1978.)

Figure 11.2   Characteristics of the Gulf Stream.

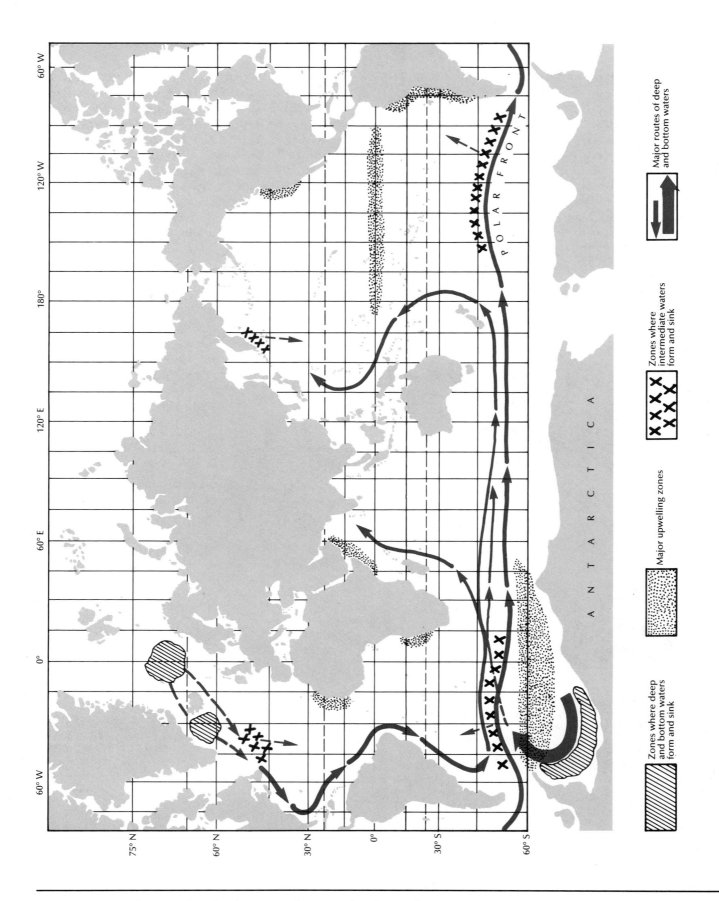

Figure 11.3 Circulation within the deep ocean layers. (After Stommel, 1958.)

The Nonsymmetry of Equatorial Currents

As indicated in Figure 11.1, the South Pacific Equatorial Current at times actually straddles the geographical equator, but the North Pacific Equatorial Current is displaced to a zone between 9 and 18°N. Meteorologists have long known that the trade winds of the hemispheres are not symmetrically disposed about the equator, especially in the Pacific; they describe this phenomenon as a "displaced meteorological equator." It is not a fixed feature of the tropical atmosphere: during the Northern Hemisphere winter the zone of covergence of the northeast and southeast trade winds is found at about 5°N; during summer it shifts to perhaps 10°N. Thus, Costa Rica is subjected to monsoon-type rains during July to September, and Columbia receives its heaviest rains in December to February. It is the displaced meteorological equator that is responsible for the shifted pattern of equatorial ocean currents.

The North Atlantic also has a displaced meteorological equator, but not to the extent found in the Pacific. A similar displacement in the Indian Ocean is closely related to the monsoon seasons there, and the seasonal wind shifts are responsible for the annual buildup and decay of a major coastal current called the Somalia Current (see 30 in Figure 11.1). The Somalia flow appears along the shores of Kenya and Somalia during the summer months June to August; in the winter it virtually disappears, or even reverses direction. Only in recent years have oceanographers begun to explore the Somalia Current, which, when fully developed, is equivalent to the Gulf Stream in speed and transport and is therefore an important feature of ocean surface circulation.

Subpolar Gyres

The subarctic and subantarctic gyres are found on the poleward sides of the main basin gyres. In each case the rotation sense is opposite to that of the adjacent main gyre, because subpolar gyres are "sandwiched" between westerly winds (on the equatorward side) and easterly winds (on their poleward sides). The reader should refer back to Figures 3.1 and 7.4 for a refresher on wind belts. Only the subarctic gyres are discussed here; the subantarctic circulation is left to a more detailed discussion of polar oceans in Chapter 20.

Salient Features of the North Pacific Subarctic Gyre

The North Pacific Subarctic Gyre was previously described in connection with the migration routes of salmon (see Figure 3.4d). Here some additional features are noted. First, this gyre feeds surface water into the Arctic Ocean via the shallow Bering Strait. This is the only connection in the Northern Hemisphere between the Pacific and the Atlantic, and it is not a major one. The transport through the Bering Strait is not extensive, being only 1 to 3 million m³/s; for comparison, this is less than one-tenth of the flow in the Gulf Stream off Florida.

Second, the arc chain of Aleutian Islands forms a leaky barrier across this gyre. Surface water enters the Bering Sea through a number of interisland straits along this chain, the most noteworthy being Unimak Pass at the tip of the Kenai Peninsula of Alaska. Within the Bering Sea some water flows along the edge of the continental shelf–slope break toward the Kamchatka Peninsula where it turns equatorward to become the Kamchatka Current. South of Kamchatka another chain of islands, the Kuriles, allows surface water to enter the Sea of Okhotsk. From here flows the Oyashio Current to join with the Kuroshio and the North Pacific currents to complete the subarctic gyre. The broken topography and shorelines serve to make the North Pacific subarctic Gyre one of the most complex of all gyres. But it is an important feature of Pacific Ocean circulation. Earlier we described the connection between this subarctic gyre and the migration routes of Pacific salmon fish. Major fish resources also occur where the cold Oyashio meets the warm Kuroshio, principally in herring species. But the extensive shallow waters of the Bering Sea are the main fishing areas. Fishing craft of four nations, the United States, U.S.S.R., Japan, and Korea, take millions of tons from the Bering Sea; chances are the fish sandwich you had last night was *pollock* from the Bering Sea.

Salient Features of the North Atlantic Subarctic Gyre

The North Atlantic Current provides a substantial input to the Arctic Ocean. The Gulf Stream feeds water far toward the north, via the North Atlantic Current, into the Barents Sea off Norway and into the waters surrounding Iceland (Figure 11.1). But the circulation becomes very complex north of the Arctic Circle. The principal westward transport in the northern part of this gyre is the East Greenland Current, which loops around Cape Farewell at the southern tip of Greenland to feed surface water into Baffin Bay along the west coast of Greenland. The flow pattern then loops again, this time turning south into the Labrador Current and eventually to a

Iceberg grounded just offshore of Godhaven, West Greenland. Each year West Greenland glaciers release large numbers of such icebergs into the Labrador Sea. (Peter J. Bryant/Terraphotographics.)

closing of the gyre via connection with the North Atlantic Current off Newfoundland.

Glaciers along western shores of Greenland discharge large numbers of icebergs into the looping Labrador Current, which carries these to its convergence zone with the North Atlantic Current. It was one such iceberg that lay in the path of the *Titanic* to produce that historic event. But the most noteworthy aspect of the gyre is its abundant fish resources.

The codfish resource of the Grand Banks off Newfoundland has been fished since the time of Eric the Red's historic voyage[1] of discovery in the tenth century. This fishery occurs along the convergence between the cold Labrador Current and warm-water extensions of the Gulf Stream, a situation very much like that described for the Oyashio–Kuroshio convergence in the Northwest Pacific. Off Norway the warm surface waters of the North Atlantic Current meet the cold waters of the Arctic Ocean, where rich stocks of Norwegian herring are found in association with the sharp temperature boundaries between these water masses, called *fronts*.

[1]Major fishing for cod off Newfoundland began only after Sir Martin Frobisher's legendary sixteenth century voyage to discover the Northwest Passage. He staked his reputation on the discovery of "gold" in what today is Labrador; it was "fool's gold" of the highest caliber. It was his other report, of the bountiful, silvery cod fishery, that provided England a true prize.

As indicated in Figure 11.1, the East Greenland Current derives only a part of its flow from the North Atlantic Subarctic Gyre. Part of its transport is a continuation of the discharge of North Pacific water into the Arctic Ocean via the Bering Strait. Additional sources are the voluminous discharges of Siberian rivers that empty into the Arctic Ocean and eventually feed into the North Atlantic. More details are given in Chapter 20.

Circulation in the Deep Ocean Layer

In earlier chapters we learned that the deep ocean layer does not feel the driving force of surface winds. Instead, the unequal heating of the earth by the sun so alters the global distribution of seawater density as to generate an "overturn" circulation of the deeper waters. But we also learned that the temperature and salinity characteristics of deep water masses are acquired only at the surface where evaporation and cooling so alter the density of water that it sinks to some equilibrium level. This is a process of massive oceanic downwelling. Where it occurs intensely in certain regions, I call it the "greenhole" process to emphasize that a major breakthrough is made in the density-stratified ocean—a "hole" is punched through the green, layered waters of the ocean. Because downwelling is a loss of water from the surface layer, there must be a compensating upward mass flow from deeper levels into the surface layer. Such is the flow we call "upwelling." The system of deep circulation is charted in Figure 11.3.

Major Known Downwelling Zones

The most intense alteration of temperature and salinity of surface water occurs in polar and subpolar regions; these are zones of major downwelling. There are three regions where major downwelling occurs, and interestingly the processes that create the downwelling are somewhat different in each.

The Antarctic Polar Front. At the Antarctic Polar Front waters of intermediate density form from the mixing of relatively warm Central and Subantarctic surface waters with the cold and low-salinity surface waters leaving the ice-melt zone of Antarctica. Sometimes called the Antarctic Convergence (Figure 9.5) or the Polar Front (Figure 9.6), it is the origin of the Antarctic Intermediate Water. The reader will recall from Chapter 9 that a mixture of two water masses having the same density, but different temperatures and salinities, is denser than either and must sink to a deeper level than that at which the actual mixing takes place.

The Arctic Polar Fronts. There are several areas in the Northern Hemisphere where dense water masses are formed. Along the confluences of the Oyashio and Kuroshio currents in the Pacific and the Labrador and Gulf Stream extension in the North Atlantic, the mixing of cold and warm water masses produces some intermediate-density waters, but not nearly as vigorously as at the Antarctic Polar Front (Figure 11.3).

But in at least two areas located near Greenland, a very dense water mass is formed through extreme rates of evaporation and cooling (Figure 11.3). As described earlier, when cold, dry air masses sweep over the oceans from continental land masses, the density of surface water can be enhanced enough to create vigorous downwelling—greenholes—through which the altered surface water spills down to fill the deepest ocean basins. These are the sources of North Atlantic Deep and Bottom waters, at a rate of about 5 million m^3/s. It is important to note that the source of surface water for these downwelling zones off Greenland is the high-salinity water carried northward by the Gulf Stream and its extension. (The importance of this is demonstrated in the next section, which discusses coupling between upper- and lower-layer circulation.)

Antarctic Bottom Water Formation. The densest water in the world's oceans is formed along the periphery of the Antarctic continent, for the most part in the Weddell Sea (Figure 11.3). Antarctic Bottom Water forms during the austral winter, and the process is the exclusion of salts from seawater as it crystallizes to ice (see also Figure 10.1b). The water layer below the ice is "loaded" with excluded salts to the point of sinking to the bottom. At a salinity of 34.68 ppt and temperature below 0°C, some 25 million m^3/s downwell in this greenhole area, which is about five times the rate of sinking of North Atlantic Deep Water off Greenland.

Deep Water Currents

Horizontal currents in the deep ocean layers are quite weak in comparison to those measured in surface layers. We tend to refer to deep motion as a "spreading" effect rather than as a current. Still, oceanographers have identified some specific zones in which deep-layer flow is concentrated, and these are illustrated in Figure 11.3. The Pacific and Indian oceans are net *receivers* of deep and bottom water masses, whereas the Atlantic is a net *contributor*.

The routes by which the major transport occurs in deep flow tend to concentrate against the western boundaries of the ocean basins. We know more about the deep flow under the Gulf Stream than about any of the others (see, for example, the Gulf Stream cross section in Figure 11.2); here the deep flow is a concentrated stream located just offshore of the core of the Gulf Stream, at depths of more than 2000 m and with speeds no higher than about 10 cm/s.

Coupling Between Surface and Deep Circulation—Some Consequences

The 5 million m^3/s of North Atlantic Deep Water formed off Greenland can be treated as water that is *lost* from the surface layer. There must be a compensating flow of water *into* the North Atlantic's surface layer. About 3 million m^3/s are provided from the Arctic Ocean overflow into the Atlantic, roughly the same quantity that the Pacific Ocean feeds into the Arctic via the Bering Strait). The remainder of the loss, about 2 million m^3/s, plus a similar amount lost from the North Atlantic water through evaporation and subsequent transport to the Pacific by air (see Figure 6.5), must be replaced from somewhere. We know that coastal upwelling is not vigorous enough to be the only replacement mechanism.

The net result is that the North Atlantic imports surface water from the South Atlantic via a trans-equatorial transport of a part of the South Atlantic Equatorial Current. The connection is outlined clearly in the surface circulation of Figure 11.1. This rather large cross-equatorial flow is unique to the Atlantic. The North Equatorial Current, together with the Gulf Stream and its extension into the North Atlantic Current, feeds the imported water toward the Greenland greenhole zones to complete the surface layer budget.

The total North Atlantic budget is not balanced, however, until we account for the 5 million m^3/s of water injected into deeper layers. From our previous work, it is clear that this volume must also be transported across the equator, from the Northern to the Southern Hemisphere—the mechanism is the deep flow of North Atlantic Deep Water across the equator toward Antarctica.

Upwelling as Compensation Flow for Downwelling

The requirement that there be mass balance between upper and lower ocean layers, as well as between the North and South Atlantic, leads to a final explanation of two facets of ocean behavior. Clearly, there must be sufficient upwelling throughout the Atlantic basins to return to the surface layer both the downwelling off Greenland as well as that associated with bottom water formation off Antarctica, a total of about 30 million m^3/s (again, roughly the transport in the Gulf Stream off Florida). If coastal and equatorial upwelling is not adequate to the return task, what is?

Massive Upwelling of North Atlantic Deep Water around Antarctica. Major upwelling of North Atlantic Deep Water occurs in the Southern Ocean off the Antarctic continent. This, the principal mechanism we seek, is not coastal upwelling. As is clearly demonstrated in Wüst's diagram (Figure 9.6), a massive upwelling of North Atlantic Deep Water provides the missing link to the overall coupling between upper and lower ocean layers. It is this regional upwelling that sustains the rich and widespread plant production around Antarctica, and with this the prodigious production of krill, penguins, seals, birds, and, of course, baleen whales.

Diffuse Upwelling Throughout the Oceans. For many decades oceanographers were perplexed by the existence of the permanent thermocline. Often found at depths of some 300 m in the tropics and up to 800 m in central waters, the permanent thermocline is a widespread, ubiquitous feature. What sustains it? In the face of continuous solar heating, why does the downward diffusion of heat not gradually erode the thermocline?

Stommel offered the hypothesis that the steady downward diffusion of heat is at some level exactly compensated by the upward diffusion of cold water, and that the process is as widespread as the thermocline itself. His model of deep circulation, on which Figure 11.3 is based, included the idea that the spreading of deep waters across the ocean depths was sustained in part by a continuous *upward* mixing of deep water. In a sense, it is a *model of diffuse upwelling everywhere.*

How rapid might such diffuse upwelling be? It is a straightforward exercise to divide the greenhole downwelling (30 million m^3/s) by the area of the oceans to get a gross estimate of diffuse upwelling, about 1 cm/day!

Stommel's model was quickly put to use by biologists who were seeking explanations of how open oceans could sustain even minor primary production in the face of the inevitable depletion of nitrates. The continuous upward diffusion of nutrient-rich deep water, albeit slow, provides a partial answer.

Summary

Sooner or later, the reader needs to study the -ography of the oceans (or "oceansography"). In a crude sense, the oceanographic equivalent to the geographer's memorization of names and locations of continents is the memorization of name and locations of water mass structures (Figure 9.5). Further, memorizing the countries spread across continents translates in oceanography to memorizing names and locations of currents. Therefore we justify the work of memorizing the surface features of the world oceans.

One extreme difference sets oceanography apart, however. In oceanography we must master a three-dimensional world. This is why the deep ocean is included in this chapter. How much more difficult is the geographical work of oceanographers who can neither see nor visit the deep provinces, and who must describe in terms that do not change as the medium itself changes! It is necessary, absolutely necessary, that both surface and deep layers of the oceans be described; otherwise we could not describe the coupling between them, or massive regional upwelling, or any number of oceanic events or processes.

Within each basin the surface water takes from 1 to 3 years to circulate around its gyre. The deep layer needs hundreds of years to cover its routes. Surface currents move enormous quantities of water, as they must to carry out their job within the heat engine of the earth. The Gulf Stream is moving about 60 million m^3/sec of seawater at the point that it departs from Cape Hatteras. Some currents are strong even though completely submerged; the Pacific Equatorial Undercurrent moves 30 million m^3/s. The largest of all currents, and one that is both surface and deep, is the Antarctic Circumpolar Current, which transports water at about three times the rate of the Gulf Stream. This great "river of the sea" flows through all the major basins. In the Drake Passage south of Cape Horn, South America, it moves surface water from the Pacific into the South Atlantic, thus providing the return flow of surface water that was first transferred from the North Atlantic to the North Pacific in the form of moisture carried across the Central America land bridge.

It is the coupling of surface to deep circulation that gives us penetrating insight into ocean behavior. In the Atlantic alone some 30 million m^3/s of surface layer water sink to join the deep realm, most of it in the Weddell Sea of Antarctica. It is the replacement of this water that drives the unique, transequatorial flow of surface water from the South into the North Atlantic basins, and forces the massive regional upwelling of North Atlantic Deep Water to the surface of the Southern Ocean, thus sustaining the world's largest animal population, the Antarctic krill, as well as its largest living mammal, the blue whale.

Study Questions

A classic approach to learning the major ocean currents is via the "drift bottle" question. If one throws a corked bottle into the ocean, where will it drift to and via what route? The following are examples.

1. Michael Rockefeller, son of Nelson Rockefeller, disappeared while engaged in a photography expedition to New Guinea. Would it be possible for a cannister of his film to wash up on the beach of a Rockefeller estate on Long Island? Trace the path(s) the cannister might take and name the currents involved.

2. A couple honeymooning in Hawaii tossed two champagne bottles with messages corked inside into the Pacific Ocean. One was recovered on the beach at Yokohama, Japan, and the other floated into the bay at Juneau, Alaska. When both had been returned, the couple noted that the bottle recovered from Yokohama had a noticeably greater growth of barnacles and algae attached to it. Can you explain why? Trace and name the major surface currents of the North Pacific that enter your solution.

3. There are two major reasons why surface waters from the South Atlantic cross over the equator and enter into the surface circulation of the North Atlantic. Can you describe the global processes responsible for this transequatorial flow in the Atlantic? Why is there no comparable situation in the Pacific? Hint: See Figures 6.5 and 11.3.

4. The argument for calling the Antarctic Circumpolar Current a great "communicator" current, as we did in Chapter 9 based on water mass structures, is even clearer after studying the charts of deep ocean circulation, as in Figure 11.3. Explain.

How do you organize a study of the marine biosphere?
(Robert Abrams/Bruce Coleman.)

THE MARINE BIOSPHERE

In which is revealed the outlines of an elegant organization of living communities, yet seen so dimly through the vast and darkened waters.

CHAPTER 12

BASIC FACTORS IN THE MARINE BIOSPHERE

Overview for Part V

The first four parts of this text present a set of broad perspectives to marine science and a demonstration of the physical laws that govern the behavior of the fluid world. You now have the framework needed to build a respectable understanding of *what* the marine ecosystem is, *how* it is organized, and *why* the system behaves the way it does. An observation by the noted British marine scientist Alister Hardy is appropriate to the next step:

> Some people think that the word ecology is just a modern name for what they have always called natural history . . . I believe there is a valuable distinction between the two, and . . . it is well to be clear as to what we mean by it.
>
> I would define ecology as that branch of science which deals with the relations of living things and their surroundings. Natural history is the description of nature. Ecology goes further; as a *science* it records organic inter-relationships in *quantitative* terms—it measures them. When we record that a particular kind of fish is confined to warm oceanic water and feeds upon various kinds of shrimps, that is simple natural history. When we can determine the temperatures and degrees of salinity which limit its distribution, and by a number of post-mortem observations work out the average percentage constituents of its food at different times of the year and at different ages, we are then beginning to learn a little of its ecology.
>
> The aim of ecology, however, is not simply to express the inter-relationships of organisms with their environment in numerical terms, but . . . in time, step-by-step, to discover more of the *laws operating in the world of living things* (Hardy, 1965).

The Open Sea

The idea that there are specific laws operating in the world of living things is the essence that distinguishes ecology from the other sciences. Physiology, biochemistry, and biophysics are laboratory sciences that describe the intricate mechanisms of life itself, but these are based on the behavior of electrons, atoms, and molecules in organic substances.

Ecology becomes a science in its own right because it adds the essential ingredient, the *quantitative* interactions of organisms as living entities.

Figure 12.1 illustrates one way to construct a framework for the marine biosphere. Embodied here are the main facets of the fluid world that the earlier parts have described. To them are added the following additional ideas to be developed in Part V.

1. ***The mechanics of fluid ocean.*** Unequal solar isolation drives the winds and currents and the overturning of the seas. Except in high latitudes, the permanent thermoclines and haloclines create a two-layer stratification, a stable barrier across which mixing is slow, diffuse. It follows that the biosphere is also divided into two distinct groups, the photic and the benthic.

2. ***The food chains.*** Energy from the sun powers the webs of life. There are two webs, one within the photic zone and the other occupying the lower fluid layers. These are connected by the detrital link. Food chains are organized into trophic levels (not to be confused with depths). At the top are the autotrophs, plants capable of organic synthesis, followed by successive levels of heterotrophs beginning with the herbivores and continuing through several carnivore steps.

3. ***The abiotic cycle.*** A continuous supply of chemical elements is introduced by rivers and volcanic emissions and submarine springs. Excess salts are chemically removed via sediment accumulation.

4. ***The decomposition factor.*** The recycling of organic material is essential to every ecosystem that does *not* have access to unlimited nutrients. Bacteria provide this function everywhere.

It is interesting that our studies in ecology may have progressed further in the marine world than in our terrestrial sphere. This is at least partly because we have domesticated so much of our common terrestrial food supply, both plant and animal. In fact, many domestic species have been so altered by genetic selection, and then bred in such prodigious quantities, thereby excluding the natural fauna and

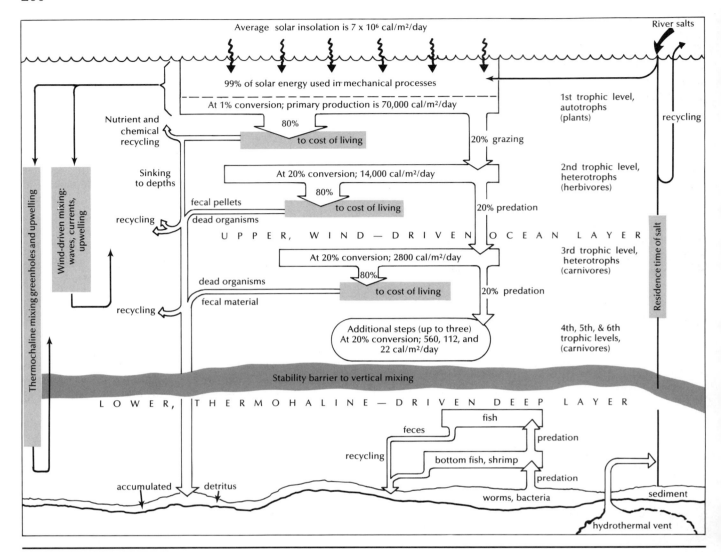

Figure 12.1 Ecosystem Ocean—An Overview.

flora from the choicest habitats, that we may never learn the natural ecology of the remaining continental species. For example, we rely heavily on corn, but corn as we know it today is so genetically altered from its predecessors that it is in turn completely dependent on us for its continued existence. We must harvest the ears, separate the kernels, and reseed these individually; modern corn left to its own would not be able to reproduce successfully.

The elemental structures in Figure 12.1 are, of course, generalizations of a quite complex system. Nevertheless, it is true that the marine ecosystem is more tractable than the terrestrial. The number of distinct biogeographical provinces is quite small, as is developed in these next chapters. Speciation in the marine biosphere is remarkably less than that of the terrestrial biosphere. Marine ecology is develop-

ing well for another reason. The marine biosphere is amenable to analyses by grouping its component organisms; for example, we can treat the phytoplankton by methods akin to those of solution chemistry. It is the fact of the fluid matrix that makes ocean ecology progress more readily than terrestrial ecology.

Terms and Definitions

At the outset, new terms that are essential to a description of the marine biosphere are defined.

Ecosystem. Any part of the natural environment in which both living and nonliving matter are in continuous interchange, with exchanges energized by a unidirectional flow of available energy.

Strictly speaking, "in continuous interchange" means that the system is *closed* to external supply of everything except energy. In contemporary ocean studies we are forced to depart from strictness—the system is too vast and our sampling is too sparse. We therefore divide the system into smaller pieces and forego for the present the neatness of being able to account for *all* factors. For example, we can study the open-ocean photic biosphere without taking into account the impact of river runoff.

Community. A group of populations of different species that are found together throughout a geographical area. Because these populations occur jointly, biological factors such as predation or competition are common throughout the community. Usually physical factors are also common, for example, in the rocky shore intertidal community.

Population. All individuals in a species whose behavior is tied to a geographical area. It is important, however, to note that individuals of one species may belong to different populations. For example, a single species of Atlantic herring is found throughout the Gulf of Maine, but the individuals are grouped into distinct populations according to the area in which they spawn, in Georges Bank, the southwest Nova Scotia Bank, or another area. The recognition of populations within a species can be difficult, particularly where these become mixed during certain seasons, as in the feeding grounds.

Component. An identifiable function within an ecosystem. Commonly a system is said to be composed of four components: the *autotrophs,* species capable of synthesizing organic compounds out of abiotic materials and light energy; the *heterotrophs,* species whose energy sources are the chemical bonds of organic compounds synthesized by autotrophs or other heterotrophs; the *abiotic* materials, nutrient and mineral elements essential to life; and the *decomposers,* faunal species that break down organic compounds into basic element form.

Niche. The job that a specific organism carries out within the community. For example, bacteria perform the job of decomposing the bodies of dead plants and animals.

Habitat. The physical location of the organism in the geographical area occupied by a community. For example, hake fish occupy a depth zone of around 140 m along the continental shelf.

Trophic level. A level in the pyramid of predator–prey sets that make up a food chain (see Figure 12.1). The second trophic level consists of herbivore animals that graze the plant populations; the third trophic level is the first level of carnivores, and so on. In the ocean, there is a strong correlation between the number of the trophic level and the size of the organism. Trophic levels are not to be confused with the physical depths of the ocean at which species exist. Decomposers are found at every trophic level, including the first one consisting of plants.

Food chain. A specific pathway along which food is transferred among the several trophic levels in the food web; for example, diatom, copepod, euphausid, and whale.

Food web. A pyramid of trophic levels containing multiple, complex pathways, the food chains. Complexity increases as predation behavior changes; for example, prey species are substituted as populations wax and wane.

Dominant species. Within a *trophic level* there is usually one species whose numerical density exceeds that of all other species. For example, among the marine plants of the first trophic level, one species of dinoflagellate can make up the overwhelming mass of individuals. Dominance is often a seasonal phenomenon, such as the massive buildup of the dinoflagellates which make up the so-called "red-tides." Dominance is also associated with *living space* in a community. For example, the mussel (a bivalve mollusk) will outcompete other sessile species in the mid-tide zone of rocky shore communities.

Predator–prey. Any specific pair of organisms through which organic energy transfers between trophic levels.

Competition. The competitive interplay among individuals for a needed resource that is itself limited. Competition may be for food or for territory, although these are usually intermixed.

Community succession. The process of community change, in which populations and predator–prey pathways change with time. Overall, the community itself remains but its makeup changes. Given a geographical area that is stable over long time spans, the final community is called the climax. If seriously disturbed, called *perturbation,* the climax community may revert to an earlier stage, one through which it normally progresses enroute to the climax stage. But the final result will once again be the same climax community.

Perturbation. An abnormal stress applied to a community. Everyday stresses are normal and therefore do not qualify as perturbation.

Primary production. The amount of *biomass of plant material produced per unit time* measured in any convenient units, for example, tons per day.

Biomass. The amount of living protoplasm that exists at *any instant in time.* Generally, we speak of the biomass of a given trophic level, for example, the herring. A hypothetical way to learn the biomass numbers for a given volume of the ocean would be to filter it through a series of progressively finer-mesh nets. The mass of all animals taken in the net of coarsest mesh would be the biomass of higher trophic levels. The biomass captured in the finest filter would be that of the phytoplankton, the first trophic level.

System Factors

How is the marine biosphere organized? If we could discover the basic laws that govern the behavior of living things, the biosphere's organization could be predicted. Conversely, we might develop models of system organization and test the models against observations in nature, hoping to discover clues from which to formulate the basic laws. As we have shown repeatedly, the factors that govern the physical behavior of hydrospace are so different from terrestrial factors that the needed models and studies

are far from complete. The technique of using models, however, is the same in all ecological analyses.

The whole idea is shown in diagram form in Figure 12.2. Here are listed the controlling physical and biological factors that must first be described in detail before a model of the marine biosphere can be designed. From physics we need adequate descriptions of all the ocean motions, from the smallest scales of turbulent diffusion to the grand scales of ocean gyres. From biology we require concise descriptions of population dynamics. All these must then be "matched" to the *observed* patterns of system behavior via the process of sampling.

The sampling process is the stumbling block in contemporary studies. We need numbers, for example, on concentrations of species, but these are difficult to obtain because swimming animals try to avoid our nets. A good measure of how organisms are dispersed is possible only after the avoidance problem and other gear constraints are solved. In the end, the high cost of ships and cruises forces us to develop models from insufficient data bases. Overall, these limitations in sampling *obscure* the organization of the behavior patterns that we do observe.

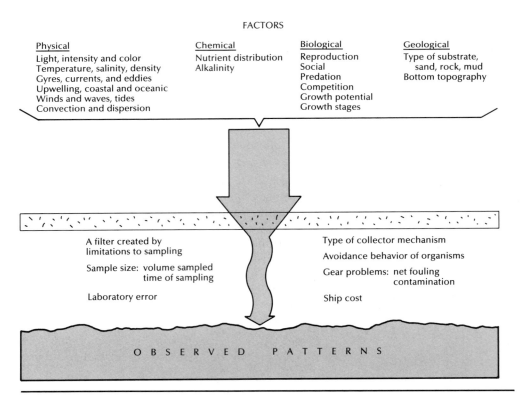

FACTORS

Physical
Light, intensity and color
Temperature, salinity, density
Gyres, currents, and eddies
Upwelling, coastal and oceanic
Winds and waves, tides
Convection and dispersion

Chemical
Nutrient distribution
Alkalinity

Biological
Reproduction
Social
Predation
Competition
Growth potential
Growth stages

Geological
Type of substrate,
    sand, rock, mud
Bottom topography

A filter created by
limitations to sampling

Sample size: volume sampled
            time of sampling

Laboratory error

Type of collector mechanism

Avoidance behavior of organisms

Gear problems: net fouling
                contamination

Ship cost

OBSERVED PATTERNS

Figure 12.2 Factors involved in the organization of the marine biosphere.

Systems Model of a Living Community

In simple form, any activity that converts an *input* to an *output* can be modeled as a system. The "system" in Figure 12.3 may represent an entire marine community, a specific food chain within it, or only a selected trophic level; the principles of operation are the same.

Feedback

An element basic to system operation is what is commonly called the "feedback loop." It has three characteristics: sign, positive or negative; magnitude, the percent of output signal returned to the input; and delay time. There is always a delay, for biological systems do not have "awareness" and cannot anticipate as humans do.

From the gross view of applying feedback to living systems, feedback can be likened to a "trial by error" solution to a problem. The problem is to evolve a stable community; the unsuccessful trial solutions attempted in nature are just that, trial communities that did not succeed and simply vanished. We "see" only the successful ones.

Let us consider negative feedback. A system is given an input; for example, the combination of spring sunshine and available nutrients in surface waters give rise to a *bloom* of plants. Initially nutrients are plentiful and the bloom proceeds. But as the population of plants increases, the concentration of nutrients available for further production decreases. This is a negative feedback—it serves to "damp out" population growth. Predation is clearly a negative feedback controlling the population density of prey.

On the other hand, positive feedback is a strong promoter of growth. We may use the concept of threshold concentration of prey as an example. When a predator switches from one preferred prey to another, the biological "stress"—the negative feedback on the former prey population is sharply reduced. Here the reduction in magnitude of negative feedback has the same impact as an increase in positive feedback.

The natural control of feedback plays a key role in explaining the diversity of niches. Fish are subject to parasites, a stress factor that if left unchecked, could destroy abnormal numbers of fish and destabilize the community. Some systems have reduced this negative feedback by evolving species of "cleaner shrimp" that feed on the parasites. Diversity is increased and the system is stabilized.

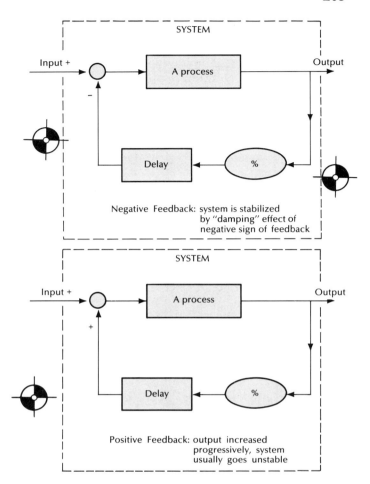

By definition, biological *stress* is a negative feedback signal within a biological system.

Examples: Predation is a stress upon a prey species.

The depletion of nutrients is a stress factor.

Excessive vertical mixing within the upper photic zone can be a stress.

Positive feedback signals are a bit more difficult to define.

Examples: Excessive stress upon a predator species becomes a positive feedback to the prey species.

The creation of "cover" can be a positive influence on a system's overall production.

The percent circle refers to the *strength* of the feedback signal in relation to the strength of the output.

Example: If a signal must cross over between adjacent trophic levels, the magnitude of the signal is diminished by the naturally low conversion efficiency between trophic levels.

The "delay" box usually refers to a delay between a change in the output and the time when the impact of the change is *felt* at the input.

Example: An increase in plant production *can* feed an additional number of grazers, but until such additional grazers are actually created by reproduction, grazing pressure does not change very much.

Figure 12.3 Feedback signals within a system—the essence of why a system works the way it does.

Magnitude (%) of Feedback

In both of the examples just given, negative feedback moderates the amplitude of variation within the living system. Nutrient depletion moderated the peak in phytoplankton production (along with grazing pressure by herbivores, to be sure) and an increase in parasite population transferred into an increase in shrimp. In general, an increase in diversity within food chains leads to a decrease in the magnitudes of feedback within specific loops, hence to an increase in the overall stability of the community. Stated another way, with high diversity, a single change does not alter the system drastically.

Conversion Efficiency and Feedback

A feature common to all food chains is the unavoidable loss of chemical energy during conversion to higher trophic levels; the conversion "efficiency" is low, about 20% in the marine system. It follows that a perturbation that occurs at *the base* of a food web will not propagate through the entire web at constant intensity. Such an event is "smoothed out" by the conversion loss factor. But a perturbation that occurs at *the top* carnivore level can cascade down the web with surprising results.

The results of two Canadian experiments on the population dynamics of young sockeye salmon contrasts the two perturbations. In one experiment, 90% of the natural predator of the young salmon were cropped from the lake, increasing by 300% the biomass of young sockeye that year. At 20% conversion

efficiency the impact on the prey species of the young salmon must have been several times more severe. In a second experiment, the amount of dissolved nutrients in the lake was increased 100% (doubled) and the result was only a 30% increase in young salmon biomass (suggesting a 30% conversion efficiency in this case). Comparing these two events yields a drastic, tenfold difference in a midtrophic-level species, depending on whether the system was perturbed from the top or the bottom.

Timing (Delay) in Feedback

The characteristic delay in feedback is as important to system stability as is the magnitude or sign. Recall the size specificity behavior of the marine biosphere discussed in Chapter 2 with the example of the need of oyster spat for food of the "right" size during the initial phases in its life cycle. In the pelagic world, the potential for right-size food is assured by the different physical sizes of the many species of phytoplankton. but each phytoplankton species has its own characteristic set of ambient conditions that must exist before the species "blooms," so that dominance among species of algae changes rapidly during the course of a spring season. A graph of overall primary production (Figure 12.4) consists of a series of distinct blooms by different species.

For nature to succeed in raising oyster spat, the highly variable bloom components in phytoplankton must "interlock" in time with the timely needs of the young oyster. It is possible that a single species may "fail" to bloom at all, or be delayed in its growth

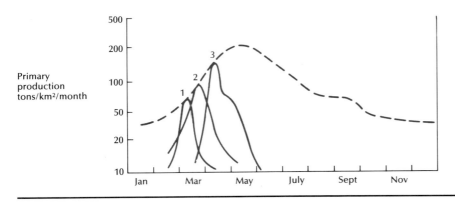

Figure 12.4 The annual cycle of total primary production. Each plant species blooms under its own preferred conditions of the primary variables of temperature, available nutrients, and light, as indicated here by the successive blooms of species 1, 2, and 3. The total curve, the dashed line, is a composite of such blooms. It is possible that a given species may fail to bloom at all, or it may be delayed in its growth pulse. Such events can have a strong impact on the food chain, particularly if the larvae of specific animals depend on this bloom for food of the *right size.*

pulse. Such events can have strong perturbation effects on the food chain, particularly if the larvae of specific animals depend on that bloom for food of the "right size" during critical early periods of larval development. Such is the complexity of "feedback" in the ocean system.

The Annual Production Cycle

Figure 12.4 examines the annual cycle of plant production as a set of contributions by many species. It is a hypothetical representation, because there are many factors that contribute to the *shaping* of the plant production curve. Principal among these are the cycle of light, the intensity of predation, and the depletion of dissolved plant nutrients.

Factors That Control Plant Production

Figure 12.5 gives the annual plant production, together with the production at the next two trophic levels, herbivore (zooplankton) and the first carnivore (herring). It is divided into several time periods so that we can describe the specific mechanisms controlling its shape.

Period a. During period *a* there is little light and many winter storms. In the fall season, November and December, cooling of the surface waters reduces the density stratification that summer heating built (see seasonal thermocline in Figure 8.3). This allows the energy of winter storms to mix the surface layer quite deeply (50 m in the case of NH-105). Plant nutrients in the deeper water are then mixed into the photic zone, a positive feedback mechanism. The state is set for the spring bloom.

Period b. Light increases in intensity and days grow rapidly longer in February and March. Plant production responds with an exponential growth rate. But during April its rate of growth diminishes; growth reaches its peak in period *d*. Two factors account for this slowing of growth. First, the bloom has *used up* a sizable share of available plant nutrients; nutrients are still available but are more difficult to find because solar heat is rapidly creating a new, seasonal, thermocline barrier to vertical mixing. Second, the population of zooplankton grazers has by now increased markedly, so predation pressure is mounting. Predation is negative feedback to plants.

Period c. Grazing pressure is the dominant cause of declining plant production in period *c*. Nutrient depletion continues; much of the nitrate originally available at the start of spring blooms has been sent to the bottom sediments via the fecal pellets of grazing copepods. Copepods also divert chemical energy into the eggs and larvae of offspring. Reproduction is positive feedback.

Period d. Periods *b* and *c* together create the peak in period *d*, that is, period *d* does not itself represent a peak *process*. But it is useful to consider how the graph might look were nutrient resupply not the limiting factor it is, for example, in upwelling. The falloff of the plant production curve in period *c* would be much less drastic, at least until the fall season and its restriction on light.

Period e. In mid-latitudes the first storms of the fall season arrive in September. These supply energy to vertical mixing and may "lift" a small amount of nutrients upward from below; with sunlight still available in this season, a short, modest *fall bloom* occurs.

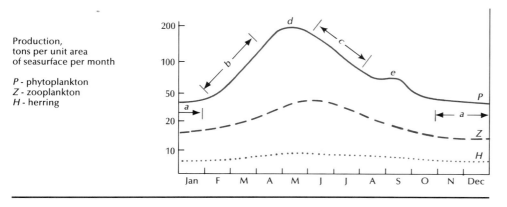

Figure 12.5 Annual production cycles of the first three trophic levels: phytoplankton, P; zooplankton, Z; and first carnivore, here taken as herring, H.

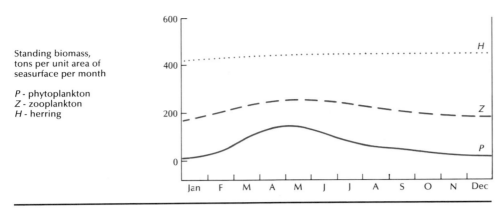

Figure 12.6 Standing biomass at three trophic levels: phytoplankton, P; zooplankton, Z; and first carnivore, herring, H. Biomass is measured as the total weight of all individuals of a given species, (or trophic level), within a stated volume of seawater at a chosen time.

Production and Biomass

It is useful at this point to tie the definitions of production and biomass to the annual production curve of Figure 12.5. We infer here that the zooplankton are dominated by copepod species. Herring, being clupeoid fish, can feed directly from the copepod level by straining them from water with special "gill rakers." The three curves here are *matched* in terms of direct conversion, one to the other. For example, during May the production by plants is about 200 tons; during the same interval zooplankton production is about 40 tons, that is, 20% of plant production goes into building new biomass in the copepods. Similarly, May production of herring biomass is about 8 tons, or 20% of the copepod production.

Now refer to Figure 12.6, which graphs the standing biomass in these same trophic levels during the annual cycle. Why is the order of its curves inverted from the order of curves in Figure 12.5? The biomass comparison indicates that there is more biomass in the herring than in copepods, and more in copepods than in phytoplankton! Yet the production curves are clearly correct in that there *must be* more tonnage of plants produced than tonnage of copepods, and so on.

This apparent *inverted order* of the biomass curves reflects several factors.

1. First, unicellular plants reproduce very rapidly. Given favorable conditions, a new generation is created every 24 hours (Table 12.1), and the biomass is doubled! Each 24 hours 50% of the biomass can be cropped, leaving the same number of plants as the day before to reproduce again, and so on. Even though the tonnage of plants at any instant (biomass) is limited, the daily production is an equal tonnage.

In comparison, the reproductive period for copepods is from 2 weeks to 2 months, with individuals living for several months. In this case, we measure production in terms of the *net* accumulation of mass within the population over the set period of time. A similar statement holds for the herring; here the reproductive period is on the order of 1 to 2 years, with individual fish living to 2 or 3 years of age.

2. The metabolism rate decreases as the size of the body increases (Table 12.1). Metabolism is de-

Table 12.1 Reproductive cycle times and metabolic rates for some marine organisms of different sizes

Organism	Size	Reproductive Cycle Time	Metabolic Rate (Relative to 1000 for Phytoplankton)
Diatom	50 microns	1 – 2 days	1000
Copepod	0.2 cm	2 weeks to 2 months	40
Herring	10 cm	1 – 2 years	2
Salmon	150 cm	3 – 4 years	1

fined as the sum of chemical changes in living cells by which energy is provided for all vital processes and activities. As a rough rule, metabolism decreases in proportion to the 0.7 power of body weight. For example, a 100-g herring will have a metabolic rate about 20 times smaller than that of a 0.005-g copepod. The rule holds both between and within species; a small herring metabolizes at a higher rate than a large herring. Larger organisms expend lower amounts of energy per unit mass in their vital activities.

It is useful to work through an example involving just diatoms and copepod.

Step 1. From the relative sizes given in Table 12.1, we calculate that the total mass of *one* copepod is the same as that of about 105,000 diatoms, assuming that both have spherical bodies. This means that for a single copepod to reproduce one offspring, which then grows to full size, 525,000 diatoms must be consumed and converted at 20% efficiency.

Step 2. By doubling every day, a single diatom yields 524,000 offspring in 19 days. This time period falls neatly within the 2 weeks to 2 months reproductive cycle of the copepod; the example is "matched" up to this point.

Step 3. In order to feed a copepod a daily ration that would in 19 days reach a total of 524,000 diatoms, its daily predation rate must be 27,600 whole diatoms.

Step 4. Now let us start the experiment in an aquarium with 27,600 diatoms and 1 copepod; after the first day's doubling of plants, 27,600 have been cropped and 27,600 remain as before. At the end of the nineteenth day, 27,600 diatoms still remain, but the copepods are increased to 2, one of which is now cropped by the next trophic level, leaving 1 copepod plus 27,600 diatoms.

Step 5. We now calculate that if the standing biomass in the aquarium is sampled at any point in time, the ratio of the weight of diatoms to that of copepods will be *1 to 4* at best (27,600 diatoms are one fourth the mass of 1 copepod). But over the 19-day cycle of production, the diatoms have outproduced the copepods by a ratio of *5 to 1* (524,000 diatoms are 5 times the mass of 1 copepod).

Returning to the curves of Figure 12.6, we conclude that the standing biomass of the larger organisms exceeds that of the smaller. If we could capture a volume of seawater and filter it through successively finer mesh-size screens, we would find that the tons of herring exceeded the tons of zooplankton, and the latter exceeded the tons of phytoplankton.

Stability of Food Webs

Here we bring together some ideas previously discussed separately. All the factors listed in Figure 12.2 as controlling the functioning of the system can also be incorporated in a discussion of stability in living communities.

"Stability is the ability of a system to maintain itself after a small external perturbation" (Hurd et al., 1971). In general, violent oscillations in any of its factors are bad for any ecosystem, and those that are unstable will experience increasing amplitude of oscillations until destruction. We can identify some characteristics that accompany stability.

Diversity in Predator-Prey. A tropical oceanic ecosystem is usually noted for complexity in its food webs. The depletion of one prey is the signal for a predator to turn to another source for sustenance. Diversity in predator–prey is a definite stabilizing factor. In effect, it switches various feedback loops on or off: by diverting its predation, an animal initiates positive feedback to the old prey but negative feedback to the new.

Cover. The marine world abounds in organisms whose adaptation to specific niches is made possible by cover. Some popularly known ones are schools of small fish that seek protective cover from predation by darting into coral reef formations, and marine worms and small porcelain crabs that obtain protection from breaking waves by living within the inter-shell spaces in mussel beds in rocky shore intertidal zones.

There are many novel examples in the marine world. Within the hard shell of sea urchins can be found a small pea crab; the crab enters the host's shell when very small and eventually grows too large to escape. For this pair cover is also a part of a symbiotic relationship. Some fishes develop a special slime coating that allows them to move into the tentacles of cnidarians such as the Portuguese man-of-war, without danger of being paralyzed by the stinging cells lining these tentacles. Cnidarians capture their prey with their stinging cells (see Plate 15.8).

Cover is also important in the open ocean. In the Sargasso Sea of the North Atlantic, floating seaweed beds provide cover for a number of species. In 1966, while on station during a cruise north of Hawaii, I observed a floating wooden orange crate. Darting in

The submerged parts of offshore platforms offer excellent cover for marine communities. Large populations of various animals from fishes such as these to demersal species such as flounder, aggregate to these structures. The structure itself becomes encrusted both with plants, mostly the blue-green algae, and with animals, mostly hydroids, barnacles, and sea anemones. (John D. Cunningham/Visuals Unlimited.)

and out of the cover provided by the crate were a number of small fishes; cruising slowly some 10 m distant from the crate was a large mai-mai (a dolphin fish), which may have claimed this microcommunity as its own special pasture for hunting.

The existence of cover, be it rock or animal or the potential for camouflage, allows the diversity of life forms to increase and extends the scope of food webs, factors that support community stability.

Migration Patterns. We do not know the underlying reasons for most of the migration patterns observed in marine species. We do observe the migration of predator species to more productive feeding grounds during set seasons. It is equally likely that movement patterns of prey species are mechanisms for avoiding predation, as hypothesized for vertical migration of copepods in the deep scattering layer case. Many other factors enter into the analyses of migration patterns: the search for food of the correct size and concentration, the avoidance of disease or parasites; or the massing of individuals for reproduction.

We can view migrations as behavior patterns that successfully survived changing conditions in the ocean. The European eel continues to migrate to the Sargasso Sea for reproduction even though the North Atlantic basin has been expanding for 200 million years, so the community in which it functions continues to exist.

Threshold Concentrations. The amassing of large numbers of individuals into a small volume is often observed. Schooling fishes such as the menhaden are highly visible examples: here biologists conclude that the schooling has a "survival" benefit in that the predator becomes confused when confronted with multiple target choices and is less likely to succeed in capturing an individual. In effect, the prey attempts to modify the feedback mechanisms that govern behavior of the predator itself.

Thresholds are presumed to function in the opposite direction as well. When the concentration of a preferred prey falls below some desired level, the predator may. switch feeding tactics to another species, perhaps in another region. The diminished population of the original prey is thus protected from overly severe depletion in numbers. In effect, the system monitors itself, converting a predation factor (negative feedback) to a protected factor (positive feedback) as needed.

Patchiness. We often find that marine organisms "clump" into patches; uniform distribution is the exception, not the rule. Such patchiness is usually related to two distinct processes. One is deliberate behavior by organisms which aggregate for reasons of spawning or for defense. But external causes can also produce patchiness. For example, a concentration of nutrients along a frontal zone between adjacent masses of water can result in high concentrations of plants relative to those in waters away from the front. Wind patterns can cause convergence or divergence of surface water; if the species are relatively nonmotile, their concentrations will also converge or diverge. Turbulence within the water is known to occur on a wide range of space scales, so turbulent eddies might also cause the concentration of individuals to vary. Whatever the cause, we find patchiness among phytoplankton and zooplankton and pelagic fishes throughout the world oceans. We conclude that patchiness plays an important role in community stability.

Statistical Probability of Encounter. Another way of discussing the "patches" phenomenon is based on the probability of encounter between predator and prey. This probability is clearly affected by patchiness. Much of the current literature in marine biology is based on statistical analyses, but which statistics to use is not always clear.

An "uncertainty" principle is involved here. We assume that our data set represents accurately the living biosphere throughout the volume of seawater sampled. In fact, it is true on some scale of time and space but untrue on others. Moreover, we apply statistics routines that are based on mathematical assumptions learned in the terrestrial world, but are these also applicable to the ocean world? The entire marine biosphere is wrapped in a maze of space scales that vary from a few centimeters to ocean wide distances of thousands of kilometers, and time scales that vary from a few seconds to the years required for particles to navigate entire ocean basins. All in all, we work in a "gray" area, heavy with uncertainties.

Dispersion. In an environment that is itself in motion, the "dispersion" factor assists in stabilizing an ecosystem. The dispersion of eggs and larvae tends to ensure that at least some offspring will develop in "suitable" habitats. Those that remain or are carried into waters of optimum T,S character will survive and propagate the species. Thus the system can adapt to gradual changes in the climate by shifting to new geographical locales. In the case of coastal animals of the benthic zone, many of which are sensitive to changes in the character of the substrate, dispersion serves a similar survival role. For example, the gradual silting of an established oyster bed is countered by dispersion of the pelagic oyster larvae over a larger geograpical area, thus increasing the probability that some progeny will settle upon suitable substrate.

Perturbations: Types, Origins, and Severity

What constitutes a perturbation to an ecosystem? Broadly defined, it is anything that upsets the "norm" of interaction among system members or of the energy source that drives the system. A norm is the *average* of physical conditions throughout the geographical area occupied by the system; under average conditions, the system functions in a normal way, with average production rates, seasonal peaks in activity of organisms, and so on.

Recall from the preceding discussion of statistics that scales of both space and time are an integral part of describing animal behavior. The same statement holds for describing a perturbation event. These ideas are best developed by studying specific cases.

Dungeness Crab Harvest on the Oregon Coast

Crab fishermen off Oregon find a high variability in the numbers of crabs taken from one year to the next (Table 12.2). The year-to-year change from 1969 to 1978 varied from a low of 53,498 pounds (1970–1971) to a high of 11,768,354 pounds (1976–1977); in contrast, the *average* annual catch over this ten-year period was 9,718,342 pounds. Although it is true that these numbers should be corrected to account for the fishing effort in each year, the data clearly show that something strongly perturbs the annual crab production in these waters.

After the small catch of 1973, Oregon State University marine scientists were asked to study the problem of diminished returns. Out of this study came a conclusion that an abnormal coastal wind can

Table 12.2 Oregon coast Dungeness crab harvest by years. (From Brown and Berry, 1979.)

Year	Pounds	Year-to-Year Change, Pounds
1969	9,783,998	
1970	14,929,347	
1971	14,875,849	53,498
1972	6,762,259	
1973	2,349,645	
1974	3,917,625	
1975	4,026,937	
1976	8,134,065	
1977	19,902,419	11,768,354
1978	12,501,274	
Total	97,183,418	Average 9,718,342

Lobster fishermen setting out from Cumberland Bay.

Hand-hauling a lobster pot from 100-m bottom depth.

The pot traps are hand-crafted of local woods.

Measuring the length of the carapace of a spiny lobster, *Jasus frontalis*. The fishermen belong to a cooperative which markets the catch.

An artesenal fishery based on pot traps, Juan Fernández Islands, Chile. One of these islands, Más a Tierra, was the location of the cave dwelling of Alexander Selkirk, the castaway sailor whose adventures on a remote island were the inspiration for Daniel Defoe's classic, *Robinson Crusoe* (Courtesy Steve Neshyba, Oregon State University.)

cause a high loss rate in juvenile Dungeness crab and thus reduce the commercial catch several years later. The crab has a pelagic larval state, as do most marine animals; the young larva undergoes a series of molts that terminates in a megalops stage in late spring. It is during this stage that the juvenile leaves the surface waters and takes up its young adult life on the floor of the continental shelf.

As we learned in previous chapters, northerly winds along the north–south-oriented coast of Oregon produce an upwelling regime; a part of that dynamic system involves the offshore transport of a thin layer of surface water. If such a wind occurs during the short time interval in which the Dungeness crab larvae are in the megalops stage, the larvae are moved offshore along with the surface water, and, as they begin their migration to the bottom, many individuals end up in offshore waters too deep for their normal adult niche in the coastal ecosystem. In a sense they are lost.

Analyses of wind patterns showed that such anomalous winds did occur in 1970. The study concluded that the mortality of Dungeness crab that year showed up in subsequent years as a drastic reduction of commercially harvestable crabs.

Now the question is whether such an event is a perturbation. The answer is no. Such perturbations are themselves a part of the "norm" of conditions for the coastal ecosystem off Oregon. Stated another way, the community of organisms we find here today is precisely composed of the species interactions that together form a stable ecosystem in a geographical area over which occasional anomalous winds are an integral part.

Mt. St. Helens 1980

In May 1980 Mt. St. Helens erupted with tremendous force. Ash falls covered large areas of southeast Washington State. The heat produced during the eruption melted accumulated snows and this water, mixed with ash, produced huge mud slides. Rivers in the vicinity of the volcano were choked with debris and ash sediment. Was this event a perturbation to the ecosystem to which the anadromous salmon belong? (Anadromous fishes migrate from fresh to salt water and back again.)

The answer would seem to be yes. Former gravel beds along the Toutle River in which the salmon spawned were wiped out or covered with silt and ash. Both the acidity and turbidity of the waters of the Columbia River were also altered, impacting in-

vertebrate animals as well as fishes. The ash clouds were carried eastward across Idaho and Montana, and fallout impacted a large swath of agricultural lands. There the addition of new minerals served somewhat as a fertilizer if it fell in moderate amounts. The watershed of the Columbia River received a widespread chemical disturbance.

But volcanoes of the Pacific Northwest have been erupting in similar fashion throughout geologic time. On the scale of space in the near vicinity of Mt. St. Helens, an eruption is a local perturbation; community succession is initiated and the system proceeds in an orderly manner toward its climax state. On the space scale of the entire Pacific Northwest with its many volcanic sources, a single eruption event is not a perturbation; rather it is the *norm* of physical events within this larger context. Taken as a whole, the oceans of the Pacific Northwest and the associated geologic activity of the Cascade mountain range are together an ecosystem of specific character. Its character would be different were it denied the input from sporadic volcanic eruptions.

The Olympic Oyster

When the Pacific Northwest was first occupied by Western people in the mid-1800s, the shallow inland waters of Puget Sound abounded in beds of native oysters, particularly along the Olympic Peninsula whence the oyster's name was derived. This species

A riverbed choked with ash and silt from the eruption of Mt. St. Helens, May 1980. (Arthur Grace/Sygma.)

CASE IN POINT

PERTURBATION OF THE CAROLINA ESTUARINE
SYSTEM

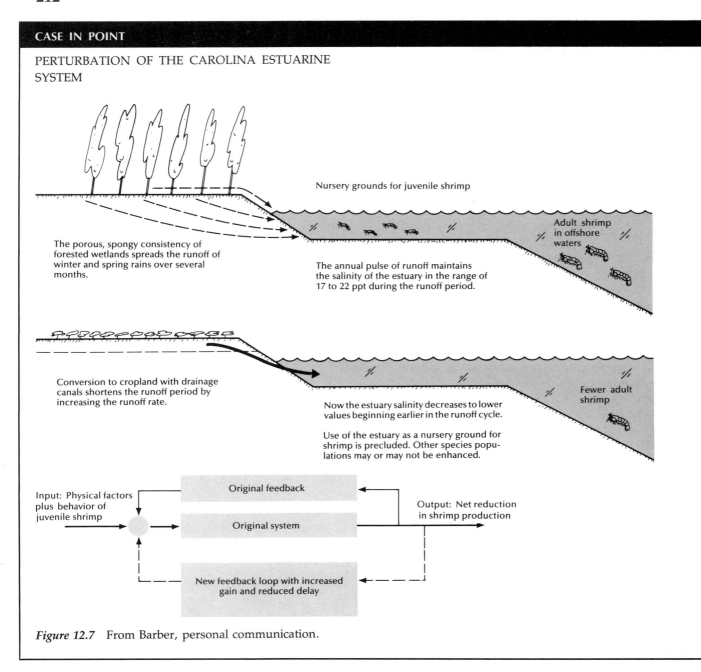

Nursery grounds for juvenile shrimp

Adult shrimp in offshore waters

The porous, spongy consistency of forested wetlands spreads the runoff of winter and spring rains over several months.

The annual pulse of runoff maintains the salinity of the estuary in the range of 17 to 22 ppt during the runoff period.

Conversion to cropland with drainage canals shortens the runoff period by increasing the runoff rate.

Now the estuary salinity decreases to lower values beginning earlier in the runoff cycle.

Use of the estuary as a nursery ground for shrimp is precluded. Other species populations may or may not be enhanced.

Fewer adult shrimp

Input: Physical factors plus behavior of juvenile shrimp

Original feedback

Original system

Output: Net reduction in shrimp production

New feedback loop with increased gain and reduced delay

Figure 12.7 From Barber, personal communication.

yields a small-size adult with succulent flavor, and many boatloads were exported to San Francisco and other coastal settlements. As production fell off after overharvesting, new oyster seed was imported from Japan in experiments to find a more rapid-growing oyster to sustain the market demand. Unfortunately, the new seed also carried the seed of a deadly predator snail, *Ocenebra japonica*, which was not native to the Puget Sound waters but thrived in these waters. The Olympic oyster was particularly hard hit by the invasion of these snails, which drill holes in oyster shells and consume the flesh *in situ*.

Was the event a perturbation to the Pacific Northwest coastal ecosystem? Clearly so. Further, once established, the drilling snail could not be eradicated and the system was permanently altered. Although some natural production of Olympic oysters remains today, the equilibrium levels between oysters and drilling snails are such that major production will probably never again occur.

Wetlands Drainage along the Carolina Coasts

The shallow, extensive shelf waters of the Carolinas are rich in marine life. Geologically, the shelf is a continuation of the lowlands and marshy forests of this section of the Atlantic seaboard of the United States. Many estuaries exist here, with their upper reaches in the natural drainage canals that thread the coastal lands.

Among the many marine species that depend on estuary habitats during some phase of their life cycle, shrimp species such as *Paneaus setiferus* are important. These shrimp spawn at sea, but the juveniles migrate into the estuaries during seasons of the year in which the runoff of coastal waters is high; after several months in these rich nursery waters, the young adults return to the open-shelf waters. The salinity of the estuarine waters in which they spend their nursery phase is reduced by river runoff, but the organic content of the waters is increased by the particulate matter washed off the land. Studies have shown that the salinity of the estuaries is controlled by the *rate of runoff*, which is in turn controlled by the percolation rate of water through the grassy marshes and porous beds of adjacent forests (Figure 12.7).

Recently, large farms have been constructed from these forests. Trees were clear-cut, the ground was contoured for more rapid drainage, and new canals were dug to increase the runoff rate. These modifications have altered more or less permanently the annual cycle of salinity change within the estuaries: now the pulse of spring runoff is much more rapid, and the estuarine waters are more turbid with silt and less saline than before. The result has been disaster for the offshore shrimp population, which is now denied nursery grounds of the specific character to which these animals were adapted. The program produced perturbations of a lasting nature, not only in the shrimp population but in the established human population of the coastal region whose livelihood was dependent on the shrimp catch.

Summary

The chapter begins Part V, The Marine Biosphere, with a list of terms and definitions pertinent to biological systems. Biological oceanographers find that their work is hampered to an extreme degree by the lack of adequate data. This deficiency holds for virtually all types of biological data, from inadequate samples of animal concentrations to multiyear variations in phytoplankton bloom successions. Two principal limitations exist: one is the very high cost of taking systematic measurements over a long enough period to gain statistical confidence in the results; and the second is the excessive amount of time required to analyze captured specimens, especially when such analyses require the use of microscopes.

From the view of collecting specimens, biologists look through a narrow window of success in sampling the full range of marine organisms. Among the first trophic levels are plants one-millionth of a meter in size, with predators that are themselves only ten times larger and still microscopic. At the large end of the size spectrum are animals that can evade capture. A biological system, like any system, can be modeled if the appropriate feedback loops are known, but these must *all* be known for the model to be a realistic one. To take "adequate" samples from a biosphere with such a huge range of sizes is a major constraint to successful modeling.

The marine biosphere differs in still another respect. With the plants so small, but with very short reproductive cycles, it is possible that the standing biomass of plant material in the ocean is less than that of the herbivores. This factor makes sampling much more difficult because *time* and *timing* add additional constraints.

Within any biosphere it is instructive to examine the elements that make it stable. In name, these are similar to those found in terrestrial systems, but their application to the fluid world can be quite different. For instance, *cover* is a well-studied part of, say, the mouse versus owl interactions, but in the marine world cover does not exist, except for some benthic species. On the other hand, dispersion plays much the same role of supporting species survival in the marine world as it does on land.

Having looked at the factors that organize the marine biosphere, we also examine factors that might disrupt it. Several case studies are given of disruptions that are not true perturbations and some that are.

Study Questions

1. The overview for Part V is summarized in Figure 12.1, which shows the ecosystem ocean. Briefly discuss the four major facets of the overview using information taken from the figure.

2. According to Hardy (1965), the aim of ecology is to discover what?

3. Consider a food chain with four trophic levels and in which the top carnivore population is decimated by a severe disease. Estimate the impact (type of feedback) and its severity on each of the first three trophic levels.

4. One way to understand how feedback works in biological systems is to contrast the overall stability of a food *chain* with that of a food *web*. Explain why *commercial* fisheries that take meat for human consumption can be more destabilizing than those that take fish for the fishmeal industry.

5. The text explains the difference between production and biomass. Show via a short essay that you know the difference.

6. What two factors are most influential in causing the biomass of herbivores to exceed that of the plants themselves in the open ocean?

7. Write a brief explanation of why the primary productivity curve of Figure 12.5 should *not* be considered as the way *each* species of plants varies its production over a year.

8. What are some of the stabilizing factors that you find most interesting in the marine community?

Are these applicable to your own community stability?

9. It seems that the more diverse a community is, the more stable it is. Defend.

10. In one sense, discussing stabilizing factors is not very useful because we never have the chance to observe ecosystems that *became* unstable. True or false?

11. Review again the connection between the *chance encounter* factor and the physical characteristics of seawater. Would marine systems be different *if the organism could see for long distances?* Essay.

12. At first look, it would appear that the *dispersion* factor that is a part of a moving, often turbulent habitat like the interior ocean would be a *destabilizing* factor. Defend or oppose.

13. Construct a time plot of the Oregon Dungeness crab harvest by years (Table 12.1). (For a guide, look at the way Figure 12.5 is plotted.) Does the graph show a cyclic nature? If so, over what span of years does the cycle occur?

14. The media during 1983–1984 spoke often of the El Niño phenomenon off Peru and California. Both 1972 and1976 were years of strong El Niño currents. What do you make of the connection between these and the variation of Dungeness crab harvest.

15. From the examples given in the text concerning different kinds of perturbations to an ecosystem, which type of perturbation is the most damaging, a Mt. St. Helens or a man-made type such as the coastal Carolina farms?

CHAPTER 13

ORGANIZING THE MARINE BIOSPHERE

The previous chapter developed an overview picture of the marine biosphere, using a very broad brush. For most of us, however, interest and excitement about marine life increase directly as the object we view becomes more and more detailed, more specific, and certainly more and more intricate. As proof of this trend, take a quick advance look at the teeth of a copepod in Figure 13.2; I for one simply marveled at the magnification of the tiny teeth of a tiny oceanic vegetarian, only a few millionths of a meter in size! Even more, each tiny tooth has a tinier duct through which the grazer injects digestive juices directly into the crushed case of the plant cell! The fact is that we become utterly fascinated by the special structures and features of marine life forms. Which brings up a para-dox. If specialty is pleasing and generality is boring, how then should one organize the subject of the marine biosphere to "capture" the reader at the highest intensity of interest possible?

Clearly, one scheme would be to discuss the communities according to the provinces outlined in Chapter 2; that would be repetitious. Other schemes are more versatile. For example, for a living system that is founded on microscopic-size plants, it is an interesting exercise to organize the entire food web on the basis of feeding types, or feeding mechanisms. We discuss here some of the principal techniques used by marine biologists in dividing up the biosphere into parts that are easy to analyze and easy to understand.

The Taxonomic Approach

Traditionally, marine biologists have organized marine life using a taxonomic scheme in which the many species are grouped into the several phyla according to their complexity and evolution through geologic time. The rationale is that a biosphere *can be described* adequately *only if* its members are named according to an accepted set of rules. We owe the origin of our currently used taxonomic scheme to Carolus Linnaeus (1707–1778), an eighteenth-century botanist. It consists of a hierarchy of levels, called taxa, into which organisms are grouped.

Taxa	*Example: The purple sea urchin*
Kingdom	Animalia
Phylum	Echinodermata
Class	Echinoidea
Order	Echinoida
Family	Strongylocentrotidae
Genus	*Strongylocentrotus*
Species	*S. purpuratus*

The primary purpose of a taxonomic method is to assign each species its own unique name. Another goal is to group organisms into natural units whose sequence follows a genesis pattern, that is, a family tree.

The kingdom Plantae contains all the autotrophs, both marine and terrestrial. Marine plant species are usually assembled into a subkingdom Thallophyta; these are plants without roots, leaves, or vascular tissue and include phytoplankton as well as seaweeds. Table 13.1 lists the several phyla of marine plants and their characteristics and gives examples. All are members of the Thallophyta, distinct from the other subkingdom Embryophyta in their reproductive strategy. The Embryophyta retain the female sex organ until after fertilization and development of the embryo and include all the terrestrial plants, with only a few marine species such as the commonly known eelgrasses and mangroves.

The kingdom Animalia contains the heterotrophs. Marine animals are in general just as complex as ter-

Table 13.1 The phyla of marine algae. (From Fell, 1975.)

Phylum	Characteristics	Examples
Cyanophyta	Minute floating one-celled plants that lack a distinct cell nucleus (the genetic material being dispersed in the cell); often colored blue-green because the pigment phycocyanin is present.	Blue-green algae
Chrysophyta	Minute floating or attached one-celled plants, having a distinct nucleus and the cell body enclosed in a two-valved secreted silica capsule; usually brown or yellow pigments present, plus chlorophyll.	Diatoms
Pyrrophyta	Extremely minute floating and swimming one-celled plants (requiring a magnification of at least x 500 to reveal detail); having one or two threadlike flagella used as swimming organs and usually several cellulose plates forming a jointed capsule.	Dinoflagellates
Phaeophyta	Conspicuous brown seaweeds attached to rocks on coasts and to hard substrates of the continental shelf; some have hollow flotation bladders enabling them to float if their attachment is lost; contain brown pigments plus chlorophyll.	Kelp Sargasso weed
Chlorophyta	Green seaweeds, some minute and one-celled or of a few cells, others larger and resembling moss or lettuce, the larger kinds attached to the seabed; chlorophyll the main or only pigment.	Sea lettuce
Rhodophyta	Red or purple seaweeds, usually a few inches long, never massive, containing red pigments as well as chlorophyll; sometimes also containing lime (in which case they resemble encrusting coral or paint spills on rocks); usually attached to substrate.	Dulse Corallines

Table 13.2 The most conspicuous phyla of marine invertebrates. (From Fell, 1975.)

Phylum	Characteristics	Examples
Protozoa	Minute one-celled animals, marine forms often with a limy or silica capsule; differ from minute algae by lacking chlorophyll; floating or benthic.	Forams Radiolarians
Porifera	Multicellular anchored forms with no mouth or central digestive cavity; grow on seabed.	Sponges
Coelenterata	Multicellular anchored or floating forms with a radially symmetrical body and a mouth leading into a central digestive chamber, but without a coelom; solitary or colonial.	Jellyfishes Sea anemones Corals
Echinodermata	Radially symmetrical coelomates, mostly free-roaming on the seabed, sometimes attached or floating; surface of body commonly spiny; solitary.	Sea urchins. Starfishes
Platyhelminthes	Bilaterally symmetrical, flattened, wormlike forms, with mouth and digestive cavity present in free-living forms but lost in some parasitic forms.	Turbellarians Fish flukes Tapeworms
Aschelminthes	Body cylindrical, wormlike, tapering at tips, with mouth and digestive cavity; coelom imperfectly developed, no obvious segmentation; often parasitic.	Roundworms
Annelida	Body elongated, wormlike, flattened or cylindrical, with well-developed segmentation, mouth, and digestive tract and a segmented coelom; mainly benthic.	Bristle worms
Arthropoda	Body bilaterally symmetrical, covered by an external chitinous skeleton, and having well-developed, jointed, paired limbs, mouth with jaws, and complex gut.	Shrimps Crabs Lobsters
Mollusca	Body bilaterally symmetrical (though sometimes coiled in a helix), usually secreting a protective limy shell of one, two, or several valves; mouth and gut well developed, but coelom secondarily reduced.	Sea snails Clams Octopuses

restrial animals, so there are members in nearly all of the same phyla. Biologists usually specialize their studies along two general lines that are not in themselves taxa: these are the invertebrates, organisms that lack a backbone, and the chordates. In Table 13.2 Fell has summarized the characteristics of the more conspicuous of the marine invertebrates. They are mostly small in size (with the exceptions of some

Table 13.3 Classes of marine invertebrates in the phylum Chordata. (From Fell, 1975.)

Class	Vernacular Names	Skeleton	Body Characters	Remarks
Cyclostomata	Lampreys and hagfishes	Cartilage only	Body eellike, no paired fins, mouth suctorial, without jaws	Aquatic animals swimming by tail motion, fins acting as stabilizers, and breathing by means of gills in water
Chondrichthyes	Sharks and rays	Cartilage only	True fishes with paired fins and paired jaws (one or both pairs of paired fins occasionally lacking)	
Osteichthyes	Bone fishes, codfishes, flounders, herrings, eels, perches, and so on	Bone and cartilage		
Reptilia	Marine crocodiles, marine turtles, and sea snakes	Bone and cartilage	Cold-blooded, egg-laying, scaled tetrapods	Originally terrestrial tetrapods, readapted to life in the sea; using limbs as paddles and breathing by means of lungs in air
Aves	Oceanic and shore birds, penguins, gulls, terns, and so on	Bone and cartilage	Warm-blooded egg-laying, feathered tetrapods	
Mammalia	Dolphins and whales, seals, manatees, and so on	Bone and cartilage	Warm-blooded, viviparous, haired, milk-secreting tetrapods	

truly large species such as the giant squid). The smallest are the single-celled protozoans, which include orders such as the Foraminifera. The larger are animals with which we are familiar, such as shrimps, lobsters, and jellyfishes. In size, the jellyfishes are among the largest, some species reaching 2 m in diameter. Among the Arthropoda are the commonly called Crustacea, such as shrimps, copepods, and crabs, which incorporate an exoskeleton that covers the entire body. Also present are the Mollusca, such as the chambered nautilus and the common clams and oysters, which incorporate a hardened shell but which expose their flesh.

Finally, there are the classes of marine vertebrates within the phylum Chordata, as listed in Table 13.3. These cover all large animals from fishes to marine birds to mammals such as whales and porpoises.

Organizing by Feeding Types

The most abundant species of living organisms on this earth are the single-celled marine plants that are microscopic in size and float within the oceanic soup. These are the producers in the marine world, and so it follows that among marine life forms the mechanisms of feeding cover a much *wider dynamic range* than in the terrestrial sphere. We can use the diversity in feeding types as a way of organizing our study of marine life.

Suspension Feeders

The relatively small density difference between water and protoplasm, either living or dead, makes the ocean a *suspension* of tiny bits of protoplasm scattered throughout the water column. In the surface layer are mostly the unicellular algae. Deeper in the water column we find small chunks of surface animals that have died and are slowly sinking while simultaneously being attacked by bacteria, which act to break up the original animal body. Near the ocean floor currents are continuously moving the sediment and resuspending material that accumulates there. It is not surprising then that many heterotrophs have evolved to feed by straining this suspended organic material from the water. We describe typical animals from each of these levels within the water column.

The Copepod–Alga Case in Surface Waters. Of particular importance to understanding the ecosystem ocean is the fundamental "grazing" of microscopic plants by the herbivores. Only in very recent years have we made substantial progress in describing the details of this important process. For the first time marine biologists have been able to ask whether copepods "select" algae, and if so on what basis. Electronic "particle counters" now enable us to measure the numbers and sizes of particles in laboratory feeding tanks; by successive measurements during the feeding cycles of copepods, we discover which particles are being eaten and how fast.

High-speed photography (movies taken at 1000 frames per second) has revealed the motion of the appendages of copepods during the feeding process. Figure 13.1 diagrams what has been learned in one

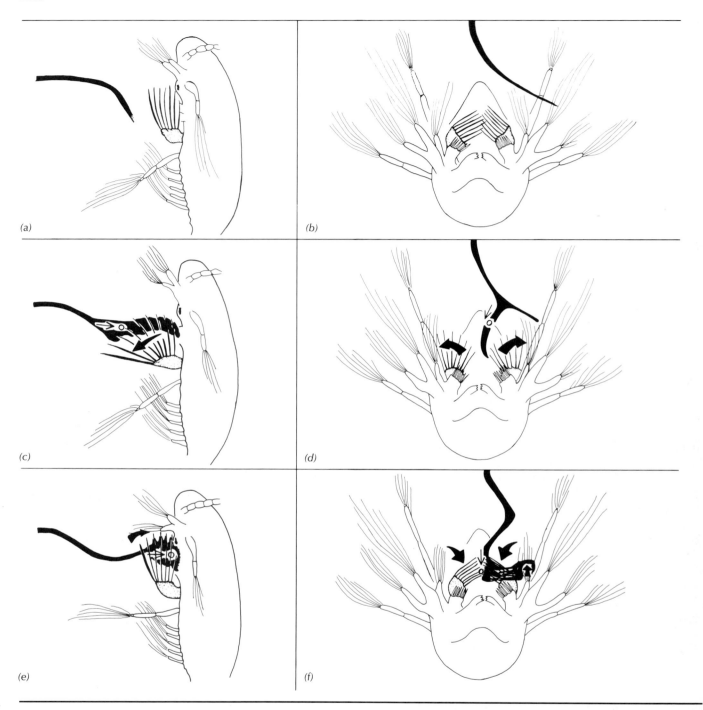

Figure 13.1 Time sequence of a copepod engaged in feeding. Black streaks are dye streams from a micropipette. Heavy arrows indicate movements of the second maxillae, and of a first maxilla in part *f*. Circles represent positions of algae and fine arrows their movements. In the first column the copepod is viewed from its left side and the first maxilla has been left off for clarity. In the second column the copepod is viewed from its anterior end. *(a,b)* Feeding currents bypass the second maxillae until an alga nears them. The alga is captured by *(c,d)* an outward fling and *(e,f)* an inward sweep of the second maxillae. (From Koehl and Strickler, 1981.)

such experiment. Water is a viscous medium at the size scale of the appendages and mouthparts of a copepod. Therefore, water that is moved by the feeding appendages has no residual momentum—it does not "coast" once the animal stops pumping. Water is shown to be moved in a series of "steplike" advances toward the mouth by the alternate flipping action of appendages.

> Filter-feeding crustacea such as copepods or euphausids extract the particulate plants from the water with a meshwork of setae (hair-like exoskeletal projections). Taken in order, the second antenna and the maxilliped move forward and backward, respectively, advancing water one "step" toward the ventral side from the surrounding volume. A ventral-to-dorsal sweep of the mandibular palp then advances it laterally and dorsally, just past the end of the second maxilla. This sequence repeats, moving water toward and around the body in a sequence of definite, individual steps. It is abruptly stopped when a particle of food approaches the animal. Then the second maxillae swing laterally, with spreading of the setae. The particle "steps" between the extended tines of the second maxillae, which then close over it. The tines are pulled closer and closer together and to the body wall, squeezing the viscous water out through the spaces between the setules (small hairlets on the hairs)—There is also a reverse sequence of

moves that push water out of the space between the second maxillae, which is used to reject particles that prove unwanted. Desired captured particles are raked from the second maxilla by the medial, proximal lobe of the first maxilla which is armed with stout spines for the purpose. These swing through the medial space between the labial palps that close the posterior side of the mouth (Miller, 1982).

The copepod is apparently able to detect particles before they touch any part of the animal itself. In this sense, copepods are not "direct filter" feeders but choose or reject suspended particles as potential food. Size of particle is but one of several selection criteria but clearly an important one: for example, copepods have been found to eat particles of two distinct sizes and manage not to eat particles of intermediate size. Electronic microscope pictures of the jaw structure of copepods reveal their complex and highly structured teeth, presumably needed to crush the silicate shells of diatom shells (Figure 13.2).

Fleming (1939) carried out an interesting experiment on the predation efficiency of copepods. He created a solution (not nutrient limited) of diatoms and copepods in which the steady-state biomass of diatoms remained constant, that is, copepod density

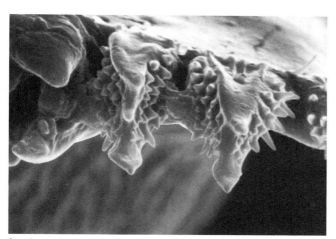

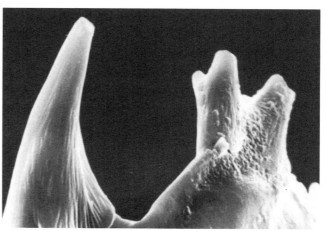

Species:

Calanus plumchrus
Body length (adult): 5 mm
Detail of two dorsal teeth of the left mandible.

These photographs, obtained with a scanning electron micrograph, show extraordinary details of the teeth of two species of copepods. It should come as no surprise that the copepod has evolved with such teeth. Recall that they graze on phytoplankton such as the diatom with its hard siliceous shell about 20 microns in length. The teeth shown are about 5 microns across; the copepod can apply about four teeth across the diatom's shell to crush it. In addition, these calcium teeth, which are much like our own, have interior ducts through

Species:

Acartia longiremis
Body length (adult): 2 mm
Detail of the ventral end of the right mandible

which they eject a toxic fluid into the protoplasm of the ruptured diatom. This fluid is an enzyme similar to our saliva and aids in the digestive process. As might be expected, the teeth of the older individuals are considerably worn, a consequence of grinding shells with the consistency of sand (SiO_2). (Photographs courtesy Barbara K. Sullivan, University of Rhode Island. From "A SEM Study of the Mandibular Morphology of Boreal Copepods" by Barbara K. Sullivan, C. B. Miller, W. T. Peterson and A. H. Soeldner.)

Figure 13.2 Copepods have teeth!

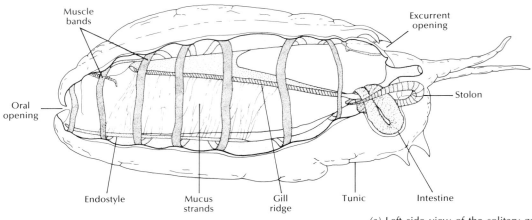

Muscle bands

Excurrent opening

Oral opening

Stolon

Endostyle

Mucus strands

Gill ridge

Tunic

Intestine

(a) Left side view of the solitary generation of the thaliacean *Thalia democratica*. Commonly called salps, these animals have a barrel-shaped body surrrounded by a transparent cellulose "tunic." The circular muscle bands produce contractions that drive water through the body, providing both food and propulsion. Small phytoplankton cells in the incoming water are filtered out by strands of mucus secreted by the endostyle. These strands collect on the gill ridge and are then passed to the intestine. (From Metcalf, 1919.)

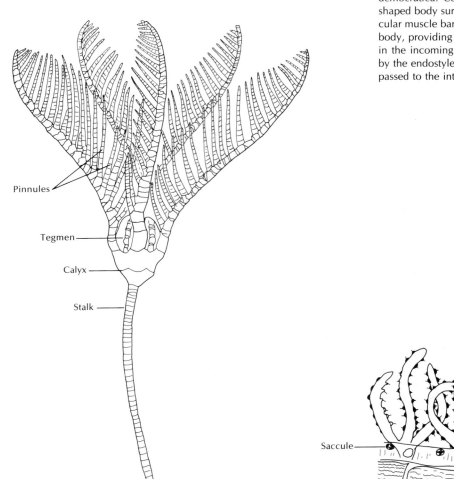

Pinnules

Tegmen

Calyx

Stalk

Podia

Water canal

Saccule

Ossicle

(b) The stalked crinoid *Ptilocrinus pinnatus,* sometimes called the sea lily. (From Barnes, 1980.)

(c) The crinoid arm, showing podia and the mucus-filled canal which transports captured detrital material to the mouth. (From Hyman, 19, 1963.)

Figure 13.3 Examples of the use of mucus by filter feeders.

was just that required to crop new daily production by diatoms. He then doubled the copepod density and measured a drop in diatom biomass of about 75% in 5 days, clearly an *overgrazing* situation. Increasing the copepod density to five times its steady-state value, wiped out the entire population of diatoms in just 5 days, testimony to the *predation efficiency* of the copepod's technique!

Mucoid Filter Feeders. The use of films of mucus to gather suspended particles is very widespread among marine animals. The technique is used by species in virtually every phylum. Distribution of mucus-using animals follows closely the distribution of suspended particle density. Thus many examples are found in the surface layer, where particulate density is highest, and again near the ocean floor, where agitation by currents and animal activity itself tends to stir up bottom sediments. Mucoid feeding is not prevalent in the mid-ocean depths, where the particulate count is low; this zone is dominated by active predator–prey behavior.

In near-surface waters salps are found, a filter feeder several centimeters in length and having a barrel-shaped body (Figure 13.3a). Through the use of muscular bands around the body the animal pumps water through its interior and across a cone-shaped sheet of mucus that filters particles out of the water stream. The sheet is a filter that is continuously generated at one side and consumed at the other. Electron microscope photographs of the sheet of mucus show highly regular and uniform-size openings on the order of 1 micron (one-millionth of a meter) square. The salps are members of the subphylum Urochordata, class Thaliacea, and are a commonly encountered open-ocean animal.

Also found in the surface layers but at deeper depths as well are the pseudothecosomes. These animals generate a huge bubble of mucous film up to a meter in diameter; the animal itself is but a few centimeters in size. When suspended in water this bubble is so large that it is literally dragged around by the motion of the currents and captures falling detritus on its sticky surface. Periodically, the animal consumes the entire structure and then spins out another.

Still other animals coat appendages with a mucous film and periodically wipe off the film with their captured particles, or they pump the enriched mucus along their appendages toward the mouth. The crinoids of the phylum Echinodermata, class Crinoidea, pump mucus toward the mouth. They are the most ancient and in some respects the most

primitive of the living classes of echinoderms, which include the starfishes and sea urchins. Some species of crinoids, called sea lilies, are attached to the bottom substrate and extend stalklike into the currents above the bottom (Figure 13.3b). They feed by means of podia, short appendages attached to the stalked arm, which contains canals coated with mucus. When podia contact a suspended particle, they "toss" the particle into the adjacent canal. Cilia within the canals beat in wavelike patterns to move the mucus along the canals toward the mouth structure (Figure 13.3c).

Some worms also employ mucous membranes across which water is pumped. A well-known example is the *Urechis caupo* worm found in shallows and mud flats. It digs a u-shaped burrow, with inlet and outlet, and secretes a mucous membrane across the inlet opening. Then, by muscular contraction, the worm pumps water through the burrow and periodically consumes the mucous film upon which suspended material has adhered.

Among the commonly encountered suspension feeders are animals we ourselves eat, for example, oysters, clams, and mussels, all members of the phylum Mollusca and all members of the benthic fauna. Further, all these animals employ some type of selection–rejection mechanism to remove the *organic* matter from the total of suspended material filtered out of the water. Human activities can severely impact these filter feeders: dredging of ship channels tends to resuspend much fine silt, which reduces the clam growth rate because (1) silt clogs their filter mechanisms; (2) the animal uses more energy in periodic flushing of the clogged filters; and (3) the added time used for flushing reduces "productive" feeding time. (Herbert F. Frolander, personal communication).

Deposit Feeders

The accumulation of organic debris on the ocean floor gives rise to a food chain that, at least in the depths of the open ocean, is strikingly different from the chain that occupies the photic zone. The energy input to the benthic chain consists of material already "dead" but still containing large quantities of chemical potential energy.

Bacteria are found in large numbers in the ocean sediment. Here these play the role of decomposers. Their activity tends to break down large-size organic debris, for example, fish bodies or dislocated kelp fronds. The small particles then become a part of the suspensoid upon which bottom-dwelling filter feed-

Raking for oysters in Chesapeake Bay. Oysters feed by filtering out particles of organic material suspended in water (Everett C. Johnson/Leo de Wys.)

ers depend. In fact, a substantial portion of the suspensoid is the total of bacteria bodies themselves.

After bacteria, the most numerous organisms in the benthic chain are worms. The largest fraction of these individuals are burrowing types that work through the accumulated sediment layer. These feed by passing "mud" through digestive tracts that en-

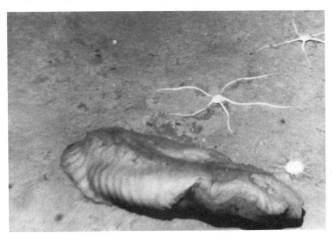

Echinoderms at 1800-m depth, photographed from the research submersible ALVIN. Two brittle stars are present, along with a sea urchin and the sea cucumber *Peloptides gigantea;* all feed on organic detritus that collects on the ocean floor (Fred Grassle, Woods Hole Oceanographic Institute.)

zymatically break down the organic constituents. Worms in turn become an important food source for many larger animals, including the many species of "flatfish" such as flounder, sole, and other members of the plaice fish.

Moving over the accumulated sediment rather than through it are many species of deposit feeders; they attack the larger-size particles in the debris. Notable among these are the brittle stars; with the advent of undersea cameras and the resultant accumulation of photographs of the ocean floor, oceanographers have found these animals to be widespread over all the world's oceans, from tropics to polar regions. They feed by ingesting the sediment mixtures, concentrating on spots with high organic content, and are thus somewhat selective. Brittle stars belong to the class Ophiuroidea, within the phylum Echinodermata.

Much less selective among deposit feeders are those that simply "sweep" the surface of the sediment layer. Typical examples are the snails (mollusks) and sea cucumbers (echinoderms). The genus *Cucumaria* has developed specialized tentacles that are first coated with a film of mucus and then swept back and forth as the animal crawls along the bottom. Periodically, the animal cleans a tentacle with its mouthparts to remove the collected organics and to recoat it with mucus.

Direct Predator–Prey Feeders

Predator–prey interaction requires no amplification here. We commonly think of oceanic animals as feeding by this means, and it is found throughout ocean space. It is *not* the most commonly encountered means of feeding, however.

Symbiotic Feeders

Just as in the terrestrial biosphere, the marine biosphere abounds in examples of symbiotic relations among feeders. Classical analysis recognizes three combinations.

Parasitism. Straight parasitic associations are those in which the host is alive but *harmed* by the presence of a parasite. An example is the bacterial disease *Vibrio* in salmonids, which can kill the host; the disease is similar to cholera in humans. Parasitic worms and flukes in the flesh or internal organs of marine animals are common pathology problems for marine biologists.

Commensalism. Commensalism is an association that is beneficial to the parasite but of no apparent consequence to the host. An example is the finding of a small pea crab inside the shell of a living clam—the crab obtains important *cover* with no apparent harmful impact on the bivalve mollusk.

The gonads of the sea urchin are considered delicacies in many maritime human cultures. When we open the test of the urchin, we may often find a small crab that entered the interior chamber of the urchin while a small larva but has grown too large to escape. Although the association is not clearly understood, humans obtain a direct benefit. A neat way of telling whether the urchin is fresh is to see whether the associate crab is still alive.

Mutualism. Mutualism describes an association in which both host and parasite benefit. There are many examples in the marine world. An example both novel and widespread is that of living algae cells within the transparent tissue of some animals. The polyps that comprise much of the population of coral reefs contain zooxanthellae, a symbiotic algae. On exposure to sunlight the algae synthesize and multiply, using as raw materials the metabolic by-products of the animal tissue, nitrogen (in urea or ammonia), phosphates, carbon dioxide, and others. The animal in turn consumes some of the algae as food and benefits also from the oxygen by-product of photosynthesis.

Even some very large animals contain symbiotic algae. The giant clam *Tridacna* found on the Great Barrier Reef off Australia spreads its mantle tissue over the edges of its calcareous shell to give its symbiotic algae exposure to light—the many colors seen in the mantle tissue are those of the embedded algae. In some cases, such as that of the worm *Convoluta roscoffensis*, the initial symbiotic relation changes to total dependency of the host on the al-

Mutualism at work. Various reef fishes gathered at a "cleaning station" where smaller wrasse fish, the dark ones with light stripes, clean them of parasites, dead tissue, and so on. (Dave Woodward/Taurus.)

gae. This animal actually becomes more and more transparent in its adult phase as it "suns" upon beach sands to give the algae needed light.

Organization Based on Special Food Chains

Some specialized, usually regional food chains are quite distinct from the normal, open-ocean types. There can be strong justification for organizing biological oceanography on this basis. Several of the more distinct are the following.

Eastern Boundary Communities. One feature that is common to the eastern boundaries of all the major oceans is coastal upwelling. In Chapter 15 we shall study the biological parts of one of these, the California Current system. Figure 15.4 shows that the higher trophic levels of the four eastern boundary communities all involve the same genera. This commonality extends to the lower trophic levels as well. Details of this community and its food chains are deferred to that chapter.

The Southern Ocean Food Chain. The Southern Ocean community is a very special one in many ways. First, it is perhaps the shortest food chain that is at the same time so extensive. Briefly, it consists of diatoms and protozoa at the microscopic level, the euphausid *Euphausia superba* as the principal grazer, and whales, seals and various birds as top carnivores. Second, the three-trophic-level chain has immense production capacity. Third, it is so very far away and covered by ice pack during so much of the year that we are unlikely to solve its intricacies soon (see Chapter 20).

The Bering Sea Saprophytic Chain. Another special community occurs over large areas of the shallow Bering Sea Shelf (Figure 13.4). The material that forms the base of the food chain in this sea is organic material derived from the continent itself. Vast areas of tidelands and floodplains occur at the deltas of the Yukon and Kuskoquim rivers. These are the sources of the eelgrass that is carried to the shelf during spring runoff. Bacteria decompose the dead plant material into suspensoids of organic debris, including the dead bodies of bacteria, and all this becomes food for filter feeders like clams. The Bering Sea system has one additional factor that makes it unique. The entire shelf is covered with pack ice in winter and returns to open water in summer. Ice

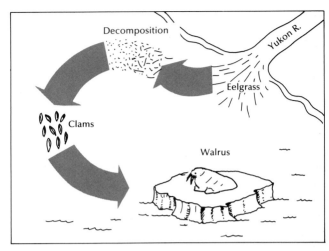

The Bering Sea Shelf, an enormous area some 500 km wide and 700 km long, with an average depth of about 75 m, is a highly productive area. During the spring months, when the ice recedes toward the north, wildlife observers have counted up to 40,000 walrus along the edge of the ice pack. Each adult walrus consumes about 2 bushels of clams per day. The animal uses its two long tusks to dig into the bottom to uncover the clams.

What is the food chain that supports the production of such a large population of clams? The delta provinces of the Yukon and Kuskoquim rivers are extensive and low in elevation, and tide ranges are high in this region. This accounts for the extensive tide flats and marshlands on which grow heavy stands of *eelgrass*. During the late-spring crush of ice floes as the winter-frozen rivers break up, the beds of eelgrass are heavily scoured. Thousands of tons of the grass are dislodged and carried out onto the coastal fringes of the Bering Sea Shelf. Here bacteria begin to decompose the dead plant material into particles small enough to become part of the suspensoid to these waters. Together with the dead bacteria bodies themselves, the suspended plant particles become the principal food source for filter feeders such as the clams. A food chain in which the *initial* material is dead organic stuff is called a *saprophytic food chain*.

Figure 13.4 The saprophytic food chain in the Bering Sea: eelgrass–bacteria–suspensoid–clams–walrus.

floes serve as resting platforms for large numbers of walruses which dig the clams with specially adapted tusks.

Organization Based on Motility

Still another way of classifying marine organisms is on the basis of their motility. The divisions are the nekton, plankton, benthon, and neuston.

Nekton. The nekton encompass animals capable of sufficiently strong motion to give them independence from ocean currents. Many animals are large. They have skeletal structures to which muscle is attached and against which the body shape is contorted to achieve a net propulsion force. Examples span many phyla: fishes, birds, marine mammals and reptiles, and some of the cephalopod mollusks such as squid. Fishes comprise the major fraction and achieve motion in two ways. The dominant way is a side-to-side warping of the body in relation to a longitudinal axis—swimming. A few species move by manipulating fins rather than the body itself; these have enlarged ventral or dorsal fins that they move in oscillatory fashion, as do those of the ocean sunfish *Mola mola* (which, incidentally, is not small—adults weigh over a ton). The most rapidly swimming species such as tuna and dolphin have adaptive recesses in the body into which fins are retracted to reduce body drag. Eels and marine reptiles have bodies elongated to an extreme degree, so that their side-to-side warping has a wavelike pattern along the body length.

Plankton. The word plankton is derived from the Greek *planktos* meaning freely-drifting. Plant and animal displacement at these tiny size scales is mostly determined by the motion of the fluid itself. For the smallest sizes, body form and other techniques are used to prevent sinking (see Chapter 9). The larger plankton generally do have some primitive means of locomotion; for example, the dinoflagellates use hairlike flagella (two) to achieve some vertical motility (see Plate 15.2). Copepods also swim. Most marine animals incorporate larval stages in their life cycles, and these vary in motility, depending on their molt stages. Barnacles, for example, have several stages of molting exoskeletons during their larval phase, each progressively larger and more motile than the preceding one. Medusae achieve vertical motion by contracting their bell-like structure to jet water downward.

Benthon. Animals of the benthon group usually move by gaining a purchase against the solid or muddy substrate, that is, they "walk" rather than swim. Examples are numerous, for there is much diversity in this group. The multilimbed arthropods, such as crab and lobster, walk. Worms burrow. Echinoderms, such as starfish, move by means of many small "tubefeet" arranged on the undersides of arms. Mollusks such as the scallop achieve limited motility by rapidly closing their bivalve shells and jetting out seawater. There are many sessile species among the benthonic fauna.

Neuston. The neuston is a special category of "sailors by the wind." The cnidarian *Physalia* uses a gas-filled membrane projecting above the sea surface to gain motility from wind pushing against the "sail." *Velella* represents another type (see examples of both in Plate 15.8).

Organization by Biogeography

The biogeographical approach is perhaps the most interesting, because we can begin marine biogeographical organization by noting that *pelagic marine species of plants and animals are far smaller in number than terrestrial organisms*. This forces us immediately to search for the contrasts between marine and terrestrial biospheres that result in the large speciation difference.

First let us look at the plants. The number of pelagic marine plant species (Fryxell, 1983) is estimated to be between 7000 and 10,000, primarily composed of diatoms, dinoflagellates and coccolithophores (excluded are macrosize plants of coastal zones). In contrast, the number of Spermatophyta (Spector, 1956) is estimated at over 200,000 and only a very few plants in the coastal regime are marine spermatophyta. Thus the ratio of blue-ocean plant species to terrestrial species is on the order of 1:20. What does this tell us? How do we attempt biogeographical organization of the marine biosphere where the speciation is so greatly reduced from that of the terrestrial biosphere? We must hypothesize either that speciation has been greatly subdued in the evolution patterns for marine plants, or that extinctions of marine plants have been extraordinarily heavy in comparison.

A look at the numbers of pelagic marine animal species in contrast to those of terrestrial species only confirms the picture gained from plant comparisons. The estimated total number of recent animals is placed somewhere around 1.5 million. Of this total, marine species comprise a small fraction: within the holoplankton (plus the meroplanktonic medusae, coelenterates that are usually caught in plankton nets) we have estimates of about 2000 species *ocean-wide* (McGowan, 1971); the number of deep-sea and benthic fish species is also placed at about 2000. If we add estimates of all the other nekton groups—squid, shrimps, mysids, and mammals—the total number of animal species in the marine world may be on the order of 6000 and is not expected to exceed 10,000 by the time the ocean is completely sampled. The ratio of marine to terrestrial animal species is therefore about 1:150.

How does this overwhelming fact of diminished speciation in the marine world affect biogeography organization? What are the critical facets of the marine world that account for this scarcity of pelagic species?

Is It the Constancy of the Marine Environment That Subdues Its Species Diversity?

With the development of biological science has come the generalization that habitats that are more uniform will contain more specialized adaptations, hence more species and greater diversity. This does not help solve our marine–terrestrial contrast. The fact is that the highly constant marine environment has *not* led to great speciation, even though we have observed that the *relative* constancy of tropical ocean environments has yielded significantly more speciation than have polar oceans.

Is the Extended Vertical Living Space in the Marine World the Key to Less Speciation?

In the marine world the rapid change of ambient pressure with depth appears to influence profoundly the makeup of species. They have adapted to life at the various levels, surface to bottom. Of all the environmental variables, pressure undergoes the widest dynamic range, from 1-atmosphere pressure at the sea surface to more than 1000 atmospheres in the deeper trenches. Still, our samples from throughout the water column indicate that organisms do *not* adapt to *narrow* pressure intervals; the deep diving by the sperm whale and the migrations of the deep scattering layer demonstrate that most marine animals function over a wide pressure range. Nevertheless, abyssal animals do not function well at near-surface pressures, as we have discovered many times. Recent studies have shown that pressure changes of 100 atmospheres can alter markedly the functioning of specific enzyme systems, which means that protein chemistry is pressure-dependent. But the point is that a dynamic range of 100 atmospheres is a *violent* change, emphasizing the remarkable capacity of marine animals to function over pressure ranges that would cause catastrophic stress to terrestrial animals, including humans.

One other observation supports the argument that pressure, or vertical living space, does not foster speciation. Net tows made at the mid-level depths of 500 m in central ocean waters contain some of the same organisms that are found at near-surface levels but at higher latitudes. The inference is that the do-main of a given species is determined more by temperature than by pressure, for the temperature at 500 m in central waters is the same as surface water temperature at some higher latitude.

Distinct Water Masses— The Key to Biogeography

In Chapter 9 we analyzed oceans in terms of water mass structures and learned that there are surprisingly few of these over the world oceans. The Pacific, for example (Figure 9.5), has but eight major sectors with different water mass structures. If we compare the locations of these Pacific sectors with the schematic of major surface currents (Figure 11.1), a strong correlation stands out: the Pacific Central Water masses occupy the same zones as its Central Gyres. We are led to conclude that *if* water masses play a large role in biogeography, the fact that each water mass occupies a great volume and area of the world ocean must mean that each biological community adapted to the water mass *also* covers large areas. In turn, the number of different communities must be quite small. This factor is the key to the relatively small number of species within the oceanic flora and fauna.

Tolerance Limits of Marine Species

Marine biologists have long recognized that most marine organisms are sensitive to small changes in ambient conditions. Salinity and temperature conditions stand out in this sense. In some species, particularly open-ocean types, the *range of tolerance* to change in these two variables is relatively small, and such species are classified as being *stenohaline* or *stenothermic* or both. It is also important to note that *tolerance* to changes in temperature and salinity may differ for the animal's reproductive, larval, or adult stages; adult stages are usually more tolerant than the early development stages. An animal may be stenohaline at any selected portion of the range of oceanic salinities, or stenothermic at any portion of the temperature range; thus we find steno-types in low-salinity, cold surface waters of the Arctic Ocean as well as in high-salinity, warm surface layers of Central Ocean Waters.

In contrast, coastal animals are usually more tolerant of variations in salinity or temperature of the water. Clearly this is an adaptation to the wider dynamic ranges of temperature and salinity in coastal waters. These waters are subject to large changes in salinity through mixing with river runoff. They also undergo the large thermal day–night changes typical of shallow waters where the bottom itself may receive and absorb sunlight. Animals that tolerate large changes in water salinity and temperature are termed *euryhaline* or *eurythermal* or both.

Estuarine animals exhibit extremes of tolerance. This is expected because estuaries are the transition waters between fresh and salt water. Still, different animals have different tolerance limits. For example, it is no accident that successful oyster farms are located in parts of estuaries where the mean annual salinity is on the order of 17 ppt. Oysters tolerate this amount of seawater dilution, but their principal predators, the starfishes, cannot; starfishes have a salinity tolerance on the order of 19 ppt. The oyster farm is *dilution-protected.*

Water Mass Uniformity versus Species Continuity

We can expect to find a strong correlation among species types found in samples taken from a single water mass, regardless of the latitude or longitude at which the water samples were taken. As long as the characteristic temperature and salinity of the water mass are maintained, so also is the "community" of planktonic plants and animal species characteristic of that water mass. A factor of fundamental significance in the study of the geographical extent of a given species is the system of circulation of oceanic waters *and* the rate at which a given water mass is mixed with adjacent, also distinct water masses. *We conclude that speciation of marine fauna is subdued principally because a water mass can be distributed over a vast area of the ocean.* Thus distinct habitats in the marine realm are far fewer in number than those of the terrestrial sphere.

The other side of this marine biogeography coin is of like importance. Once the physicist is aware that distinct faunal communities are correlated with distinct water masses, biological data become a valuable aid in the interpretation of circulation. The case described in Figure 13.5 illustrates how interdiscipline cross talk in marine science yields greater knowledge than either discipline alone.

Indicator Species

Is a particular type of animal or plant more useful than others as an indicator species for a water mass? The answer is yes. Looking first at plants, we find that they are generally not used as indicators because their reproductive span is so short and their

population numbers respond so rapidly to small changes in conditions, for example, to vertical wind mixing with the associated nutrient resupply. Further, plants are found only in the first tens of meters of surface water and thus would not be useful in tracing deeper water masses.

Although the distribution of benthic organisms does in an indirect way reflect the productivity of the overhead surface community, the nature of the substrate also influences greatly the composition of the benthic community. Whether the bottom sediment consists of mud, sand, or other dominant constituents depends as much on geographical location and water depth as it does on the character of the overlying water mass. Therefore, benthic communities do not contain the most useful indicator species for pelagic biogeography. Neither are the nektonic fauna a useful source of indicator species simply because of their ability to migrate.

Therefore, we look within the zooplankton to find the most useful indicator species. Copepods, chaetognaths, and euphausiids are usually used. They have a relatively long life span, are relatively easy to capture without being mutilated in the process, are widely distributed vertically, and have limited motility. Moreover, the taxonomy for zooplankton is relatively well established.

In addition, the species that have narrow tolerances in their reproductive phase allow the biogeographer to place additional constraints on the community being studied. Certain species literally must have a "home area" in which to spawn, and this means that the water mass movement must have within itself the circulation features to which the indicator species has become successfully adapted. Often, this factor is simply the semienclosed circulation of a major gyre, where the influx of "new" water into the water mass at one point is just balanced by an outflow somewhere else. Mature adults of an indicator species are sometimes rafted away from their biogeographical water mass. They may survive but they are incapable of reentering the home grounds for reproduction purposes. In such studies, of course, the temperature and salinity ranges within which spawning can be triggered are useful knowledge gathered from laboratory experiments on the animals themselves.

By definition, then, a biogeographical area is one in which all phases of development of the indicator species are present. If the larval state is very short we expect to find adults and larvae simultaneously. If it is relatively long, however, the adults may tend to aggregate in areas more favorable for their feed-

CASE IN POINT

THE VARIATION IN PHYTOPLANKTON COMMUNITY COMPOSITION INSIDE AND OUTSIDE OF A DISTINCT WATER MASS

Background. Because of its remoteness from the centers of world population and transportation, the Southeast Pacific is to date a relatively unsampled ocean. We knew from existing data of a low-salinity surface water that extends westward from the coast of Chile, as shown by the contour of 33.90-ppt salinity in illustration *a*. There are two plausible explanations for this tongue: (1) rainfall at sea dilutes surface water, or (2) the heavy coastal runoff of fresh water from the Andes Mountains dilutes the surface water, and a current then carries it away from the coast and toward the west.

To test the second hypothesis, we set up a corollary hypothesis that the plankton "community" inside the low-salinity tongue should be substantially different from that outside. Further, because the water masses differ, Central Water to the north, Antarctic surface water to the south, the community on the equatorward side of the tongue should be different from that on the poleward side. On a cruise we sampled the plankton at 11 stations located as shown in illustration *a*.

Results. We identified and counted all plankton species in the 11 samples; in all some 26 species were identified. Fourteen species each contributed more than 2% of total abundance; these are listed in illustration *b* along with their relative population density at each

station. Sample 37, taken at 41°S, is the physical south boundary of the tongue.

Interpretation. South of sample 37 the dinoflagellates *Ceratium tripos* and *Ptychodiscus inflatus* are the dominant species. Within the low-salinity tongue the diatoms *Rhizosolenia bergonii* and *Nitzschia* (two species) dominate. To the north of 37° S, the north physical boundary of the tongue, the plankton community is again dominated by the dinoflagellates, this time *Ceratium* (three species).

Conclusion. The salinity differences between the low-salinity tongue water and outside water are definite biological "barriers" which prevent the mixing of plankton species. The physicists have not yet proved the existence of a westward current at this latitude, but these biological data are strong evidence that such a current exists.

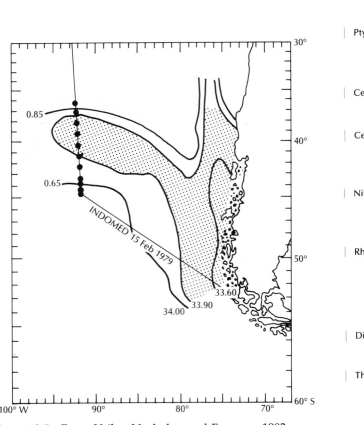

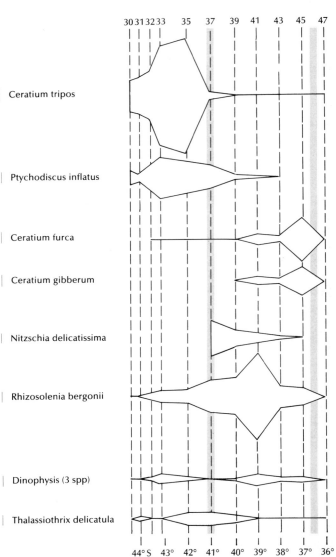

Figure 13.5 From Uribe, Neshyba, and Fonseca, 1982.

ing, and the larval stages are found in a separate locale; then the entire area is the biogeographical "home area."

Water Masses and Marine Species Distributions

The development of marine biogeography follows along the lines defined by three questions:

1. What are the principal patterns of species distribution and abundance?
2. What factors maintain the observed patterns?
3. What are the communities?

Here we shall discuss species distributions only for the Pacific Ocean; similar patterns, but involving different combinations of species, typify all the major oceans.

Water Masses

The water masses of the Pacific are outlined in Figure 13.6*a*. Broad categorization divides the waters into four groups.

Subpolar. In the North Pacific, Subarctic Water occupies the zone poleward of about 45°N; in the South Pacific, Subantarctic Water occurs poleward of 45°S. In both hemispheres subpolar water moves toward the equator along the eastern boundaries of the ocean basins, a reflection of the California and Humboldt current systems along the North and South American continents.

Central. Central Waters occupy the areas interior to the major circulation gyres of both hemispheres. In both hemispheres, there are two subdivisions that distinguish waters of the western and eastern sectors.

Equatorial. Pacific Equatorial Water straddles the geographical equator but is displaced in latitude slightly into the Northern Hemisphere, a reflection of the similar displacement of the "meteorological" equator in the Pacific.

Transition. Major transition zones occur between the subpolar water and the Central Waters, zones that the physicists have long labeled as "convergence" zones between these major water masses. As we shall see in a later section, biologists define these as transition zones because of the distinct species distributions found along the boundaries *between* major water masses.

Typical Examples of Species Distributions

Subarctic. Figure 13.6*b* shows the subarctic distribution of the chaetognath *Sagitta elegans*. This animal

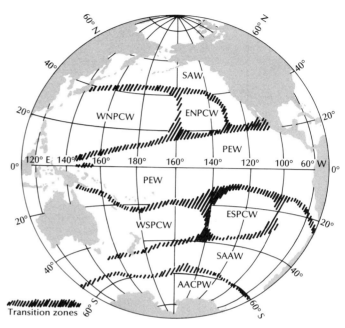

(a) Water masses of the Pacific Ocean (reproduced from Figure 9.5.)

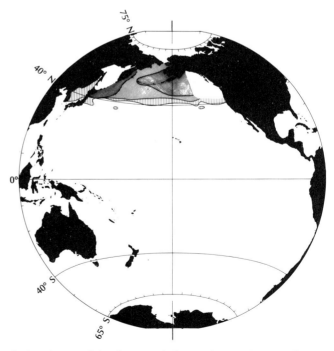

(b) Distribution of the chaetognath *Sagitta elegans*. (From McGowan, 1971.)

Figure 13.6 Biogeography of a chaetognath in relation to water masses.

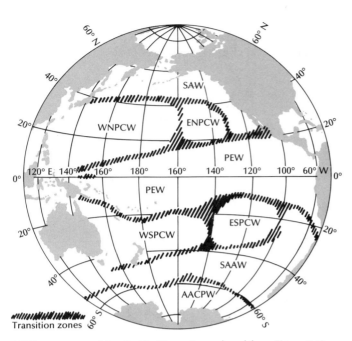

(a) Water masses of the Pacific Ocean (reproduced from Figure 9.5).

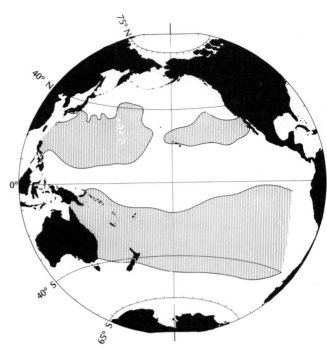

(b) Distribution of the pteropod *Cavolina inflexa*.

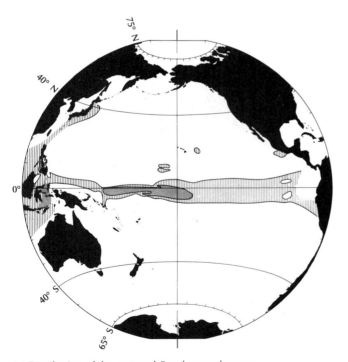

(c) Distribution of the copepod *Eucalanus subcrassus*.

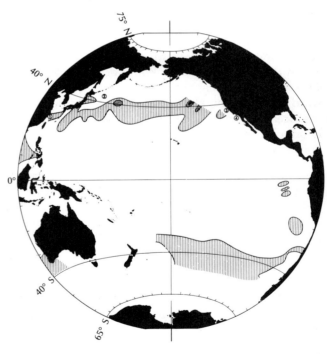

(d) Distribution of the copepod *Eucalanus elongatus hyalinus*.

Figure 13.7 Biogeography of a pteropod and two copepods in relation to water masses. (From McGowan, 1971.)

occupies the northernmost part of the broad West Wind Drift Current that we call the North Pacific Current and extends poleward into the Bering Sea. If we compare the circulation chart of Figure 11.1 with the *S. elegans* distribution, it is immediately clear that the home area of this animal coincides closely with the semienclosed Subarctic Gyre. This gyre is defined by the northern portion of the North Pacific Current, the Alaska Current along Canada and Alaska, the Alaskan Stream along the Aleutian Island chain and into the Bering Sea through the passages in that chain, and the Kamchatka Current, which completes the gyre.

Central. Figure 13.7*b* shows the distribution of *Cavolina inflexa*, a pteropod of the phylum Mollusca. In the Northern Pacific, this animal occupies both the eastern and western sectors of the Central Water mass. Further, the species occurs in both Northern and Southern Hemisphere central waters; the term "bitropicality" is used to describe such distributions of a single species. Many species exhibit bitropicality.

Equatorial. The distribution of an equatorial species is shown in Figure 13.7*c*. The animal is the copepod *Eucalanus subcrassus*.

Transition Water. Both the Subarctic and Subantarctic Water masses are separated from Central Waters by a zone of transition of water characteristics. This zone is noted particularly for its strong horizontal temperature gradients. Physical oceanographers call this zone the Subtropical Convergence. Figure 13.7*d* shows the distribution of another species of copepod, *Eucalanus elongatus hyalinus*. This herbivore occupies the transition zone, both in the Southern and in the Northern Hemisphere.

The transition zones occupy latitudinal strips between about 34 and 45° north as well as south. The species distribution shown in Figure 13.7*d* is bitropical.

Evaluation

Basing our studies on known distribution patterns of plankton like these, we can make several important generalizations about them.

1. The distribution boundaries of most of the species of all the taxa have the same position and shape.

This is an important point. Figure 13.8 superimposes the boundaries of oceanic areas occupied by numbers of central, equatorial, and transition zone

animals. The agreement is striking; the principal patterns of species distribution in the open ocean are repeated from taxon to taxon and at all trophic levels from primary producer to top carnivore.

2. For a large number of species in each group, the boundaries of the area each occupies are almost identical with the physical boundaries of water masses.

But water mass boundaries are not *exclusive* rigid boundaries to distributions of *all* species found within a given water mass. We must add three other generalizations.

3. A number of species may occur at highest abundance within a water mass, but their distribution boundaries may extend beyond the boundary of the water mass per se.

4. Many species are found throughout several water masses. Such species are termed "cosmopolitan."

5. Other species are found only within certain parts of some water mass areas.

What Factors Maintain the Observed Biogeographic Patterns?

First of all, the ocean is apparently a much less complex domain than is the land. There are far fewer niches per unit space.

Second, the physical processes that maintain the shape of species patterns must be related to the processes that maintain the shape of water mass structures. We can demonstrate this for the North Pacific using the *T–S* nomogram technique. Figure 13.9 plots the water mass structures as revealed by hydrostations from both the East and West North Pacific Central Waters (ENPCW and WNPCW in Figure 13.7*a*). The part of the *T–S* domain occupied by the Central Waters community, that is, the ten species whose geographic boundaries are shown in Figure 13.8*a*, is indicated by the line shading. Note the strong correlation between water mass structure and the zone occupied by one community. A similar correlation is evident between the *T–S* zones occupied by the subarctic community (dot-shaded) and the water mass graphs from two stations in the subarctic region. Note that the occupation zones of central and subarctic communities do not overlap; rather there is an open *T–S* space between them. If we were to plot the *T–S* zones occupied by transition communities, these would *fill* the empty spaces.

Let us further explore the connection between

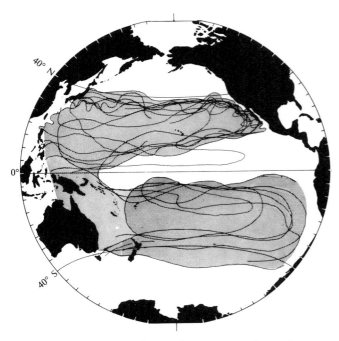

(a) Central water mass distributions for 10 species of oceanic organisms.

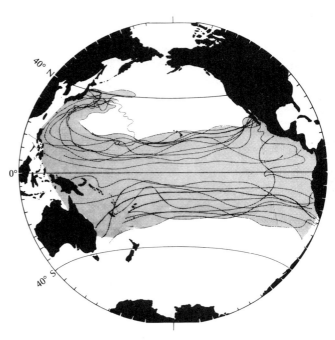

(b) Equatorial water mass distributions for 11 species of oceanic organisms.

transition zones and their associated species. By contrast, the idea that circulation *inside* of gyres is strongly coupled to the larval-spawning–adult-feeding cycles of animals endemic to gyres does not at first seem applicable to transition zones. Clearly, the physical character of transiton zones is one of relatively strong turbulence, brought about by the intermixing of two water masses of differing current speeds and directions. If the zone is one of converging streamlines of the two water masses, it must also be a zone of downwelling or of intense interweaving of the two water masses via turbulent eddies. If the zone is associated with a divergence of surface winds, there may be strong upwelling along the frontal zone. What we have just stated, in a real sense, is that *transition zones have highly specialized mixing processes* as well as different time scales for mixing. What the observations show (Figure 13.8c and 13.9) is that some animal groups have adapted their life cycles precisely to such physical processes. Transition zones are unique in both physical and biological ways.

Thus far our analysis has focused on distributions of planktonic forms. Analogous study of the distribution patterns of large nekton shows that many large forms are also found in greatest abundance within the same gyre–water mass structures as the planktonic species, even though such animals are capable of long-distance swimming. In fact many an-

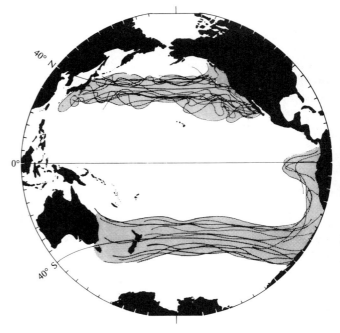

(c) Transition zone distributions for 12 species of planktonic organisms.

Figure 13.8 A superposition of the boundaries of oceanic areas occupied by several species of organisms found in the same geographical areas. Overall, a remarkable coincidence exists between the boundaries of these communities of species and those that define the limits of water masses.

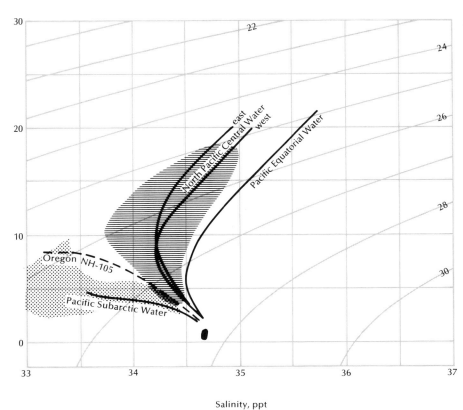

Figure 13.9 Domains occupied by Subarctic Water and Central Water communities of species superimposed over the *T–S* nomographs for these oceanic regions. (Based on Johnson and Brinton, 1963.)

Salinity, ppt

nually migrate over long distances spanning two or more water masses. Why do the abundance data for such nekton conform so well to distribution patterns of plankton? It cannot be argued that the populations of large and powerful animals are conserved by the comparatively slow-moving currents.

The answer must be that the large species are interacting with the species lower in the food web whose populations *are* conserved by the physical processes that dominate the structure of the water masses. We may call this a coupling by *interaction linkages.* Thinking in an ecological sense, we say that a given species is *adapted* to the presence and behavior of another species, and that such adaptation is an integral part of the overall feedback mechanism that controls living oceanic biogeography.

Summary

There are many ways to organize a study of the marine biosphere.

1. The taxonomic approach emphasizes that every community is composed of individual species, and each has a specific name that is internationally recognized.

2. Classification of the various feeding types is another way to organize marine studies. There is much to be said for this scheme. Because seawater and living protoplasm have almost identical densities, the debris of organic material remains in suspension. This is why suspension feeders are found among so many phyla. The use of films of mucus is a key element of suspension (filter) feeders. It is an elegant way to capture organic bits and pieces, and organisms that use the technique range in size from microscopic radiolarians to animals many centimeters in length.

3. Some biological oceanographers specialize by studying a specific but entire food chain. This is actually an ecosystem approach.

4. Another scheme for study is to organize organisms on the basis of motility.

5. The most illuminating method of all, from the view of general ocean studies, is the biogeographic method. Here we discover that communities are not randomly distributed. Rather there is a clear correlation between the physical processes that shape the distribution of species *and* entire communities. Species are found to exist over areas that are also uniform in water characteristics. Moreover, the species that do so occur across the lists of taxa.

Last, we ask the most penetrating question of all. Given that organisms of the first two trophic levels are dispersed throughout the volume of one water mass, why does it also happen that the larger nekton remain associated with the same community, when they can obviously migrate across the transitions between water mass structures? There are three reasons. First, if we consider that organisms have evolved, it follows that the links between organisms, such as predator–prey interactions, also evolve. Given this, the connections within a community are unbreakable; wherever the least of the species go, so also go the mighty. A second reason, perhaps trivial, is that the associations are maintained simply by habit.

A third reason is based on the body chemistry of the larger animals. Having evolved without complex regulatory systems for temperature control, they are adapted to a narrow range of water characteristics. Because they incur problems on crossing water mass boundaries, they do not.

Study Questions

1. (a) There are few blue-green algae in the open-ocean, deep-water system; most phytoplankton are of the Chrysophyta and Pyrrophyta. What are their common names?

   (b) Biologists classify animals into two major types that are *not* taxonomic. What are they?

   (c) Which phyla of invertebrates contain the major commercial species?

   (d) Which class of vertebrates contains the major commercial species?

2. Another rather useful way of classifying marine animals is on the basis of how they feed. Describe the five major feeding types and give an example of each.

3. Many animals in the marine world use the suspension feeding technique. Briefly, why is this method not found in terrestrial, nonaquatic systems? Many marine animals employ films of mucus in their feeding mechanisms. Why don't land animals?

4. What is meant by the term *saprophytic* food chain? How does this type differ, if at all, from the one we ourselves use?

5. Cows have teeth, but why should copepods, only 2 mm in average size, need them? The experiment in copepod grazing efficiency by Fleming was classical in the sense that it showed that *over*grazing of microscopic plant cells is possible. True or false?

6. Most copepod species do not produce male and female offspring; they are mostly hermaphrodites. Why should this method of reproduction be advantageous for them?

7. Briefly describe the organization scheme based on motility of marine organisms.

8. By far the most interesting scheme of organizing the marine biosphere is on the basis of biogeographic zones. Develop this theme, using two different lines of approach.

9. Question 8 can be stated another way. Explain the connections between the very large spatial extent of water mass structures (Chapter 9) and the relatively meager species diversity within the marine biosphere.

10. If it is true that the key to understanding biogeography in the oceans is in the *rafting* of phytoplankton by the circulating waters, why not use phytoplankton species as the best indicators of biogeographic provinces and their boundaries?

11. Develop the idea that it might be better to use *dead* marine plants as biogeographic indicators than live ones.

12. Describe, using at least one example, how the distribution of an indicator species actually conforms with the area covered by a specific water mass structure.

13. What is the meaning of the term *cosmopolitan* attached to an indicator species? What is wrong with this question?

14. Based on your knowledge of ocean circulation, sketch the boundaries of major biogeographic provinces in the North Pacific.

15. Write a short essay developing the theme that a biogeographic province does not *have* to occupy an entire oceanic gyre. For example, Figure 15.4 shows that the eastern boundary communities of *all* the major oceans are similarly organized, and are distinct from communities in other parts of the gyre. Does the satellite photograph in Figure 14.8 offer any clues as to how the eastern boundary community may be a separate, *subprovince* of the larger gyre community?

CHAPTER 14

LIGHT AND COLOR IN THE SEA

There remains to be studied one physical property of water that has such a great impact on how life in the sea is organized that it warrants its own chapter. The property is the capacity of water to propagate rays of light, called its transparency. It is here that we discover what I call the GRAND COINCIDENCE—that the fluid H_2O has its clearest window to electromagnetic radiation at almost precisely the same color as that at which the sunlight is the most intense! There is absolutely no reason why this should be so: the transparency of water to different colors of light depends only on its molecular structure, not on the temperature of the sun. Conversely, the temperature of the sun in no way depends on how the water molecule is structured. Yet it is the 6000 K surface temperature of the sun that forces its peak radiation intensity to occur at the same color, blue-green, as that to which water is most transparent! I am forced to believe that a great "tinkerer" must have designed the original system, saw that it would work better if this or that property of substance was "tweaked" a bit, and so adjusted the emission temperature of the sun to match the window of the sea.

Light from the Sun

We begin with a description of the light that reaches the surface of the sea. Traditionally it is called *white* light, because the total package contains all the colors of the rainbow mixed together in proportions that we sense as the color white.

Actually, sunlight on the ocean surface consists of many colors, all arranged in a scheme we call the *spectrum* (Figure 14.1). The central band, that in which sunlight is most intense, is the band of visible colors. There are many other colors not visible to the human eye; some energy exists at colors redder than red, called infrared, as well as more violet than violet, called ultraviolet. The infrared section is further subdivided into the near- and far-infrared. Keeping track of colors can rapidly become tedious, so we use instead an equivalent scale that is marked off in terms of the *wavelength* of the light, also shown in Figure 14.1. The unit of measure for wavelength is the nanometer (nm); 1 nanometer is one-trillionth part of a meter.

A brief analysis of the sun's spectrum places its most intense radiation within the visible band, 400 to 700 nm. The peak intensity occurs at about 500 nm. Intensity drops off rapidly toward the shorter-wavelength end, with very little energy below about 300 nm, the ultraviolet range. If we could look at the sun's spectrum at the top of the earth's atmosphere, we would find much more radiation energy in the ultraviolet. Most of this is absorbed by the ozone layer before the sunlight reaches the sea surface. The spectrum intensity tapers off much more gradually toward the long wavelengths of infrared, as clearly seen in Figure 14.1.

One additional point should be made: this spectrum has been smoothed. In reality, there are some rather specific and deep "holes" in the spectrum, particularly in the infrared bands. Such holes are due to the selective absorption of energy at specific wavelengths by specific gases in the atmosphere; for example, virtually all sunlight at wavelengths greater than about 6000 nm is absorbed by the CO_2 and H_2O vapor in the atmosphere. Further divisions of incoming solar energy are detailed next.

A Radiation Budget for Earth

Weyl (1970) gives a succinct summary of the earth's radiation budget in Figure 14.2a. Setting the total so-

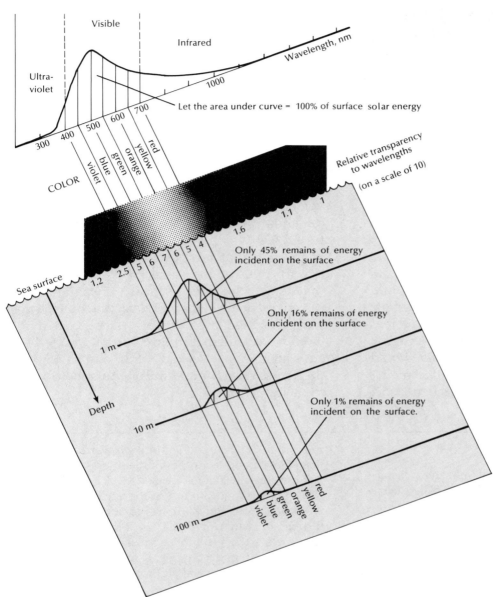

Figure 14.1 How the energy in sunlight decreases with depth in the ocean.

lar energy arriving at the top of the earth's atmosphere equal to 100%, he proceeds:

. . . let us see how, on a global average, this energy is disposed of. The ultraviolet, comprising 3 percent of the 100 units of incoming sunlight, is absorbed mostly by the ozone of the upper atmosphere. About 40 of the remaining 97 units interact with clouds. Of these, 24 are reflected back into space, two are absorbed by the clouds, and 14 are scattered, reaching the earth's surface as diffused radiation.

The water vapor, dust, and haze in the atmosphere interact with 32 units of incoming radiation. Thirteen (13) of these units are absorbed, seven (7) are reflected back to space, and 12 reach the earth's surface as dif-

fused sunlight. Out of the original 100 units, then, the surface receives 25 units of direct sunlight and 26 units as scattered, diffused light. Of this total, four (4) units are reflected from the earth's surface back to space. Thus the total amount of sunlight reflected back to space equals 35 percent of the incident sunlight. Of the 65 percent absorbed by the earth, 3 percent is absorbed in the upper atmosphere, 15 percent in the lower atmosphere and 47 percent by the surface of the earth, both land and sea.

Note that the 47% of total solar energy that passes through the atmosphere to be absorbed by land and sea must also be given off by land and sea if the earth is to remain in thermal equilibrium. This *budget*

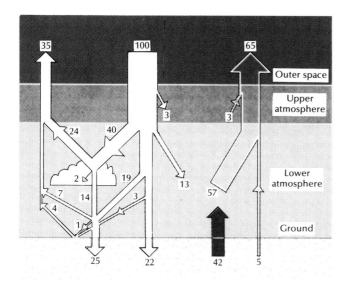

Figure 14.2 Tracing light rays and their energy, from the top of the atmosphere to the sea surface and into the ocean interior.

(a) A radiation budget for earth (From Weyl, 1970).

(b) The behavior of light rays incident on the ocean surface.

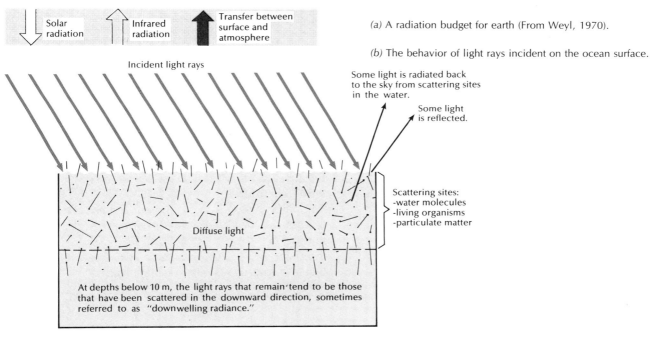

approach is different for land and sea. Readers will recall that we studied the heat budget of the ocean in Table 6.2. There the table set the incoming short-wave radiation equal to 100%, exactly the same *quantity* of energy as the 47% calculated by Weyl as the amount of total solar energy that actually reaches the surface.

Definition of Terms

We have used some terms whose meaning may not be clear. Some definitions follow.

Color. Color is not a trivial term to define. It refers strictly to a nerve sensation that we have when electromagnetic radiation of a specific wavelength enters our sensors.

Wavelength. We consider electromagnetic radiation as a wave, and so the distance between successive wave crests is the wavelength.

Photon. We can also treat light rays as a stream of particles traveling at the speed of light. The elemental particle of light is called a photon. It is the radiation equivalent of energy. Photons at different wavelengths which are seen as different colors, have different quantities of energy. (In quantum mechanics terms, photons possess different quanta of energy; whether a photon is absorbed by a chlorophyll molecule, for example, will depend on the "match" of quanta between photon and the absorbing molecule).

Intensity. A convenient way to quantify intensity is in terms of the number of photons that pass through a unit area per second.

Absorption. When applied to electromagnetic radiation, absorption means that one or more of the incident photons are captured and their energy absorbed. In marine biology, it refers to the capacity of a plant pigment, for example, chlorophyll, to absorb the energy of a photon and thereby energize the photosynthesis process. Conversely, a plant cell's photosynthetic capacity is measured in terms of the minimum number of photons per second needed to activate its synthesis chemistry.

Scattering. A photon can be scattered instead of absorbed. The direction of scatter can be forward, backward, or in any direction.

Attenuation. Attenuation combines the effects of both absorption and scattering. It is a measure of how rapidly light is *lost* as it propagates through seawater.

Transmissivity. Transmissivity is the opposite of attenuation. It is a measure of how much light passes through a given path length in seawater.

Light Propagation in Seawater

The capacity of light waves to pass through pure water is governed solely by the structure and arrangement of water molecules. In seawater, light is also absorbed or scattered by suspended and dissolved particles. In absorption, photons are captured by molecules and their energy is absorbed; in scattering, the photons are deflected away from their original propagation path to a new direction, without absorption. This concept is illustrated in Figure 14.2. Light that is reflected from a medium, for example, the sea surface, is scattered in the direction of the viewer. Below the sea surface, the points from which scattered light arrives at a viewer's eyes are distributed in three dimensions. A consequence is that the total field of light in seawater is much diffused, as those readers who have ever dived into the ocean and opened their eyes can testify; an equivalent example is the diffuse daytime light field under a cloud-filled sky.

Combined absorption and scattering, or attenuation, accounts for the extinction of light; the inverse of extinction is the transmissivity of light. In practice, it is easier to measure the rate of extinction than the rate of transmission. Oceanographers measure extinction by lowering a white, flat disk below the surface to the depth at which the disk can no longer be seen. Called the "Secchi disk depth," this is a measure of the rate at which total sunlight is attenuated.

We know, however, that different colors attenuate at different rates. Blue-green light propagates the longest distance, and the ultraviolet and infrared wavelengths are the most severely attenuated. Figure 14.1 illustrates this phenomenon by depicting seawater as a window of variable transmissivity. The clearest part of the window occurs at the wavelengths seen as blue-green. Fortunately, as we have indicated, this clearest section of the window aligns almost precisely with that part of the sun's spectrum at which solar energy has its greatest intensity. It is worth reiterating that this alignment is *not* an evolved characteristic. So far as we know, the water molecule has always existed in the exact form it has today.

The Distribution of Light in the Sea

Because light waves of different colors are attenuated at different rates, it follows that the total light field changes with depth below the sea surface. The reader can verify this statement by noting how in Figure 14.1 the field of sunlight is altered both in intensity and in color makeup at successively deeper levels in the purest open-ocean water.

At the 1-cm level (not shown), 15% of all solar energy has been absorbed. Primarily, it is the energy at the far infrared wavelengths that is lost to absorption in this uppermost "skin" of the ocean.

At the 1-m level, 55% of all energy has been absorbed, including all the infrared waves and a major fraction of the ultraviolet.[1]

At the 10-m level, 84% of all energy has been absorbed. Only the blue and green photons remain to

[1] The absorption of ultraviolet photons has been a concern of oceanographers ever since the discovery that these energetic photons are harmful to living tissue. If life originated in the sea, it must have done so at depths below which the ultraviolet rays were absorbed. I mention this only because you may sometimes read that "life began in the shallow primal seas"—to which you should add "not too shallow."

propagate to still greater depths. If you were to encounter a "Yellow Submarine" below this level, chances are you would be unaware of its yellow color simply because too few of the yellow photons

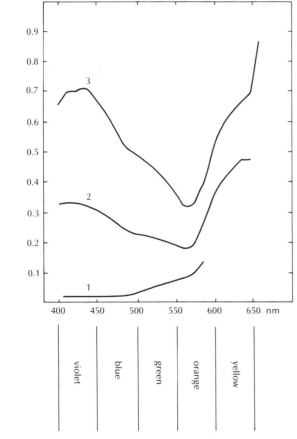

1. *Sargasso Sea.* Waters of the Sargasso Sea in the central part of the North Atlantic are among the "purest" of ocean waters in terms of their clarity to light waves. The colors violet and blue are the *least* attenuated; observers sailing over this sea therefore see its water as a deep indigo blue.

2, 3. *An upwelling region near the African coast.* The attenuation of light is much greater for African coastal waters than for the Sargasso Sea; in water column 2 attenuation is some three times greater, and in water column 3 it is four to seven times greater. In addition, the color band that is *least* attenuated here is the orange. In active upwelling zones with abundant nutrients, plant population can be quite dense. This increases the absorption of light photons and therefore the attenuation of light energy. In addition, such productive waters also contain heavy concentrations of dissolved organic substances, which add a "yellowish" tinge to the "green" of waters containing large amounts of chlorophyll. This explains the shift in color of least attenuation.

Figure 14.3 A comparison of how light is attenuated at different rates in different water masses. (From Jerlov, 1976.)

remain to activate the corresponding color receptors in your eyes.

At the 100-m level, only 1% of the sunlight incident at the surface remains. In color, this residual light is blue-green, centered at about the 470-nm wavelength.

The purest open-ocean water to which this description applies is found, for example, in the central waters of the North Atlantic, far from land masses. In waters close to shore both the concentration of suspended particles and the amount of dissolved organic substance are much greater. Prime sources of particles are microscopic plant cells, plentiful because the supply of nutrients is good, and the heavy suspensions of organic particles, including fecal pellets. Consequently the attenuation rate is greater. Figure 14.3 compares light attenuation in upwelled water off Africa with that for the Sargasso Sea. In African coastal water, light energy falls to 1% of incident surface value at a depth of only 10 m instead of the 100-m value quoted earlier for the purest oceanic water. Much greater amounts of organic substances are also dissolved in coastal waters, and their effect is to shift the clearest portion of the transmissivity window toward the green-yellow part of the spectrum. In Figure 14.3 the sharp minimum in the attenuation curve is at about 570 nm for African coastal water, compared to a minimum attenuation at 450 nm for the Sargasso Sea.

Vertical Distribution of Light versus Primary Productivity

Considering the rapid decay of light energy with depth, we might expect plants to grow best at the surface of the sea where light is the most intense. We do not find this to be the case.[2] Observations revealed long ago that phytoplankton are the most concentrated somewhere in the range of 7-m to 17-m depth. What is the explanation for this apparent contradiction in plant demands and light supply?

[2]It is intriguing to play the game, however—What if marine plants grew best in the most intense light? If optimum growth and reproduction occurred just at the surface, any cell that, by virtue of turbulence, happened to move below the surface would be shaded by the overlying layer of plants and therefore be severely disadvantaged. Further, herbivore forms would probably be quite different from that of the copepod: with plant cells distributed in a thin layer, a more efficient grazer would "skim" the surface rather than pump water as does the copepod. Finally, such a distribution would require the mixing system to pump nutrients all the way to the top layer. All in all, the present scheme seems best.

CASE IN POINT

HOW DO WE MEASURE PRIMARY PRODUCTIVITY IN THE OCEAN?

Primary production must be measured *in situ*. It is almost impossible to duplicate real oceanic conditions of light intensity and color, pressure, and controlled temperature in a laboratory, much less aboard ship. The procedure used at sea is at follows.

1. The problem is to determine primary production at the 5-, 10-, 15-, and 20-m levels.

2. With the ship on station, extract samples of water from each depth desired. Filter all samples to remove the copepod predators; the plant cells remain in the sample.

3. Transfer the samples to sets of three equal-volume flasks; one is clear glass to admit light; the second is black to obstruct all light; the third remains on deck for measurement of its oxygen content.

4. Lower the clear and dark bottles in pairs to the same depth levels from which the samples were extracted.

5. After a suitable period of time, at least two hours, recover all flasks and immediately *(a)* fix the contents of the clear and dark bottles chemically to prevent further physiological activity—in other words, kill the sample; *(b)* measure the oxygen content of each flask.

6. The difference between the amounts of oxygen in the dark bottle and the control flask on deck is the total respiration of the population of phytoplankton. Since no light was available to the dark bottle, no photosynthesis occurred.

7. The amount of oxygen in the clear flask, *plus* the amount respired in the dark one, is the total primary production at each depth for which samples were run.

One other method of estimating primary production is frequently used. It is based on the assumption that the chlorophyll content in a water sample is proportional to the production. The chlorophyll molecule is known to fluoresce. A sample of water is subjected to a flash of light to "energize" these molecules, and then a sensitive light de-

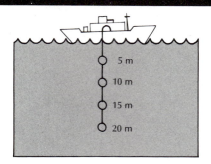

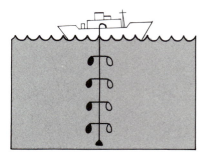

tector measures the photons emitted by the molecules as they fluoresce back to a normal state. It is not difficult to rig such a device to yield continuous measurement as the ship is underway, usually by pumping seawater directly through a fluorometer in the ship's lab. All methods are subject to considerable error. Thus research continues to devise ways of measuring primary production, hoping to find the definitive one.

An explanation must include the physical factor of vertical motion in the near-surface layer. Recall that convective cells driven by surface cooling and evaporation result in vertical mixing; additional mixing is forced by wind waves that break and create turbulent motion in the near-surface waters. All motion carries the minute phytoplankton with it, so some type of vertical distribution of plant cells is inevitable.

On the other hand, nutrient supply comes from below the mixed surface layer. Over the great majority of the ocean where upwelling does not occur, the only mechanism for nutrient replenishment is the slow diffusion upward of nutrient-rich water into the surface layer. Turbulent motion in the surface layer assists by actively entraining water from below. We call this turbulent diffusion.

Therefore, *the condition for optimum plant production should exist at some level between the surface with highest light intensity and the bottom of the mixed layer where*

nutrients enter. The 7- to 17-m range in which we find the densest concentration of plants is, on the average, the compromise level selected by the plants themselves. In fact, laboratory experiments have revealed that many species of phytoplankton are *photoinhibited*, that is, they reduce production rate when exposed to high light intensities. Photoinhibition is not a universal characteristic, however; there is a large variation among species in terms of the light intensity at which each obtains optimum production. Dinoflagellates, for example, seem to have greater tolerance to strong light levels than do the diatoms, so their contribution to total production is much greater in tropical waters. Diatoms predominate in the colder waters at high latitudes where, because of the sun's angle, peak light intensities are lower.

The ability to predict primary production is a goal long sought by biological oceanographers. A useful tool is the simulation model, in which the factors thought to govern plant behavior are specified and the model itself is a set of mathematical equations believed to simulate the plant response. Once a model is found that simulates accurately the plant behavior in a real ocean, the model can be run again and again, each time with a different set of weights assigned to the factors. One such model is illustrated in Figure 14.4. It begins with a uniformly distributed plankton that is then subjected to variations in light and nutrient supply, predation, and mortality. The

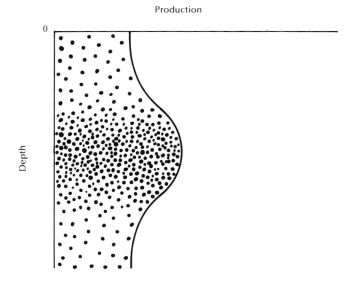

(a) Initially, seed phytoplankton are uniformly distributed throughout the depths of the photic zone, together with an ample and also uniformly distributed supply of dissolved nutrients

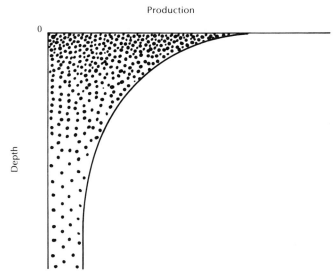

(b) After some time, the phytoplankton are heavily concentrated in the uppermost layer.

(c) After still more time, the heavy initial production at the surface has depleted nutrients; peak production is now found in the middle range.

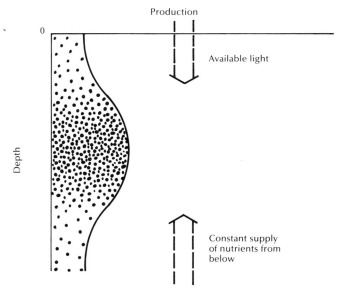

(d) In the final steady state the densest concentrations of phytoplankton are found in the depth zone between 7 and 17 m. But overall production has fallen below the peak of the middle depth.

Figure 14.4 A model of how phytoplankton production changes with time at different depths in the photic zone. The reader should know that a ''model'' is simply a set of mathematical equations that defines the *inputs* to a system. Here inputs are (1) the available light and nutrients, (2) an initial supply of plant cells, and (3) the rates at which the plankton respond to stimuli. Eventually, the model reaches a steady state of highest production, at a level where the needs for light and nutrients just balance.

result is a final steady-state condition that accurately predicts the average plant cell distribution. This type of oceanographic research is continuing, since the universally applicable model has not yet been developed.

Mixing Depth versus Compensation Point

Given that in general the rate of photosynthesis falls as the available light intensity decreases with depth, marine biologists have devised the concept of a *compensation point*. Because the respiration rate of a plant cell is relatively constant and independent of whether the plant is producing or not, there must exist a point at which the oxygen consumed in res-

piration equals the oxygen being generated in photosynthesis. This is the compensation point. The depth at which it occurs is the *compensation depth.* Plant cells at greater depth may still synthesize, but their lot is an ever-poorer one; their state of "health" will steadily degrade, and with this their capacity to reproduce or to store the fats whereby buoyancy is maintained (Figure 14.5a).

When coupled with active vertical mixing, however, the compensation point takes on a different meaning. Discounting for the moment all other factors, the planker must spend at least as much time above this level as below. Alternatively, it may adapt to "use" its time above compensation point to synthesize more rapidly, and then simply rest or carry

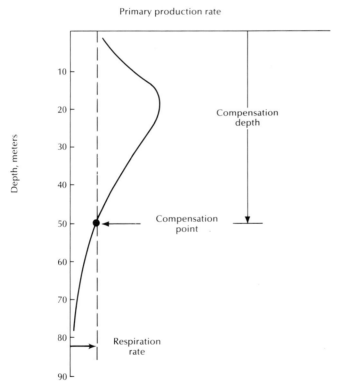

(a) *Compensation point* is the condition for which the *rate* of oxygen generation within the plant cell during photosynthesis just balances its *rate* of oxygen consumption in body metabolic processes. The *depth* at which this occurs depends on how rapidly the intensity of available light decreases with depth; thus clear water will yield a deeper compensation point than will the greenish waters of coastal areas. Plants can live at depths below these levels, but they must lose energy to the work of maintaining adequate oxygen intake.

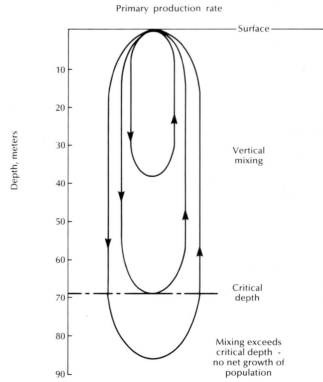

(b) *Critical depth* pertains to the entire population of phytoplankters *and* the depth down to which the outside forces of winds and waves cause the surface layer to mix. Plankters near the surface and high light intensity can synthesize more new organic chemical energy than they consume in body metabolsim. Plant cells mixed deep into the water column suffer from the lack of adequate light; they tend to produce less new chemical energy than they consume. When vertical mixing is confined to depths shallower than "critical," the entire population produces a net surplus. But mixing to greater depth results in a net loss, with overall consumption outstripping overall production.

Figure 14.5   The definitions of compensation point, compensation depth, and critical depth.

out other necessary cell functions during its stay below, such as the assimilation of nutrients for later use.

The Concept of "Critical" Mixing Depth

Compensation depth loses its meaning when we refer to a *population* of phytoplankton. For this we invent a different concept, called the *critical mixing depth*. This is the depth to which turbulent mixing can distribute the entire population before "net" production falls to zero. This idea (Figure 14.5*b*) has important implications. If the active mixing within the ocean surface layer occurs to depths below the critical depth, a "bloom" in plant production cannot occur because the overall, composite demand for chemical energy in metabolic processes exceeds the production. If mixing is restricted to lesser depths than critical, a "net" production occurs, that is, a "bloom."

Clearly this is a trade-off situation. Mixing must occur if there is to be a continuous, adequate resupply of nutrients. The deeper the mixing depth, the greater is the rate of resupply, and production can increase if adequate light is available. But with greater mixing depth, the plankters spend proportionately more time at depths below their individual compensation levels, and overall production can drop.*

Still another factor is involved. As we shall study shortly, the sensitivity of the pigment that the plant cell is using to capture photons is wavelength-dependent. A plant with a pigment that captures photons at shallow depths can lose efficiency at greater depths because the photon's wavelength does not allow it to penetrate with sufficient intensity. Plants with multiple pigments covering a wider range of absorption wavelengths have a greater potential to survive.

Light Wavelength versus Pigment Sensitivity

Within the tissue of the plant cell are specialized areas called chloroplasts; they contain the chlorophyll and other pigments necessary for photosynthesis. The process of photosynthesis is outlined briefly. The chlorophyll molecule absorbs one or more photons of light, specific wavelength or energy

levels, and becomes "energized" in the sense that one or more electrons are pumped into higher-energy states within the molecule. These high-energy electrons are passed into a sequence of other molecules whose function is to create adenosine triphosphate (ATP), which in turn serves as the energy source in a chemical reaction from which sugars, amino acids, and other needed tissue molecules are produced (Figure 14.6*a*).

A critical point in this complex process is the "tuning" of the chlorophyll molecule to the precise energy levels of photons; a mismatch here means that the photon is not absorbed, and consequently no energy to drive the life system is produced. Strangely, research into the wavelength sensitivity of chlorophyll shows that this molecule absorbs best the photons associated with violet (440 nm) and red (660 nm) (Figure 14.6*b*). From our previous discussion of transmission of light in water, recall that the natural "window" for transmission occurs at 470 nm in open-ocean water and at 500 to 550 nm for coastal and upwelling water. Because only a few meters of water are required to absorb the violet and red portions of the available light, it would appear that plants dependent solely on chlorophyll pigments would be at a disadvantage when their depth exceeds about 5m.

Figure 14.6*b* shows the wavelength sensitivity of fucoxanthin, an accessory pigment found in many marine plants. It absorbs light energy in the portion of the available spectrum where chlorophyll itself is a poor absorber. How did this adaptation come about and why? One hypothesis is based on the idea that the early oceans were shallow seas in which the violet and red wavelengths were available in ample supply to sustain plant needs. Chlorophyll was therefore an adequate pigment by itself. Still, why not evolve a pigment capable of direct absorption in the portion of the spectrum at which the greatest amount of photon energy is available? The answer probably lies in the limitations of chemical bonding of elements into molecules. There are simply not an infinite number of molecular combinations that can do the job that chlorophyll does, and chlorophyll may have been the best answer at that time and place. With deepening of the ocean waters, the limitations imposed on the plant system adapted to chlorophyll were offset by the evolution of the accessory pigments. But it is important to note that the chlorophyll chemistry has been retained; in fact, the accessory pigments are not capable of performing the entire sequence of molecular interaction needed to produce sugar and other tissue needs.

---

*This explains why there is no winter bloom, even though winter storms have mixed nutrients up from below. The bloom is triggered only when daily increases in the springtime solar radiation support it.

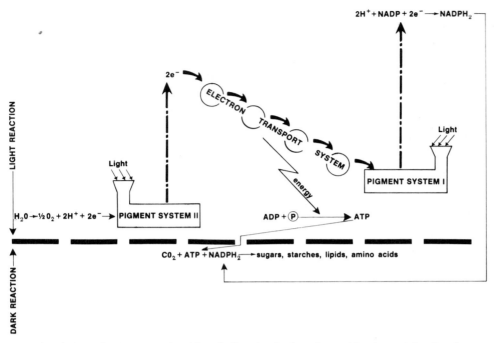

(a) In the photosynthesis process the chlorophyll molecule alone is capable of energizing the electron transport system and generating the sugars, amino acids, and other substances needed for protoplasm construction. The auxiliary pigments are capable only of feeding the electron transport chain.

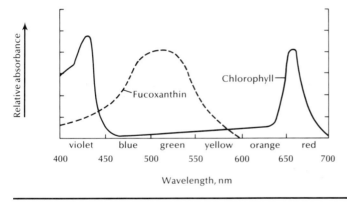

(b) Chlorophyll and other pigment molecules are sensitive to only certain parts of the color spectrum of sunlight. Chlorophyll is best able to absorb photon energy in the violet and orange-red color bands and is weakly sensitive to the blue-green photons, which are the most abundant. To compensate for this apparent mismatch, plant cells have evolved auxiliary pigments, such as fucoxanthin, which has greater ability to absorb blue-green photons, and which passes the absorbed energy on to the chlorophyll pigment molecule for the actual conversion into biomass.

Figure 14.6 The photosynthesis process and the sensitivity of pigments. (From Sumich, 1976.)

The Index of Refraction of Seawater: Some Consequences

Most of us have seen the apparent "bend" in the shape of a stick that is partly immersed in water. This is a consequence of the difference between the index of refraction of water (1.33 for pure water; up to 1.35 for seawater) and that of air (1.00).

This idea is illustrated in Figure 14.7a. A ray of light that impinges upon the sea surface at an angle *i* is bent toward the vertical on passing through the air–sea interface; in the water the ray direction is at an angle *r*, which is less than *i*.

One consequence of the larger index of refraction is that light rays penetrate to greater depths in the ocean than they would if permitted to continue direction along the incident angle *i*. Stated another way, if the mean path length through water is fixed by attenuation, the light ray oriented as vertical as possible penetrates to the greatest depth.

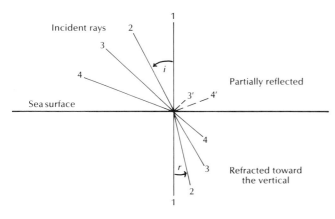

(a) Light rays are refracted toward the vertical when passing from atmosphere into the ocean; *i* is the angle of a ray incident on the surface, *r* the angle to which it is refracted.

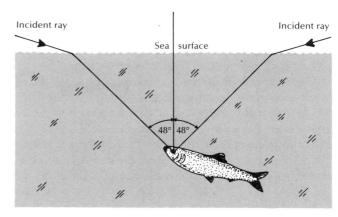

(b) A "fish's-eye" view. Because of the higher index of refraction of water, a fish can see light rays almost from horizon to horizon when scanning a field of view of only 96°.

Figure 14.7 Effects related to the index of refraction of seawater.

A further consequence of the high index of refraction of water is that a fish near the surface can "view" the entire hemisphere above the air–sea interface, horizon to horizon, by capturing all the rays of light within a field of view of only 96° (Figure 14.7*b*). This phenomenon has been optically copied in some special photographic lenses called fish-eye lenses. This capability of marine fishes is of little known consequence to behavior patterns except that such fishes are aware of the approach of seabirds

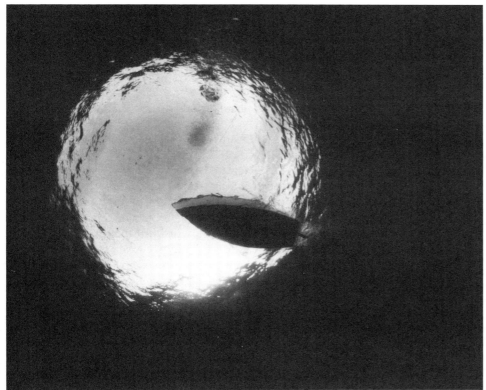

A "fish eye" view looking upward toward the sea surface and the underside of a boat, demonstrating how the index of refraction of water allows the fish to see almost horizon to horizon while itself submerged. (Tom Stack & Associates)

from virtually any direction.[3] For trout fishermen of inland streams, the consequence is well known— one approaches a trout-bearing stream carefully to avoid being silhouetted against the horizon.

The fact that light rays tend to propagate more nearly vertically downward may have important implications in the shapes and colorations of the marine fishes inhabiting the near-surface layers. We find many species with distinctly lighter shades of colors on the lower part of their torsos. This is interpreted as a defensive measure against detection by predators from below. Because much light is also scattered backward from water molecules and from particulates in seawater, there is considerable "upwelling" radiance. Such light is reflected from the underside of fish more efficiently if this part of the body is light-colored, thus blending the fish against the background of downwelling radiance.

Light Scattering in Seawater: Optical Signatures of Water Masses

During the past twenty years a relatively new science, "hydrological optics," has emerged. The objectives in studies of the optical properties of seawater are many, but two results have overshadowed the others.

Nature of the Scattering Particles

It is axiomatic that the true nature of suspended particulate matter in seawater can only be studied *in situ*. Attempts to gather microscopic-size materials from the ocean and then resuspend them in laboratory tanks have been only moderately successful in shedding new light on the role of passive particulates, which tend to clump or undergo chemical change once gathered in mass, or on the role of living but microscopic plants and animals. Sophisticated optical devices have been invented that can

measure the index of refraction of these living and nonliving particulates in their natural state. Further, optical oceanographers can now measure the angles of reflection and scattering from such particulates and correlate these data with types of suspended materials as well as with species of living organisms. The result has been great progress in developing theoretical models of light scattering in the sea.

One important by-product of this research is the ability to correlate the light scattered back into space with the nature of the constituents in surface and near-surface seawater. Satellites now routinely chart the oceans in terms of backscattered sunlight. By analyzing the backscatter radiance for its "green" content, we obtain a chart of the chlorophyll concentration (Figure 14.8). In other color bands, for example, yellow, satellite sensors can chart the concentration of suspended silts off river mouths. Such research might be able to determine how man-made pollutants are dispersed along coastal margins.

Optical Signatures of Water Masses

Once oceanographers knew how to correlate optical measurements with the biological fraction of suspended particulates, they had a new tool with which to classify water masses. From biogeography we learned of the connection between water masses and their communities of living animals. Using optics we can explore the suspended, microscopic material in ways that are denied to oceanographers with nets that capture only the larger particles. In this sense, optical oceanography can *extend biogeography into the finer-size fraction of the plankton world*. More and more, optical data supplement the information gathered by both biologists and physicists.

Figure 14.9 shows two cross sections of the upwelling region off Oregon. The first, the vertical distribution of temperature, shows a tongue of low-temperature water extending some 10 km offshore and traceable to depths of about 40 m. The physicist infers that water upwelled very near to shore is initially transported westward but eventually begins to sink. The second chart is a vertical cross section of the turbidity of water as sensed by an optical device. These *turbidity data show the tongue much more clearly* than does the temperature. Together, these provide an illuminating picture of circulation in an upwelling zone.

Generally, each major water mass originates through the process of air–sea exchange of heat, water, and gases; further, the region of origin is usually well defined. It follows that the region of origin is also a specific biogeographical zone with its well-de-

---

[3] I know of one particular fish individual to which this ability served for naught. While a visiting scientist in Valparaiso, Chile, I was given use of an office with a window directly over the sea. There I could look down from a 5-m height and watch the waves crashing upon rocks, a pastime often indulged. On one occasion a cormorant dived to about 1-m depth and gave chase to a small fish. The rapidity with which that bird could maneuver and chase the fish astounded me. Within a horizontal area not greater than 3 m², the predator–prey pair maneuvered through at least half a dozen hydrobatics involving turns up to 180°. During my watch, the distance between the bird and fish did not change perceptibly. I never learned the outcome of that matched performance, but I shall never forget it.

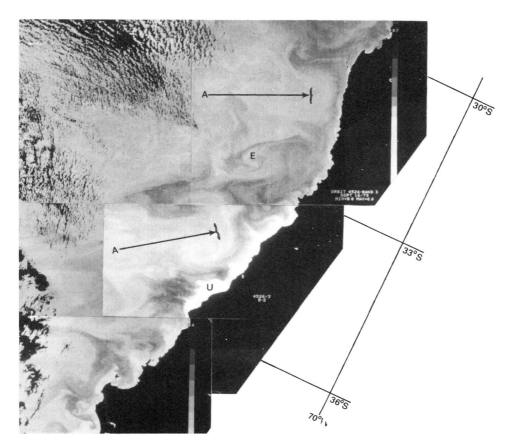

Figure 14.8 Images of the sea surface off the west coast of South America (From Espinoza, Neshyba, and Zhao, 1983.)

The *NIMBUS-7* satellite carries a radiation sensor that scans the sea surface in the green color band, 540 to 560 nm. Sunlight that scatters back from the uppermost 5 to 10 m of surface water in this color range is proportional to the amount of chlorophyll and other plant pigments in this layer and is thus a direct indicator of the quantity of plant production in the sea. In this mosaic of images, the black parts are either land or clouds. The whitest part, *U*, is a strip of coast water only about 20 km wide, from which the radiation is most intense; this strip is a coastal upwelling zone of great productivity. Also seen in this mosaic, which covers about 700 km of coastline and up to 500 km out to sea, are large-scale, swirling patterns. These can be interpreted as surface water movement. Points *A* mark two intrusions of oceanic-type water inward toward the coast. Between them lies a plume of darker water, with low chlorophyll content; it extends hundreds of kilometers to sea. Point *E* marks an "eddy" that is perhaps 50 km in diameter.

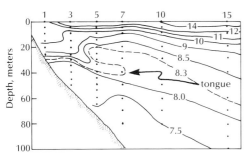

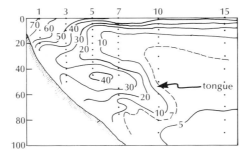

The temperature field. These data were taken with standard hydrographic thermometers. The objective of the measurement program was to learn whether the cold water that is upwelled near to shore is warmed fast enough by sunlight to prevent its sinking back down to some equilibrium depth. The tongue of relatively warm water could be interpreted as the return flow of water that was not warmed fast enough to maintain neutral buoyancy in the surface layer. But temperature data alone are insufficient to "prove" the interpretation.

The turbidity field. These data were taken with an optical *transmissometer* which measures the turbidity of water *in situ.* They also show the shape of a tongue, coincident with the warm-temperature tongue. Because turbidity in water is a function of the concentration of suspended particles, and is fully independent of the water's temperature field, we conclude that these optics data confirm the hypothesis first postulated on the basis of temperature alone. Because high turbidity is known to accompany the high productivity typical of upwelling zones, we also conclude that the water in the descending tongue had shortly before been exposed to high productivity in the surface layer.

Figure 14.9 An example of how the optical characteristics of seawater are used to study its motion patterns. (From Pak, Beardsley, and Smith, 1970.)

fined suite of plants and animals, and that a water mass will carry with it a *biological signature*, which can be sensed optically as well as taxonomically. To this can be added the idea that windblown particles of terrestrial origin that eventualy settle into the sea also have different characteristics in different oceanic regions. These will add to the overall signature of the water mass. The data gathered by optical instruments are just as useful in the open ocean as in the coastal example of Figure 14.9, with one exception. The intensity of optical signatures of the open-ocean and interior waters is generally less. Nevertheless, we find increasing utility in the data that optics provide.

Bioluminescence

Every ocean-cruising marine scientist has come face to face with the phenomenon of bioluminescence in marine animals. My first exposure was as a student on a cruise into the Gulf of Mexico. We were required to take nighttime bucket temperatures. We scooped water from the sea surface in a bucket and took the water's temperature by swishing the thermometer bulb around in it to get a quick equilibration of the mercury column. Each swish of the thermometer in *my* bucket produced in the water a multitude of sparkles of pinpoint light—I was fascinated, hooked on bioluminescence. Very probably the light was produced by a genus of dinoflagellate, *Noctiluca*, known to inhabit tropical and subtropical waters. Later, during field work in which I tested a hypothesis that using intense light pulses as the electromagnetic source for underwater radar might elicit troublesome responses from luminescent marine creatures, I devised a sensitive receiver, which was lowered to depths of 500 to 700 m. There it recorded the ambient luminscent response of sea life to flashing lights of various colors. What I discovered was that the absolute darkness often considered characteristic of the deep ocean simply does not exist. Instead, the flashing of countless bursts of light from deep-sea organisms maintains the light in deep waters at a somewhat constant, albeit very low, light level. I also learned that introducing a series of man-made light pulses—here xenon light flashes—excited the underwater world to respond in kind: ambient light level increased 10,000-fold under pulsed light stimulation, from a background level of about 10^{-7} μwatts/m^2 to about 10^{-4}! One set of the recordings I obtained is reproduced in Figure 14.10.

Bioluminescence must have been noticed even by primitive humans. The phosphorescence of dead fish, seen in the marketplace in the dark of night, was known to Aristotle. Today we recognize the luminescent bacteria on decaying flesh as the source of this light. The chemist Robert Boyle (1627–1691), when experimenting with vacuums, noted the extinction of light from fish flesh when air was removed from the vacuum jar. At that time (1668) oxygen had not yet been isolated as an element. We now know that oxygen must be present for the chemical reaction to occur and produce light. It was not until 1853 that the source of light was traced to a bacterium. Since then, many luminescent bacteria have been identified and the chemical reaction common to these and higher organisms as well is now established.

Common Features of Bioluminescence

1. The energy required to produce photons of light is obtained chemically from the ATP molecule. When the ATP molecule reacts with the enzyme luciferase, energy is released to a different molecule, called luciferin, an oxidizable compound, in the form of high-energy electrons. Within the luciferin the electron falls to a lower energy state, and a photon of visible light is emitted—hence chemiluminescence of "cold" light. No other external source of energy is required. In fact, bioluminescence is the *reverse* chemical process from photosynthesis by plants!

The participating compounds were so named early in the history of research when the work of the devil Lucifer was thought to be involved. When the bodies of animals capable of luminescence were ground up and cold water filtered through the mixture, a compound called luciferin was leached out. With hot water, a second compound, luciferase, was obtained. Light is produced when the two extracts are combined in the presence of oxygen.

2. The pelecypod *Cypridina* (a small bivalve mollusk) has been used as a source of such extracts. Extremely low concentrations of extract to water, 1 to 100,000,000,000, make light that is visible to the eye.

3. Luciferins obtained from different species are chemically different. This leads to the conclusion that marine bioluminescence is a "convergent" evolutionary trait. Luciferin mixed with luciferase from different species will not "light up" unless the species are closely related.

4. The intensity of the light is generally low; in color, it falls into the visible spectrum, predominantly blue-green coincident with the transmission "window" of water (readers will deduce that this is an "expected" selection).

STIMULATING BIOLUMINESCENCE FROM MARINE ORGANISMS AT DIFFERENT DEPTHS IN THE OCEAN

Experiments were carried out in the Gulf of Mexico during dark nighttime to detect the response of marine organisms to the introduction of man-made pulses of light. The instrument transmitted short, intense flashes of light from a xenon source and recorded the post-flash light field with a sensitive photo-multiplier receiver. Results were as follows. *(a)* At 700-m depth, during dark of night, the background light level is extremely low; seen against this background are occasional weak bursts of light attributed to bioluminescent animals. Once a series of xenon flashes is begun (at time t_0), the background level rises by a factor of over 1000; the occasional flashes seen before stimulation are now buried in a "flood" of bioluminescence. The animals responsible for the effect are not known.
(b) With the instrument at the 100-m level, background light rises by a factor of 10 over that found at 700 m. Even in starlight some light reaches deep into the sea! Here the beginning of the flash series also produces a rise in background activity, much as that found at 700 m, but probably caused by different animals. The effect is brief, with responses falling markedly between adjacent flash pulses.
(c) At 30 m starlight is quite intense. The introduction of xenon flashes into the water elicits no systematic response as it did at deeper levels.

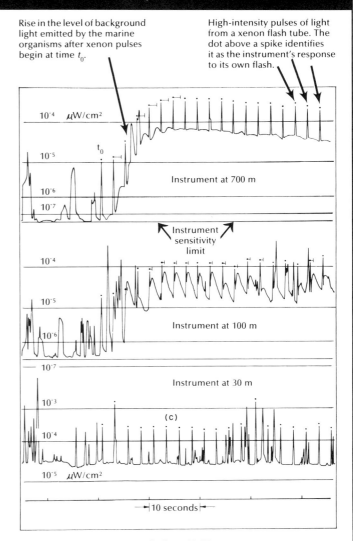

Figure 14.10   From Neshyba, 1967.

5. Bioluminescence is a very "efficient" process, in that the ratio of visible light energy to total emitted light energy is high, on the order of 20 to 90%. Whatever the reason that marine animals have evolved with luminescent capability, they do so at low energy cost.

Principal Ways in Which Organisms Produce Light

1. The light-producing reaction occurs within the individual cell. This type of reaction is quite rapid and is the type used in the flagellate *Noctiluca*. Microscope study of these small protozoans reveals that each contains a number of specific sites within the cell protoplasm at which the reaction occurs.

2. The reaction may occur exterior to the organisms. The luciferin and luciferase are excreted from epithelial cells on the skin, and when brought together produce their light reaction.

3. The animal may store luminous bacteria in a symbiotic relationship. The small fish *Anomalops* has a rotating organ adjacent to the eye; the organ is rotated to expose the symbiotic bacteria toward the outside or rotated inward to "extinguish" the light. The fish *Photoblepharon* has similar sites in which the bacteria are cultured, but it has a dark curtain that it can raise or lower to expose or cover the organ.

The Distribution of Bioluminescence

Perhaps the most striking biological fact regarding the emission of light by animals is the great number of totally unrelated and diverse organisms which have developed this ability. No clear development of luminosity along evolutionary lines is detected, but rather a cropping up of luminescence here and there, as if a handful of damp sand has been cast over the names of various groups written on a blackboard, with luminous species appearing wherever a mass of sand struck. The Ctenophora have received the most sand . . . it is probable that all members of this phylum are luminous. The Cnidaria also contain many luminous species scattered among certain of its orders. Another peculiarity . . . is the almost complete absence of luminous species in fresh water. The most striking instance of this rule is to be found among dinoflagellates in which only the salt water species can emit light (Hardy, 1952).

Protozoans. Many genera of dinoflagellates luminesce; the most outstandng example is the *Noctiluca.*

Ctenophores and Cnidarians. Of the closely related phyla Ctenophora and Cnidaria, the ctenophores are the most shining examples. Also called comb jellies or sea gooseberries, they have eight rows of luminescent sites. Some species are even reported to have luminous egg stages. Among the cnidarians many hydroids and jellyfishes will luminesce, as will almost all the siphonophores and deep-water sea pens; only among sea anemones and corals is luminosity absent.

Annelids. A fair number of the segmented worms of the phylum Annelida are luminescent; most of these are polychaetes.

Mollusks. Among the Mollusca luminescence is widespread in the cephalopods; in fact, some squid and octopi have the most complex of luminous organs, incorporating lenses, reflectors, and pigment screens. One species of squid is even capable of producing light of different colors. Many squid are deep-sea forms, and it is among these that the number and complexity of luminous organs are the greatest.

Luminosity is generally absent among the gastropods, with the exception of some nudibranch species, and rare among the pelecypods.

Arthropods. In the phylum Arthropoda, which accounts for over 75% of all species in the animal kingdom, bioluminescence exists mostly among the crustacea. The euphausids are so named because of their widespread luminosity. Among the ostrocods, members of the genus *Cypridina* are well suited for laboratory study of luminescence because they are widespread in occurrence, easily caught, and produce copious amounts of light and enzyme material. Some 3 to 4 mm long as adults, these animals are bottom dwellers by day and come out at night to feed.

Numerous species among the copepods are luminous. In the decapod group a number of pelagic shrimps carry photophores scattered over parts of their bodies; this is particularly true of the species that participate in the vertical diurnal migration of the deep scattering layer. Some forms of shrimp project luminous clouds of "light" into the water; this is found for the most part in deep-sea shrimps.

Chordates. Many species of fish are luminous. We find this particularly true of the deep-sea species; their assortment of luminous organs is as varied as among the squid and raises fascinating zoological questions.

Bioluminescence in fishes increases with depth to perhaps some 1500 m, then decreases only to increase again in those living near the bottom. It is best developed among species that inhabit the region below 500 m. In this "twilight" zone fishes also have big eyes. In the mid-depth range of 2000 to 3000 m, where the number of bioluminescent species diminishes, so also do the eyes tend to diminish to a state of near blindness. Eye size again increases among the bottom dwellers.

Species that have been well studied are the myctophids, the lantern fishes which make up the higher carnivore levels in the deep scattering layer. Some species have rows of photophores along lateral lines on each side; hence the origin of their name. Whatever the pattern of location, the position of the photophores and their number are specific to the species and are a significant aid in taxonomical classification. Frequently there is a slight difference in the position of photophores in male and female, thus sustaining the idea that in deep-sea fish these have reproduction roles.

Reasons for Bioluminescence

Proof of reasons why bioluminescence is so widespread among marine species has eluded biologists since the earliest days of discovery. It is relatively simple to list some of the more obvious reasons that animals in the sea might have evolved the function. But no single purpose has been demonstrated that could tie together the diverse forms and mechanisms among the many luminous phyla and species.

Species Recognition or Sex Recognition. Species or sex recognition is a clear and potential function. Studies of the flashing of light by the terrestrial firefly have verified that the timing and duration of pulse sequences are a part of the mating procedure. In fact, further research has even shown that predator species have evolved with the capacity to simulate the mating signals and thereby lure the males of the prey species—clever, but altogether depressing.

It is difficult to monitor bioluminescent behavior of marine species *in situ*. When the process is observed, more questions arise. For example, the light of the fish *Anomalops* has been observed to be constantly turned on and off, 10 seconds of light and 5 seconds of dark. The *Photoblepharon*, on the other hand, has not been so observed in its natural state, but it is known to flash its light in laboratory captivity. Why should these two fishes, which are closely related, have developed different means of exposing and covering their light sources if the ultimate function is similar in both species? It is worth pointing out that the culture of luminous bacteria has a definite "cost," for the light organ itself is richly supplied with host blood so that the bacteria can receive oxygen and nutrients.

Feeding Lure. We have identified some species, for example, the deep-sea species collectively called anglerfish, that use the light organ as a lure in attracting their food. The photophore in these animals is placed at the end of a filamental appendage, which is actually a modification of the first dorsal fin ray. In *Ceratias holbolli*, which may reach 100 cm length as an adult, the lure can be extended or contracted muscularly to protrude up to 30 cm in front of the mouth or be drawn back to a position immediately above the mouth. That the species is most prevalent in the deep sea attests to its use of a light lure in a dark environment. But shore forms of anglerfish, called marine frogs or "fishing frogs," have been known since classical time. Therefore, an attempt to link the use of the light organ as a lure to a specific part of the environment, whereby specific adaptation can be evoked, fails.

An Emergency Device. Prominent among reasons for the development of bioluminescence is its use to foil predation. The excretion of clouds of luminous material by squids is one example. In deeper waters

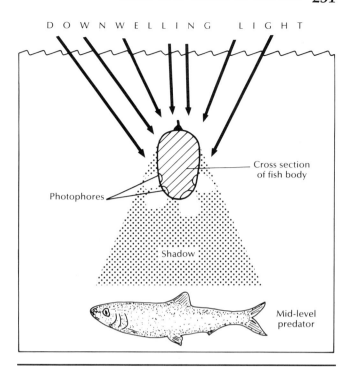

Figure 14.11 Countershading is a possible reason for the evolution of bioluminescence in some marine fishes. The light emitted by photophores located along the underside of the fish "fills" the shadow created by its body in the natural downwelling radiance from the surface. (From Miller, 1983.)

with little ambient light (see Figure 14.10), the flashing of a light some 10,000 times brighter than the background could be interpreted as a defensive measure to confuse the predator momentarily. Such an application of bioluminescence is supported by the finding that flashing or glowing can be triggered by adrenalin, but this is not a universally used trigger mechanism.

Countershading—An Example of Evolved "Cover." We have learned in recent years that some bioluminescence is apparently used to counter the "shadow" of animals living near the surface; Figure 14.11 illustrates the mechanism.

Note that many fish, squid, shrimp, even other forms, have their photophores on their ventral sides, while mid-water fishes have their eyes fixed in an upward-looking orientation. These (mid-water) predators must cruise along and spot their prey silhouetted against the cold blue of the downwelling light. The evolutionary response of filling in the silhouette would surely be richly rewarded in terms of survival. Hence the ventral location (and downward focusing lenses) of the photophores.

There are tests. The countershader must match the light coming down to good precision, and it must be able to vary its countershading intensity over about four orders of magnitude. Direct tests of matching capability show that shrimp, squid, and fish all match to within very narrow limits the downwelling light. Making the match requires not only that the animal can measure downwelling light (eyes), but that it can measure and compare its own bioluminescence. They do that by having a photophore associated with the eye. All photophores respond to systemic (neural) controls in the same way. Thus the eyes look at both downwelling light and at bioluminescence, directing the match. If the eye photophores are covered or excised, the capability for a match disappears. The experiments are very clean and convincing. It is clear that bioluminescence, even in forms that definitely countershade, has communication, luring, and possibly other functions as well (Miller, 1983).

(a) Their shading as might be seen during daylight:

(b) Their patterns of bioluminescence as might be seen during hours of darkness.

Luminous fishes of the ocean surface layer. An *Astronesthes* is shown pursuing two myctophids (lantern fish). (From Beebe, 1926)

Concluding Remarks on Bioluminescence

There remains much to be learned of animal behavior in the sea, but nothing excites the imagination more than the role of luminosity. The most extensive observations of deep-sea animals, and particularly the myctophids, were reported by William Beebe (1925), based on his bathyscaphe observations during the cruise of the *Arcturus* in 1925 as well as on observations made in darkened aquaria on the *Myctophum coccoi.*

> Scattered over the body are many small round luminous organs, which we may divide into three general sets. First, 32 ventral lights on each side of the body, extending from the tip of the lower jaw to the base of the tail; second, about 12 lateral lights arranged irregularly along the head and body, and third, a series of three to six median light plates or scales, either above or below the base of the tail.
>
> The lower battery, when going full blast, casts a solid sheet of light downward, so strong that the individual organs could not be detected. Five separate times when I got fish in a large, darkened aquarium, I saw good-sized copepods and other organisms come close, within range of the ventral light, then turn and swim still closer to the fish, whereupon the myctophid twisted around and seized several of the small beings.
>
> . . . Perhaps the best distinction between various species of lanternfish is the arrangement of the lateral light organs, and in the darkroom in absolute darkness, I could tell at a glance what and how many of each species were represented in a new catch, solely from their luminous hieroglyphics. When several fish were swimming about, these side port-holes were almost always alight, and it seems reasonable that they may serve as recognition signs, enabling members of a school to keep together, and to show stray individuals the way to safety.

Color Change in Animals

Although not an active factor of light in the sea, as are sunlight and bioluminescence, passive coloration and color change in marine animals are well known as a specific adaptation and warrant a brief discussion. Many marine animals have structures called *chromatophores,* which are used to alter color. These specialized skin cells may contain a variety of pigments, each with its specific absorption color band.

Color changes are found to be responses to one of two stimuli.

1. *Neural stimulus.* An action stimulus via a nerve transmission exposes or closes the chromatophore by activating or relaxing controlling muscles; color change is rapid.

Octopus seen against a rock background. Although this photograph is taken at night, the octopus has retained the mottled skin coloration it adopted for camouflage during daylight hours. (Carolina Biological Supply Co.)

2. *Humoral stimulus.* The trigger is activated by an endocrine, of which there are two types. A water-soluble chemical is carried by the bloodstream or within the lymphatic system to activate the control muscles; or a fat-soluble chemical moves through tissue by diffusion. The color changes are slow.

Color Change in the Octopus. The color change in the octopus takes place in individual cells of the skin; muscular contraction causes the cell to be flattened and thus exposes more area to the outside. Some octopi also have iridophores, layers of reflecting tissue located beneath the outer layers of the chromatophore; this tissue increases the amount of scattered light. Each chromatophore organ is supplied by nerves from a central nerve trunk; such neurologically controlled color change is very rapid. If the nerve trunk is dissected, the controlling muscles relax, the color cells close, and the animal's skin blanches.

Color Change in Crustacea. To many species of crustacea it is important to adapt to day or night conditions of ambient light. The objective is to match the color of the surrounding environment. Most animal species will go dark by day and alter color to a pale hue during weak lighting; such behavior is commonly found among shrimps, crabs, and the hermit crab regularly seen on beaches day or night. The eyes of the animal are important links in the chain of command that induces color change. Removing the eyes of an animal causes it to turn pale. Control is by means of neurohumor; injecting blood taken from a dark crab into a pale crab of the same species causes it to turn dark.

Color in Deep-Sea Animals

A perplexing observation made by biologists who study animals of the deep, dark sea is their often vivid coloration. Benthic amphipods, for example, may be bright red on arrival at the surface from depths of thousands of meters. Most deep-sea exploration by camera has used black-and-white film, so we lack extensive observations of color of the benthic fauna *in situ*. Still, the question of why red is intriguing. One explanation is that bioluminescent light produced by predators is blue-green in color, which is where the window is. Because a red substance does not absorb blue, a red prey would be relatively invisible in otherwise dark water.

Summary

A surprising connection exists between the transparency *window* of water and the temperature of the sun. Both produce their maximum effect at the same wavelength of electromagnetic radiation. Because of its 6000 K surface temperature, the sun emits its energy with greatest intensity just at the greenish color of light, about 470 nm. Water has its maximum transparency at this wavelength.

The earth's surface receives about 47% of the solar energy incident at the top of the earth's atmosphere. This is the energy that heats the oceans at the top of the water column; for a balanced budget the ocean must lose the same quantity. But it is important to recall that the ocean budget is almost *never* in balance on a *local scale*; a given locality will export heat if tropical, import heat if polar, or both, in the sense that upwelling imports heat at depth and exports it at the surface.

On a vertical scale, light intensity falls exponentially with depth. Only 45% of incident light remains below the 1-m level, and only 1% reaches to 100 m. Light that penetrates the deepest has a radiation of the same wavelength as that of the window, that is, a blue-green color in open-ocean water or a yellow-green in coastal water.

The vertical distribution of light is a fundamental factor in how life is vertically distributed in the ocean. Biologists identify several aspects.

1. An individual plant cell has a *compensation point* at some depth in the sea, defined as the point where respiration needs are just balanced by oxygen production.

2. There is a depth to which turbulence mixes the upper ocean layer and its plankters as well. When mixing occurs beyond a *critical depth*, overall productivity decreases because the population of plant cells spends too much time in the *dark* part of the water column. If mixing is too shallow, productivity again falls because nutrients are not replenished adequately by entrainment of deeper water into the *light* part of the column. It follows that an optimum productivity rate exists somewhere between these limits, in the range of 7- to 17-m depth.

Optical oceanography has come into its own as a valuable specialty in marine research. Among other uses, it allows us to *classify water masses* according to optical signatures, independently of the temperature and salinity of the water itself.

But it is the research on the bioluminescence of sea creatures that has some of the most fascinating aspects and questions. Why is it that the capacity to luminesce is so much more prevalent in the marine world than in the terrestrial, given that light is extinguished much more rapidly in the ocean? The answer is that we can guess, but we do not know.

Study Questions

1. Figure 14.1 presents the idea of color-dependent (wavelength-dependent) *transparency* of seawater by means of a filter whose transparency changes with wavelength. Figure 14.3 presents the same information in a different way; it shows that the *attenuation* of light rays in seawater changes with wavelength but also with the type of water. Which of these two methods is more useful and understandable to you?

2. What does it mean to you that there is a nearly perfect alignment between the *window* of water and the wavelength at which sunlight is the most intense?

3. Is it merely a coincidence that our own eyes, as well as those of marine animals, are most sensitive to blue-green rays of light?

4. Is it a coincidence that bioluminescence *color* is of the same wavelength at which peak energy is emitted from the sun? Or of that of the water window?

5. If the average depth of the oceans were only 100 m (instead of 4000), how might the open-ocean photic zone be different? How might it be the same?

6. There is the question of the *sensitivity* of the important chlorophyll molecule to different colors of light. Nature has thrown us an improbability, namely, that chlorophyll is *not* most sensitive to the blue-green color photons. Is this a disadvantage to marine plants? If so, what does the mismatch mean in terms of plant growth?

7. A partial answer to question 6 is that the world of plants has auxiliary pigments. What function do these perform?

8. Defend the statement (made earlier in the text) that the oceans are heated from the top down, using the absorption characteristic of energy from the sun.

9. In Figure 14.3 we find that the attenuation of light by different types of oceanic water is also

different. In what type of water is attenuation the greatest and why?

10. One of the more interesting exercises in oceanography is to attempt an explanation of *why* we observe the maximum concentration of plant cells at around 5- to 15-m depth instead of at the surface. Try it.

11. It is essential that the student understand the meaning of compensation point and compensation depth and be able to differentiate this concept from that of the so-called *critical depth.* Note that it is not at all necessary to introduce the factor of *vertical mixing* when discussing compensation point.

12. In Figure 14.8 the term "turbidity" of seawater is introduced. For the particular case described, what *causes* increased turbidity in waters that have been upwelled?

13. What is the stated basis for the assignment of the words luciferin and luciferase to chemical extracts from the tissue of marine animals known to luminesce?

14. Is bioluminescence an "efficient" process of producing light?

15. Research into means of propulsion for interstellar spacecraft has yielded the idea of using laser light. A laser aboard the spacecraft would be energized to emit photons of light from the rear of the craft, which should move the craft forward because photons have mass. Could fish have evolved with light emitters as an efficient means of propulsion? At what speed does light travel in seawater?

16. In a general sense, where among the various phyla of marine organisms is bioluminescence most likely to be found?

17. Write a short essay on the theme that bioluminescence in the marine biosphere must have the same purpose for all the different plants and animals that use it, because it seems to have evolved separately among the various phyla.

18. Of what earthly use would it be to produce *luminous* eggs?

CHAPTER 15

PLANTS AND ANIMALS OF THE OPEN-OCEAN PHOTIC ZONE

In a sense, this chapter begins with John Steinbeck and Cannery Row, a location first made famous by the expanding sardine fishery of the 1920s and 1930s, then famous as the location of Steinbeck's novel, and finally as the scene from which the Pacific sardine disappeared in the early 1950s.

The collapse of the sardine canning industry promoted what has been one of the world's great detective stories. What happened to *Sardinops sagax?* The industry began in 1916 with the first recorded landings of about 30,000 tons in California. It expanded rapidly to British Columbia, Washington, Oregon, and Mexico, to a peak of 791,000 tons in 1936–1937, and then crashed to a mere 25,000 tons in 1952! In 1967 California banned all directed sardine fishing, allowing only small catches incidental to other fishing. The niche once occupied by the sardine is now filled by various other fishes, such as the anchovy and jack mackerel.

The search for the reason that the sardine disappeared led to the creation of the California Cooperative Oceanic Fisheries Investigations program (CALCOFI). This systematic project began about 1949 and has continued since. A valuable data bank has been created and is summarized in the series of CALCOFI atlases from which much of this chapter is taken.

A Look at the Marine Community in the California Current

The chart in Figure 15.1 gives the basic station plan of the CALCOFI program in which each station is visited at timed intervals to sample all sorts of factors, from physical to chemical to biological. The objective is to learn as much as possible about this ocean environment, its living community, and the life cycles of specific species that interact with the sardine. I have selected just one of these stations as the locale for the following *show and tell* pages; Station 60.90 is about 90 nmi west of Monterey Bay and Cannery Row.

We start with a set of names of typical plants and animals that CALCOFI scientists find at Station 60.90 itself or in the nearby surface waters (Figure 15.2). There are five subdivisions to the set: the *phytoplankton,* the *protozoa,* the *zooplankton,* the *micronekton,* and the *nekton.* These divisions also mark the trophic levels.

Before proceeding to the marine life charts and photographs, however, the reader should be reminded of some limitations to the data.

1. There is no way that a single cruise can accommodate a full suite of sampling. One cruise may tow only for zooplankton, and then only between selected depths; a different cruise would study the primary production; and so on. The reader must not be led to think that at one instant in time, during one specific cruise, each of the plants and animals shown here could actually be taken at Station 60.90.

2. Because we already know how variable the populations of species can be, we should expect that the dominance by one species today would change to a dominance by a different species tomorrow.

3. These species are only *representative* of the water masses within the California Current.

Phytoplankton in the California Current

Diatoms

General characteristics

• Sizes of diatoms range by a factor of 10, from about 10 to 100 microns.

• These plants make shells of silica (SiO_2) in a

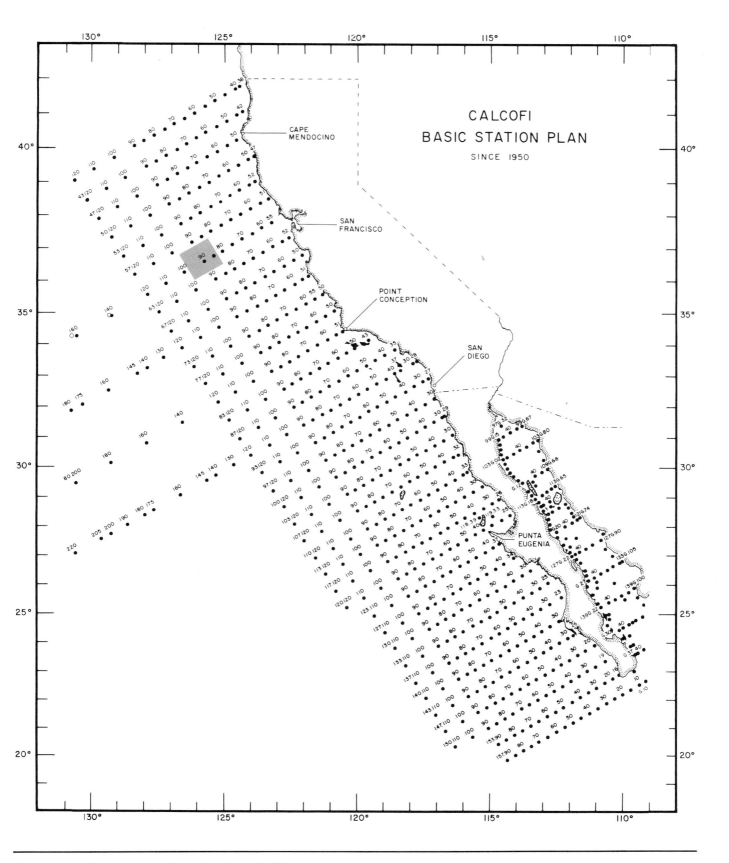

Figure 15.1 The basic sampling plan of the California Cooperative Oceanographic and Fisheries Investigation Program (CALCOFI). The plants and animals described in this chapter are the dominant forms that have been sampled from the boxed Station 60.90 (Courtesy J. L. Reid)

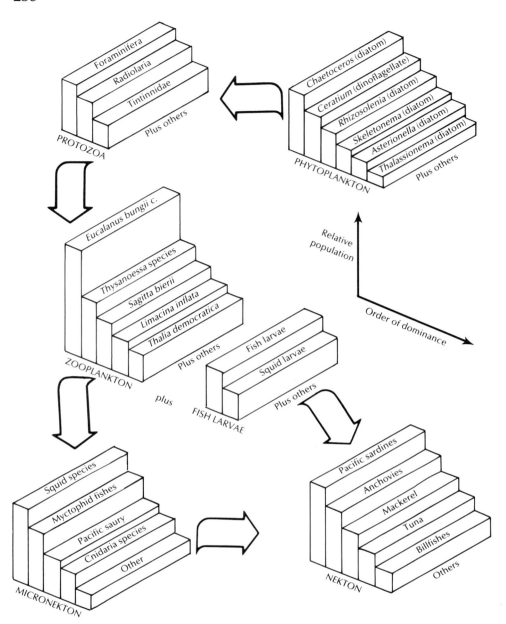

Figure 15.2 The five subdivisions of the open-ocean plant and animal community at CALCOFI station 60.90, arranged in food-chain order. (Compiled from various sources.)

wide variety of shapes, with one of two symmetries.
Centric. The shell has radial symmetry, as in *Chaetoceros* (Plate 15.1).
Pennate. The shell is elongated with bilateral symmetry, as in *Rhizosolenia* (Plate 15.1).

• Both centric and pennate types may combine into "chains," particularly when the supply of nutrients is abundant and the cells grow rapidly, as in an upwelling region. Chaining is one of the factors that allow filter-feeding fishes to predate directly on the primary producers. The plant cells by aggregating form a *large* particle, one that these fishes can

easily strain out of seawater. For this reason herring, anchovy, and menhaden schools are always found in coastal waters and together supply roughly a third of the total worldwide fish catch. (This is also the explanation for the "1.5" trophic level statement by Ryther in Table 1.2.)

• Diatoms flourish in colder waters, as seen in the distribution of siliceous sediments in Figure 5.11; SiO_2 is highly insoluble in seawater.

• Diatomaceous sediments, where geologically uplifted and exposed, have been mined for industrial uses; as a filler agent in paints and toothpaste;

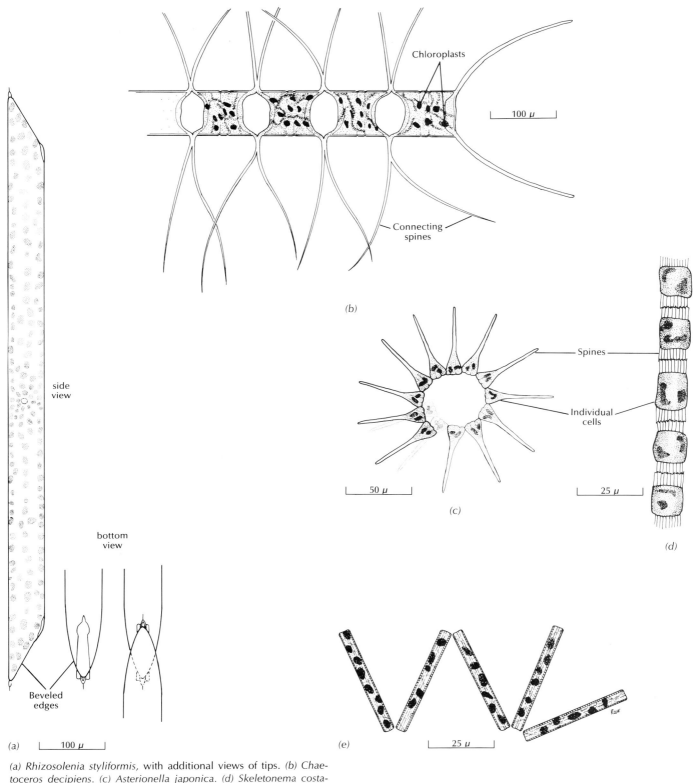

Chloroplasts

100 μ

Connecting spines

(b)

side view

Spines

Individual cells

50 μ

(c)

25 μ

(d)

bottom view

Beveled edges

(a) 100 μ

(e) 25 μ

(a) *Rhizosolenia styliformis,* with additional views of tips. (b) *Chaetoceros decipiens.* (c) *Asterionella japonica.* (d) *Skeletonema costatum.* (e) *Thalassionema nitzschiodes.*

Plate 15.1 Diatoms in the California Current

as a carrier agent in insecticides; and in dynamite, as a neutralizer to the unstable explosive nitroglycerin. (The invention of dynamite by Alfred Nobel today supports the activity of the Nobel Prize Committee.)

• Reproduction is by cell division. The shell itself divides into two half shells, each with its own cell nucleus and share of protoplasm. Each then forms a new half shell, called a *frustule*, to complete the case.

California Current examples (Plate 15.1)

• **Chaetoceros.** A number of *Chaetoceros* species are found at Station 60.90; together they can make up as much as 40% of the individual plankters in a single net haul. They are centric, chain-forming species; a pair of long, thin spines at the end of each frustule fuse with those of neighbor cells to form the chains. *Chaetoceros decipiens* is common in the California Current cold water.

• **Rhizosolenia.** *Rhizosolenia robusta* and *R. styliformis* are common in the fall season. These elongate species have tapered ends that cause the shell to flutter as it sinks downward, thus slowing the descent rate. These cells can also unite in chains. Individuals are quite long, more than 100 microns.

• **Skeletonema.** *Skeletonema* are lens-shaped with parallel spines around the edge by which multiple cells link up into straight chains. Cells are small, about 10 microns. *Skeletonema costatum* is a common California species; it appears to prefer water of only moderate nutrient concentration, so that it blooms at different times than a nutrient-hungry genus like *Chaetoceros.*

• **Asterionella.** The *Asterionella* genus competes well with *Chaetoceros* in that the two are usually found together. Individuals are about 50 microns in size and have asymmetrical shells that are lumpy at one end. Members of species *A. japonica* form clusters by fusing their lumpy ends, as shown in Plate 15.1.

• **Thalassionema.** *Thalassionema* has a pennate form some 40 microns long; the ends of members' shells join to form zigzag composites.

Dinoflagellates

General characteristics

• Dinoflagellates have two flagella for locomotion. These hairlike filaments are arranged in a characteristic way: one lies in a groove that circles the body like an equator (Plate 15.2, the *Peridinium*), a second groove generally intersects the lateral groove and contains the second, whiplike flagellum, which provides forward propulsion.

• The cell wall is usually made up of several cellulose plates glued together. Because the cellulose is dissolved after cell death, dinoflagellates do not contribute to bottom sediments. Plates have many pore holes, like diatom shells.

• Species range in size from 25 to 500 microns. Spines are typical, as in the *Peridinium,* and both horns and spines are common in various species, particularly in species of *Ceratium.* In general, dinoflagellates do not form chains as do the diatoms.

• Reproduction is by longitudinal cell division. Each new half cell quickly adds plates to form a complete shell.

• Dinoflagellates are sometimes classed as animals, because some species do ingest small particles of food. Photosynthesis is a major function and takes place in the chloroplast structures inside the body.

• Dinoflagellates flourish in warm waters, where production might exceed that of diatoms.

California Current examples (Plate 15.2)

• **Ceratium.** Ceratium appears to be the dominant genus of dinoflagellates; many species occur. These dinoflagellates prefer warmer waters and bloom profusely when surface waters warm in the summer or when warmer Central Gyre water is mixed into the California Current. *Ceratium extensum* is a common species; its very long spines can extend hundreds of microns in length.

• **Peridinium.** Because members of the genus *Peridinium* also prefer warmer water, their presence indicates either absence of upwelling or mixing of the California Current with Central Gyre water. Their size is about 50 microns and many species occur, most with the usual one spike up and two spikes down.

• **Gonyaulax.** Members of *Gonyaulax* are normally present in only small quantities and thus do not contribute much to total production. *Gonyaulax catenella* is a common species.

Special Remarks. Unlike diatoms, many dinoflagellates are highly bioluminescent, particularly in tropical waters. *Noctiluca* species really shine out. If you pick up a bucket of bay water from a tropical sea on a dark night, you will likely see brilliant tiny flashes of light each time the water is agitated. The wake of a ship is strikingly "flashy."

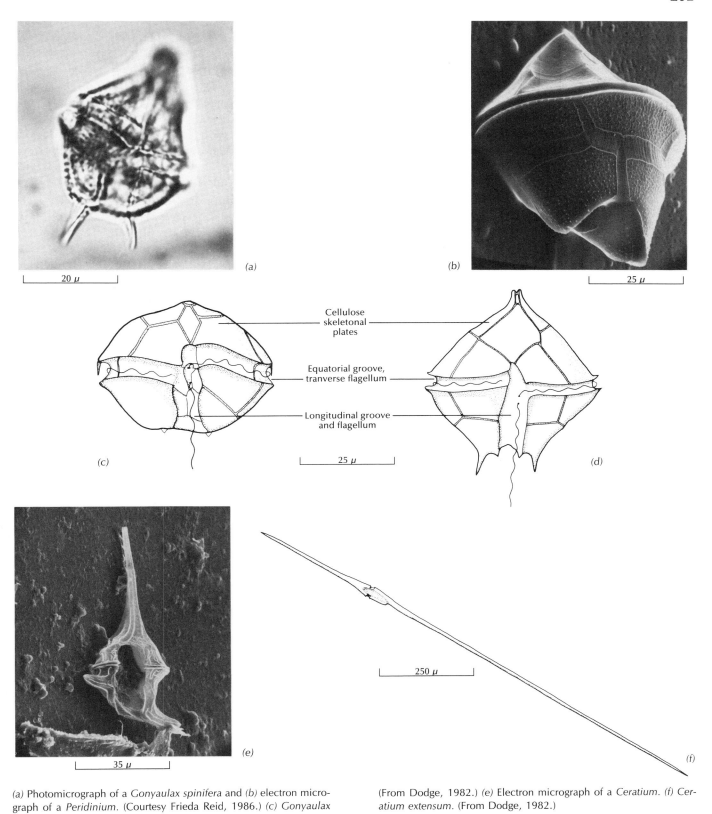

Cellulose
skeletonal
plates

Equatorial groove,
tranverse flagellum

Longitudinal groove
and flagellum

(a) Photomicrograph of a *Gonyaulax spinifera* and (b) electron micrograph of a *Peridinium*. (Courtesy Frieda Reid, 1986.) (c) *Gonyaulax catenella*. (From Whedon and Kofoid, 1936.) (d) *Peridinium granii.* (From Dodge, 1982.) (e) Electron micrograph of a *Ceratium*. (f) *Ceratium extensum*. (From Dodge, 1982.)

Plate 15.2 Dinoflagellates in the California Current

Under certain conditions that are still not well understood, some species of dinoflagellates develop an "explosive growth" or "bloom." Concentrations of cells can become so high, 7000 or more per cubic centimeter that the sea surface becomes colored; this is the "red tide" phenomenon. Many of the species that do this also contain toxins: *Gonyaulax catenella* is the toxic species common in the California Current. Because some toxins are lethal to the plankton-filtering fishes, extensive fish kills can occur in connection with red tides. Other toxins concentrate in the flesh of invertebrates, such as the mussel *Mytilus californianus*, without killing them; however, a human who eats the mussel can become violently ill.

Coccolithophores and Sargassum (Plate 15.3)

General characteristics

● *Coccolithophores.* Although not found in abundance in the Pacific, coccolithophores are small, 10- to 15-micron plant cells that do add to overall production in the North Atlantic Gyre. Because most of the chalk deposits worldwide are attributed to coc-

colithophores, it is thought that these may have had a more dominant role in times past. They form calcareous shells by gluing together a number a very small calcium carbonate disks called coccoliths.

● *Sargassum.* Plants of the genus *Sargassum*, North Atlantic plankton, are notable because of their kelplike appearance in both color and size. The plant is multicellular, with small gas-filled flotation bags that provide buoyancy. This seaweed may form extensive mats in regions where winds and currents cause a surface convergence.

● *Nannophytoplankton.* Last, but not at all the least, are the *extremely small* plants called nannophytoplankton. In size, they range downward to perhaps 1 micron. Little is known of the role these plants play in the open ocean, partly because it is so difficult to collect and study them *in situ*. We infer their existence from the evidence that the salp mucus feeders (see Plate 15.7) spin webs of mucus with a mesh size of 1 micron and that some grazing animals are themselves only abut 10 microns in size (for example, the Tintinnidae, Plate 15.4) and must therefore graze on much smaller-size plants.

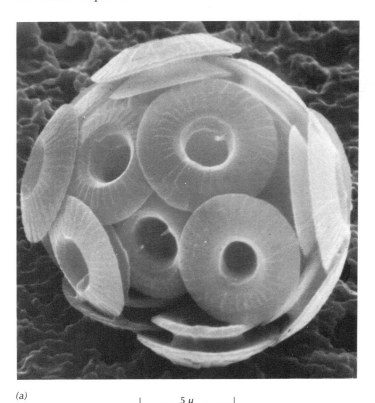

(a)

5 μ

(a) Electron micrograph of an *Umbilicosphaera sibogae*. (Courtesy Frieda Reid, 1986.)

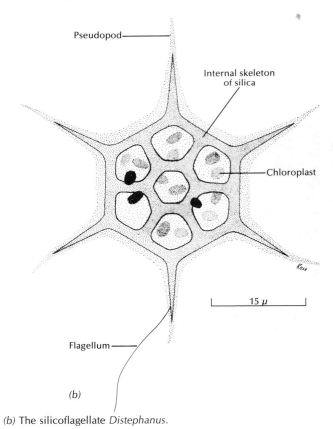

(b)

(b) The silicoflagellate *Distephanus*.

Plate 15.3 A coccolithophore and a silicoflagellate from the California Current

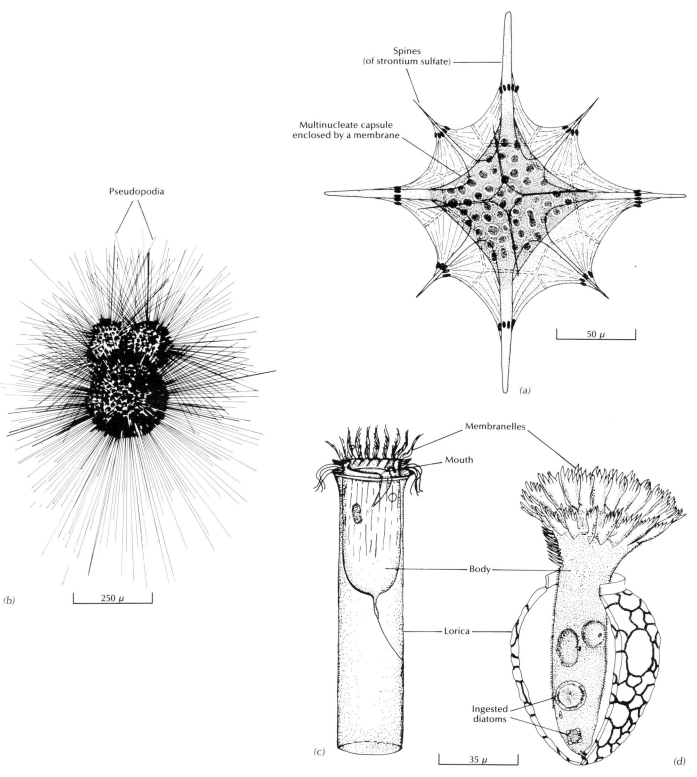

(a) The radiolarian *Acanthastaurus purpurascens*. (From Schewiakoff, 1926.) (b) The foraminiferan *Globigerina bulloides* (From Brady, 1884.) (c) The tintinnid *Tintinnus fraknoii*. (From Sleigh, 1973.) (d)

Cutaway drawing of the tintinnid *Stenosemella ventricosa* (Campbell, 1954)

Plate 15.4   Protozoans in the California Current

Open-Ocean Zooplankton

Broadly defined, *zoo*plankton are small, hetero-trophic animals. They occupy the second trophic level and are generally divided into four categories: *grazers,* (herbivores); *predators,* (carnivores); *holoplankters,* animals that spend the entire life cycle in the plankton; and *meroplankters,* animals that spend but a part of the life cycle in the plankton. A single animal may fit into three of these four categories: for example, a herring fish larva will initially feed on algae, then predate on larger-size organisms including herbivores, but is still a planktonic organism until large enough to swim against the current, at which point it enters the category of nekton.

Holoplankters and Meroplankters

• Examples of holoplankters are the many crustacean species of copepods, mysids, euphausids, jellyfish (except those that sail), and some worms and mollusks.

• Size itself is not an adequate means of distinguishing between holoplankters and meroplankters. For example, some jellyfishes may reach 2 m in diameter but still not swim against currents although they can and do move vertically.

• The advantages of a meroplankton stage in the life cycle of an animal are many: (a) the parents need not "look after" offspring; (b) being small helps disperse the species to new territory, as eggs and larvae are swept along by currents; (c) food in the photic zone is both the "right size" and plentiful; and (d) the presence of large numbers of larvae, all also subject to predation, diversifies food webs and greatly stabilizes the ecosystem.

• The disadvantages to the individual species are; (a) larvae are themselves a major food "stock" in the photic zone, so that (b) large numbers of eggs must be produced to overcome survival statistics.

Protozoan Zooplankton in the California Current

The next organisms we study in the CALCOFI region are the Protozoa. These are single-celled animals of very small size, ranging from 100 microns (one-tenth of a meter) to thousands of microns (several millimeters). They are grouped in three categories: the Foraminifera, the Radiolaria, and the Tintinnidae.

Foraminifera

General characteristics

• Foraminiferans secrete calcareous shells. On death these sink rapidly to the bottom to form extensive calcareous sediments; Figure 3.10 shows that these types of sediments cover most ocean basins. Typical species are the *Globigerina* spp.

• Animals exude protoplasm from the many holes in the shell to entrap bacteria or other food particles. Their size ranges up to several millimeters.

• Although the number of pelagic species is only several hundred, they are found everywhere but are separated into warm- and cold-water species. Because the shells usually become permanent parts of the sediment layers, paleontologists can analyze sediment cores to (a) reconstruct photic zone conditions in ancient seas and (b) to date the sediment layers by analyzing the carbon-14 content in the shells.

• By fixing carbonate (CO_3) into shells and sediment, these animals provide a rapid route for the oceans to remove excess CO_2 from the atmosphere.

California Current examples (Plate 15.4)

• *Globigerina bulloides.*

• *Globoquadrina pachyderma.*

Radiolaria

General characteristics

• Radiolarians are of the same general size as the foraminiferans, that is several millimeters.

• Animals secrete an internal skeleton of silica (SiO_2), usually in a spherical pattern; hence the name radiolaria; the skeleton is also very porous.

• Because radiolarians are generally cool-water organisms, siliceous sediments from these animals are often found in upwelling regions, such as the equatorial upwelling in the Pacific (see Figure 5.11).

• Feeding, like that of the foraminiferans is by encounter with bacteria and particles. The siliceous skeleton often has numerous spines, along which the animal exudes mucuslike protoplasm to "catch" food particles.

California Current examples (Plate 15.4)

• *Stylatractus* spp.

• *Acanthastaurus* spp.

Tintinnidae (Plate 15.4)

• Tintinnids are very small animals, on the order

of 10 to 50 microns. Most have bands of cilia around the mouth opening, these hairlike growths provide both mobility and a capture mechanism. Tintinnids usually develop a thin siliceous shell and in general are not well studied.

• The reproduction time of the small tintinnids can be as short as one day, making them able to expand rapidly to "eat a bloom."

Zooplankton in the California Current

Passing now to the group of next largest animals, the zooplankton of Figure 15.2, we find that only a few species dominate the California Current community. The top three species alone make up 50% of all the individuals taken at this station, and all are crustacean copepods. Of the 74 species identified, 22 alone make up 90%; 14 are copepods (70% of total), 3 are euphausids (9%), 3 are chaetognaths (7%), and 2 are mollusks (4%).

The copepods and euphausids are both members of the phylum Arthropoda. These are more commonly called the crustaceans. Their more familiar members are the lobster, crab, and shrimp, but these highly edible groups are coastal and not found in the open sea. Euphausids are often called the *krill*, but the small crustacean *copepods* are the dominant animal. Most of the species listed as zooplankton in Figure 15.2 are herbivores, but some are voracious predators, particularly the chaetognaths, commonly called arrowworms.

Copepods

General characteristics

• Copepods are the single most important zooplankton group.

• Adults range in size from 1 to 4 mm. Herbivorous copepods use modified appendages to strain smaller phytoplankton cells (see Figure 13.1).

• Copepods have many stages in a life cycle: up to six stages as *nauplius* larvae, followed by six additional stages as *copepodites* (juveniles). Being crustaceans, they have a chitinous exoskeleton, which they molt at each step to the next life stage.

• Because each stage of copepod is progressively larger in size, and because there are so many individuals, a grazer of the *right size* is always present to take whatever size plant cells happen to be in bloom at the time.

California Current examples (Plate 15.5)

• *Eucalanus bungii californicus* (similar to *E. elongatus*).

• *Metridia pacifica.*

Euphausids

General characteristics

• Euphausids are grazers; see Figure 20.11 for a description of their filtering mechanism.

• Euphausids of the tropical and temperate waters are generally 1 to 2 cm long and are often bioluminescent.

• They are an important animal of the sea and rank second to copepods in abundance.

• Even though euphausids are second in abundance, they are larger animals than the copepods and so can rank *first* in biomass. For the upper 1000 m of the Pacific Ocean, one biomass estimate ranks euphausids first at 35 to 50% of pelagic animal biomass, fishes at 20 to 45%, and shrimps at 15 to 25%.

• The fact of large biomass, plus their tendency to collect in large swarms, makes euphausids economical to harvest directly. Japan harvests *Euphausia pacifica* for feedstock for aquaculture farms, and California fishermen take them for sale as aquarium food.

• Euphausids of the open ocean migrate vertically in a diurnal cycle, coming near to the surface at night to feed and returning to depths of about 400 m by day. (See also the discussion of the deep scattering layer in Figure 2.2.)

California Current examples (Plate 15.5)

• *Thysanoessa gregaria* is one of many prominent euphausids in the Pacific, a species notable for the sharp spine protruding forward from the carapace.

• *Nematoscelis difficilis.*

• *Euphausia pacifica* is a cosmopolitan species found throughout the Pacific.

Chaetognaths

General characteristics

• Commonly called arrowworms, chaetognaths have translucent bodies 2 to 5 cm long.

• They feed for the most part on copepods and other smaller zooplankton. Spinelike attachments around the mouth are used to grasp prey. They are swift-moving and direct their predation on individ-

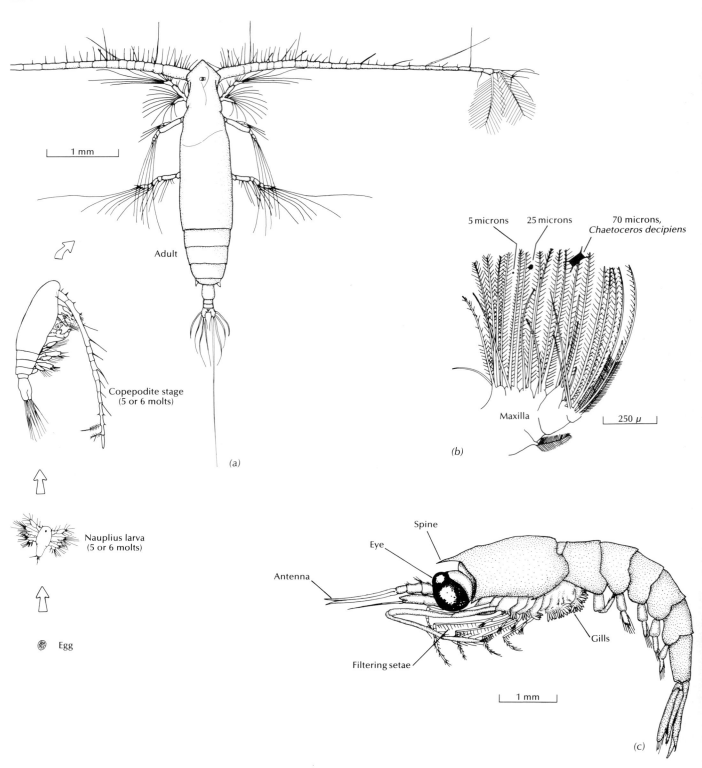

(a) A calanid copepod with life cycle stages, nauplius larva–cope-podid–adult. (Adult from Geisbrecht, 1892.) (b) Comparison of setae on the feeding maxilla appendage of a calanid copepod with three different-sized particles: 5 microns; 25 microns; 70 microns, the size of the single cell *Chaetoceros decipiens* (From Barnes, 1980.) (c) The euphausid *Thysanoessa gregaria* (From Brinton, 1967.)

Plate 15.5 Crustacean zooplankton in the California Current:

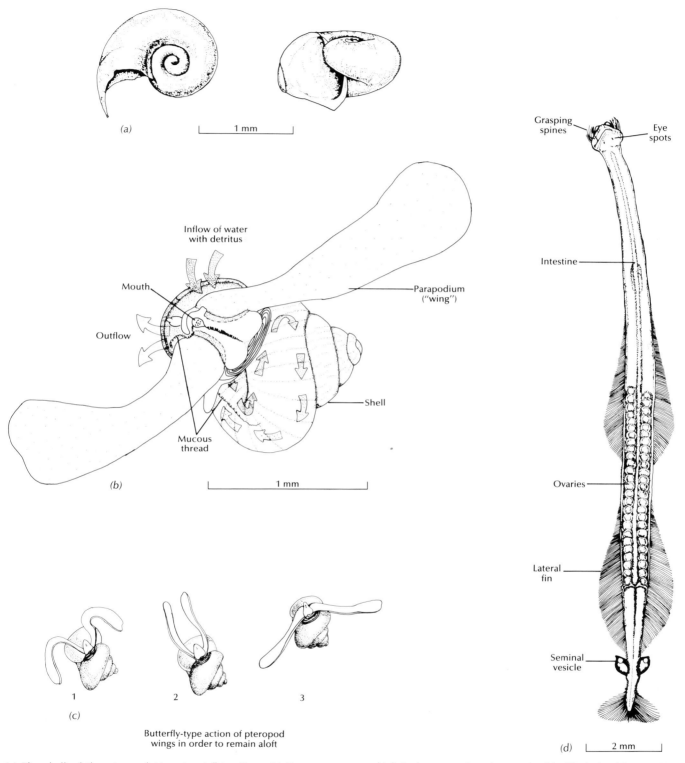

(a)

1 mm

Inflow of water
with detritus

Mouth

Outflow

Parapodium
("wing")

Shell

Mucous
thread

(b)

1 mm

1   2   3

(c)

Butterfly-type action of pteropod
wings in order to remain aloft

Grasping
spines

Eye
spots

Intestine

Ovaries

Lateral
fin

Seminal
vesicle

(d)   2 mm

(a) The shell of the pteropod *Limacina inflata*. (From McGowan, 1968.) (b) *Limacina retroversa*: in the cilia-driven flow of its feeding process particles are trapped by mucus and formed into a string, which is then passed to the mouth. (Modified after Morton, 1954.) (c) The swimming motion of the *Limacina*. (Modified after Morton, 1954.) (d) The chaetognath *Sagitta bierii*. (From Alvariño, 1965.)

Plate 15.6   Zooplankton in the California Current:

ual prey. In turn, chaetognaths are a common food for jellyfishes. Next to euphausids, they may have the largest biomass of planktonic animals.

California Current examples (Plate 15.6)

- *Sagitta bierii.*
- *Sagitta euneritica.*
- *Sagitta scrippsae.*

Pteropods

General characteristics

- Pteropods belong to the phylum Mollusca, which consists of bivalve invertebrate animals. Only a few species spend entire life cycles in the plankton. Sizes range from 3 to 5 mm (3000 to 5000 microns). Most species are filter feeders.

- In polar waters the genus *Clione* provides a major food source for the right whales. In temperate waters the genus *Limacina* is common (Plate 15.6). Animals move by beating a pair of flaplike wings, which are actually modified halves of the gastropod "foot" commonly found in snails. When threatened, animals will retract the wings and sink rapidly, which means that they expend considerable energy in swimming just to stay in the photic zone.

- *Limacina* has evolved a complex feeding process. Water is ingested into a cavity within the shell and moved across the mucus-lined cavity walls; cilia provide the pumping power while at the same time forming the particulates into a "thread" of food, which is then passed out of the shell, over the neck, and into the mouth, as shown in Plate 15.6.

California Current example (Plate 15.6)

- *Limacina inflata,* a filter feeder.

Salps (Thaliacea)

General characteristics

- Salps are relatively large in size, from 0.5 to 10 cm, and are weak swimmers; most of their mobility is used simply to move seawater through their hollow body where particulates are extracted by means of a mucous "web."

- Salps can reproduce by budding off new individuals. It is not unusual to find five individuals per cubic meter in California water, with density sometimes reaching 500 per cubic meter, a very heavy concentration of filter feeders.

California Current example (Plate 15.7)

- *Thalia democratica.*

Micronekton in the California Current

We now pass to larger animals found in the California Current, the micronekton group shown in Figure 15.2. Most prominent among these are the squid, large medusae, shrimps, and the vertically migrating species of lantern fish.

Cephalopods (squids)

General characteristics

- Squid belong to the Mollusca phylum, but with the "foot" evolved into a series of eight short arms plus two tentacle arms. The two tentacles are used to sense prey and accomplish the initial capture; after drawing the prey into the reach of the shorter arms, these take over the task of feeding the prey toward the beaked mouth.

- Squid beaks are able to inject toxin into the flesh of the prey, rendering it unable to resist.

- Reaching some 10 cm in length, the California Current species is an important member of the micronekton. Individuals move rapidly by ejecting jets of water. There is no skeleton, but a semirigid interior "bone" gives some rigidity to the animal.

- Squid in general represent a relatively untapped food resource for humans. The problem is that they are difficult to catch, for they can outswim a towed net and do not swallow hooks. At night, however, they do tend to be attracted to lights on the surface, where fishermen can "jig" (snare) them.

California Current examples (Plate 15.7)

- *Abraliopsis felis.*
- *Loligo opalescens* is a commercially important species.

Cnidaria (formerly called Coelenterates)

General characteristics

- The Cnidaria phylum contains a variety of quite interesting organisms, including the sea anemones and corals (both described in Chapter 16), the medusa-type jellyfish, and the neuston or "sailing" jellyfish.

- Almost all members are unique in that they employ a "stinging cell" called a *nematocyst* to capture and neutralize prey animals.

- One type of stinging cell houses a coiled tube-like filament of protein, the inside of which contains a strong toxin. When activated by neuron sensitization, the cell expels the filament, which inverts itself

Muscle bands pump water through the body for feeding and propulsion.

→ → → → → → → → → → → →

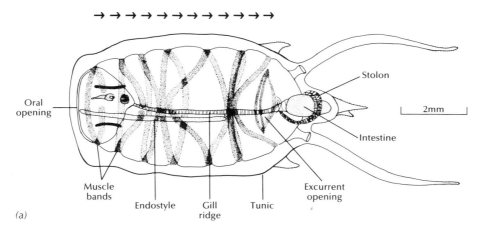

(a)

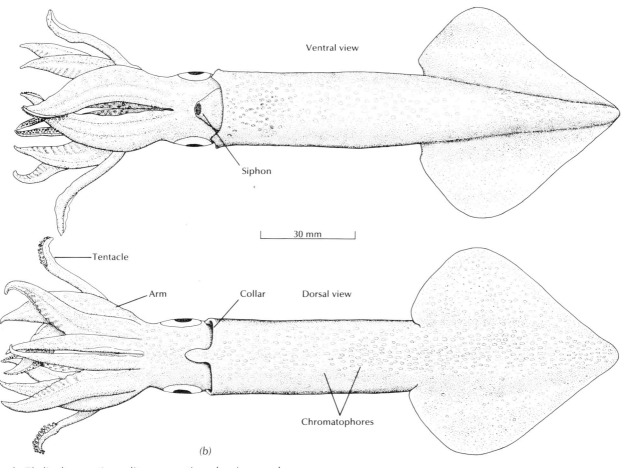

(b)

(a) The salp *Thalia democratica,* solitary generation, showing muscle bands and flow of fluid through the animal. (From Berner, 1967.)
(b) The squid *Loligo opalescens.*

Plate 15.7 Salps and cephalopods in the California Current:

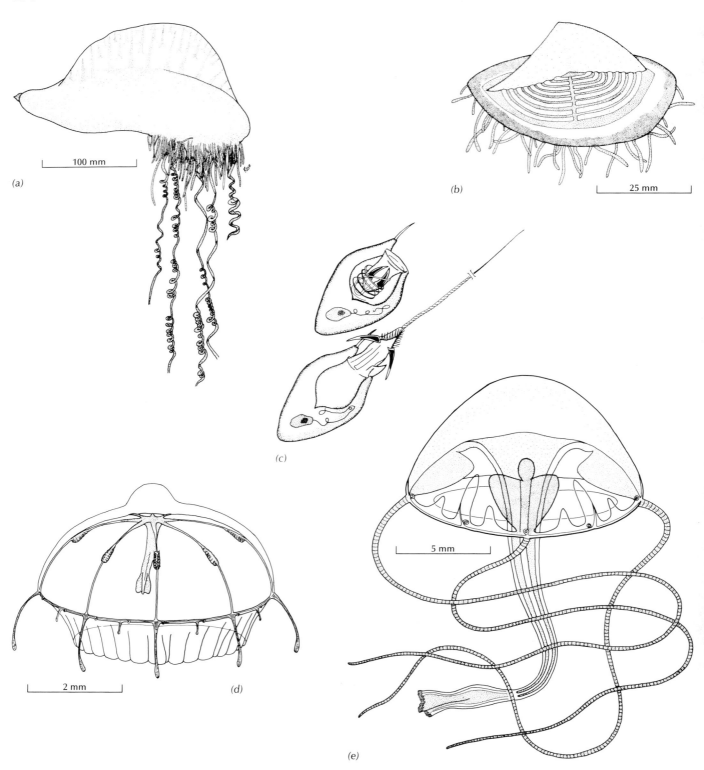

(a) *Physalia physalia*, the Portugese man-of-war, showing sail bladder and tentacles with nematocysts, which are stinging cells. *(b) Velella velella*, the sailor by-the-wind. *(c)* A nematocyst before and after dis- charge. (From Sumich, 1984.) *(d)* The small medusa drifter *Rhopalo- nema velatum*. (From Mayer, 1910.) *(e)* Another small medusa, *Lir- iope tetraphylla*. (From Mayer, 1910.)

Plate 15.8 Cnidaria in the California Current and their characteristics:

in the process, exposing the toxin for outside contact with a prey animal. Spines and barbs on the filaments help in holding the prey once contact is made (Plate 15.8c). Other stinging cells act like tiny hypodermic needles, with internal cell pressure causing the injection.

• In general, the reproduction process has two stages, or two body types: a free-swimming or planktonic *medusa* body form (Plate 15.8d) and an al-

ternating sessile *polyp* form. Some jellyfishes keep the medusa body form each generation, that is, they have eliminated the polyp state. For most corals, however, the opposite is true; most corals reproduce directly from polyp to polyp.

• Jellyfishes (class Scyphozoa) are usually treated as zooplankton because of their limited mobility. Some species, however, can reach 2 m in diameter with tentacles dangling 30 m or more below—a large

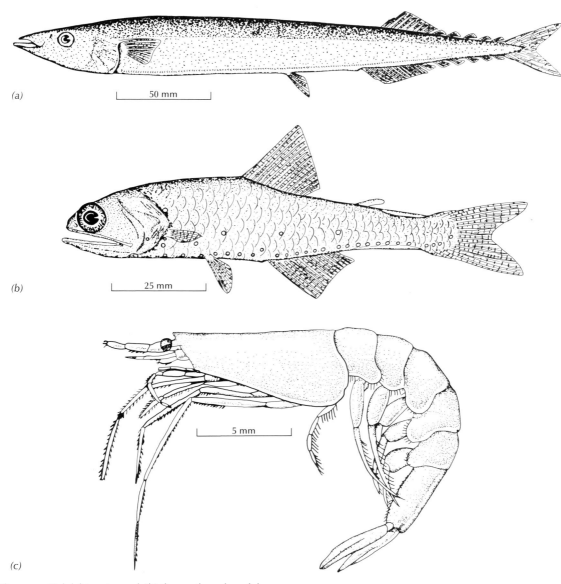

(a) The Pacific saury *Cololabis saira* and (b) the northern lampfish *Stenobrachius leucopsarus*. (From Hart, 1973.) (c) The shrimp *Sergestes* spp.

Plate 15.9 Nekton in the California Current:

animal. The medusa body is made up mostly of a gelatinous material called mesoglea. Jellyfishes swim by alternately expanding and contracting their bell-shaped body.

• Siphonophores are a unique group that also uses tentacles covered with stinging cells to capture prey. Two genera are worth noting: the large Portuguese man-of-war *(Physalia)* and the smaller sailor-by-the-wind *(Velella)*. Both maintain the body itself just at the surface by means of a buoyancy bag, and both erect membranes above the sea surface so that the wind will move them relative to the water. In this way, the dangling tentacles are moved through a much larger volume of water than are those of the wholly submerged jellyfish. These animals literally "drag" for prey.

California Current examples (Plate 15.8)

• *Liriope tetraphylla* is a rather small animal, a few centimeters in diameter. Investigators have found a strong inverse relation between the number of these medusae and the number of anchovy fish larvae present, suggesting that this medusa is a strong predator of anchovy (Alvarino, 1980).

• *Rhopalonema velatum* is also present in the CALCOFI community, this medusa being smaller and probably not an active predator of fish larvae.

Myctophids (Lantern Fishes)

General characteristics

• Lantern fishes are found over all the world oceans, including the CALCOFI zone. Together with another small fish, the Pacific saury, they form the very important link between the small zooplankton and the high-level carnivores, as shown in Figure 15.3. Most species are about 10 cm in length.

• Lantern fishes migrate vertically to follow the diurnal migration of copepods and euphausids. In fact, the gas bladder of the lantern fishes produces the echo sounder signal whose diurnal depth migration was investigated and assigned the name "deep scattering layer" (Figure 2.2).

• As we have learned, the myctophids are named "lantern" fish for their conspicuous arrays of bioluminescent organs along the ventral sides of the body. Each species has its own arrangement of organs *or* its own pattern of energizing the organs (see also Chapter 14 for Beebe's visual description).

• In the California Current community, the larvae of the myctophids exceed in number *all* other fish larvae combined. They are key species in the food chains.

California Current examples (Plate 15.9)

• *Triphoturus mexicanus.*
• *Stenobrachius leucopsarus.*

Saury

General characteristics

• Unlike the lantern fishes, the Pacific saury does not migrate vertically.

• It preys upon large copepods, amphipods, euphausids, and other fish larvae, and so it forms a trophic link between the zooplankton and high-level carnivores, as shown in Figure 15.3.

• Because the saury is not a schooling fish, commercial harvest is by trawling within the upper 150 m. The adult reaches about 15 cm in length.

California Current example (Plate 15.9)

• *Cololabis saira* is commonly called the Pacific saury.

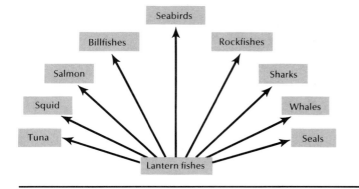
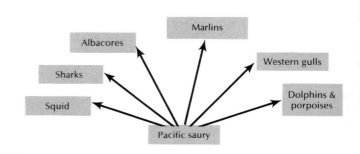

Figure 15.3 Some known predators of lantern fishes and the Pacific saury in the California Current system.

A street scene today on Cannery Row in Monterey, California. (J. Hackett/Leo de Wys.)

Shrimps

General characteristics

• It is common to think of shrimps as coastal or estuarine animals that feed as scavengers over the shallow bottoms, but some species inhabit the open ocean. In the California Current *Sergestes similis* is a pelagic shrimp carnivore. About 2 cm long in its adult stage, it preys upon whatever it can catch, most often the larvae of many fish species in these waters. In turn, the shrimps are a favorite prey species for the squid *Loligo opalescens*.

Large Nekton in the California Current

Figure 15.3 shows the principal types of large nekton in the California Current system. Some of these are important commercial fishes. Tuna, for example, has long been a prized fishery, but from Figure 3.4*e* we know that the feeding grounds for tuna extend northward only to about San Diego, California. Only during years when the El Niño phenomenon moves warm waters farther northward do the tuna follow into central California waters. Salmon, on the other hand, are subarctic species, and original salmon stocks occupied California waters south to San Francisco Bay. The billfishes such as marlin are warm-water species and are taken only in the southern reaches of the California Current. Whales, particularly gray whales, migrate the entire length of the

Pacific coast of North American from Baja to the Bering Sea.

These large nekton are essential parts of the complete California Current system biosphere. Many of the species are sought in sport fishing, especially billfishes, salmon, rockfishes, and sharks. But the bulk of commercial value in California Current fisheries belongs to a specialized few.

The Commercial Role of Eastern Ocean Boundary Communities

Now that we have looked at sample plants and animals of the Eastern North Pacific, an appropriate question is "How typical of the world's ocean biospheres is the California Current system?" The answer is that, at the level of the most commercially important species, the California Current community is very similar to each of the other major eastern ocean current communities. Figure 15.4 shows that the same commercial fishes dominate the Peru (South Pacific), Canary (North Atlantic), and Benguela (South Atlantic) current communities.

It is important to note, however, that the lists in Figure 15.4 are not necessarily in order of dominant species. Certainly the California Current community in the time of Steinbeck and Cannery Row was dominated by the sardine, but since 1950 the anchovy has assumed the dominant position. In the Peru Current community, after the collapse of the an-

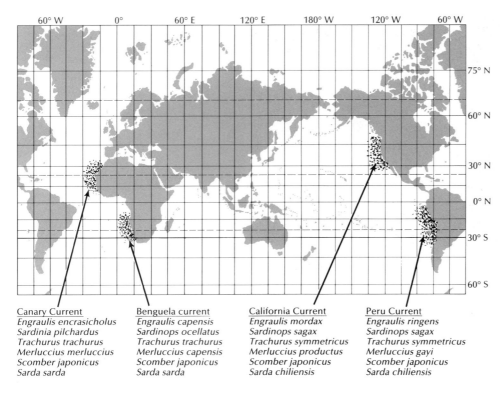

Figure 15.4 The major commercial fishes are common to each of the major eastern boundary currents of the world's oceans. *Engraulis spp.* are anchovies, *Sardina spp.* pilchards, *Trachurus spp.* horse mackerel, *Merluccius spp.* hake, *Scomber spp.* mackerel, and *Sarda spp.* bonitos. (After Bakun and Parrish, 1980.)

Canary Current
Engraulis encrasicholus
Sardinia pilchardus
Trachurus trachurus
Merluccius merluccius
Scomber japonicus
Sarda sarda

Benguela current
Engraulis capensis
Sardinops ocellatus
Trachurus trachurus
Merluccius capensis
Scomber japonicus
Sarda sarda

California Current
Engraulis mordax
Sardinops sagax
Trachurus symmetricus
Merluccius productus
Scomber japonicus
Sarda chiliensis

Peru Current
Engraulis ringens
Sardinops sagax
Trachurus symmetricus
Merluccius gayi
Scomber japonicus
Sarda chiliensis

chovy-dominated fishery in 1972, mackerel species became dominant in the north and sardines in the south.

Summary

After examining all the drawings in this chapter, the reader may have the impression that the marine community of the California Current is well known. That would be a false impression. The fact is that even after more than 30 years of systematic sampling and analyses by the CALCOFI groups, and even though this marine community is the most thoroughly sampled of any in the deep, open ocean, we have only sketchy views of how it works. (The North Sea community has been studied and harvested for much longer, but it is a shallow sea between England and the European continent.) For example, it is not possible in Figure 15.2 to list the order of dominance of phytoplankton in the same way as is done with the zooplankton. The task of working on microscopic samples taken from a vast ocean is too vast. Instead, biologists resort to measuring the amount of chlorophyll and other pigments that are part of the photosynthesis; this variable is a good measure of total primary production, but it tells almost nothing about the work of individual plant species.

Nonetheless, the drawings of this chapter document many characteristics of species that live in the different trophic levels: how they stay afloat, how they move about, how they capture prey or filter seawater for particulate food particles, and how human society has impacted the natural system. Readers are encouraged to read further in the many volumes of work that have been published (and, if you can, start by reading *Cannery Row* or *Log From the Sea of Cortez*, so that you begin with the "taste" in your mouth).

But there is one more technique for summarizing pictorially this discussion of the California Current community. Selected plants and animals can be put on a single chart with their sizes in relative scale. Figure 2.3 has already done this. The upper part shows the principal players in the California Current food web in correct size scale beginning with the lantern fish. The problem is that the other end of the food web is seen only as a series of tiny specks. The circular inset is a ×25 magnification of the region around the copepod and shows the *Chaetoceros* diatom (plant), the much smaller carnivore (and omnivore) tintinnid, and again a small speck for the unknown nannoplankton. The marvel of this marine world is how it can be so minutely "sized" and yet so elegantly "organized" to produce the bountiful harvests of the Northeast Pacific.

Study Questions

1. The tendency for many species of diatoms to form "chains" by aggregating together is the basis for the term *food chain*. True or false?

2. In fact, the aggregation of phytoplankton into chains is a behavior seen for the most part in upwelling zones. True or false?

3. Chained diatoms are one reason that clupeoid fishes such as herring, menhaden, and anchovy can *bypass* one trophic level and feed directly on plants. True or false?

4. If the major fisheries of anchovy occur in cold, upwelled water regions, it is because of the food size and abundance, not because of the cold water. True or false?

5. The reason we don't find sediments of dinoflagellate shells is that shell material dissolves quickly on death of the plant. True or false?

6. *Gonyaulax species* are the source of toxins that are found in the red tide. True or false?

7. Tintinnid animals are a special problem to marine biologists because they are so very small (10 microns) that we do not know what they eat. True or false?

8. Foraminifera, like Radiolaria, exude protoplasm out of holes in their porous shell to trap particles, which are then passed through the holes for digestion inside the animal. True or false?

9. Foraminifera, like Radiolaria, reproduce by means of eggs. True or false?

10. Crustaceans such as copepods may pass through as many as ten molts enroute from egg to young adult. True or false?

11. In the California Current, the majority of zooplankton individuals are copepods. True or false?

12. In terms of evolved methods of staying afloat, the pteropod is perhaps the most bizarre, using two appendages as a pair of wings. True or false?

13. The grasping spines of the euphausids are used to capture individual prey. True or false?

14. Salps feed by first stunning their prey, using their toxic *stinging cells.* True or false?

15. A squid *holds* its prey with two of its tentacles while it bites the prey and injects paralytic toxins. True or false?

16. Neuston is a type of jellyfish found only in Texas bayous. True or false?

17. Write a short essay exploring the physical, chemical, geological, or meteorological conditions that could explain why the eastern boundary zones of all the major ocean basins have such similar sets of commercial fishes.

18. Write a short essay developing the theme that we may still not know the true composition of the California Current community.

PART VI

COASTAL OCEANOGRAPHY

The specialized study of nearshore environments from coral reefs to estuaries

CHAPTER 16

MARINE PLANTS AND ANIMALS OF THE OPEN COASTS

Human beings first began studying plants and animals of coastal zones over a million years ago when our ancestors walked the shores of ancient lakes in Africa; their early studies were prompted by the need for food for survival. Today, we still go to the shore, but now for recreation.

For most of us, the marine shore community is the only part of the living marine world that we can observe closely. And because the chance to observe is also linked to the cycles of tides and waves, the observer quickly notices that there is an "organization" to the marine community. Whether the shore that we visit is rocky, or a sandy beach, or a mud flat, we find that its organisms occur in distinct patterns. A careful study of the *how* and *why* such patterns exist is one of the more fascinating recreations that "weekend oceanographers" can have.

We focus first on the rocky shore. Here the zones occupied by the different plants and animals are clear to see and the easiest to explain in terms of the forces and stresses that help to shape its zonation. Next we examine the sandy shore, although not in as much detail. A final section looks at the special conditions of the tropical coral reefs. We leave to other courses the equally interesting subjects of marine life along the quiet waters of estuaries and tide marshes.

The Rocky Shores—Factors Determining Distribution of Organisms

Suppose you are walking on a rocky shoreline during a low tide, observing how different organisms appear to be "banded" into different zones above water level. What factors could you list as the most influential in creating the zonation? Surely the rise and fall of the sea must be an important factor. Perhaps the slope of the rock faces, whether they are steep or shallow, is influential, and also the direction in which the rock face opens—north or south, seaward or landward, sunlight or shadow. The energy of breaking waves is another physical factor. There are also irregular shapes in the rocky shore itself, shapes forming pools and tiny coves in a wide range of sizes. These are influential *physical* factors.

The marine ecologist knows that another set of factors are at work, factors that are not visible but are just as important as physical ones. These are the *biological stresses* imposed on each species. Perhaps the most active stress is the competition that the individual plant or animal meets while claiming its own place on the rocks. Just as important is the direct stress of predation of one species upon another; in predation the uppermost level at which a predator can function sets the lower limit to which the prey species will survive. The temperature of water in tide pools exposed to sunlight will heat up more during the day than does the water in protected pools. Rainfall can dilute the salinity of a pool while evaporation can cause salinity to rise. Both cause biological stress by rapidly changing the environment itself.

The impact of all these factors is summarized in a set of general rules that control the zonation of species within the intertidal range of a rocky shore.

1. The upper limit of a zone occupied by a specific organism is set by the stress of physical exposure out of water, of predation from land animals, or of both.

2. The lower limit is set by the exposure tolerance of the organism's oceanic predator; the organism itself can live and reproduce at lower levels, but predation keeps it from doing so.

3. The concentration of individuals within a zone is determined either by interspecies competition or by symbiotic association among species.

Tide-Related Exposure

Most open coasts experience tides that are a mixture of diurnal and semidiurnal components. Figure 16.1 shows the typical daily sea level rise and fall with two high-water and two low-water marks over the diurnal tide period of 24.8 hours. The exact levels of each of the highs and lows change from day to day, principally because of the spring–neap tide cycle of 14.75 days (see Figure 17.13). If we refer all levels to the mean lowest low (Lo-Lo) as the 0 datum level, the average Hi-Lo mark might occur at about 0.7 m above datum, the average Lo-Hi mark at about 1.4 m above datum, and the Hi-Hi at perhaps 2.0 m. The entire zone between highest and lowest water is called the *littoral zone.*

Each of the high and low tide levels is important from the view of biological stress. Between the Lo-Hi and Hi-Hi marks is the *high intertidal zone;* a plant or animal attached to the rock in this zone is exposed to air for most of the tide cycle. Between the Hi-Lo and Lo-Lo marks is the *low intertidal zone,* where attached organisms are covered for most of the tide cycle. In the central or *midintertidal zone* there is a twice-per-cycle exposure–cover sequence.

Two other zones are included in Figure 16.1. Above the Hi-Hi is a somewhat indefinite upper limit that is wet with the spray from breaking waves, called the *splash zone.* Although treated as a zone here, it is very irregular because the distribution of wave spray along a rocky shore is quite irregular. Below the datum line the shore is covered at all times except during the occasional very low tide; here begins the *sublittoral zone,* which can extend to sea for a considerable distance. (Some consider the entire shelf bottom to be an extended sublittoral zone).

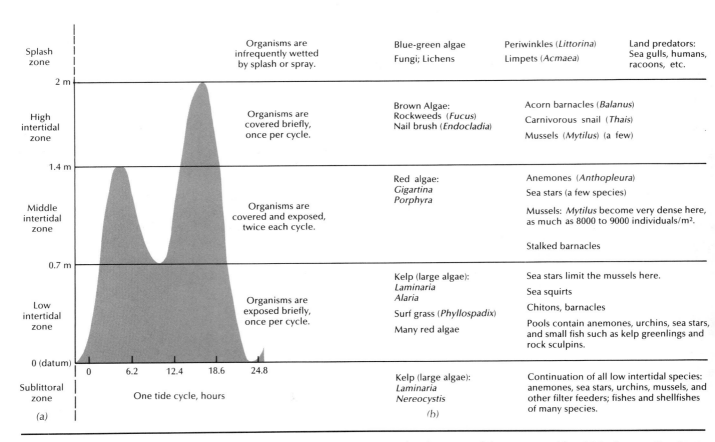

Figure 16.1 Physical and biological description of the intertidal zones. *(a)* Classifying the intertidal zones according to tide levels. The tide range is subdivided according to the degree and frequency with which the sessile plants and animals are exposed to air. *(b)* Examples of dominant plants and animals found in each intertidal zone.

Typical Organisms of the Intertidal Zones

We begin with Plate 16.1, page 304, photographs of intertidal fauna and flora on the rocky Oregon coast. Because of the great similarity of intertidal communities throughout the global coasts, the following descriptions are an oceanwide representation. Each of Plates 16.2 to 16.5 shows one section of the main photograph of Plate 16.1, corresponding to one division of the littoral zone described in Figure 16.1, with its typical plants and animals.

Organisms of the Spray and High Intertidal Zone

(Plate 16.2). In its uppermost parts the spray and high intertidal zone is briefly covered, with most of its wetness caused by waves breaking on the rocks. It is a zone of algae and fungi and the animals that graze upon these plant films and mats. Here are found the limpets, small gastropods that move slowly over the rock faces, scraping algal growth with a radula mouthpiece. When disturbed or in danger of being flushed off by surging waves, the animal uses its large foot as a suction device to hold on. During low tides it seeks out depressions in the rock face in which to "bed down" so to speak; by holding its shell edge close to the rock, it avoids desiccation as well as reduces the chance that a predator sea gull can pry it off. Limpets reproduce by releasing eggs and milt into the seawater during high tides. Coloration of the individual offspring govern their survival; those best camouflaged against the rock face are most likely to be overlooked by predatory birds and land animals.

Some acorn barnacles are found in these upper intertidal zones. The offspring larvae of barnacles, which are crustaceans, live in the shore waters through a number of molt stages; when they reach the *nauplius* stage, they seek a suitable substrate to which the individuals cement themselves permanently. As an individual grows it enlarges its shell. A pair of calcareous flaps across the shell opening protect it against drying out during long hours of air exposure. A typical *Balanus* species is shown in Plate 16.2. A major predator of both barnacles and limpets is the whelk, or gastropod *Thais* spp., a larger animal. Its predation holds the barnacles to an average life of about two years or less. The barnacles that survive beyond two years are then too large to be taken by the whelk and may live many more years. In the Southern Hemisphere, a species of very large barnacle called "picoroco" is the main ingredient of a native soup dish.[1]

[1]The dish, called "sopa marina," is absolutely delicious!

Some larger algae often form thick mats in the intertidal zone. Species of *Fucus*, a brown algae that grows to lengths of 10–15 cm, are common. They attach firmly to the rock face to avoid being stripped away by breaking waves. Other plant types include the wiry *Endocladia*, sometimes called nailbrush. It is also limited to about 10-cm length. It is thought that these plants exist here only because the predators of *Balanus* barnacles keep enough space open for the plants to attach and grow. Otherwise the barnacles would soon cover the entire area and exclude other life forms.

Organisms of the Mid-Intertidal Zone (Plate 16.3)

The dominant organism in the mid-intertidal zone is the mussel. On the West coast of the United States the dominant species is *Mytilus californianus*, and on the East coast *Mytilus edulis* is dominant. The major predators of mussels are starfish, but these are limited to the low intertidal zone because they tolerate only limited exposure to air. Therefore, starfish predation sets the lower limit of the "mussel" zone at about the level of the mean highest low water tide (mean Hi-Lo). The highest level to which mussels dominate is set by their exposure to air, so few mussels live above the mean lowest high watertide (mean Lo-Hi). Within this zone mussels reach a density of some 8000 to 9000 individuals per square meter! They form a multilayer mat that can be up to 30 cm thick; the individuals are linked loosely by the network of byssal threads secreted by each animal as its "tie down." The mat itself becomes a special habitat for many other organisms such as the tiny porcelain crabs, worms of various species, and, of course, the acorn barnacles.

Sharing the mid-intertidal zone with mussels are gooseneck barnacles (Plate 16.3). These animals often form clumps. The individual barnacle secretes a shell made up of a number of calcareous plates, held together with flexible chitinous material. The shell is the outermost extension of a flexible stalk also made of tough, chitinous material. This barnacle, like the related acorn barnacle, feeds by extending a set of hair-covered tentacles into the water to capture small suspended particles.

Numerous plant species occur in the mid-intertidal zone. The red algae *Porphyra* are common. Some authors even describe the mid-intertidal as the "red-algae belt." The plants gain nutrients during the twice-daily inundation needed by these plants. *Fucus* species also thrive in this zone.

Clumps of the sea anemone *Anthopleura* are found in rock crevices or in shaded overhanging rock for-

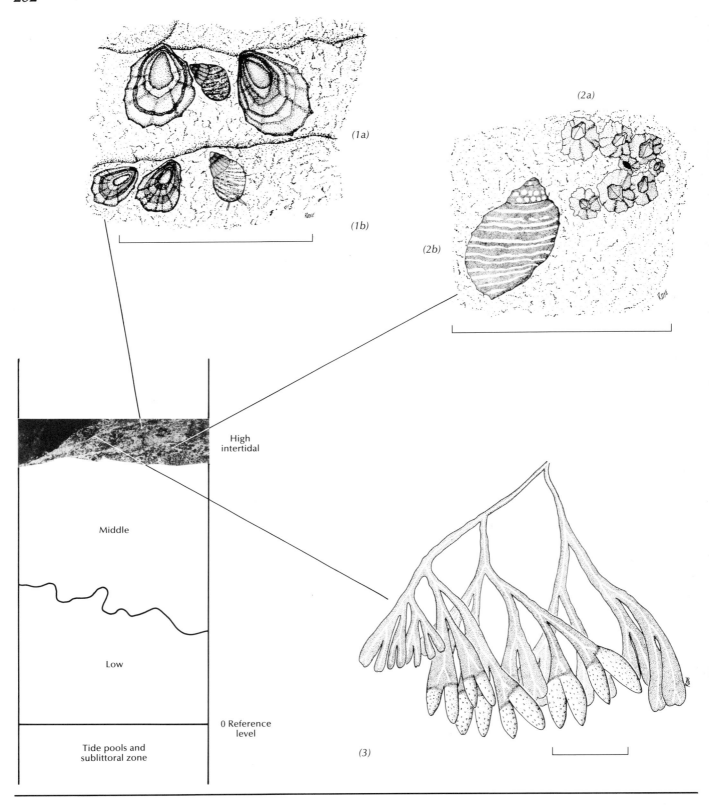

Plate 16.2 Organisms in the high intertidal zone (1a) the limpet *Colisella digitalis;* (1b) the snail *Littorina planaxis;* (2a) the acorn barnacle *Balanus glandula;* (2b) the predatory whelk *Nucella* (formerly called *Thais) emarginata;* (3) the brown alga *Fucus distichus.* Each bar is 5 cm long.

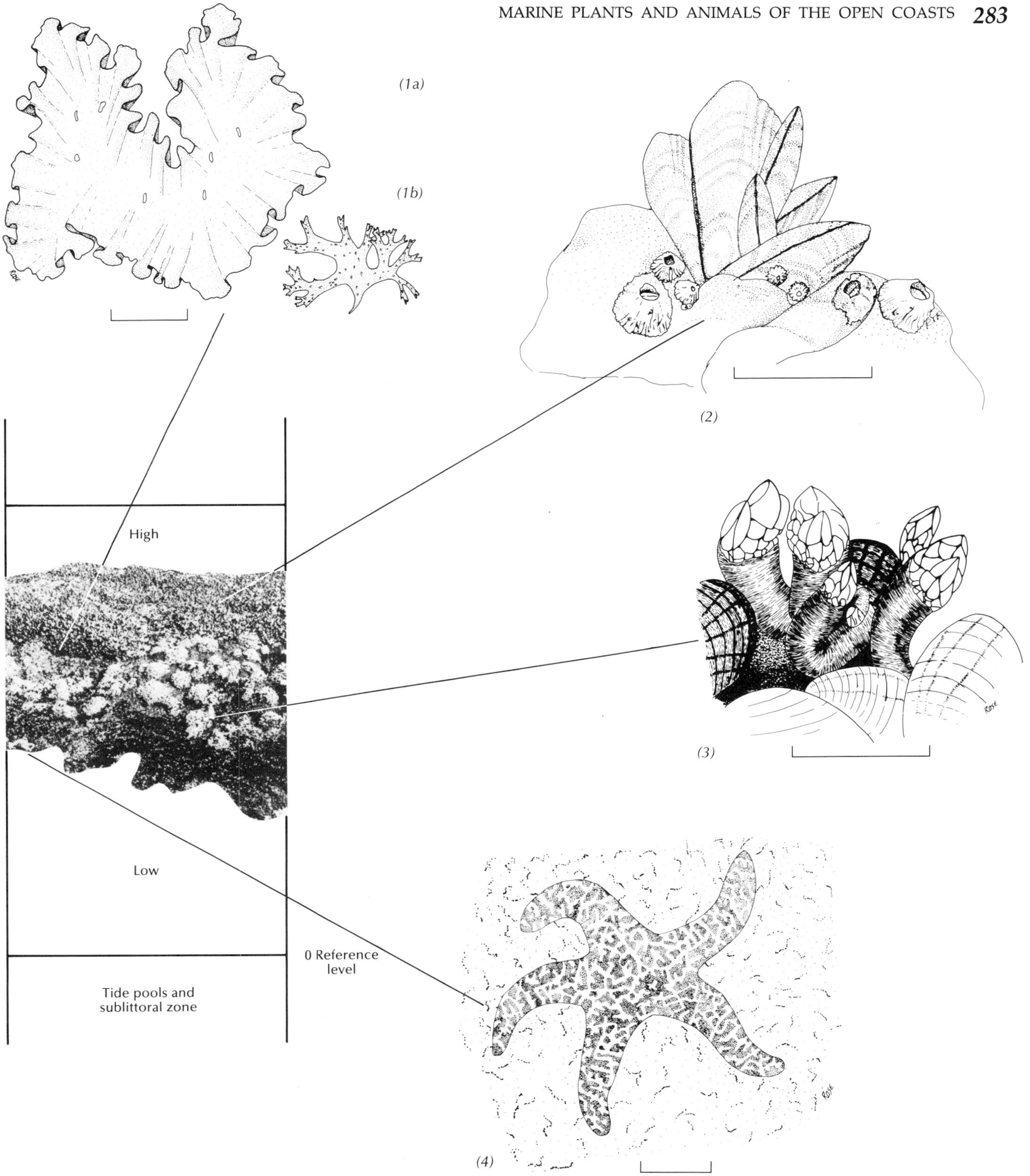

(1a)

(1b)

(2)

(3)

High

Low

0 Reference level

Tide pools and sublittoral zone

(4)

Plate 16.3 Organisms in the mid-intertidal zone (1a) the red alga *Porphyra perforata*; (1b) the red alga *Gigartina papillata*; (2) a mussel, *Mytilus californianus*, with attached barnacles; (3) a goose-neck barnacle, *Pollicipes polymerus*; (4) a sea star, *Pisaster ochraceus*. Each bar is 5 cm long.

mations. These animals reproduce by division, so all individuals within a cluster are clones. Studies have shown that one clone group will actively "fight" another, using special tentacles that carry chemicals toxic to alien clone groups. The fight is over territory, the most limited of all resources in the midintertidal zone.

Organisms of the Lower Intertidal Zone (Plate 16.4) The immediate impression given by the lower intertidal zone is that it is dominated by large algae. Plate 16.1 shows this in two ways: first, in the growths of surfgrass and *Laminaria* species in Plate 16.1*a*, and second, in the character of the sublittoral zone exposed in Plate 16.1*c*. Several examples of large brown algae are detailed in Plate 16.4; these are *Laminaria* species common to both Pacific and Atlantic coasts of the United States.

Still larger species of brown algae, the sea kelp, dominate some select offshore areas. *Postelsia* are the large brown plants in Plate 16.1*c*. Still farther offshore we may encounter *Macrocystis,* a very long-stemmed plant that grows in waters up to 50 m deep by extending multiple fronds to the surface, each buoyed by small pneumatocysts, buoyant bags, along the stem. Another genus is *Nereocystis;* it differs from others in that it develops a single buoyancy bag just at the surface, with multiple fronds attached at this point. Pneumatocysts are also seen on the brown alga *Egregia* in Plate 16.4(3).

Many animals can be found under the dense kelp fronds of Plate 16.4. Typical are the large chitons, gastropod animals that graze upon algae. Limpets are found as well, some with attached kelp growing to their shells. Tube worms can be seen by lifting the fronds hanging off rock faces; they are filter feeder animals. Also to be found are worms of various types, crabs, sea slugs (these are gastropod snails without shells), starfishes, and many others.

During most of the day the lower intertidal is covered with water. Fishes are active during these times, and large crabs such as the blue crab may enter with the high-level water to scavenge among the rocks. Flatfishes roam the bottoms in search of invertebrates.

Organisms of Tide Pools and Nearshore Sublittoral (Plate 16.5). Visitors to the rocky shore are always fascinated by the different life forms found in the tide pools. Here are what appear to be loose rocks and shells strewn across the pool floor, but closer examination reveals that many of these "loose" materials are actually attached to the numerous sea ur-

chins and sea anemones distributed around the pool.

Sea urchins are echinoderms; they have the characteristic five-point radial symmetry typical of this phylum, as well as the tube feet used for locomotion. In addition, urchins have developed numerous spines arranged in rows around the calcareous test. The spines have a defensive role. In between spines are the tube "feet" with suction cups at their tips; these are used to grab pieces of algae and move them around and under the body to the mouth, which is underneath the animal. During immersion periods urchins may move about seeking algal food, but they always return to their preferred depression in the rock to wait out exposed periods.

Other echinoderms, the sea stars, are found in almost every tide pool. A particularly active genus is *Pycnopodia*, named after the numerous feet each animal possesses. These starfishes keep tide pools virtually free of bivalves such as clams and mussels.

A few species of fishes occupy tide pools, and additional fishes, such as the kelp greenling, live in the offshore kelp regions.

The Sandy Beach Community

The most likely place for the reader to "touch the living organisms of the sea" is at the beach. In contrast to the abundant life along the rocky shore, the first impresion given by a sand beach is its apparent absence of life. Appearances to the contrary, there is life *in* and *on* the beach sands.

Factors Determining Its Distribution of Organisms

What factors control the distribution of "beach" life? Some of the more basic ones are the following.

1. Beaches are inherently areas of deposition. The most obvious thing that is deposited is sand. Just how much sand, and what the grain size and texture of the sand are, depend on the source of the sand and the amount of energy available from breaking waves by which the sand is sorted, winnowed, and otherwise moved about. (Chapters 17 and 18 treat these subjects at greater length.) In turn, the character of the sand substrate—its grain size, how firmly it packs, its porosity to percolating water, and the mineral from which the grains are derived—determines which organisms are found where on the beach.

2. Because the substrate sand is always being shifted about, there are no sessile plants or animals

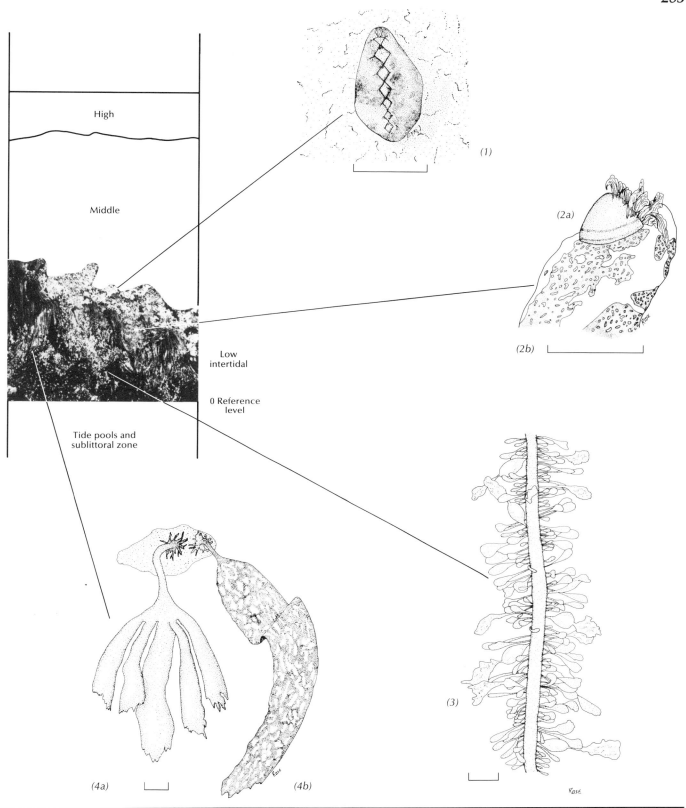

Plate 16.4 Organisms in the lower intertidal zone (1) the chiton *Katharina tunicata;* (2a) the limpet *Acmaea mitra* with seaweed attached and growing on its shell; (2b) an encrusting, coralline alga, *Lithothamnion;* (3) the brown alga *Egregia menziesii;* (4a) the brown alga *Laminaria andersonii;* (4b) the brown alga *Laminaria farlowi.* All bars are 5 cm long.

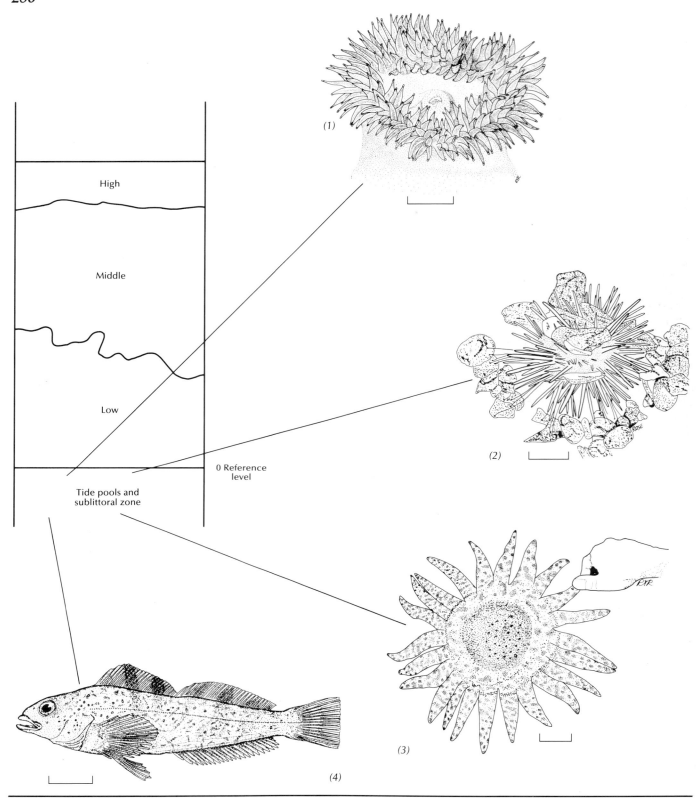

High

Middle

Low

0 Reference
level

Tide pools and
sublittoral zone

Plate 16.5 Organisms of the tide pools and nearshore
sublittoral zone (1) the green sea anemone *Anthopleura xantho-
grammica* with symbiotic zooxanthallae; (2)a sea urchin, *Strongylo-
centrotus franciscanus;* (3) the sea star *Pycnopodia helianthoides;* (4)
a kelp greenling fish. All bars are 5 cm in length.

on the beach face. Therefore, there is no local primary production to sustain a conventional food chain. Instead, the organic "foundation" of the beach community is "imported" in two general forms: (a) suspended particles, the most likely source being pieces of surfgrass and other detritus exported from a rocky headland nearby, and (b) large chunks of *Macrocystis,* which wash up on the beach and are stranded there. These materials are distributed *along* the shoreline by the prevailing *littoral* current (see Chapter 18.)

3. The distribution of both organic detritus and fluid water over the beach is fixed by the tide levels and by the breaking waves. Beach sand is usually very porous. Where wave action is vigorous, the beach sands are coarse and do not retain much fluid water from one high tide to another. Where waves are moderate, the beach sands have a smaller grain size and will retain large amounts of fluid. Water in the swash zone percolates downward through beach sand, which acts as a filter to remove particles carried by the swash flow (Figure 16.2). The result is that the surface layer of beach sand carries a high concentration of organic matter, and a corresponding variety of animals are adapted to feed and live there.

4. The percolating seawater is saturated with oxygen when it enters the surface of the beach sand.

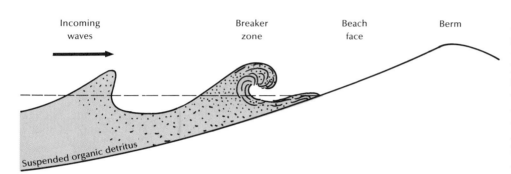

Incoming waves • Breaker zone • Beach face • Berm

Suspended organic detritus

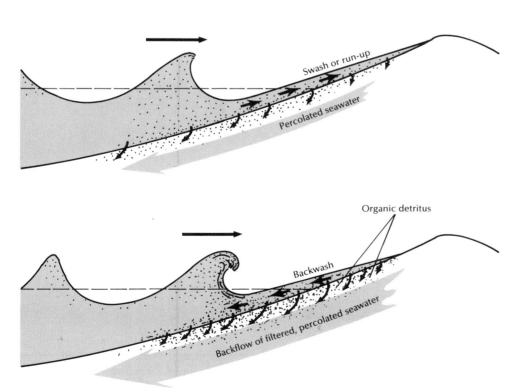

Swash or run-up

Percolated seawater

Organic detritus

Backwash

Backflow of filtered, percolated seawater

Figure 16.2 How organic detrital material enters the domain of the sand beach community. Seawater holding suspended organic material is thrown up the beach face by the breaking wave. As this swash water percolates downward through the sand on its return path to the sea, the organic detritus is filtered out within the uppermost few centimeters of sand.

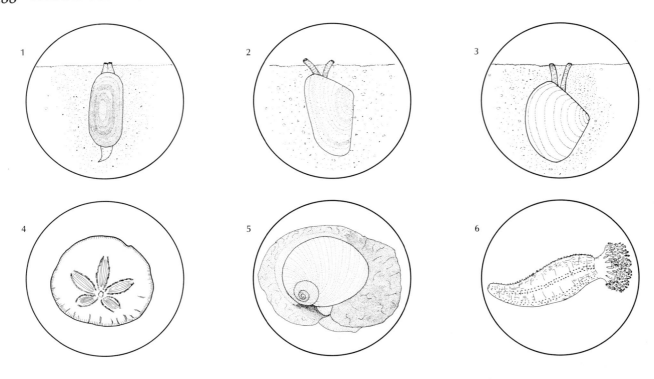

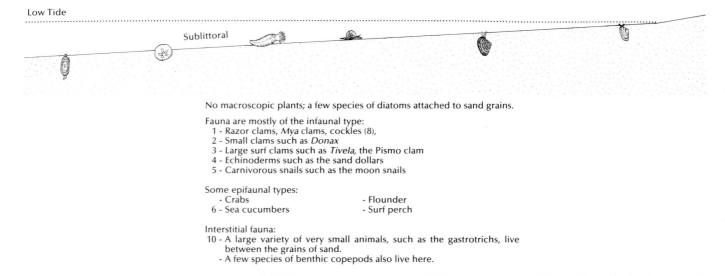

High Tide

Low Tide

Sublittoral

No macroscopic plants; a few species of diatoms attached to sand grains.

Fauna are mostly of the infaunal type:
1 - Razor clams, *Mya* clams, cockles (8),
2 - Small clams such as *Donax*
3 - Large surf clams such as *Tivela*, the Pismo clam
4 - Echinoderms such as the sand dollars
5 - Carnivorous snails such as the moon snails

Some epifaunal types:
- Crabs   - Flounder
6 - Sea cucumbers   - Surf perch

Interstitial fauna:
10 - A large variety of very small animals, such as the gastrotrichs, live
between the grains of sand.
- A few species of benthic copepods also live here.

Figure 16.3 The distribution of organisms in the sand beach community. The high- and low-tide levels mark major changes in the biota, but many of the organisms living in the sandy beach intertidal are mobile and move up and down the beach face according to tide level and wave activity.

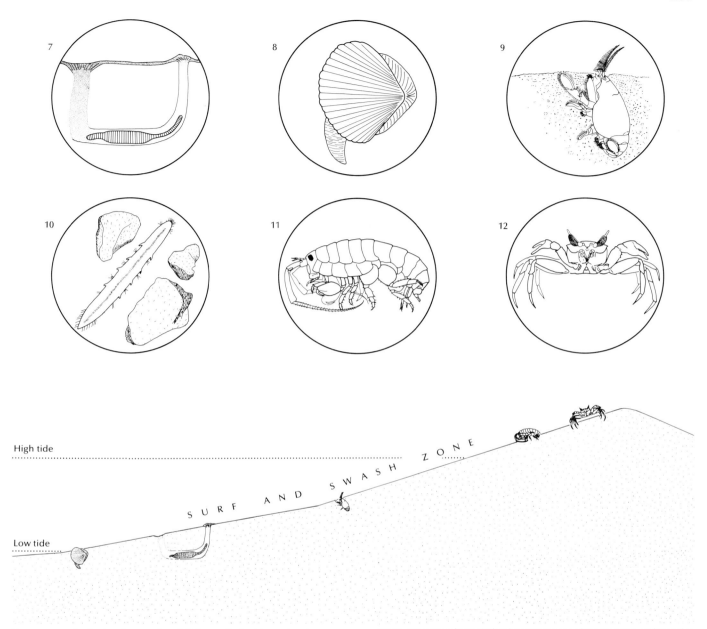

High tide

SURF AND SWASH ZONE

Low tide

No macroscopic plants; a few species of diatoms attached to sand grains.

Fauna are mostly of the burrowing types:
- Various amphipods and isopods
7 - Lugworms and burrowing polychaetes
- Burrowing shrimps such as ghost shrimp
1 - Razor clams
8 - Cockles
9 - Mole crabs, such as *Emerita*

Interstital fauna:
10 - A large variety of very small animals, such as the gastrotrichs live between the grains of sand.
A few species of benthic copepods also live here.

No macroscopic plants. Only a few species, mostly scavengers, make up the macroscopic fauna:
11 - Beach hoppers (amphipods)
12 - Ghost crabs (crustaceans)
- "Pill bugs" (isopods)

This is the source of oxygen for the various organisms living within the sand. In areas of low wave action and fine-grained sands the percolation flow is much less, limiting the oxygen supply to only the top surface of the sand, as are the animals.

5. Last, sand exposed to sun heats up rapidly; burrowing species need to dig deeply to avoid excessive daytime temperatures.

Adaptation in the Sand Beach Community

Because salt water is denser than fresh water, rain falling on an exposed beach does not penetrate or mix with the seawater retained by the sand, and because most beach sands are highly reflective, solar heating does not penetrate deeply into the sands. The net result is a fairly constant environment below a few centimeters into the beach surface. The organisms that obtain food by filtering particles out of the seawater burrow deep enough for protection against salinity and temperature changes or wave erosion but still shallow enough to permit extending their filtering mechanisms into the overlying fluid. The various clams shown in Figure 16.3 illustrate how syphons can be extended just above the sand surface to ingest water with its suspended particles.

Some species have adapted to feed best in the swash zone. A classic example is the sand crab *Emerita*, (Figure 16.3) which burrows into the sand only far enough to cover its body. It stays covered while the swash moves up the beach face but extends its hair-covered tentacles into the returning backwash; the meshlike tentacles capture small organic particles from the moving stream. The tentacles are then pulled back, cleaned of captured food particles, and made ready to reinsert in the next wave's backwash. This crab must move up and down the beach face as the tides change, and it must also be able to burrow rapidly. It rides a swash upward to a new level as the tide is rising, but quickly burrows down to avoid being washed back out in the backwash; it reverses this procedure during ebb tide. By staying in the swash zone it avoids the crashing waves of the breaker zone.

Being able to burrow rapidly into the sand is a successful adaptation of other organisms. The razor clam *Siliqua* (Figure 16.3) has a shell that is quite thin, but it protects itself against damage in the breaker zone by burrowing rapidly down into the protective sand. It too can change its position up and down the beach face, and it feeds by pumping water through its syphon and internal filter to remove suspended organic particles.

Clams without the ability to burrow rapidly obtain protection against the crashing surf by excreting heavy-walled shells. The California Pismo clam (*Tivela*) (Figure 16.3) has a shell about 1 cm thick, and its considerable weight also helps the clam to hold its position in the sand during times of heavy wave erosion.

An unusual adaptation is found in the sand dollar *Dendraster*. Sand dollars belong to the same family as starfishes and urchins, all of which have the five-point symmetrical designs seen on the familiar empty shells of these animals. They retain the tube feet of the starfishes and the spines of the urchins, but both features are diminished to very small size. The sand dollar is found below the levels of breaking waves. It feeds by burrowing through the surface sands, collecting food particles that fall through its covering of spines into mucus-lined channels, which then convey the food to its mouth. Some species can be found with part of the body projecting above the sand into the moving seawater, filtering particles from it in the same way.

Other expected members of the beach community include various types of worms. One type that is well adapted to feed on the organically richer surface sands deposited by percolating swash water is the lugworm *Arenicola* (Figure 16.3). Making a U-shaped burrow, it takes in organic-rich surface sands that are washed into one opening by the swash waters and digests these through its gut. It ejects the residue via a seaward downsloping opening.

Interstitial Animals and Carnivores in the Beach Community

A very important part of the beach community are the tiny animals that live in the *interstitial spaces between sand grains*. Sometimes called psammon, (or meiofauna), they are at most a few millimeters long, with a narrow body shape so that they can move easily between the grains of sand; typical sizes range down to 0.1 mm (100 microns).

Many of the invertebrate phyla are represented among the psammon. The Protozoa are present with many species of ciliated body shapes. Small flatworms are also present, as well as large numbers of nematodes, which are roundworms. Especially well represented are the gastrotrichs and the annelids, and species of benthic copepods represent the crustaceans. In addition, the juveniles of some larger invertebrates occupy benthic roles within the sand. The concentration of individuals can be large, up to about 1 million/m^2 and together weighing perhaps 5

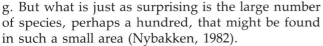

(a) A rocky shore is actually a rich habitat for sessile plants and animals. (Timothy Eagan/Woodfin Camp.)

(b) At high latitudes the flora and fauna of a coastline must cope with the freezing of seawater and with the abrasion caused by blocks of ice. (Michael S. Renner/Bruce Coleman.)

(c) At low latitudes protected bays and estuaries are usually lined with mangrove swamps, which provide a rich habitat for both marine flora and marine fauna. (Mark Boulton/Photo Researchers.)

g. But what is just as surprising is the large number of species, perhaps a hundred, that might be found in such a small area (Nybakken, 1982).

Because the adults of these interstitial animals are small in size, they tend to reproduce by brooding a few offspring rather than casting many eggs. With all the filter feeders straining the overlying water, reproductive success seems to lie in the sand rather than in mass dispersion.

With the interstitial world defined, we now understand better how carnivorous worms survive in the sand world. Nematodes move through the sand seeking living specimens in the interstices between grains. The small benthic crustacean copepods move through the sand searching for benthic diatoms attached to the grains. In turn, larger meiofauna, for example the cnidarian hydroids, prey on passing copepods. Some of the worms and gastrotrichs "lick" grains of sand to ingest the organic film coatings.

Finally, larger animals burrow through the sand seeking living prey. Moon snails will search out clams, and flatfishes scour the bottoms looking for invertebrate forms.

The Coral Reef Community

Coral reefs are a special kind of shoreline or shallow bottom habitat found in latitudes between 30°N and 30°S. A unique combination of conditions allows them to support an extremely complex biological community.

1. The myriad of shapes and crevices of the hard coral structures offer a variety of covers to animals.

2. Coral reefs are located in sunlit, warm waters which support rapid growth rates.

3. Nutrients are rapidly recycled because the vertical dimension of the habitat is but a few tens of meters.

4. The primary productivity base includes heavy production of blue-green algae, which fix dissolved nitrogen gas, and algae that are symbiotic with the coral animals.

Corals grow wherever surface seawater temperatures exceed about 20°C throughout the year. They flourish wherever a shallow, hard substrate is kept silt-free by prevailing currents and waves. Corals do not survive in sediment-laden water, such as near a river discharge. Their dependence on intense sunlight restricts the depth of most living coral communities to less than about 50 m, although some corals will grow at depths to 100 m. The conditions of coral reefs translate into factors that make them a prime target for research in marine ecology … shallow water accessible to divers, warm waters hospitable to human submersion, high ambient light, and rich faunal diversity sustained by a nonconventional nutrient base.

The Indian and Pacific oceans have the most extensive coral reefs. These include waters of the Philippine Islands, of the vast Indonesian archipelago, and of northern Australia, as well as of the Great Barrier Reef, which alone is over 100 km wide and extends 2000 km along the eastern coast of Australia. These reefs have the greatest species diversity and are likely the oldest. The second largest area of reefs is in the Caribbean Sea, from Florida to Trinidad and westward to Central America. These West Indies reefs have less diversity, partly because the Atlantic itself is a young ocean. The Eastern Pacific coasts from Mexico south to the Galapagos Islands and Peru have some coral reefs, but these are not extensive. This text draws primarily from the West Indies reefs because of their proximity to major United States population centers.

The Structure of Coral Reefs

Coral reefs are massive deposits of calcium carbonate that are excreted by individual coral animals and cemented as clusters of distinct, external skeletons. The forms taken by cluster structures, forms that we usually call the *corals*, are many and varied. Some examples are shown in Plates 16.6 and 16.7; they vary from large domes created by brain corals to the flat leaf types found at greater depths to the whip-like stalks of black corals. Coralline algae also contribute to the material structure of the reef zone by laying down encrustations of calcareous material.

A new reef is built over a solid substrate in shallow water. If the substrate is geologically stable, reef growth will eventually reach a "balanced" state in which the rate at which coral animals deposit new calcareous material is just matched by the loss of material through erosion. Reef erosion is a continuous process, owing in part to mechanical abrasion by breaking waves but also to the "chewing" by many reef animals, especially by parrot fishes and sea urchins. These animals scrape the hard corals to feed

on encrusting algae or the coral animals themselves. The erosion is evident by the large deposits of coral sands in the lagoons and beaches of tropical shores.

If the substrate is itself subsiding, as is the chain of Hawaiian volcanic islands, new reef construction will try to offset the rate of sinking. Borehole samples taken during oil exploration provide evidence that on many continental shelves reefs can build to thicknesses of thousands of meters. The atolls commonly found in the equatorial Pacific are the work of coral reefs that began as fringing reefs around new volcanic islands. As the island later subsided, the reef continued to accrete and so maintained its living surface in shallow water. It became a bank reef as the periphery of the island shrank. Eventually, all that remains is a circular reef surrounding a central lagoon.

Structurally, West Indies reefs have a systematic organization (Figure 16.4a). Close to shore is the *fringing reef*, which extends seaward until water depth reaches about 10 m, but this is variable from reef to reef. Farther from shore, at distances of 500 m or more, is the more complex outer reef called the *bank reef*. Between the bank and fringe reef structures is a *reef flat*. This flat is covered with loose sand and rubble, derived from erosion of the reef structures, and with occasional isolated pillars of small coral groups.

At its shallowest point the bank reef is 10 to 15 m in depth and of variable width. When the bank is relatively narrow, only the longest of wind waves arriving from the open sea will break there. Usually, the wind waves are forced to "top off" as they run up the seaward slope of the bank reef, but some energy remains in the moving wave, continuing it over the bank toward the fringing reef and shoreline. Where an outer reef is very wide and shallow, as is the Great Barrier Reef of Australia, *all* wave energy is dissipated by breaking on the seaward slope and over the bank reef itself. Many lagoons and nearshore waters of tropical shores are thus protected against high waves. Along the seaward edges of shallow reefs, the surging and breaking waves carve a system of *surge channels* (Figure 16.4b). If the tide range is large, such channels may extend for considerable distances up and down the forward slope. The channels serve as an avenue for sand and debris moving seaward from the inner reef flats. They also help increase the total surface area of the reef as well as its nooks and crannies; both factors serving to enhance food chain production.

Tide waves, of course, will flood and ebb over even the most extensive barrier reefs, and storm

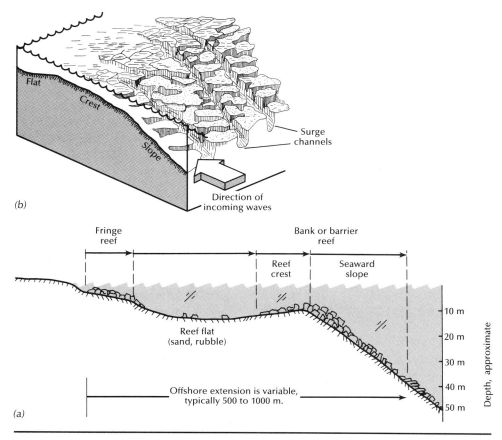

Figure 16.4 A typical coral reef structure: *(a)* cross section from the beach to the 50-m level seaward of the bank reef; *(b)* the surge channels cutting into the seaward slope by wave action.

surges may spill enormous amounts of seawater from the open ocean onto the reefs and into the lagoons that are shoreward of the reefs. Lagoons usually have at least one major "channel" to the open sea via a break in the surrounding outer reefs. Because the lagoon is a favored feeding and nursery ground for many coastal fishes, there can be heavy migrations between lagoon and open sea. Tropical fishermen know from experience that these are ideal places to set nets and gather fish at specific times of the year. When the outer reefs are very wide, tide waves and storm surges serve to "flush" the reef structures periodically, preventing waters from stagnating within the reefs.

Examples of West Indies Reefs

Plate 16.6 is a photograph of a fringe reef off the west coast of Barbados, West Indies, taken at a depth of 10 m. Dominant in the foreground are three

species of "brain" coral, *Diploria labyrinthiformis, Meandrina meandrites,* and *Colpophyllia natans.* The small pillars in the background are the mountainous star coral *Montastraea annularis.* Encrusting parts of the pillar is the sponge *Mycale laevis,* which grows on the undersurface of certain corals. Other sponges and the long-spined urchin *Diadema antillarum* are also seen. Urchins are common in fringing reefs but scarce on the outer bank reefs; in contrast, fishes are much more abundant on the outer banks.

Plate 16.7 was taken at a depth of 30 m on the seaward slope of the bank reef at the same Barbados location. Notice the contrasts with the character of the fringe reef of Plate 16.6. First, the greater depths receive much less light; at 30 m deep there is only half as much visible light intensity as at 10 m. The shapes of coral colonies at 30 m deep are much flatter, for they need to expose as much area as possible to the weaker light field. The "leaf" coral *Agaricia lamarchi* abounds throughout the area seen in the photograph. Pillar corals are not evident; the star coral *Montastraea cavernosa* builds a domed, pillar-like structure. Whiplike stalks of the black coral *Sticho-pathes lutkeri* survive at this depth, in part because water motion induced by waves is too weak to damage such structures.

The reader may get the impression that many more crevices and holes exist among the corals on the seaward slope then on the fringe reef. This is true, particularly where the surge channels are deep and corals grow along the sides of such channels as well as on the top of the reef. Plate 16.7 shows a variety of *cover* for all sorts of animals, for example, the small fish at the right center of the photograph. In contrast, part of the scene in Plate 16.6 shows an open, sandy "patch." We expect then that life is more diverse on the bank reef than on the fringe reef.

Coral Animal Biology and Reef Primary Productivity

Animal and plant biology seem to be more tightly coupled in the reef trophic system than in other marine systems. We have long known that coral reef communities are high producers, even though tropical ocean surface waters in general are nutrient-limited and have low productivity. Nybakken (1982) estimates primary production on coral reefs to be in the range 1500 to 3500 g carbon/m^2(yr), a rate that is 30 to 70 times *higher* than the 50 g carbon/m^2(yr) given in Table 1.2 for open-ocean water in general!

Coral animals belong to the phylum Cnidaria, along with sea anemones, hydroids, and jellyfishes, but the individual animals are smaller in size than these. The adult form is that of a polyp permanently attached inside a limestone cup called a *corallite*, which the individual secretes and expands in height as it grows. Figure 16.5 shows the individual polyp and its cup in cross section, and Plate 16.6 shows how colonies of polyps can be arranged over the surface of a dome of brain coral. Live polyps are found only on the outside surface; in normal growth these extend their cups, by secreting calcium carbonate, at rates from 1 to 10 cm/y (Goreau et al., 1979). The cementing together of the corallites of colonies of polyps creates the massive, calcareous reef structure at growth rates that are able to "keep up" with a sinking substrate or rising sea level.

Like other cnidarians, coral polyps have tentacles with stinging cells. Tentacles are arranged symmetrically around the animal in multiples of six or eight. Depending on the species, polyps range in size up

to 20 cm in diameter, large enough to capture small reef fishes. Most polyps are smaller, from 0.1 to 1 cm in diameter, and feed on particles ranging in size from bacteria up to large zooplankton. They are nocturnal feeders, for it is during night hours that the waters above the coral reef come alive with zooplankton and larvae of many types and species. The myriad of nooks and crannies within the reef structure itself provide a great deal of cover for small animals and larvae during daylight hours, making the biomass of the zooplankton, mostly meroplankton, very large.

The association of the coral polyp with the symbiotic algae *zooxanthellae* makes the coral community unique. The alga is a modified dinoflagellate and is embedded in the outer skin layer of the polyp (Figure 16.5). This symbiotic relationship, in which the polyp *cultures* its own "autotrophic garden" inside its body tissue, explains how coral communities can be so productive while the surrounding seawater is at the same time relatively barren. The mutualism makes good logic. The animal produces phosphates, urea, and CO_2 during metabolism, and these and other chemicals diffuse through the body tissue to become "nutrients" for the algae. In turn, the algae located in the skin of the animal get enough sunlight to photosynthesize new organic material. We still do not know exactly how the animal then gets nourishment from its plant cells—it does not *eat* them unless the cells stop producing or die—but tests show that as much as 40% of the plant's production can show up in the tissue of the animal itself. The adult polyp passes "seed zooxanthellae" to its offspring. Polyps may reproduce by asexual budding, with the new individual taking its place alongside the parent in the colony. At times corals release a planktonic form, the *planula*, which is able to start new colonies.

Recalling that the coral animal is definitely a carnivore predator, how much of its food is gained by predation and how much by the work of symbiotic algae? Although answers are not yet clear, the animal is estimated to gain from 10 to 20% of its nutrition by predation and by capturing particulates on its mucus-lined skin and mesenterial filaments (Figure 16.5). The polyp's efficiency as a filter feeder is one reason reef waters are kept clean of suspended particulates.

The Contributions of Blue-Green, Coralline, and Macroalgae. Most of the primary production in coral reefs comes from benthic forms of algae, a sharp contrast to the open-ocean system, which de-

Living coral polyp

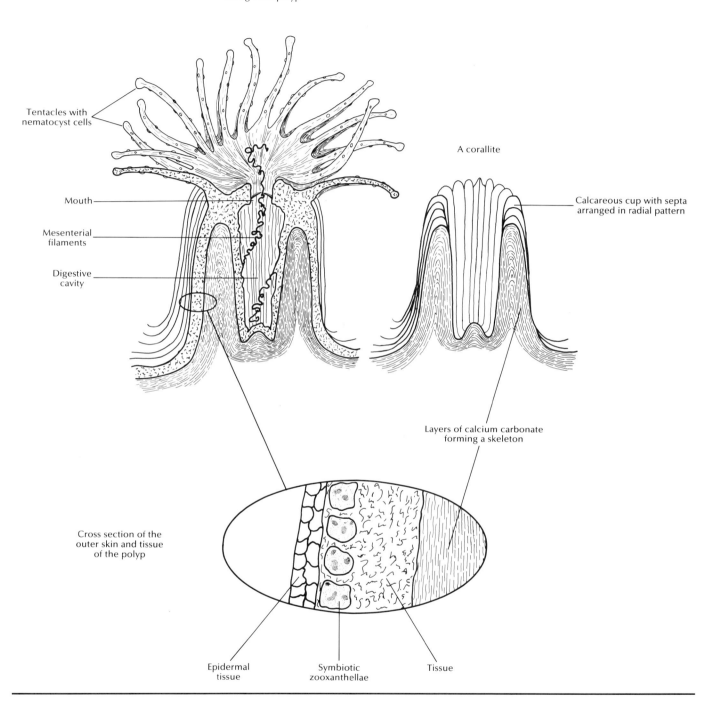

Tentacles with
nematocyst cells

A corallite

Mouth

Calcareous cup with septa
arranged in radial pattern

Mesenterial
filaments

Digestive
cavity

Layers of calcium carbonate
forming a skeleton

Cross section of the
outer skin and tissue
of the polyp

Epidermal
tissue

Symbiotic
zooxanthellae

Tissue

Figure 16.5 The coral animal and the calcareous cup that it secretes. Tentacles capture larger particulates and feed these to the mouth and digestive cavity. Mucus-covered filaments aid in digestion, but they are also used to clean the animal of organic matter that settles on the surface. The outer "skin" of the polyp carries the symbiotic zooxanthellae, which are dinoflagellate-like algae. Polyps range from 1 mm to many centimeters in diameter.

pends primarily on phytoplankton. Moreover, it appears that most of the nitrogen needed in photosynthesis by the benthic algae comes from the atmosphere in the form of nitrogen gas dissolved in seawater. Because surface water in contact with air is saturated with nitrogen gas, and because the shallow reef waters are turbulently mixed top to bottom, benthic algae have a plentiful supply of nitrogen. Thus the reef community is *not* nutrient-limited by nitrogen depletion, another sharp contrast to the conventional open-ocean system where nitrate is usually depleted, limiting primary production.

In 1971 a group of scientists studying seawater around Enewetak Atoll in the Pacific discovered that ocean water became noticeably richer in nitrogen compounds *after* it moved across a shallow reef. They traced the source of the nitrogen to a number of species of N_2-fixing algae growing on the reef; the filamentous blue-green alga *Calothrix crustacea* was the dominant species (Weibe, Johannes, and Webb, 1975). Overall, this reef community "exported" nitrogen at the rate of 3 kg/(hectare)(day), a rate equal to or better than nitrogen production from a terrestrial alfalfa field!

The filamentous algae forms a mat up to several millimeters thick on exposed rock surfaces. Rocks and dead corals that have been scraped by algae-eating fishes and urchins are rapidly covered with new growths of the algae. Other species grow on the sand flats at the base of the reef structures. The algae also contribute nitrogen compounds to the water column directly by releasing peptides and free amino acids. In addition, wave action may tear pieces of the filamentous algae free, and these then become floating particulates consumed by coral polyps or other filtering organisms.

Coralline algae, for example, *Lithothamnion*, also contribute significantly to the overall primary production. These are found as encrustations over rocks and dead corals and in greatest concentration on the crests of the outer reefs, where the forces of breaking waves keep other organisms from attaching. These algae are also grazed by herbivores. But one of the more important roles they play is that of "cementing" together fragments of reef rubble to form durable structures.

Macroalgae provide substantial primary production to the reef community. These are also called epiphytes or "marine grasses;" *Thalassia* and *Cymodocea* are examples. Dense stands of these grasses are found on the shallow sand flats behind the reef crest or in the shallow lagoons. Shallow water, of course, means high light intensities, so that production in reef zones is much like that found in tide marshes and shallow estuaries. Reefs help to stabilize the loose sands of lagoon floors, provide a habitat for diverse organisms, and serve as spawning grounds for animal species that attach egg masses to their stems. Just as important, they are also nitrogen fixers and so contribute to the overall export of nitrogen from reefs.

Knowing the mechanisms by which the nitrogen fixed by the blue-green algae actually enters into the coral community helps in understanding how this system works. The filamentous mats are grazed directly by fishes, sea urchins, sea cucumbers, and other animals. Among the grazing fishes are the parrot fish, the *Scaridae*, which are equipped with special teeth to scrape the algal coating off the rock or coral substrate. But because these fish are somewhat inefficient in digesting the algae, about one-half the organic material is passed off in feces, which then helps sustain a wide range of invertebrates. Algal mats and the dense stands of marine grasses both sustain large populations of bacteria. Some bacteria decompose algal fragments and some are nitrogen fixers; all bacteria in turn become food for filter feeders, including the coral animals.

The Shallow-Depth Factor. Because the entire reef community, from the planktonic forms to the benthic invertebrates, is compressed into a thin layer, total production is concentrated and nutrient recycling is very rapid.

The Reef Surface Irregularity Factor. Because of the many nooks and crannies within the reef structure, together with the channels and overhangs, the actual surface area of a reef is many times greater than its simple horizontal area. Much of the reef production is area dependent, and the area can be horizontal or vertical or even the underside of coral structures. This factor is basic to high reef production.

The Reef as a "Home Base". Many of the animals that seek refuge within structures during resting periods move away from the reef for feeding. Herbivores feeding on sand flat grasses do, for example.

Reef Fishes

For sheer diversity of fishes, the coral reef community exceeds all other marine communities. About one-third of *all* fish species are associated with reefs in one way or another (Moyle and Cech, 1982). Smith and Tyler (1972) collected 75 species from a

Figure 16.6 Some carnivorous fishes of a West Indies reef community: (1) snapper; (2) Nassau grouper; (3) grunt; (4) goby cleaner fish; (5) goatfish; (6) trumpet fish; (7) barracuda; (8) mid-water damselfish; (9) butterfly fish; (10) needlefish; (11) moray eel; (12) scorpion fish. (Compiles from various sources.)

single isolated reef dome 3 m in diameter and less than 2 m high! Over 2000 species are cataloged. One way to classify reef fishes is by feeding types: general carnivores, specialized carnivores, and herbivores. Carnivores are the majority of species, but herbivores exceed the other classes in population density.

Reef fishes can also be classified by their periods of activity. Some species are nocturnal, feeding only during nighttime; crepuscular, feeding only during the dawn and dusk periods; and diurnal, feeding only during daytime. When not feeding, each group seeks shelter within the coral structures, with members of one group often moving into the same crevices just vacated by members of the alternate feeding group. This phenomenon, called "diel changeover," happens twice each day, at dawn and at dusk, and requires about 20 minutes to complete.

Fishes can also be distinguished by whether they live *and* feed on the reefs, or feed away from the reefs but return to rest in the security of cover provided by the reef structure. Some species of carnivorous grunts, for example, move off the reefs to feed at night.

Carnivores. Figure 16.6 shows some carnivorous reef fishes. Generalized carnivores are the largest specimens. A typical fish is the grouper (family Serranidae), which is for the most part a daytime predator. It cruises slowly over the reef watching for a prey fish, shrimp, or crab to make an "error" in its own defense. Piscivores, (predators of fish), seem to be the most active during periods of low-level light at dawn or dusk; examples are found in the snapper family Lutjanidae, but other snappers feed during night hours as well, usually in small groups. Still another predator is the grunt (Pomadasyidae), which feeds mostly on the sandy flats away from the coral, taking all sorts of benthic invertebrates such as crabs, worms, shrimps, starfish, snails, or chitons.

Another striking feature among reef fishes is the number of different "ambush" feeders. Notable among these are the scorpion fish (Scorpaenidae), which have evolved with such unusual body shapes and fin structures and colors as to be almost invisible when resting on rock surfaces. Other ambush predators are the lizard fish (Synodontidae), which camouflage themselves against the loose sand to await passing prey, and the trumpet fish (Aulostomidae), which may ambush from resting spots next to coral. Other specialists are the drift stalkers with their elongate bodies and long snouts full of sharp teeth. They drift across a reef with the currents; when a prey comes into "range," the quiet drifter lunges suddenly across a short space to make a kill. Examples are the barracuda (Sphyraenidae) and needlefish (Belonidae). In contrast, elongate predators such as the moray eels (Muraenidae) have small heads so that they can seek out prey that are hidden in the cracks and crevices of the coral. As defensive measures against such predation, some small fishes wedge their fin rays against the sides of a crevice to prevent their being "pulled out."

An abundant group of reef fishes are the goatfish (Mullidae), so named because of their "chin whiskers" or *barbels*. These appendages are used to probe the sandy bottom in search of small hidden invertebrates. Once found, the prey are "'sucked up'" into the mouth. This specialized technique does not require light, so both nocturnal and diurnal species of goatfish exist in the same reef community.

Still other families are specialized to locate invertebrate prey using vision rather than touch, as do the goatfish, and these are necessarily daytime pred-

ators. Being diurnal, they have evolved bright color patterns over the body and strong teeth and jaws with which to pick or crush invertebrates. Examples are the butterfly fishes (Chaetodontidae) and angelfishes (Pomacanthidae), which use their long snouts and sharp teeth to snip off tips of coral branches.

As will be discussed in the summary, a surprising aspect of the coral reef community is the high density of zooplankton that appears over the reef during nocturnal hours. These also have predators, for example, some species of damselfish (*Chromis*), which roam the reefs in dense clouds picking up zooplankters.

Herbivores. Herbivorous fishes account for only 25% of all reef fish species, but they have the most abundant populations on the reef. These are the fishes most often seen moving about the reef itself as well as on the sandy flats away from the coral structures. Three families supply the major herbivorous fishes: the parrot fishes (Scaridae), the surgeon fishes (Acanthuridae), and the damselfishes (Pomacentridae). Figure 16.7 illustrates some species.

Parrot fishes can be the most abundant of all species. They have fused teeth and strong jaws with which they scrape, bite off, and chew pieces of dead coral to get the algal growth. The coral "sand" generated this way is a main source for the sandy flats around reef structures. The contribution of parrot fishes to reef erosion was noted first by Charles Darwin in the Journal of Researches during the Voyage of the HMS *Beagle* (1845). The algal mats form only on *dead* coral substrates. Live coral animals cover their calcareous cups with a mucus membrane that not only catches particulates and keeps the animal clean but also protects them against being "smothered" by overgrowth of such algae.

Adult parrot fishes have loosely defined territories over the reef, perhaps 1000 m^2 of area. Their movement is a slow progression along the reef with occasional stops to feed. They feed also on the coralline algae, but these algae are not a major part of their diet. They also graze on the sand flats wherever algae grows. At night the parrot fishes retreat to cover inside the coral; some species then surround themselves with a mucus envelope, which is apparently toxic to nocturnal predators such as the moray eel.

Damselfishes are strong competitors of the algae-grazing parrot fishes and defend their small territories (1 to 2 m^2) vigorously against any other herbivore. Parrot fishes can graze in these territories only when moving in small groups, and then only for short periods. Recent studies show that the damsel-

Figure 16.7 Herbivores and other organisms of a coral reef community: (1) mats of filamentous blue-green algae; (2, 3) parrot fish *(Sparisoma viride* and *Scarus coeruleus);* (4) algae-grazing sea urchin *(Diadema antillarum);* (5) surgeon fish, blue tang *(Acanthurus coeruleus);* (6) damselfish *(Eupomacentrus planifrons).* (Based on observations by Oxenford, 1985.)

fish territoriality plays a unique role in the reef community by preventing both overgrazing and undergrazing of the algal growth. The consequence is that the species of algae are more diverse than they would be were there overgrazing or undergrazing. In overgrazing some plant species vanish because their reproduction falls below threshold levels; in undergrazing dominant plant species "crowd out" other species (Hixon and Brostoff, 1983).

Surgeon fishes feed mostly on filamentous algae. Some species have gizzardlike stomachs and ingest sand apparently to aid in the grinding of algae during digestion (Randall, 1968). These fishes also "browse" over the reef surface in groups, much like parrot fishes. They are also seen on the reef flats grazing on the larger macroalgae that grow there.

Other Organisms and Special Features of the Reef Community

Many other animal types live in the reef community. All sorts of benthic invertebrates are important parts of the living commuity. These include the sponges (Porifera), two species of which are shown in Plate 16.6, the plate sponge *Callyspongia vaginalis* and the pineapple sponge *Ircinia* spp. A number of burrowing organisms are found: examples are the sipunculid worms, which burrow into the hard substrate of coral, and the rock-boring urchin *Echinometra lucunther.* There are even species of rock-boring blue-green filamentous algae.

Other invertebrates scrape the substrate for algae, for example, the snail *Nerita tesselata,* which becomes geologically important for the quantity of sand it creates while feeding. The long-spined urchin *Diadema antillarum* (see Plate 16.6) was an important herbivore on West Indies reefs until 1983, sometimes reaching a density of 14,000 individuals per hectare! But the population fell throughout the Caribbean, in some places to fewer than one individual per hectare, after the tropical ocean warmed during the 1983 climatic event associated with El Niño and the Southern Oscillation.

Brittle stars (Echinodermata) live within the corals, as do sea cucumbers (Holothuroidea). Both of these animals feed on organic detritus that collects on the surface. The brittle star "grazes" on the spots of sand or mud with high organic content; the sea cucumber moves slowly over the bottom while "sweeping" it with mucus-covered tentacles, which it periodically passes through its mouth for cleaning and recharging of mucus. Their role in the ecosystem is best understood from the view of the large quantity of feces discharged in the reef area by the large populations of herbivorous fishes. The fishes that feed off the reef but return to it for security during resting periods add their feces to this organic cycle. The many irregular cracks and crevices of the reef structure catch this organic matter, which invertebrates then recycle. The unique feature of the coral commuity is the rapidity with which the nutrient cycle operates.

Still another feature of the reef community is the abundance of toxic chemicals used by animals, probably as defensive measures. Many of the fishes produce a toxic substance, for example, the venom on the spines of stonefish. Some extrude poisonous material onto their body surface, as does the parrot fish with its nighttime mucus "cocoon" (Collette and Talbot, 1972). Others have toxic substances in their internal organs, for example, the pufferfishes (Tetraodontidae), which feed on tips of coral branches. Still other fishes accumulate toxins that are first produced by certain species of dinoflagellates (see Plate 15.2) but are concentrated at each successive step in the food chain. The groupers and snappers of Figure 16.6 are examples of fishes in which the toxin accumulation becomes dangerous to humans because they are commonly fished and marketed. Eating these tropical fishes can cause an illness called ciguatera, the effect of which is neurological and can result in death. Many of the toxic reef species have marked, highly visible color patterns, perhaps nature's way of giving warning against predation. (Similar high-visibility colors mark venomous animals of the tropical jungles.)

As might be expected of a community that has

high production at elevated water temperatures, the reef system has many symbiotic and parasitic relationships. One of the more interesting of these is the "cleaning" symbiosis in which certain species of fishes and shrimps create their ecological niches by removing parasites, decayed or diseased flesh, mucus film, and the like from other fishes. In Figure 16.7 a cleaner fish common to West Indies reefs, the goby *Globiosoma evelynae,* is cleaning a damselfish. But cleaners also "service" the largest carnivores on the reef, for example, the grouper of Figure 16.6, without danger of being eaten. In fact, the fish being cleaned value the service and are often seen "standing in line" at a cleaner's service station.

Contrasts with the Open-Ocean Ecosystem

Reef ecology is unique, differing from that outlined in the open-ocean ecosystem overview given in Figure 12.1 in at least three important ways:

1. It is a very shallow system with close coupling between pelagic and benthic fauna.

2. Ambient conditions vary little over the seasons in the low-latitude tropics; the productivity cycle graphed in Figure 12.4 does not apply to the reef community.

3. The autotrophic level draws heavily on the dissolved nitrogen gas in seawater and is therefore not limited by the supply of nitrate nutrients, as is often the case in open oceans.

Figure 16.8 shows two simplified models of the coral reef and open-ocean systems. In the conventional oceanic pelagic community the phytoplankton are supplied nutrients by the upward mixing of deep waters. Nutrient cycling periods are very long, on the order of hundreds of years.

In the reef community, vigorous vertical mixing keeps a high level of dissolved nitrogen gas from the atmosphere available throughout the water column. A direct consequence has been the evolution of nitrogen-fixing algae as the main autotrophs, as Figure 16.8 indicates. The symbiotic zooxanthellae get nitrogen directly from their host polyps.

Copepods are the dominant animals among oceanic pelagic zooplankton, directly grazing on phytoplankton and in turn serving as prey for the meroplanktonic species. In reefs, primary production by free phytoplankton is a much smaller part of total productivity, so the proportion of holoplankters to meroplankters is also low. The main algal production by nitrogen-fixing species passes directly to large herbivorous fishes and then to the invertebrate benthic species via detrital feces. In turn, the large

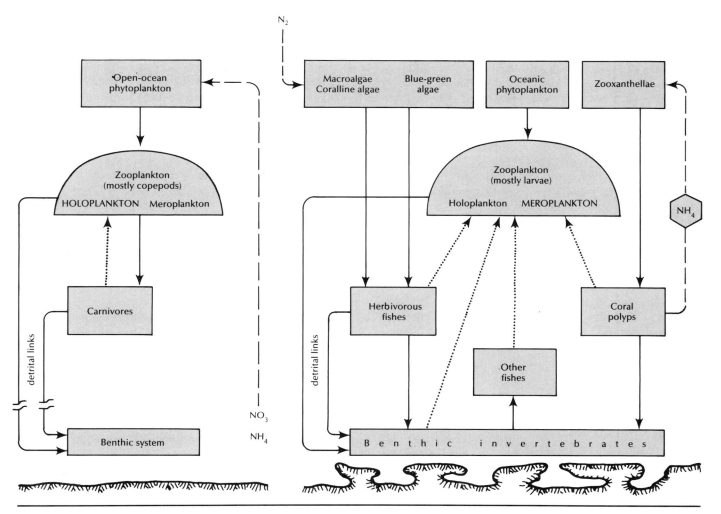

Figure 16.8 Contrasts between the conventional open-ocean community and the coral reef community. Solid lines are paths of transfer of chemical potential energy, prey to predator or grazer. Dashed lines are nitrogen nutrient replenishment paths to support algal production.

Dotted lines indicate the multitude of meroplankton derived from eggs and larvae of diverse reef animals. The zooplankton trophic level in the reef community is dominated by meroplankton feeding upon one another.

populations of herbivores and benthic invertebrates produce enormous quantities of meroplankton larvae. The dense concentrations of zooplankton found over reefs during nocturnal hours suggest that many larval-stage animals take refuge in the reef during day hours and come out at night to feed, usually upon each other! The nighttime scene is then one of vast exchanges of chemical energy among egg, larval, juvenile, and adult stages of diverse species, as suggested in Figure 16.8 by the large zooplankton box dominated by meroplankton forms.

Such a vigorous interaction among meroplankton could not happen if spawning cycles were seasonal, as is usually the case for pelagic oceanic communities. In the reefs animals spawn at frequent inter-vals, usually lunar, throughout the year. This evolved behavior assures a steady supply of organic "particulates" to the plankton in a region where conventional autotrophs contribute little to the plankton biomass.

Summary

Every reader will, sooner or later, visit one edge of the ocean. It may be a rocky shore, a sandy beach, a tide marsh, or even a coral reef. Each demonstrates the complexity of marine life of the open coast. Each has a set of physical constraints and a set of biological stresses, and many examples of specific adaptation. Each has its own fascination.

On the rocky shore, physical constraints are the exposure during low tides; the rapid diurnal changes in temperature and salinity of shallow water; the crashing of waves, with the concomitant shifting of rocks and other coast debris; and, in polar seas, the freeze-up in ice. Biological stresses are competition for space on the rocky shore; the danger of desiccation; predation from both land and sea; and, because of human habits, exposure to pollutants.

The sandy beach community is entirely different in the physical sense. Sand is a shifting habitat; thus there are no exposed life forms, only those that can burrow into the safety of the sand. Plants do not thrive here; the organic particles are swashed up the beach face and filtered out as the seawater percolates back. This organic film is home to the specially adapted interstitial organisms and their burrowing predators. Surprisingly, the environment a few centimeters below the sand surface is quite constant in temperature and salinity.

A special type of open coast is the coral reef, one of the oldest communities in the ocean. As such, it has a greater diversity in species than any other marine community. Physical constraints are the considerable dissipation of wave energy, the high light intensity, and shallow depths with well-mixed waters. There are also intriguing chemical constraints; for example, the amount of dissolved nutrients available for phytoplankton is very small, definitely not enough to account for its high food chain production. One specially adapted organism is the filamentous alga: it attaches to the coral rock and extracts nitrogen gas from the water, a nutrient in virtually unlimited supply from the air above. This food base is directly cropped by large fishes, so the role of the oceanic copepod grazer is minimal. Fish feces feed the invertebrates and benthic community. Plankton of the coral reef are mostly meroplankton, existing in larvae and eggs. The coral animal itself plays a key role in that it has evolved with a symbiotic algae. It is able to start new reefs because of this self-sufficiency and because it reproduces for the most part by budding, thereby avoiding the problems of dispersion from wave-swept shallows. The coral reef is the ocean's marvel community.

Study Questions

1. Write a short essay developing the role that physical factors have in shaping the zonation of rocky shore marine communities. As a conclusion, comment on whether the same physical forces are present worldwide, and if so, are there similarities among the rocky shore communities the world over?

2. Write a short essay describing *how* the sandy beach community receives its initial organic input, understanding that macroalgae do not live on sandy beaches themselves.

3. The brown alga *Fucus* lives in the upper intertidal zone. Why do these macroalgae need seawater at all? The largest species of macroalgae, such as *Macrocystis* and *Nereocystis,* grow in offshore waters to 50 m depth. Why do these need water except for the buoyancy that holds them near the light?

4. Explain why the mussel reaches its maximum density only within the mid-intertidal zone, even though it is a filter feeder and presumably would grow faster if it was submerged *all* the time.

5. Why are coralline algae called coralline algae?

6. Sea urchins are a major predator problem for large sea kelp. What animal controls the urchin population?

7. Write a short essay developing the factors that control the distribution of life forms in the sandy beach community.

8. Describe the feeding technique of the sand crab *Emerita.*

9. When you pick up the shell of the dead sand dollar *Dendraster,* you will usually see part of its highly specialized feeding mechanism exposed. Describe it and identify the part that is visible on the empty shell.

10. Why do you think so many of the *beach dwellers* burrow into the sand?

11. Write a short essay describing the microworld of interstitial organisms of the sandy beach.

12. The coral reef communities are perhaps the most significant marine community to be studied. List as many features as you can that make it so unique.

13. From the list in question 12, pick the two factors that you consider to be the critical ones in shaping the character of the coral reef community.

14. Compare the primary productivity of the coral reef community with that of other marine communities (use Table 1.1 as a data source).

15. The mesenterial filaments of the coral polyp serve a special role for the animal. Explain.

16. Write a short essay developing the critical role played by nitrogen-fixing algae in the coral reef community.

17. Write a short essay developing the role played by the symbiotic algae zooxanthellae in the coral reef community.

18. Comment on how fishes that use the reef as a home base, but roam afar to feed, contribute to the organic food base of the reef community itself.

19. Comment on the role of *cover* as a stabilizing factor that has allowed the coral reef community to diversify extensively.

20. Write a short essay developing the idea that the coral reef community has a "short circuit" built into it, by which plants support directly a large fish population without the need for the dissolved nutrients required by open-ocean phytoplankton.

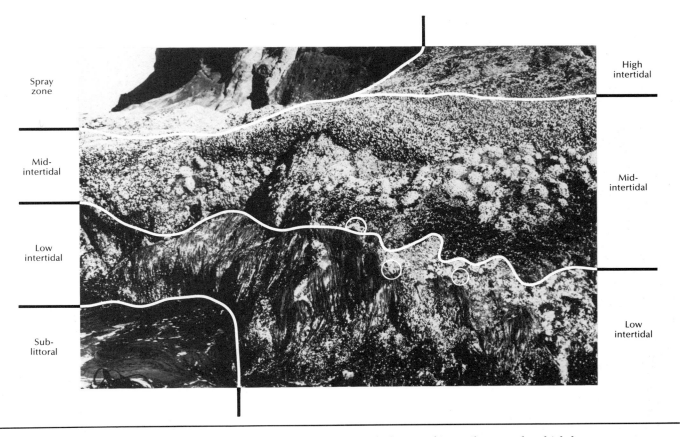

Spray
zone

Mid-
intertidal

Low
intertidal

Sub-
littoral

High
intertidal

Mid-
intertidal

Low
intertidal

Plate 16.1 Photographs of a rocky shore (a), some animals of the sublittoral tide pool (b), and extensive beds of surf grass and sea kelp extending seaward from the base of the rocky shore (c), exposed only during extremely low tides. Note that the zone limits tend to meander unevenly over the face of the rocks. This is to be expected because the irregularity of the rocky shore channels the breaking waves unevenly, leading to large local variations in the degree of wetness or exposure to air.

The upper intertidal is but sparsely represented in this photograph. The mid-intertidal, however, is rich in zone detail. Its uppermost part extends shoreward as a flat area, covered with a thick mat of mixed mussels and gooseneck barnacles. Its mid-section presents a more vertical face, and is occupied by clumps of goosenecks mingled with algae-covered rock, while its lowest part carries

a dark mat of juvenile mussels which have not yet acquired the usually attached acorn barnacles whose presence gives a whitish hue to mussel beds. Note carefully the presence of at least four starfish along the transition between mid- and low inter-tidal zones, shown above in white circles. No mussels or goosenecks occur below the transition because of starfish predation.

The low intertidal is marked by heavy mats of attached plants. Brownish-green leaves of surfgrass are seen here; below these are mats of various laminaria. The mats harbor a variety of animals, such as chitons, crabs and tube worms. The tide pool shown is a special part of the sub-littoral sea world below the lowest intertidal. It carries its own community of plants, including the corraline algae, as well as a host of animals; sea anemones and sea urchins are commonly found here.

a

(Courtesy Jefferson J. Gonor.)

b

c

Plate 16.6   A fringe reef near Porter's Location, Barbados, West Indies (Courtesy Hazel Oxenford)

Plate 16.7   A bank reef at Porter's Location, Barbados, West Indies (Courtesy Hazel Oxenford)

Euricea sp. (soft coral)

Yellowhead wrasse (adult male)

Callyspongia vaginalis (vase sponge)

Diploria labryinthiformis (grooved brain coral)

Agaricia agaricites (encrusting coral form)

Montastraea annularis (star coral)

Colpophyllia natans (hard coral)

Meandrina meandrites (butterprint brain coral)

Diadema antillarum (long-spine sea urchin)

School of larval fish

Agaricia lamarchi (leaf coral)

Stichopathes lutkeri (sea whip, a black coral)

Colpophyllia natans (hard coral)

Montastraea cavernosa (cavernous star coral)

CHAPTER 17

WAVES AND WAVELIKE MOTIONS

More has been written about waves on the foamy brine than about any other aspect of ocean behavior. That is a natural consequence, for people have always had to cross the wave-swept borders of the seas to get onto the main ocean itself, and the crossing is not always easy. If one goes down to the sea in ships, the crossing is almost always made at the entrance to bays or estuaries—a furious place, full of foam and fast-moving waves.

Much of the mystique of oceans in the folklore of humankind may have come about as people watched the borders of oceans from the beaches. The mystical nature of the ocean is reinforced in each of us when we watch the strange process in which an ocean swell seems to appear suddenly quite close to shore, steepen as it approaches the beach, and then hurl itself in noisy turbulence in a final surge up the face of the beach. Whence does the ocean obtain this fury? Why does the behavior of the sea surface change once we move away from shore out over the open sea? And what is the source of the great power that causes the tides to surge up over the coasts, only to retreat again and again?

An Overview of Wavelike Ocean Motion

A great deal of the motion of ocean fluid occurs in ways that appear *wavelike,* so it is important at the outset to identify where waves occur and what wave motion does, and to describe the wave motion itself.

1. As a general statement, whenever work is done on an ocean fluid within a given region, at least a *part* of that energy will move out of the region as wavelike motions of the fluid itself. Indeed, it is this one aspect of the physical behavior of large bodies of fluid that makes it difficult at times to explain its *local* behavior; some of what happens locally is the result of energy released locally although generated far away, or it results from energy just "passing through" a local area enroute to some other province.

A remarkable example was discovered during the 1976 large-scale experiments on the upwelling phenomenon off Peru. Physicists installed a number of current meters to measure how the nutrient-rich deeper water moved onshore in response to local Ekman wind forcing (see Figure 10.8). But analyses of recorded currents showed that a very large *wavelike motion* accompanied the upwelling, and this wave was *not correlated with the local wind;* instead, it appeared as a passing wave generated by very large-scale wind cycles. Only a part of the local fluid behavior was a response to local forcing (Smith, 1978).

2. Motion in the form of waves is one way that energy is transferred from one point in a fluid to another point. It is *energy* that is transferred; there need be no net transfer of mass during wave motion, as indicated by two important examples. (a) Wind storms over the Southern Ocean near Antarctica transform large amounts of energy into large surface waves that then travel across the expanse of both the South and North Pacific oceans to dissipate eventually along the shores of our hemisphere. The breaking of these waves in the surf zone of Oregon beaches releases energy that moves sands up and along the shoreline, energy originally generated in wind waves thousands of miles away. (b) A seismic disturbance to the ocean floor off Chile inserts energy into a tsunami wave, which then travels across the Pacific basin at speeds up to 450 miles per hour, delivering this energy to the shores of Hawaii in walls of water surging up the beach. In neither case

Rough seas under winds of 30 knots (Leo de Wys.)

is seawater "moved" from the ocean of one hemisphere to another.

3. Wavelike motions in fluids are most intense where there are sharp changes in fluid density (Figure 17.1). At the water–air interface the energy source is the friction between surface winds and the water, which creates ordinary wind waves. The "concentration" of this energy is greatest at the interface itself. Energy in wind waves decreases rapidly with depth below the surface; a submarine cruising at 50-m depth will scarcely feel the wave forces that are at the same time tossing a ship about on the surface.

Does this mean that, wherever there are sharp changes in seawater density—as in the zone of the thermocline in a two-layered ocean—waves move along this density interface? The answer is yes (Figure 17.1), and these are the "internal"-type waves. An observer at the surface may be unaware that a few tens of meters below, the thermocline itself may be heaving up and down in a wavelike motion that slowly propagates along the thermocline. One of the still unanswered questions in oceanography is *how* this type of internal wave is generated. We do know that storms at sea generate trains of internal waves with amplitudes that can exceed a hundred meters

or more. Other internal waves are found near isolated islands in patterns similar to those of lee waves in clouds sometimes seen downwind from high mountain peaks.

The coastlines of oceans are also sharp discontinuities, here between the ocean fluid and the continental rock. Special types of surface waves occur here, called "edge waves." We cannot "see" these waves—their length of several kilometers is too great—but sensitive instruments can detect a periodic rise and fall of the level of the sea. Again, the amplitude of the wave is greatest at the shoreline, tapering off with distance offshore. The wave propagates parallel to the shore; hence its name "edge" wave. The wavelike motion mentioned in connection with coastal upwelling is a very large scale wave that travels along the continental margin as a "coastal trapped wave"; its crest-to-crest distance alongshore can be hundreds of kilometers (see also the discussion of El Niño in Chapter 10).

4. There is one other way in which large fluid bodies support wavelike motion. This is called the "seiche" or standing wave. All of us know how these work; they are the same as the waves generated by sloshing back and forth in the bathtub. Regardless of how such oscillations are generated, their

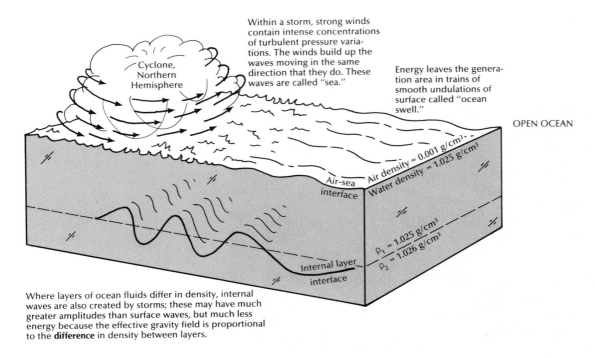

Within a storm, strong winds contain intense concentrations of turbulent pressure variations. The winds build up the waves moving in the same direction that they do. These waves are called "sea."

Cyclone, Northern Hemisphere

Energy leaves the generation area in trains of smooth undulations of surface called "ocean swell."

OPEN OCEAN

Air density = 0.001 g/cm³

Water density = 1.025 g/cm³

Air-sea interface

$\rho_1 = 1.025$ g/cm³

$\rho_2 = 1.026$ g/cm³

Internal layer interface

Where layers of ocean fluids differ in density, internal waves are also created by storms; these may have much greater amplitudes than surface waves, but much less energy because the effective gravity field is proportional to the **difference** in density between layers.

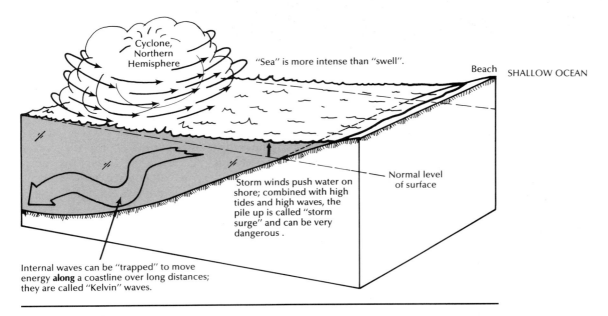

Cyclone, Northern Hemisphere

"Sea" is more intense than "swell".

Beach   SHALLOW OCEAN

Normal level of surface

Storm winds push water on shore; combined with high tides and high waves, the pile up is called "storm surge" and can be very dangerous.

Internal waves can be "trapped" to move energy **along** a coastline over long distances; they are called "Kelvin" waves.

Figure 17.1   Waves move energy.

cyclic *rate* is a direct function of the size of the container. In the ocean basins, the tide cycles are the main source of excitation energy, and so it is that every ocean basin and sea has a seiche at either the half-day tide (semidiurnal cycle) or the daily tide (diurnal cycle). In very special locations, for example, the Bay of Fundy, Nova Scotia, the extreme range between high and low tide marks of over 15 m is a consequence of the bay wave being "in resonance" to the open-ocean tide cycle.

Wave Types and Characteristics

Waves and wavelike motions occur in the oceans over a tremendously wide range of wavelengths, their distance crest to crest; and periods, the time interval required for two successive wave crests to pass by an observer. The smallest are the capillary surface waves with lengths measured in but a few centimeters and at periods that are fractions of a second. The longest waves are the tides, reaching crest to crest across half the circumference of the earth, about 20,000 km. But tide waves are not the longest in period. That distinction goes to slow internal waves, which require months to traverse the ocean basins.

One way to describe the wide ranges of lengths and periods of wave motions in the ocean is by us-

ing a spectrum graph, as given in Figure 17.2. Along one scale of the graph are described the various *types* of waves, classified according to the *period* of time required to observe one cycle. Capillary waves move very fast, considering their crest-to-crest distance, so their periods are the shortest of all. In the midrange from a few minutes per cycle to a few hours per cycle are the tides and tsunamis. These waves also move very rapidly, hundreds of kilometers per hour, but have such long wavelengths, crest to crest, that the time required for one cycle to pass an observer is quite long. At the far end of the spectrum are phenomena that have extremely long periods. For example, the slow changes in ocean currents that result from seasonal changes in winds can be treated as wavelike behavior with an annual cycle. Other long-period wavelike behavior may take several

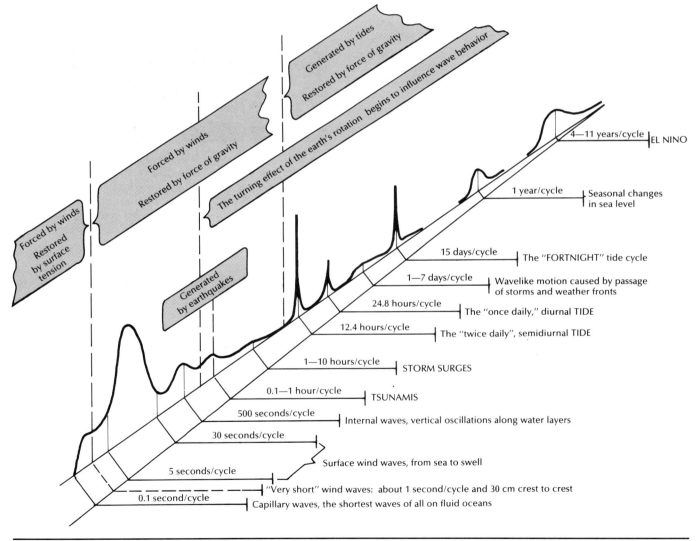

Figure 17.2 A spectrum diagram of how energy in ocean waves is distributed among the many different types encountered in the oceans.

years per cycle, as does the El Niño–Southern Oscillation phenomenon.

Figure 17.2 also charts the mechanisms that generate wave motion *or* oppose it *or* influence its movement. For example, because the water in the crest of a wind wave is actually above the mean sea level at that point, the wind has done *work against gravity* to put it there; wind, then, is a generating force. For tides, the generating forces are the mass attractions between planets. For tsunamis, the generator is a seismic event on the sea floor, usually an earthquake, but large turbidity flows can also generate tsunami waves. Again, the generating force does work against gravity, so it is gravity that provides the main force to restore the sea surface to its undisturbed state. In generating capillary waves, however, the wind works *against* the surface tension of the water surface itself, so surface tension is also the principal restoring force. For very long period waves, the Coriolis force enters the picture by forcing a deflection in the direction of wave travel. In propagating internal waves along density interfaces such as the thermocline, the generating mechanism must again work against gravity. The difference in fluid density above and below the thermocline is very small, however, less than one ten-thousandth part of the difference in densities of fluid air and fluid seawater across the air–sea interface. For this reason an internal wave can have an amplitude of hundreds of meters and still not carry as much energy as a small wind wave. Oceanographers are just now unraveling the mystery of internal ocean motion.

The *height* of each curve in Figure 17.2 represents the total amount of energy that exists in waves of that specific wavelength at any time, summed over all the oceans. For example, there is more energy in wind waves than in any of the other types of waves. This does *not* imply that *one wind wave* as we might see it or play in it at the beach is the most energetic of all waves. Rather, it means that at any one instant there is more energy stored in the total of countless wind waves across the vast ocean surface than in any other kind of wave, including the tides. We now describe different types of waves in more detail.

Capillary Waves

The *shortest* waves on the sea surface are generated by the friction between fluid wind and fluid water. These waves are the very first distortion of the surface when a wind begins to blow. If we stand on a high shore above a still lake early in the morning, as

An excellent photograph of a sea surface with a wide range of wave types. The smallest capillary waves are responsible for the tiny, glinting spots of light reflected from the sun. Next larger in size are the very short gravity waves, which give the surface its "leathery" appearance. Still larger waves are ordinary "sea." (Jerome Wexler/Photo Researchers.)

the first breezes of day begin to stir, we see sudden patches of "riffled" water surface—sometimes called "cat's paws." These are patches of capillary waves, with wavelengths from only about 2 to 5 cm. The friction with moving air distorts the water surface into successive wavelets while the surface tension of water continually attempts to restore the surface to its least-energy, initially smooth state. Thus capillary waves dissipate their energy of motion by passing energy directly into heat via the molecular viscosity property of water.

At sea the capillary waves are often overlooked in the presence of other waves ranging up to hundreds of meters in length. But they are there whenever the wind speed is greater than a few meters per second. In fact, because the winds at sea blow most of the time, to witness a "glassy smooth" ocean surface is a rare occasion. I have seen such an ocean but once and will testify to its eerie nature; long swell waves contort the glassy surface, but no riffles are present.

Wind friction is not the only generator of capillary waves. A *very* careful observer at sea will note the sudden appearance of capillary waves just in front of the crests of very short gravity waves that have grown to be sharply peaked, almost unstable. But instead of breaking forward, as do large wind waves, the unstable fluid at the crest "slumps" forward to create a set of wrinkles. These too are capillary waves; they are an important link in the dissipation of energy from large waves because they pass such energy directly into heat, via viscosity, and so act to "short-circuit" the continued accumulation of energy into "big waves," as described next.

Very Short Gravity Waves

As the surface wave grows to lengths of 5 to perhaps 30 cm, the force of gravity takes on more and more importance in controlling shape and behavior, relegating surface tension as an important force to the highly curved part of the wave near the peak. Reaching periods of up to one second, these waves travel quite slowly, much more slowly than the typical surface waves. Consequently, we see these short waves riding up and over the *crests* of the faster-moving sea and swell. It is during this time that the very short waves become most peaked and lose energy via the wrinkling capillary wave route just described.

During a storm at sea, the rough peaking of waves combined with strong winds brings about much breaking of waves, known as *white caps* or *combers*. Energy cannot be dissipated fast enough via the short waves and capillary route to reduce the

sharp crests of large sea waves and to prevent the breaking. Ancient mariners knew this from experience. In rough seas, the helmsman would sail the ship in the same direction as wave travel and allow the waves to overtake the ship from the stern. To prevent the breaking of waves over the stern, a dangerous occurrence, the crew would trail from it canvas bags filled with oily rags. This had the effect of reducing surface tension, hence lowering the drag of the slow-moving short and capillary waves on the large wave, which would then pass under the ship without breaking. From this practice comes the soothing adage "Cast ye oil upon the troubled waters"; today we would need an environmental permit to do this.

Wind Waves

As stated earlier, the ocean stores more energy in wind waves than in any other type; in Figure 17.2 it is the large, broad peak of energy between 5- and 30-s periods. Most of the energy built into these waves across the vast ocean area eventually arrives at some coastline and dissipates in the turbulent surf zones. We will learn in Chapter 18 that the concentration of energy in a wave increases as the *square* of its height, H^2. The amount of such energy is staggering—a wave with a 10-s period and a 2-m height carries enough energy in *each 1 meter of crest* to light up 250 light bulbs, each bulb using 100 watts. Such energy is not evenly distributed over the worlds oceans, however. Because winds are the generators of these surface waves, we expect to find the most energetic waves in the same belt patterns as those of surface westerly and easterly winds. The most dramatic example (Figure 17.3) is the association of the

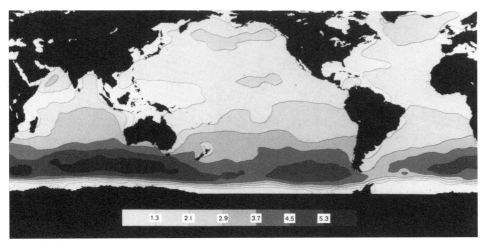

Figure 17.3 The global ocean contoured for the average height of surface wind waves during the austral winter season of 1978, based on altimeter data from the *Seasat* satellite. Clearly, of all oceans the Southern Ocean has the most concentrated wave energy, derived from the winds of the roaring forties and furious fifties, (the latitudes of the westerlies). The gray scale at the bottom groups the waves according to their heights. (Courtesy Jet Propulsion Laboratory; Chelton, Hussey, and Parke, 1981.)

ocean's largest wind waves, in the belt of Southern Ocean latitudes 40 to 50°S, with the westerlies of that region, sometimes called the "roaring forties and furious fifties." Here, because the wind blows unimpaired around the globe, we find the longest of all surface waves—some reach wavelengths of over 500 m and speeds of over 50 knots. Small wonder that the ordeal of early sailors in "rounding the Horn"—sailing around the tip of South America, latitude 54°S—was world-renowned as a test of strength and courage and worthiness of ship.

How Winds Generate Waves. To begin a description of *how* waves are generated, we first describe the character of the wind itself (Figure 17.4). Near the surface, winds are *not* uniform and "smooth"; rather there is a "roughness" in the pattern of speed, sometimes higher or lower, and the pattern is *random*, that is, turbulent in nature. There are two other important statements to be made about the wind's pattern of speed.

1. The pattern is not stationary—it moves along with the mean wind speed.

2. The mean separation distance between the most energetic cells in the pattern stays the same as long as the mean wind speed stays the same; in this way the cell *pattern* can efficiently transfer energy to the ocean wave pattern.

These statements imply that there is a "structure" to the wind near the surface. This is precisely the point, and the structure can be mapped to reveal a random field of pressure fluctuations, as in Figure 17.4. To map such a field accurately, we would need to make simultaneous measurements of atmospheric pressure over a wide area of ocean, a feat that has not yet been done. But we can infer the existence of the structure by *timing* the variations in pressure that are "rafted" past a single measuring point, for example, a ship or buoy station.

The analogy of wind patterns over a Kansas grain field might help the reader understand this moving, random, turbulent structure. Those who have observed large grain fields will recall the movement of patterns of light reflecting from patches of grass that are slightly more bent over by the wind than are other patches. In a large field, we can see a number

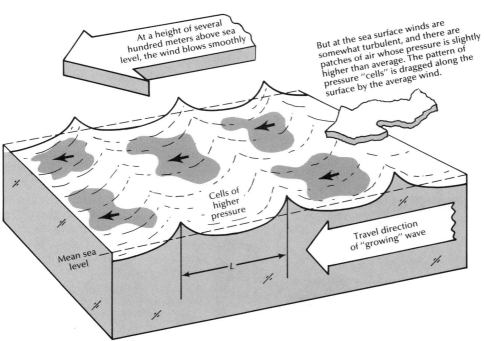

Figure 17.4 Winds impart energy to surface waves on a selective basis. The wind at the sea surface moves at some average speed, but imbedded within this mean flow is a pattern of small pressure variations, randomly distributed, but with a maximum spacing that depends on wind speed. A "steady" wind will maintain this spacing and build waves of similar length. Higher wind speeds space out the pressure variations at longer intervals, building longer, and higher, surface waves.

At a height of several hundred meters above sea level, the wind blows smoothly

But at the sea surface winds are somewhat turbulent, and there are patches of air whose pressure is slightly higher than average. The pattern of pressure "cells" is dragged along the surface by the average wind.

Cells of higher pressure

Travel direction of "growing" wave

Mean sea level

L

Within the ocean, waves will continue to extract energy from the turbulent wind and grow in amplitude **if**
(a) the crest-to-crest distance *L* "fits" the surface pressure pattern of air cells so that the patches of higher pressure coincide with troughs and the areas of lower pressures coincide with crests, and
(b) the growing waves are moving in the same direction as the mean wind and with about the same speed, so that the pressure pattern can insert energy into the wave on a continuous basis.

of such patches simultaneously, and all moving in the direction of the mean wind. Over the oceans, these correspond to the "cells" of Figure 17.4. It is important to note that if the wind field were suddenly lifted away from contact with the grain field, the patch structure would vanish immediately—no "waves of grain" would continue to propagate once the forcing wind was removed.

In contrast, waves that are generated in the fluid ocean continue to exist after the forcing winds disappear. They are simple water waves moving along the surface in the direction of the wind that originally generated them. Therefore, if such waves move in the general direction of a forcing wind, and at speeds similar to that of the average wind, they continuously grow more energetic; the turbulent wind cells continue to "do work" on them as the two move more or less coherently in the downwind direction.

We now make a definitive statement on how energy transfers from the winds to ocean waves. Beginning with the energy of motion in the mean wind field above the ocean, turbulent cells are generated in a layer of air just above the surface. Through a process of "matching" high-pressure disturbances in the turbulent cells with the troughs of moving sea waves, and low pressure disturbances with the crests, the wind field pushes and pulls the sea surface into ever greater wave heights. The final wave field depends on (1) wind duration, the period of time over which the transfer process works; (2) the speed of the wind during this time; and (3) the "fetch" distance over which the wind and wave fields stay coupled.

Role of Wind Duration and Fetch. Intuition tells us that there must be a finite limit to the size to which a wave field can grow during the energy transfer from the wind field. One such limit is described by a maximum in the ratio of wave height H to wave length L, the H/L of Figure 17.5. This ratio is called "wave steepness," and we find that wind waves commonly break over when the steepness ratio begins to exceed 1:7, that is, when the height of the wave grows to one-seventh its length. This breaking over of the wave crest at sea, creating the whitecaps or combers mentioned earlier, causes part of the energy in the wavelike motion to dissipate through the turbulence of the whitecap itself. The wave continues onward with its remaining energy, and if it stays coupled to the wind field will eventually regain the lost amount, become critically steep again, break again, and so on. When a sea has reached this state,

it is called a "fully developed sea," that is, fully matched to the specific driving wind field.

Whether a given wave can grow to the critical H/L steepness depends on how long it is exposed to the forcing wind. Where the wind field dies off, or its parent atmospheric disturbance moves off to other ocean areas, the sea state may not become fully developed. Stated another way, the "age" of the wave field has not reached maturity; it is referred to as a *duration-limited* wave field.

A wave field may also be termed *fetch-limited*. This occurs when the waves encounter an obstacle before becoming fully developed. One example is when the wave field encounters a shoreline before achieving its maximum sea state. Most of us are aware of this limitation from experiences on small lakes; the larger the lake, the larger the waves encountered on the windward shores.

Figure 17.3 points dramatically to the sea state that can exist when an ocean has *unlimited fetch* to a wind field—the large waves in the Southern Ocean in the latitude band where no continental barrier exists. Here the pattern of westerly winds can also provide an almost *duration-unlimited* driving force.

Role of Wind Speed in Controlling Sea State. Mariners know from experience that the energy in a fully developed wave field increases markedly with the speed of the forcing wind. The term "sea state" is used to describe the condition of the sea surface after a wind of a given speed has generated its full wave potential. The sea state is largely determined by a *visual* assessment of the condition of the sea surface. In 1934, Cornish published a collection of such visual observations summarizing the behavior of wave fields for different wind speeds (Table 17.1). Before describing this behavior, we should point out that a human observer, when asked to estimate the average height of waves at sea, is strongly influenced by the highest waves, a natural bias since these dominate the visual senses. As a consequence, the data in Table 17.1 are averages for the highest third of the waves actually present, that is, the most dominant waves that are generated by a given wind. We can draw several useful conclusions from these data.

1. The average wave speed of the highest waves is about 0.8 times the speed of the forcing wind for all the speeds observed, from 14 m/s (31 mph) to 31 m/s (68 mph). This result confirms the earlier statements that energy passes from winds to waves best when the wave itself moves at about the same speed as the pattern of pressure cells in the mean wind.

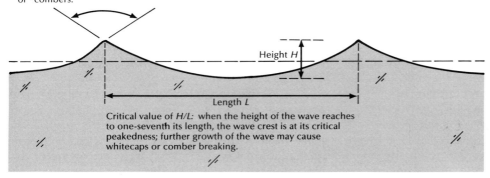

Critical angle: does not fall below 120°. If a wave becomes more peaked than this in the open sea, it breaks forward to "lose" peakedness while the remainder of the wave continues on; we call these breaks "whitecaps" or "combers."

Height *H*

Length *L*

Critical value of *H/L:* when the height of the wave reaches to one-seventh its length, the wave crest is at its critical peakedness; further growth of the wave may cause whitecaps or comber breaking.

(a) The critical condition for a wave to break depends either on a minimum angle for the "peakedness" of the wave or on a maximum height, *H,* of the wave for a given wavelength, *L,* crest to crest.

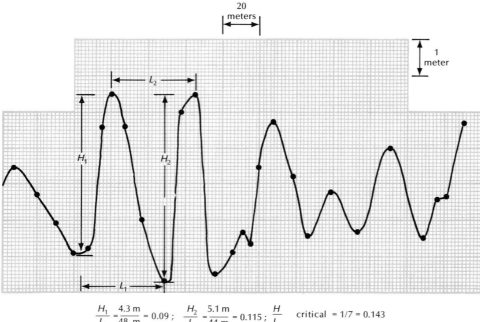

$$\frac{H_1}{L_1} = \frac{4.3 \text{ m}}{48 \text{ m}} = 0.09 \; ; \quad \frac{H_2}{L_2} = \frac{5.1 \text{ m}}{44 \text{ m}} = 0.115 \; ; \quad \frac{H}{L} \quad \text{critical} = 1/7 = 0.143$$

(b) The topography of ocean surface waves derived by interpreting aerial photographs. The technique gives a "snapshot" slice through the air-sea interface. It is used here to show how we can estimate values for the ratio *H/L* for real waves.

Figure 17.5 The concept that waves can be limited in height.

2. Both the observed average wave period and wave height increase directly with wind speed: as wind speed increased by a factor of 2.19 (from 14.2 to 31.1 m/s), wave period increased by 2.21 and height by 2.14.

3. But notice that the *length* of the average wave more than quadruples, from 78 to 384 m, as the wind speed doubles from 14.2 to about 31 m/s. This tells us that the faster-moving waves are also the longer waves, or, what is the same thing, that long waves travel faster than short waves.

4. The energy *density* in the field of longest waves also more than quadruples over the same change in wind speed. The wave developed by the 31.1-m/s

Table 17.1 Characteristics of wind waves observed at sea under different wind conditions. (From Cornish, 1934.)

Wind Speed m/s	Wave Speed m/s	Period, T, seconds	Length, L, meters	Height, H, meters	H/L	Energy, kilojoules/m²
14.2	11.5	7.0	78	6.9	0.088	59.5
16.0	12.8	8.0	103	7.7	0.074	74.1
19.2	15.3	9.5	147	9.2	0.063	106.0
22.9	18.3	11.5	209	10.9	0.052	148.5
27.0	21.5	13.5	290	13.0	0.045	211.0
31.1	25.0	15.5	384	14.8	0.039	274.0

wind carries 274 kilojoules for each square meter of ocean surface, compared to 59.5 kilojoules for the shorter wave developed by a 14.2-m/s wind, a ratio of roughly 4:1. Moreover, the longer wave delivers its energy about twice as fast. Therefore, the longer wave is some eight times more powerful! We examine wind wave energy in more detail in Chapter 19.

5. Finally, the steepness ratios for these dominant waves decline as the wind speed and other factors increase. This results directly from the fact that wavelength, which quadruples, increases much more rapidly than does wave height which merely doubles.

Figure 17.6 is a pictorial summation of the results of wave generation by winds using data from Table 17.1. It shows how the peak in the distribution of energy changes with the character of the generating wind, assuming that wind duration and fetch are adequate to allow full wave development in each case. To put this picture into perspective, compare it with the wind wave portion of the overall spectrum in Figure 17.2.

The Storm Surge: A Large-Scale Wind Effect

We should discuss the *storm surge* phenomenon before leaving the subject of winds that do work upon surface water. Although the storm surge is not strictly a wave, it impacts very severely on humans because it occurs in association with wind waves and tide waves. For this reason it is included here.

Chapter 10 contains a description of the Ekman effect: in deep ocean water a wind blowing over the surface forces a *net* transport of surface layer water in a direction at right angles to the wind. We now know that this effect is the result of (1) a transfer of momentum from wind to water via the mechanisms of capillary waves and the very short gravity waves and (2) the Coriolis turning effect. As water becomes shallower, however, the direction of Ekman transport aligns closer and closer toward the downwind direction, and therefore in closer alignment with the wind waves being generated by the same wind. This is important in the recognition of storm surges as potentially dangerous phenomena; Figure 17.1

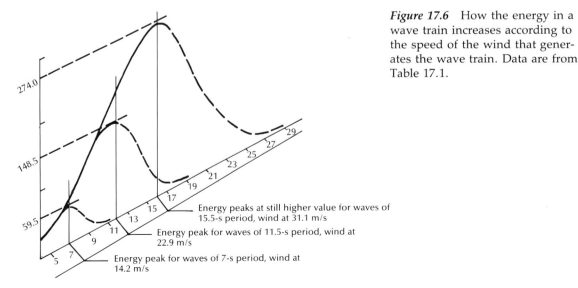

Wave energy kilojoules/m²

Energy peaks at still higher value for waves of 15.5-s period, wind at 31.1 m/s

Energy peak for waves of 11.5-s period, wind at 22.9 m/s

Energy peak for waves of 7-s period, wind at 14.2 m/s

Figure 17.6 How the energy in a wave train increases according to the speed of the wind that generates the wave train. Data are from Table 17.1.

Ordinary wind waves rarely top over seawalls as shown here. But hurricane winds can push surface waters against a shore; this surge of water can be several meters high, topping over such barriers and wreaking heavy damage on coastal installations. Notice the two streetlight posts under water. (Official U.S. Navy Photograph.)

shows how the Ekman transport can cause a surge of water against a shoreline. If this surge occurs during a high tide, the combination of surge and high wind waves can cause the ocean to breach the shoreline.

Historically, the peoples of the Netherlands have more experience with storm surges than inhabitants of most other areas because their countryside is low-lying, much of it wrested from the adjacent, shallow North Sea by the construction of dikes and pumping out seawater. Storms over the shallow North Sea have repeatedly caused damage and loss of life in the Netherlands. In the United States, the famous example is the 1900 storm surge that arose from the shallow continental seas off the coastal port city of Galveston, Texas; it caused extensive damage and claimed about 2000 lives.[1] In 1973, a major storm

surge in the Bay of Bengal inundated large sections of the coastal provinces of Bangladesh with a loss of life exceeding 300,000!

Ocean Depth: A Factor Controlling Wave Behavior

Proceeding along the wave spectrum of Figure 17.2, we encounter next a group of waves with long periods and very long wavelengths, of which the tsunami and tide waves are prime examples. We observe that the behavior of very long waves is controlled in part by the water depth, so we need to describe the *how* and *where* that waves can "feel" the influence of the ocean floor. To do this, we need to describe how the water medium itself behaves as a wave passes through it.

Figure 17.7 demonstrates the behavior of water particles as a surface wave moves along the sea surface. The initial position of the wave is with two successive crests at positions 0 and 20. These are *seconds of time*. (But keep in mind that a wave of 20-s period is unusual; the majority of wind waves have periods of about 8 to 10 s.) The goal is to track the trajectory of an individual water particle as the wave goes by, and for this we begin with the particle at the surface but located in the trough midway between crests. As the wave form progresses to the right, a dot marks the location of a "tagged" particle at each successive

---

[1]Strangely, the Galveston episode did not trigger research into causes and predictions of storm surges, but an event during World War II at the remote South Sea island of Tarawa did. The occasion was an amphibious assault by U.S. Marines on the beaches of that Japanese-held island. Japanese defenses included steel and log barriers to prevent the passage of landing craft. We knew that the average depth of water at these barriers would normally allow assault troops to disembark there and wade ashore. What actually happened, however, is that a storm surge, generated by a typhoon, increased the water depth sufficiently to make wading impossible. We lost over a battalion of marines in the assault. One immediate result was a crash program of research on predicting storm surges and other wave effects.

second. Particle behavior has three important aspects.

1. As the crest approaches, the surface particle is moving *toward* the crest and upward; when the crest is at the center position 10, the particle is at the top and moving with the crest. Therefore, the forward slope of a wind wave is a zone of *convergence* of surface water. (Recall that it is here that the very short waves can "peak out" and transfer energy into capillary waves and hence into viscous dissipation.) In contrast, the back slope of the wind wave is a zone where the surface is diverging and stretching out. Those who have been to sea know very well that the back slope of a wave is always "smoother" than the more wrinkled foreslope.

2. As the wave crest reaches position 20, the water particle returns to its original starting position, having executed a trajectory in the shape of a circle. Actually, the trajectory never quite closes on itself but instead undergoes a small net displacement in the direction of wave travel. In the deep open sea, such net displacement is trivial in relation to the maximum excursion that the particle goes through.

3. The *surface* particle trajectory has a diameter equal to the total vertical distance between crest and trough, the "height" of the wave, but particles at depth undergo much smaller trajectories. The *rate* at which the trajectory shrinks in size with depth is exponential; in Figure 17.7 trajectories are *shown to true scale relative to the length of the surface wave.* It is clear that at depths greater than about one-half the length of the wave, particle motion is negligibly small, and this leads us directly to the next section.

The Wave "Feels" the Bottom Through Friction

On the basis of point 3, we deduce that surface waves in the open ocean cannot "feel" the bottom because particle motion dies away with a depth far above the solid bottom. In Table 17.1 the longest wave, 384 m, had a period of 15.5 s; our 20-s wave may have a wavelength of about 600 m. Therefore, when ocean depth is greater than 600/2 = 300 m,

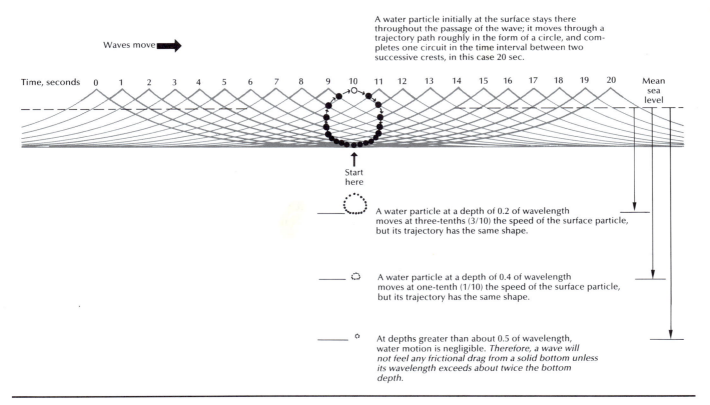

A water particle initially at the surface stays there throughout the passage of the wave; it moves through a trajectory path roughly in the form of a circle, and completes one circuit in the time interval between two successive crests, in this case 20 sec.

Waves move ➡

Time, seconds 0 1 2 3 4 5 6 7 8 9 10 11 12 13 14 15 16 17 18 19 20 Mean sea level

Start here

A water particle at a depth of 0.2 of wavelength moves at three-tenths (3/10) the speed of the surface particle, but its trajectory has the same shape.

A water particle at a depth of 0.4 of wavelength moves at one-tenth (1/10) the speed of the surface particle, but its trajectory has the same shape.

At depths greater than about 0.5 of wavelength, water motion is negligible. *Therefore, a wave will not feel any frictional drag from a solid bottom unless its wavelength exceeds about twice the bottom depth.*

Figure 17.7 The motions made by water particles at different depths during one cycle of a passing wave. The successive sketches focus on the circular motion of a single parcel of surface water as the wave propagates. The surface particle executes a near-circular trajectory; particles at greater depth have similarly shaped trajectories, but the amplitudes of the orbits of deeper particles diminish rapidly.

particle motion at the bottom is always negligible. There is no friction "drag" to slow down the wave or begin to extract energy from it.

The friction aspect changes as our wave approaches a shore, however. Recall from Chapter 5 that the average depth of water at the point where the shelf breaks into the continental slope is 135 m, that is, about 0.2 of the length of our test wave. In Figure 17.7 water particles at this relative depth still have substantial trajectories of sizes and speeds up to 0.3 as high as those of the surface particle. As the wave moves onto the shelf, the bottom now *interferes with particle motion.* Clearly the trajectory at the bottom cannot be a circle but is instead flattened out into a forward–backward excursion *along* the bottom; this is the source of friction and the mechanism by which the wave begins to "feel" the bottom. Figure 5.8 indicated the forces that move sediments across the shallow continental shelves. Over the mid-shelf region, in waters of 100 m depth or more, only the very long waves generate sufficient water motion to stir and move bottom sediments, for example, the long wave just described. Of course, *most* wind waves are much shorter than the wave of Figure 17.7 and would not begin to feel bottom until much closer to shore. (See Chapter 18 for details of wave behavior at beaches and shorelines.)

Very Long Waves: Tsunamis and Tides

Common usage in the English-speaking world assigns the word *tides* to the daily ups and downs of the sea surface along our shores. We do not commonly refer to these movements as *tide waves*, which is precisely what they are; instead we use the terms flood tide, ebb tide, slack tide, spring tide, neap tide, and so on to describe the reality of its wavelike behavior. In this text, both common jargon and the term tide waves are used in describing tide effects. It is also common to refer to the occasional catastrophic wall of water rushing over the shore as a *tidal wave.* Oceanographers and this text use the term *tsunami*, a Japanese word meaning "wave in bay" for this sometimes devastating release of energy.

In fact, these two phenomena have no common basis except that both are (1) very long waves in the ocean so that (2) the speed of each (meters per second) is governed by the depth of the ocean according to the formula

$$\text{speed} = \sqrt{10 \times D},$$

where D is the ocean depth in meters. For the ocean with an average depth of 4000 m, the speed of these waves is then $(10 \times 4000)^{1/2}$ or about 200 ms (440 mph). Tides are caused by forces that hold the planets of our solar system in their steady orbits. Tsunamis are generated by seismic movements of the sea floor itself; the energy of the movements is transferred into very long waves.

Tsunamis

In the spectrum of Figure 17.2 the periods of tsunami waves are in the range of 6 to 60 minutes, whereas tides have periods on the order of 12 to 24 hours. This great difference in characteristic periods is due to the character of the forces that shape the waves. Tsunamis are the ocean's response to a large-scale *impulse* of energy and are "free" waves once generated. Tides are always "forced waves" in that the oceans are continually under the influence of the generating forces.

We will not give much attention to tsunamis in this text, but there are some aspects of the phenomenon that should be mentioned. Let us begin with the meaning of this Japanese term, "wave in bay." Why should this meaning have been assigned to the seismically generated wave? The answer is that the tsunami wave is visually undetectable on a ship at sea. Only when the wave arrives in very shallow water at the shore is it suddenly seen, often taking the form of a huge breaker that rolls its way far up the shore. Japanese fishing villages were often devastated by these, but the fishermen themselves, if out to sea at the time, would not be aware that the catastrophe had occurred until they returned home. Since villages were always located inside bays and coves of shallow waters, the fishermen, not having witnessed the wave offshore, must have reasoned that it was created inside the bay itself.

Given the speed of the free wave as calculated at 200 m/s (720 km/h), roughly 10 times faster than a Volkswagen on the interstate, a tsunami with a 10-minute period would in the open sea have successive crests spaced 120 km apart. These crests are not very high, perhaps centimeters at the most; it is impossible to sense that a ship rises and falls 2 cm over a period of 10 minutes, since the ordinary wind waves at sea toss the ship about much more vigorously.

However, as the wave enters the shallow waters of the bay, its speed is drastically reduced, again according to the formula given. If the front part of a wave cycle slows down while the part at sea is still rapidly approaching, the total energy of the wave begins to be compressed. The wave *form* changes as

Tsunami coming ashore at Hilo, Hawaii, 1, April 1946. The structure at the left is part of a pier. The top of the wave in the upper right portion of this scene is at least as high as the roof of the already demolished pier. (NOAA.)

the crest grows higher and the troughs grow deeper. Eventually, the crest begins to spill forward like ordinary breakers in the surf zone, but unlike the breaker, tsunami crests may be tens of meters high and unleash devastating destruction as the wave swash surges inland.

The death and destruction caused by tsunamis can be avoided to a great extent by early warning. Oceanographers have constructed a network of Tsunami Early Warning Stations (TEWS) throughout the Pacific (Figure 17.8). These stations monitor sensors that can detect the tsunami signal and flash warnings across the network. Since the speed of the waves is easily calculated, the network also predicts the time that the waves will arrive at shore; civil defense agencies can then prepare local populations for the oncoming disaster. The TEWS is also linked to earthquake-monitoring seismometer stations so that the network is alerted immediately after an earthquake is detected in an oceanic area.

We have saved mention of how tsunami waves are generated to the last for a reason. The cause is thought to be the sudden movement of basin crust during an earthquake. We do not know how the moving earth puts energy into water motion: some earthquakes cause tsunamis and others do not. But if a ship happens to be just above the crust that moves, that is, at the epicenter of the earthquake, it will receive a succession of pressure waves generated by the quake. These propagate with the speed of sound in water. Seamen report that the effect is similar to a huge Davy Jones slamming the bottom of the ship with a giant sledgehammer. It is an experience I hope to avoid.

Tides

Next to wind waves, the tides are the best-known characteristic of the sea. Our language is full of metaphors about tides: "We sail with the tide." "Time and tide wait for no man." "There is a tide in the affairs of men." Implicit in most of these is the characteristic *pulsation* or regularity of tides. Not only do the pulsations occur daily, but the range between the daily high- and low-tide marks seems to pulse between a maximum, called *spring tides,* and a min-

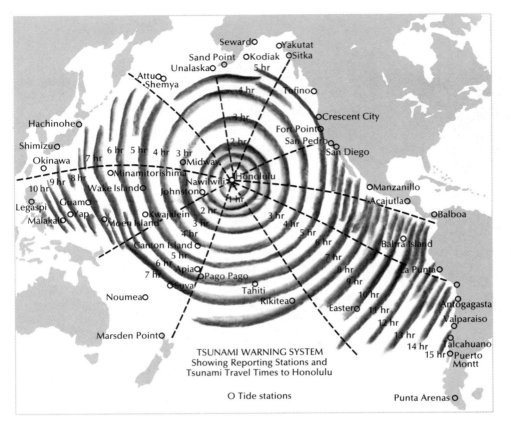

Figure 17.8 The Pacific Tsunami Early Warning Station Network. The distance between each of the contours is the distance traveled by a tsunami in one hour. A tsunami generated in the ocean off Chile would travel to Hawaii in about 15 hours, at an average speed of 700 km/h. (From Weigel, 1974.)

imum, called *neap tides*, with a 14.75-day period that is as regular as clockwork. In fact, the term for this period is the "fortnight," which comes from our heritage in the English maritime tradition.

Entire books have been written about the tides. Early mathematicians noted the regularity in tides and concluded that there must be an equation by which to calculate the tide height for any point along a coastline. The correlation between the timing of tide peaks and the passage of moon and sun overhead has been clear from early times. But early efforts failed to produce a prediction formula for the tidewater level. For although the strength and timing of the driving forces for tides could be calculated with great accuracy, based as these are on the interplanetary orbits, no mathematical formula could account adequately for the complex, dynamic response of the fluid ocean. We will return to this point later.

The following general statements can be made about tides. We follow them with detailed descriptions.

1. The double-bulge tide generating force. Common knowledge holds that ocean tides are caused by the moon; more specifically, we would say the mass attraction between the moon and the

earth. Neither is correct in detail, but the idea holds. The earth–moon pair of masses interacts to produce *one* tide-generating force, which in turn produces *two* high-water bulges around the global ocean. Distributed in a symmetric, wavelength pattern, the high-water bulges are at both the near and the far sides of the earth *at the same time.* It is a concept always difficult for students to master.[2]

2. The composite tide. Because there are many planetary bodies that pair *with the earth* and produce additional tide-generating forces, the *total* force responsible for the earth's tides is a *composite of many.* We then refer to the actual tide as a composite tide made up of a sum of the ocean's responses to many forces. Each of these is a cyclic force, with a period determined by the relative orbits of the earth and of the other planet in the pair. No two periods are the same, and they all differ in timing and relative strength.

3. The diurnal inequality. Because the spin of the earth about its axis carries an observer through the double-wave tide pattern in one revolution, each

---

[2]And not just beginner students; I have seen Ph.D. students fail to master the double-bulge concept.

wave cycle lasts for about 12 hours. But because the spin axis of the earth is inclined about 23.5° to the ecliptic plane in which it orbits the sun, the two daily *high tides* are of *unequal heights.* Thus the diurnal inequality.

4. *The local establishment.* Although each tide wave component is "forced" by its generating force, the speed at which the composite wave travels is governed by the ocean depth and is slower than that of the "force" itself. Therefore, the time at which a

Figure 17.9 Factors that govern the tide-generating forces.

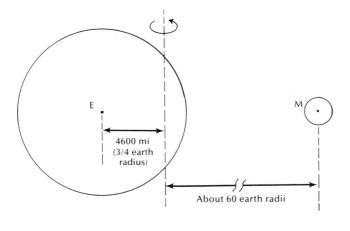

(a) The earth-moon pair rotates about a common axis.

(b) But it is important to note that the earth does not "revolve" about the common axis in the sense of a solid-body rotation; rather, it *translates* around the axis without rotation. The reader should differentiate between this *pair-body* rotation and the earth's own spin about its polar axis.

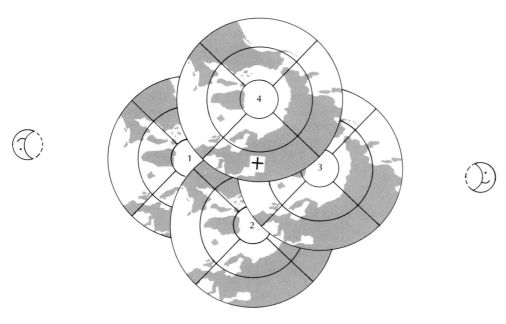

tide wave crest reaches a given point on the shore usually *lags behind* the time at which the moon or tide-producing planet passes directly overhead. This time lag is called the local "establishment."

5. *The tide currents.* Because the tide wave is of such great wavelength, one-half the circumference of the earth for each of the two cycles of the double wave, the *orbits* of water particles as the wave passes are in fact not orbits at all; rather they are completely flattened into what is best described as a "sloshing back and forth" in the open ocean. Nearshore we call these "ebb and flood" tide currents. But because the earth spins beneath the moving tide currents, we perceive the tide wave motions as currents that continually change directions, the Coriolis effect. (See Chapter 18 for a description of tide currents on a continental shelf.)

6. *The "standing" tide wave.* Most tide waves produce maximum flood–ebb currents when the tide mark is at mean sea level, not during periods of either highest or lowest water levels; tide waves are therefore classed as "standing waves."

The Double-Bulge Tide. The one aspect of tide waves that is the most difficult to understand is its twice-daily high–low behavior. Simply stated, *if the tide is caused by the moon's gravitational pull on the earth's water, there should be but one high-water mark—and on the side closest to the moon!* The fact is that there is at the same time an *equal high water* on the opposite side of the earth from the moon. How do we explain this double-bulge or double-wave characteristic of the tide? One way we *cannot* explain these double tides is on the basis of gravitational attraction alone. In magnitude, the direct "pull" of the moon on a water parcel at the earth surface and directly under the moon is only 1/9,000,000th as strong as the "pull" of the earth itself on the same parcel. For example, the value at the equator of the earth's gravity acceleration acting on a unit mass of seawater at mean sea level has been measured empirically to be 9.78049 m/s^2; by comparison, the maximum force of attraction by the moon on the same unit mass of seawater is only 0.000001 m/s^2, and hence negligible.

The key to understanding the double-wave tide is contained in four statements.

1. Each pair of astronomical bodies, one orbiting the other, is in equilibrium rotation. The axis about which the earth–moon pair rotates is about three-fourths of an earth radius from the center of the earth (Figure 17.9*a*). It is much closer to the earth's

center *E* than to the moon's center *M* because the mass of the earth is 81.5 times greater.

2. The earth *translates* around the common mass center—it does *not* rotate. Figure 17.9*b* shows this translation behavior of an earth with its own spin removed; the orientation of the earth would remain constant. Notice the position of Greenland during each of the four quadrants of revolution of the earth–moon pair; its translation about the common axis means that each particle of Greenland has the same speed and direction, hence centrifugal acceleration, at each instant, as does each and every other particle of earth at that same instant.

3. A particle at the center of the earth is in force balance (equilibrium) because it experiences precisely equal and opposite forces: its *centrifugal acceleration directed away from the moon* and its *mass attraction directed toward the moon*. But particles on the earth surface are not at equilibrium. Surface particles on the side nearer the moon experience greater attraction forces than centrigual and hence are attracted *toward* the moon; at the same time, particles on the far side of the earth experience a weaker moon attraction than the local centrifugal force, which then causes a bulging of the earth's surface *away* from the moon. This effect is illustrated in Figure 17.10.

4. Ocean waters are forced to move only when the small *imbalance* between the centrifugal and the attraction forces is directed *horizontally along* the earth's surface (Figure 17.10*a*). Even small-magnitude forces *can* move seawater along the horizontal direction, for example, along surfaces of constant density (recall in Chapter 8 the data for an upwelling zone). Therefore, these imbalances, called *traction* forces, are responsible for converging the ocean waters into two "humps," one on the near side and the other on the far side of the earth from the moon.

This last statement examines what happens during the earth's daily spin about its own polar axis. An observer at a seashore point travels through the two zones of water convergence, or bulges, during a 24.8-hour period and pronounces each a *high tide* (Figure 17.10*b*). To an observer, the alternation of high-to-low-to-high-to-low waters appears as a wave with a 12.4-hour period; that is, it occurs twice daily.

The Composite Tide. Let us examine a typical record of successive hourly tide marks at a shore station (Figure 17.11). The gross features of the record repeat at intervals of about 24.8 hours, which is called the "diurnal" tide period. Mathematicians can "dis-

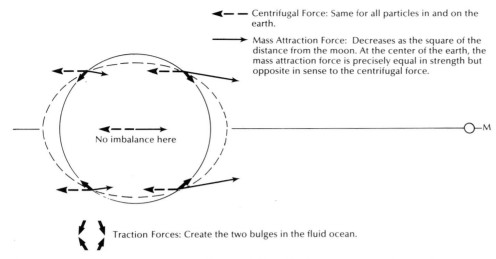

◄ ─ ─ Centrifugal Force: Same for all particles in and on the earth.

──────► Mass Attraction Force: Decreases as the square of the distance from the moon. At the center of the earth, the mass attraction force is precisely equal in strength but opposite in sense to the centrifugal force.

Traction Forces: Create the two bulges in the fluid ocean.

(a) The two forces that can move ocean fluid are balanced only at the center of earth; elsewhere the net imbalance is a small force that causes oceans to converge into two equal "bulges," as shown.

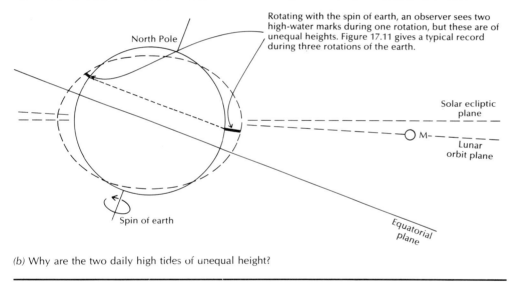

(b) Why are the two daily high tides of unequal height?

Figure 17.10 Explanation of the double-wave tide.

sect" this record into a number of components, each with its own strength and each having a different, but exact, period of repetition; these are called *partial tides*. In 1922 A. T. Doodson listed some 390 different partial-tide components; they range in period from very long, a year or more, to just a few hours. The majority have such small amplitudes that it is impractical to include all of them in a forecasting analysis; usually tide forecasting is done with the four major components.

The strongest tide-producing force is the earth–moon interaction. Its "basic period" is the time between successive appearances of the moon at its zenith above a fixed point on the earth's surface; because the moon orbits the earth in the same sense of rotation as the earth's spin, this time period is slightly more than one solar day, about 24.84 hours. Now, because a fixed point on earth will move through *two* high–low stands of sea level during one earth's spin, the double-wave tide, the basic period of this partial-tide component is 12.42 hours, called the semidiurnal M_2 partial tide. Table 17.2 lists the major partial-tide components, their exact periods, and their relative strengths compared to that of M_2. For example, the second-strongest semidiurnal component is the earth–sun interaction. It is called the

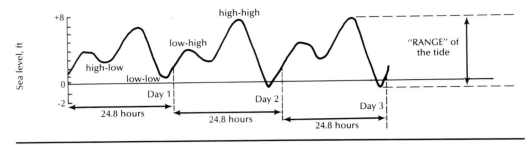

Figure 17.11 A graph recording of three diurnal tide cycles at a Pacific coastal station.

S_2 component, is 46% as strong as the M_2, and has a period of precisely 12.0 hours.

All these principal components are related to the sun–earth, moon–earth, or combined interactions. The strongest diurnal partial-tide component is the luni-solar component K_1 at 58.4% of M_2 but with a period of 23.93 hours. Notice that the four principal components, M_2, S_2, K_1, and O_1, together account for more than 68% of the total potential for producing tides.

If we know the strengths and periods of these tide-producing forces, why is it so difficult to predict the tide? First, because the speed of tide waves is controlled by ocean depth, the submarine topography in the vicinity of a given shore point controls the strength and time of arrival of each component wave at the measuring station. Each shore point has its own unique topographic vicinity, so each will have a different "set" of partial-tide amplitudes and times of arrival and each a different "composite tide" behavior.

Second, the size and shape of an ocean basin control the resonant oscillations of water in the basin. Called "seiching," this phenomenon is responsible

Table 17.2 Partial harmonic components of the composite tide

Name of Partial Tides	Symbol	Speed, degrees per mean solar hour	Period, solar hours	Coefficient Ratio, $M_2 = 100$
Semi-diurnal Components				
Principal lunar	M_2	28.98410°	12.42	100.0
Principal solar	S_2	30.00000	12.00	46.6
Larger lunar elliptic	N_2	28.43973	12.66	19.2
Luni-solar semidiurnal	K_2	30.08214	11.97	12.7
Larger solar elliptic	T_2	29.95893	12.01	2.7
Smaller lunar elliptic	L_2	29.52848	12.19	2.8
Lunar elliptic second order	$2N_2$	27.89535	12.91	2.5
Larger lunar evectional	ν_2	28.51258	12.63	3.6
Smaller lunar evectional	λ_2	29.45563	12.22	0.7
Variational	μ_2	27.96821	12.87	3.1
Diurnal Components				
Luni-solar diurnal	K_1	15.04107°	23.93	58.4
Principal lunar diurnal	O_1	13.94304	25.82	41.5
Principal solar diurnal	P_1	14.95893	24.07	19.4
Larger lunar elliptic	Q_1	13.39866	26.87	7.9
Smaller lunar elliptic	M_1	14.49205	24.84	3.3
Small lunar elliptic	J_1	15.58544	23.10	3.3
Long-Period Components				
Lunar fortnightly	M_f	1.09803°	327.86	17.2
Lunar monthly	M_m	0.54437	661.30	9.1
Solar semiannual	S_{sa}	0.08214	2191.43	8.0

for the *standing-wave* nature of tides. Each basin is configured differently, amplifying specific components and dampening down others.

But why are there so many components that make up the total tide-producing force? The answer is composed of two parts.

1. Mathematically, we can show that the strength of a tide-producing force varies inversely as the cube of distance from the earth to the body. Because the moon's orbit is an ellipse with the earth at one of the foci (Figure 17.12*a*), its tide-producing force is 18% greater at perigee (when the moon is closest to earth) and 15% smaller at apogee (when it is farthest from earth) than its mean value. Throughout the lunar month, then, the intensity of the M_2 component changes—this is the source of the lunar monthly component M_m (Table 17.2) and other, higher-frequency components related to the lunar elliptic. The earth's orbit about the sun is also elliptic with the sun at one of the foci, so the change in separation distance between perihelion, when the earth is clos-

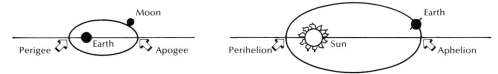

(a) The orbits both of the earth about the sun and of the moon about the earth are elliptical. But the earth's ellipse is eccentric and changes on a cycle of about 100,000 years, becoming almost circular at times.

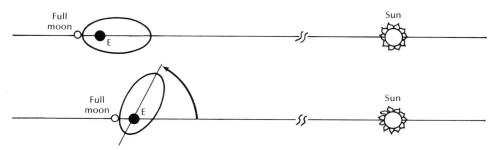

(b) The strength of the moon's tide-producing force varies greatly over the 8.85-year period during which the moon's elliptical orbit cycles. Spring tides occur just after the moon is new or full. When the full moon is closest to earth, spring tides have their largest ranges.

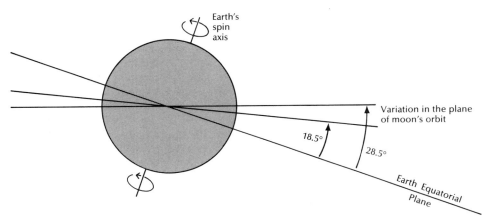

(c) The plane in which the moon orbits the earth changes or nutates from 18.5° declination to 28.5°, making a full cycle every 18.6 years.

Figure 17.12 Some examples of how the complexity of interplanet forces is responsible for the many partial tide components listed in Table 17.2.

est to the sun, and aphelion when it is farthest from the sun, is also a source of partial-tide components.

2. The elliptical orbits keep changing, for both the earth about the sun and the moon about the earth. The moon's orbit is by far the most powerful in generating partial-tide components. The axis of the moon's orbit rotates around the earth, making a full cycle every 8.85 years. The impact of this orbit change is primarily in the fortnightly partial tide, M_f, and its higher-frequency components. Figure 17.12*b* diagrams this phenomenon; when the sun, earth, and moon are aligned, as in the *full* and *new* phases of the moon, the tide-producing forces of the sun and moon are additive. They force the spring tides. But the moon's contribution to the forcing of spring tides varies over the 8.85-year cycle; it is greatest when the moon is closest to the earth, and the three align, along the short *O–E* line in Figure 17.12*b*.

In addition, there is an 18.6-year cycle in the inclination of the plane of the moon's orbit. Relative to the earth's equatorial plane, the moon's orbital plane oscillates between a minimum of 18.5° and a maximum of 28.5° (Figure 17.12*c*).

Tide Prediction. In spite of its complexity, the tide can be predicted with practical accuracy. Every seacoast village has at least one store that sells booklets of local ''tide tables'' in which the times and heights of high- and low-tide marks are given for as much as a year in advance. How are these tables made? The National Ocean Survey (formerly the U.S. Coast

and Geodetic Survey), at one time in the past, recorded the tide level at a shore point, analyzed the record for its principal components (amplitude and timing), and then set these data into a tide simulation model. When programmed into the future, this model produced the desired prediction tables. Early simulation models were mechanical, constructed of pulleys and eccentric cams of different diameters that simulated the strength of the tide components, with a cable that interconnected all to ''sum'' all components. Once set up for a specific coast station, the machine was motor-driven into ''future'' tide cycles while recording the predicted tide. Another technique is to build a scale model of a coast region, Figure 17.13, and ''flood'' it with simulated tide waves; the model then solves its own complex tide prediction problems.

Tide Waves Are ''Standing Waves.'' With crest-to-crest distances spanning half the circumference of the earth, tide waves behave as though the 4-km average basin depth is *very shallow* water. To such waves, the continental margins appear as vertical walls, and so most of the energy in the tide waves is simply reflected back and forth across the basins. Such behavior supports the formation of standing waves. The horizontal motion of the fluid stops when sea level is highest and lowest; these periods are called ''high-water slack'' and ''low-water slack,'' respectively. In between the periods of slack water are the periods of the most rapid horizontal flow: the ''flood'' tide occurs about midway between lowest-

Figure 17.13 Narragansett Bay tide simulation model. The physical model is built to a scale of 1:100 vertical and 1:1000 horizontal. An automatic tide generator reproduces normal tide waves at the entrance to the bay, and water level is then recorded at various times and places within the bay. Studies are made of the effects, under various tide conditions, of man-made barriers and channels on tidal circulation, pollution and flushing, and sedimentation. (U.S. Army Corps of Engineers.)

The tidal bore in Truro, Nova Scotia. Under certain conditions river outflow keeps the incoming tide from moving upstream until the tide wave at the mouth of an estuary reaches a height that overwhelms this local outflow. The incoming water then proceeds upstream as a single, roiling, hydraulic phenomenon called a tidal bore. (Allen Rokach/National Audubon Society-Photo Researchers.)

water and highest-water levels, and the "ebb" tide currents occur as high water moves toward the next low water.

We observe this behavior of tides almost everywhere. With so many partial-tide components available, every ocean basin is capable of producing the wave reflections that reinforce at least one component into a strong standing wave. For example, the Pacific Ocean resonates at both the diurnal and semidiurnal periods; tide records on Pacific coasts therefore have the typical "mixed-tide" appearance shown in Figure 17.11. The Atlantic, on the other hand, has a muted diurnal behavior but quite a strong semidiurnal standing wave. By comparing the periods of high tide at many stations around the Pacific basin, we can show how the crest of the standing wave appears first in the Western North Pacific and 12 hours later in the middle of the Eastern North Pacific (see Figure 18.3).

Spring Tides and Neap Tides

A discussion of tides would be incomplete without a fuller description of the spring–neap tide cycle. Early humans observed that the range of tides varied from a maximum, which they called the "spring" tide, to a minimum, which they called the "neap" tide, and that a full cycle, spring to neap to spring, required about 15 days, the period the English came to call the fortnight (Figure 17.14a). We would conclude

that tide currents are strongest during spring tides and weakest during neap tides. During spring tides, the water level will both reach higher up the beach and expose more tide pools.

The explanation for the fortnight cycle of tides is the periodic alignment of the earth, moon, and sun (Figure 17.14b). Such alignment occurs twice during each lunar month of 29.5 days. When the earth is between the moon and sun and the moon is full or new, spring tides occur because the double-wave crests generated by the earth–moon pair coincide with the double-wave crests generated by the earth–sun pair. When the sun and moon are in quarterly positions relative to the earth (shown in Figure 17.14b), the double waves are separated, and this is a period of neap tides.

If early humans noticed the spring–neap cycle regularly and in fact used the phenomenon to keep time in units of fortnights, the animals of the sea might similarly "use" the cycle to some purpose. The classic example of such an adaptation is the spawning behavior of a small (about 15 cm long) West coast fish called the *grunion*. During spring and summer these fish choose the highest high-tide point just after the spring tides for spawning far up on the slope of California beaches. Further, the fish sense the "highest" waves during the brief hour of highest high tide and ride the swash of these highest waves to gain the uppermost reaches of sand above mean sea level. There the female wriggles a depres-

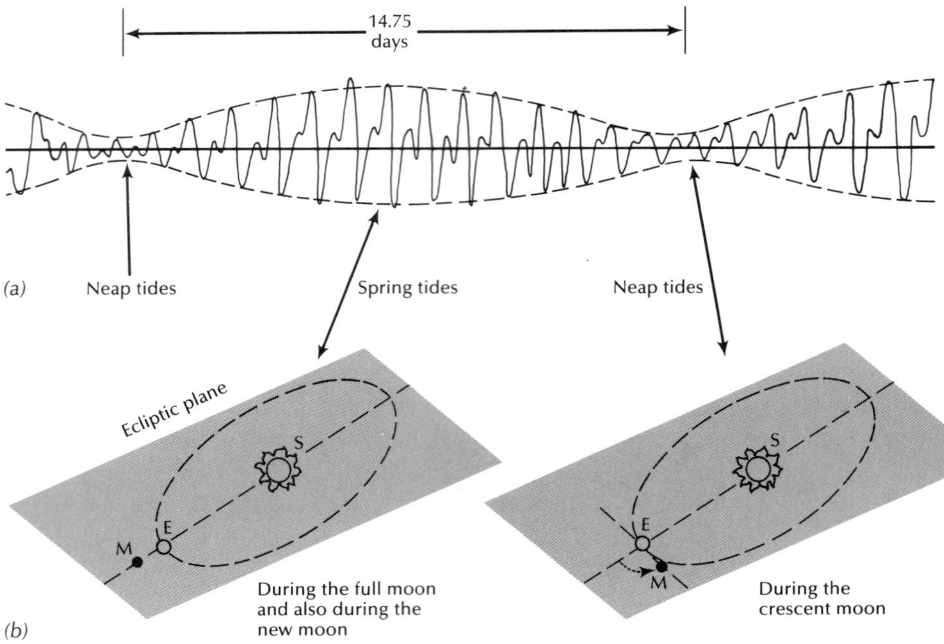
14.75
days
(a)
Neap tides
Spring tides
Neap tides
Ecliptic plane
S
E
M
During the full moon
and also during the
new moon
S
E
M
During the
crescent moon
(b)

Figure 17.14 The spring tide–neap tide phenomenon.

sion into the sand and releases her eggs, to be followed by the male, who releases his milt into the depression. As the next wave swashes up the berm, the two fish fight their way back to the safety of water. Thus the eggs are deposited safe from predators in the water, and the warm sun incubates them. Just before the incoming highest high tide and waves of the following spring tide, about 14 days later, the eggs hatch. As the new waves swash up the beach to their highest reach, the fry wriggle down to the sea to assume their aquatic life-style.

Summary

Putting the phenomena of waves and wavelike motion into perspective is a task made very difficult by the wide range of characteristics of ocean waves. We can summarize wave motion into a spectrum graph that is useful in showing which motions are wind-driven, which result from tide-generating forces, and so on. The following are the principal types of waves discussed.

1. ***Capillary and very short gravity waves.*** Capillary and short gravity waves are the very shortest of waves created by winds moving over fluid surfaces. They result from and at the same time produce friction between wind and water.

2. ***Winds waves.*** Wind waves are the waves we always thought we knew best, for the obvious reason that they are always there to be studied. Yet it required the work of many people over many years to decipher the intricate coupling between the wind field pressure pattern and the rate at which wave energy grows. The final sea state is determined by three wind–water relations: wind speed, the period of time that the wind blows, and the fetch over which it blows. A surprising finding of wave research is that long swell waves can propagate energy across 10,000 km of ocean from the point of generation.

3. ***Storm surge.*** A special wave called a storm surge can occur when the wind-driven Ekman transport over very shallow water coincides with high tides and high wind waves. These have caused extensive loss of life and property in areas like the Gulf coast of Texas and coastal Bangladesh on the Bay of Bengal

4. ***Tsunamis.*** Tsunamis are trains of waves generated by seismic upheavals on the ocean floor. Traveling at speeds over 700 km/h, they carry destructive force to distant shores.

5. ***Tides.*** Of all the waves we see, tides have always fascinated us. Once it was understood that there are a number of tide-generating forces, each created by the orbit relations between the earth and another planetary body, it was thought that future tides could be predicted mathematically. We now appreciate the degree of complexity introduced by the variable topography of coastal areas. Tides can be predicted only by building a successful model of tides in one locale and extrapolating the behavior of this model into the future.

Study Questions

1. Where within the overall depth range of a column of ocean water do we encounter waves?

2. In a simplistic sense, what is it that waves *do*?

3. What is the difference between *sea* and *swell*?

4. What are the forces that (a) generate capillary waves and (b) dissipate them?

5. What is meant by the statement "ordinary wind waves have periods ranging from 5 to perhaps 30 seconds"? In what way are very short waves of lengths ranging from 5 to 30 cm different from ordinary wind waves?

6. Very short wind waves, combined with capillary waves, play a distinct and important role in dissipating the energy that the ocean fluid gains from the wind fluid. Explain.

7. Where over the world oceans are we most likely to find the *highest* waves of all? Where over the world oceans are we most likely to find the *longest* waves of all? Briefly, tell why we are most likely to find both wave characteristics in the same place.

8. Briefly, describe how winds transfer energy to surface waves.

9. As a follow-up to question 8, describe the three inherent *constraints* to the process of wave building by winds at sea.

10. What is a storm surge?

11. Defend the statement "It is the *form* of the surface wave, a type of distortion in the sea surface, that moves *along* the surface; there is no net translation of water itself."

12. Under what conditions does a surface wave *feel* the presence of the ocean bottom, and what happens to the wave when it does?

13. What aspect of ocean tides is the most difficult to accept: (a) That it is a very *long* wave? (b) That it is a very *shallow* wave? (c) That it has *two* highs and *two* lows during a time period slightly longer than one solar day?

14. What is meant by the term *diurnal inequality*, which is used in describing the behavior of the sea level at a given point on a shore?

15. If it is true that one point of strongest tide-generating force occurs directly below the moon at the zenith point, why is it that the local highest water usually lags behind in time, sometimes up to several hours?

16. What is the difference between a *tide current* and the motion of a water particle under the influence of a wind wave?

17. What is meant by the statement "The actual tide measured at a point is a composite of several partial tides"?

18. Doodson showed that there are hundreds of partial-tide components. Why are there so many? (This is a generic question—you need not explain them all.)

19. Why is it so difficult to predict tides? After all, we can usually purchase booklets of precise times and heights of tides in every seacoast village.

20. What is a tsunami? Is it a long wave or short wave? At what speed might such a wave travel? Would you know if one passed you while you were on the high seas?

CHAPTER 18

COASTAL OCEAN STUDIES

In describing the nature of the fluid ocean in Chapter 2, we divided it into seven provinces. Three of these provinces make up the boundary area between the open ocean and the continent, an area defined more or less as that between the 1000-m-depth contour and the beach. The three coastal provinces are province IV, the margin zone, the waters over the shelf and upper slope; province V, the energy dissipation zone, the open shore between high and low tides; and province VI, the estuarine zone, where the river meets the sea. All these provinces are of special importance to us. Margin and estuarine waters provide the major part of our seafood. The energy dissipation zone is where most of us "meet" the sea for recreation, and the estuarine zone is where all rivers meet the sea.

Waters here are subject to extremely complex physical and chemical processes. The multiplicity of biological niches found in coastal marine waters accounts for much of the speciation found in them. Not only are ambient conditions complex, but their variations cover much larger dynamic ranges than those of the open sea. Perhaps marine scientists who specialize in coastal topics should be called "coastal oceanologists"; it is a course of study quite different from the oceanography we have studied to this point.

How Do Margin Waters Differ from the Open Ocean?

Physical, chemical, and geological processes not yet studied in this course, processes that in turn control the biology of coastal ecosystems, make margin waters much different from the open ocean. Among the physical processes are those that control the "energetics" of the area. The foremost of the chemical processes are linked to the "flux of organic and inorganic materials" from lands to oceans. The geological processes are those that "shape" the boundary itself. To be complete, the process list should contain anthropological factors, the ability of *Homo sapiens* to alter coastal systems using mechanisms that operate in physical, chemical, or biological ways.

Energetics of Ocean Boundary Zones

Continents as Barriers to Large-Scale Ocean Motion

The major influence of the continents on large-scale ocean motion is to *block east–west flow*, to *create the gyre patterns*, and to *force rapid global heat transfer to the poles*. Recall the *symmetry* that exists among the gyre currents in the major basins (see Figure 11.1). Each central gyre has an east-flowing current along its poleward side, tied directly to the belts of westerly winds. Examples are the West Wind Drift (WWD) of the South Pacific and the North Pacific Current in the Northern Hemisphere. These flow head on against the South and North American continents, respectively, whereupon they each divide to feed a pair of *eastern boundary currents*: the Cape Horn and Humboldt in the Southern Hemisphere, the Alaska and California in the Northern Hemisphere.

Similarly, the North and South Equatorial Currents of all the major basins are produced by the trade winds of the hemisphere. These currents feed warm surface waters toward the western ocean boundaries and into the intense poleward currents found there, for example, into the Kuroshio of the North Pacific and the Gulf Stream of the North Atlantic.

Differences Between Atlantic and Pacific Coast Studies. Western boundary currents tend to be the most intense. Currents such as the Gulf Stream are narrow and thick; (they are say 100 km wide and can reach the bottom, as seen in Figure 11.2). They flow at surface speeds of up to 4 knots. In contrast, the California Current along the eastern boundary of the North Pacific is many hundreds of kilometers wide, reaches to depths of only about 500 m, and moves at only a few tenths of a knot (0.1 to 0.4 knot; Sakou and Neshyba, 1969).

Such *intensity differences* between eastern and western boundary currents explain much of why studies of the Atlantic seaboard must differ from those of the Pacific coast. For example, biologists at the Scripps Institute of Oceanography in La Jolla, California, have learned that the community of plankton living near this coast is also present hundreds of kilometers out to sea (McGowan, personal communication, 1978). These data support the physical picture of a planktonic community dispersed throughout a broad ocean current off this shore.

Differences in Temperatures of Boundary Currents.

The fact that equatorward-flowing boundary currents are *cooler* on the average than poleward-flowing boundary currents at the same latitudes (see Figure 6.5) is another reason that coastal oceanography is so varied a science. A classic example is the impact of cool eastern boundary currents on the climate of adjacent coasts. The severe deserts along the coast of North Africa, the Namibian desert in southern Africa, and the deserts of Peru and Chile are all related to the presence of cool ocean waters along these eastern boundaries. Moreover, meager rainfall also means little river outflow, so that the nearshore waters off these desert regions are chemically more isolated from the outflux of materials in continental drainage than is typical of most coastal regimes. Because there are fewer estuaries, the biological system here must also be different from those of other boundary zones. Many species of animals found as adults in coastal waters are dependent on estuaries during some part of their life cycles (review Figure 12.7).

Coastal Margins Dissipate the Energy of Ocean Motions

A review of the basic patterns of surface winds over ocean basins (Figures 10.5 and 10.7) might make us wonder whether, with the winds continuously "spinning up" the oceanic gyres, the speeds of ocean currents should also continuously increase. The fact that we do *not* observe such a speedup means that some mechanism is now removing en-

ergy from ocean motion as fast as the winds are inserting it. In the steady state the energy gained must equal what is lost, and the place where the energy of waves, tides, and currents is dissipated, by and large, is in the shallow coastal margins.

Friction Between Continents and Boundary Currents.

The concept of how continents exert a "friction drag" on ocean motion is drawn in Figure 18.1 using the North Atlantic basin as the setting. The pattern of surface winds, from the zone of easterlies at low latitude to the westerlies around 40°N, tends to spin up the interior ocean in a clockwise sense of rotation, as marked on the sketch by the dashed circles tagged with a ($-$). When all the distributed clockwise spin-ups are summed over the ocean basin, the result is the large-scale gyre of circulation we studied in Chapter 11. And because the wind is continuous, it is continuously applying this *clockwise torque* to the North Atlantic.

Figure 18.1 also shows idealized east and west boundaries, together with their boundary currents: the slow, broad Canary Current along the eastern boundary and the rapid, narrow Gulf Stream against the western shore. Because of friction the fluid flow immediately next to these boundaries is zero. But farther offshore the flow rate increases until it reaches the rate that is "normal" for each of these currents. Consequently, the boundary friction introduces a *counter clockwise torque* into the fluid, as indicated in Figure 18.1 by solid circles with a ($+$) tag. This friction-induced ($+$) torque at the solid boundaries is the mechanism most responsible for offsetting the wind-generated ($-$) torque.

But note also in Figure 18.1 that the ($+$) torque introduced by friction along *the western boundary* is many times greater than its eastern counterpart. In 1948 Henry Stommel showed that this condition has to exist because of still another source of torque, the increase in the Coriolis force in the poleward direction. Consider this example: if we start at the equator with a column of ocean water that has no spin at all and push the column directly northward, it would begin to spin clockwise ($-$) faster and faster as it approached closer and closer to the pole. At the pole itself, the water would spin one complete revolution in one solar day, right? And the more rapidly we push the column northward, the more rapidly does the Coriolis effect "torque" or accelerate the column in the clockwise ($-$) sense; this relation between speed of northward push and torque is linear, 1:1. But if the moving column has frictional contact with a continent during its northward push, the generation of counterclockwise ($+$) torque increases

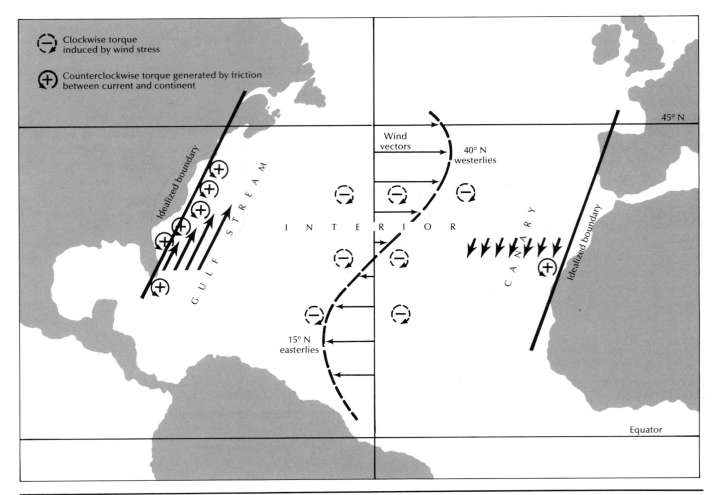

Figure 18.1 Friction between large-scale coastal currents and the rigid continental boundary generates "eddies" that spin with a rotation opposite that imparted to the ocean by the large-scale wind patterns. In the North Atlantic the Gulf Stream supplies most of the compensating "spin down" because it is the intense western boundary current for this ocean basin. The numerous eddies gener- ated in the friction zone between the boundary currents and the continent are a form of large-scale turbulence that transports river runoff waters across the continental shelves to the open sea. Each major ocean basin has its own intense western boundary current that performs this essential "spin down" function.

with speed at a much faster rate, say, 2:1. Stommel then reasoned that the ocean would organize itself with the most rapid flow on the western boundary so that the flow would not only counteract the Coriolis effect but have enough (+) torque left over to offset the wind effect as well! This, like the Ekman invention, was another "oceanmark" event because it explained why intense western boundary currents like the Gulf Stream exist.

Friction Between Continents and Tide Currents. All water particles inside the water column, top to bottom, *respond equally* to the tractive forces that gener-

ate the tides, as shown in Figure 18.2a. Because the period of a tide cycle covers many hours, we view these displacements as real "currents." Notice also, Figure 18.2a, how these currents are amplified in intensity when the tide wave sweeps up the slopes and onto the shallow waters of the continental shelves.

To get a better picture of how tide currents behave near shore, let us look at an actual data record from a current meter installed to measure shelf currents, as shown in Figure 18.2a. Typically, the machine records fluid speed and direction of flow every few minutes; later, we average the recordings over

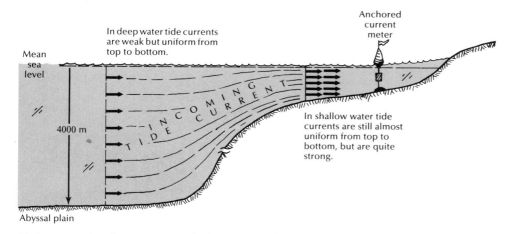

(a) A cross section showing vectors of tide currents in deep and shallow water.

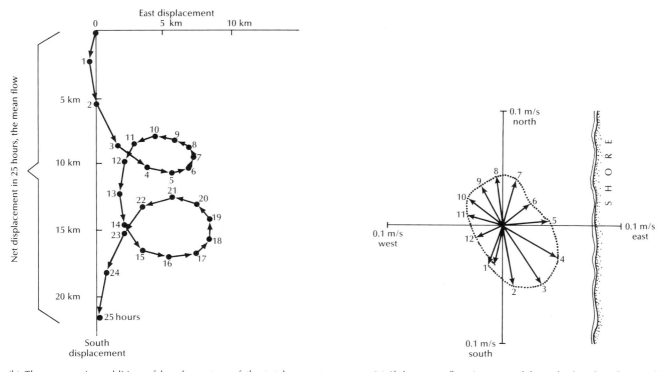

(b) The progressive addition of hourly vectors of the total current, measured by a current meter anchored over a shallow continental shelf. Numbers 1 through 25 are successive hours for one diurnal tide cycle. The loops are formed when strong tide currents are present in addition to a mean flow.

(c) If the mean flow is removed from the hourly values and the remaining numbers are plotted from a common origin, the result is the "tide ellipse" that exists at this measurement point.

Figure 18.2 An analysis of tide currents.

hourly intervals to produce numbers that we interpret as the *net* displacement of fluid during each successive hour of a tide cycle. Because tide currents ebb and flood, shifting direction during the tide cy-

cle, we expect that a plot of successive hourly numbers will indicate how the current over a continental shelf changes during a tide wave cycle. Figure 18.2b shows this behavior quite clearly; here we plot the

The extremely high range of tide at Hopewell Cape, New Brunswick. Exposed during low tide are the undercut portions of cliffs, estimated at three times the height of the persons walking. Waves superimposed over the wide

range of high-to-low water level created the undercutting. The change in shadow direction suggests about 6 hours between high and low points, a semidiurnal tide wave.

successive hourly current measurements as vectors connected one to the other, and the presence of strong tide currents is revealed as telltale "loops" in the vector diagram. Here two full loops in 25 hours shows that the tide wave is *semidiurnal*, having two cycles within this period.

We interpret the graph in Figure 18.2*b* as composed of a "looping" tide current combined with a "mean" flow. A coastal oceanographer may choose to replot these data to emphasize the tidal part alone. To do this, we subtract one twenty-fifth part of the total displacement of fluid during the 25-hour measurement period from each of the hourly vectors. The residual vectors then represent only the tide currents; when plotted alone, these produce the characteristic "ellipse" pattern of tide currents in shallow waters that is evident in Figure 18.2*c*. It indicates the *relative strength of the tide current alone,*

hour by hour. The strongest tide-produced flow occurs during the third hour of measurement; its speed is about 0.6 m/s, in contrast to the mean flow rate of 0.23 m/s. Notice that the major axis of the tide current ellipse is oriented more or less parallel to shore, which is typical of tide currents over the continental shelf. (If a similar study is made of tide currents in an estuary, however, the "ellipse" is usually found collapsed into a line diagram that represents a flood current up the estuary followed by an ebb current in the opposite direction.)

For these reasons the coastal oceanographer must include local tide effects in all studies of mixing processes over the shelves as well as within bays and estuaries. The concept described in the previous chapter, that tides differ from one shore point to another, can be expanded to a global picture (Figure 18.3), to show the extreme differences in tide behav-

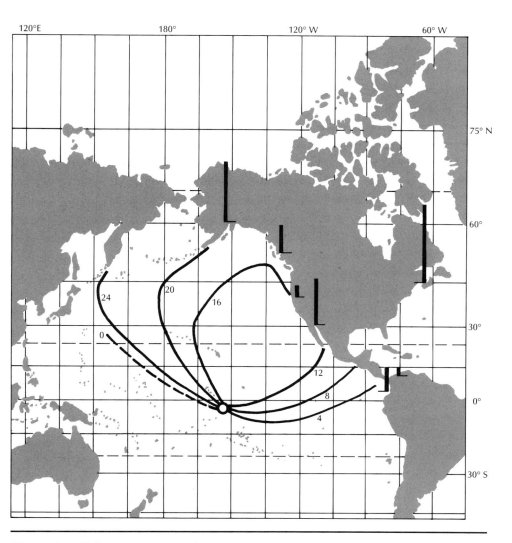

Figure 18.3   Tide waves are standing waves in ocean basins. Because of their long
wavelengths, their energy is reflected from the basin walls, and the waves themselves
oscillate back and forth within the basin, often building up amplitude with each
swing. The buildup becomes especially evident at the head of a long, shallow inlet; in
Cook Inlet, Alaska the range reaches as much as 10 m. A similar buildup at the head
of the Gulf of California makes tide ranges of 8 m, and the most spectacular of all is
found in the Bay of Fundy, Nova Scotia, where the difference between successive high
and low tides may be 15 m. Along open coasts tide ranges are much more moderate;
for example, San Francisco has a range of about 3 m.

Because tide waves are standing waves, their movement across a basin is quite pre-
dictable. A *node* usually occurs at a point somewhere within the basin, where the
range of a tide component is zero, and around which the wave progresses much like
the spokes of a wheel. Here the node for the diurnal tide is located south of Hawaii,
and the position of the crest of its wave is charted at successive 4-hour intervals along
each of a set of curves. This wave "rotates" around the Pacific basin in a counterclockwise sense. (Based on Defant, 1961.)

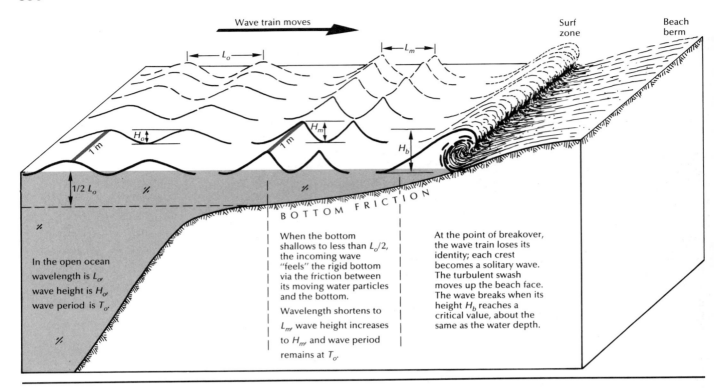

Wave train moves

Surf zone

Beach berm

L_o

L_m

H_o

1 m

H_m

1 m

H_b

1/2 L_o

BOTTOM FRICTION

In the open ocean wavelength is L_o, wave height is H_o, wave period is T_o.

When the bottom shallows to less than $L_o/2$, the incoming wave "feels" the rigid bottom via the friction between its moving water particles and the bottom.

Wavelength shortens to L_m, wave height increases to H_m, and wave period remains at T_o.

At the point of breakover, the wave train loses its identity; each crest becomes a solitary wave. The turbulent swash moves up the beach face. The wave breaks when its height H_b reaches a critical value, about the same as the water depth.

Figure 18.4 How a wave train is modified as it progresses from the open ocean toward the beach. In the open ocean the wave feels no frictional contact with the bottom. Once frictional contact begins, the wave changes in several ways simultaneously: the wave's speed drops, its height increases, and its direction shoreward becomes more and more perpendicular to the beachline. At the end, each wave crest becomes a solitary, turbulent toss of water up the beach face. Two characteristics of the wave train remain the same throughout the process: the period of time for successive crests to pass an observation point is always the same, and the rate at which energy moves toward the beach is constant.

ior over the global shoreline. Friction[1] between tide currents and the shallow floors of continental shelves, bays, and estuaries provides much of the energy for mixing coastal waters; more will be given on this later.

Dissipation of Wind Wave Energy in the Surf Zone. Most of all wave energy in the sea is in wind waves, and much of it is carried eventually to some

shore where it dissipates in the turbulent, chaotic breaking of waves in the surf zone. Here we describe how the open-ocean wind wave is *modified* as it moves into shallower water and onto the beach. In another section we study the geological processes in which the energy released by waves plays a key role.

Suppose that a train of swell waves approaches a shoreline in such a way that the crests are parallel to the beach (Figure 18.4), and we fix our attention on a portion of the train that lies between two points spaced 1 m apart along the crest lines. We can calculate the amount of energy that the wave train moves toward shore. Its *energy density E* is described by

$$E = \tfrac{1}{8}\rho \, g \, (H)^2 \text{ joules/m}^2$$

where ρ is the water density in kilograms per cubic meter, g is the gravitational constant 10 m/s^2, and H is the wave height in meters. A wave 2 m high has an energy density of 5000 joules/m^2 (see Table 17.1

---

[1]A discussion of tide-induced friction between ocean and continent would be incomplete without mentioning the classic physics problem that links this process with the phenomenon that the period of the moon's orbit around the earth is gradually lengthening. Groves and Munk (1958) analyzed global ocean tide data to show that the net torque exerted upon the earth by the moving tide waves is directed westward, slowing the earth's rotation rate, and would alter the moon's orbit period about 11 seconds per century. This amount is very close to that calculated by astronomers based on historical data collected since the seventeenth century.

for the energy density in even high waves). A good way to evaluate how much energy this is would be to calculate how many light bulbs the energy in the train could power continuously. If the wave train delivers its energy toward the beach at a speed of 5 m/s, a normal condition, 25,000 joules/s are delivered; this is 25,000 watts, enough power to keep 250 hundred-watt bulbs burning continuously, or enough to light about ten average classrooms. Bear in mind that *each* 1-m-wide strip of crest line delivers this amount of energy to the shore.

We follow this process in Figure 18.4 by first noting how the open-ocean wave train is *modified* as it approaches the beach, and then how each wave of the train becomes an individual, solitary breaker to deliver its power to the surf zone. In the open ocean, each wave in the train has a length L_o meters, a height H_o meters, a period T_o seconds, and an energy density E_o joules/m^2. Subscript o identifies a wave characteristic that is not yet modified enroute to the beach. Closer to the beach, after the wave begins its modification, subscript m applies, and in the surf zone of breaking waves we use subscript b.

Recall now from Figure 17.7 that a surface wave begins to feel a friction drag against the bottom when it enters water of depth less than about half its own wavelength, $L_0/2$. Friction with the bottom forces water particles into ever-flatter elliptical orbits as the water depth shallows. It also reduces the speed of the particles, and this in turn forces the foremost wave crest in the train to slow down, which then results in a shorter crest-to-crest length of the wave, L_m, as shown in Figure 18.4.

The total energy that each wave is delivering to the beach must stay constant, however. In the 1-m-wide strip of Figure 18.4, the total energy between successive wave crests is calculated as the product of energy density E and the wavelength L. Then the open-ocean value $E_o \times L_o$ must be equal to the product $E_m \times L_m$ for the modified wave. But because L_m is shorter than L_o, this means that E_m must be greater than E_o, and the only way this can occur is if the wave height increases from H_o to a *new and larger value* of H_m. Thus for observers at a beach the approaching crests seem to "grow upward" out of the water just before breaking in the surf zone.

Eventually, as bottom friction continues to slow the wave's speed and increase its height, the wave is distorted into an "unstable" shape. It becomes too "peaked" and begins to spill forward at the crest in the first phase of its final breakup. The mechanics of the breakup are complex. Just how the wave behaves in the final few seconds of its existence depends on several factors: the slope of the bottom and the slope of the beach itself, the shape of the wave while still in open water, and the nature of the substrate, whether rock, gravel, or sand. Studies of breaking waves are now usually carried out in ocean engineering programs and are mainly concerned with forces on structures such as piers, with the protection of shores against erosion, and with the wave-induced movement of sediments.

Chemical Aspects of Coastal Oceanography

Virtually every marine scientist who studies coastal problems becomes involved, sooner or later, with some aspect of the flux of materials, suspended, dissolved or floating, that moved from lands into the coastal seas. Here we examine the topic in an introductory way. Where does the material come from, how much of it crosses from land to ocean and in what forms, and does the human factor play a significant role?

Freshwater Runoff

The most important chemical species that crosses the coastal boundary zone is water, H_2O. Marine biologists know that the annual cycle of river runoff volume controls seasonal changes of salinity or turbidity of the coast ocean, a signal to which many marine organisms respond. It can also alter the temperature as well as the density stratification of shallow waters and in this way it affects vertical mixing and with it nutrient replenishment. In short, runoff cycles impact coast ocean production at all trophic levels.

Figure 18.5 shows where the major freshwater fluxes occur. The basin boundaries in this chart are stylized to emphasize the latitudes and sizes of river outflow. Only twelve rivers worldwide make up a full one-third of the total runoff (Table 18.1), and of these the Amazon and the Congo together have more (18.5%) than the other ten combined (15.8%). Most of the total runoff is into the tropics and subtropics of all ocean basins. It enters the Pacific and Atlantic mainly along their western boundaries, but it enters the Indian Ocean along its eastern boundary (Baumgartner and Reichel, 1975). The reader is urged to compare these data with the distribution of watersheds, Figure 4.11.

Dilution of coastal oceans by riverine fresh water can be quite strong. For example, there are stories of

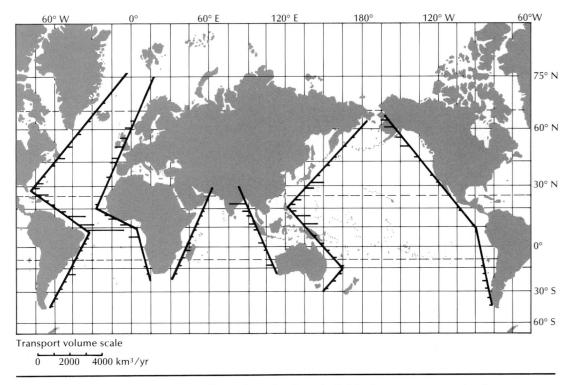

Figure 18.5 The major freshwater fluxes from lands to individual oceans summarized over 5° bands of latitude. Coastlines are idealized for simplification for each of the major basin boundaries except the North Polar Ocean. The length of the bars are proportional to the river outflow within the latitude band. (Based on Baumgartner and Reichel, 1975.)

how merchant ships *off* the mouth of the Amazon could replenish freshwater stores simply by filling buckets with surface water. The usual case, however, is that surface water at a river's mouth is already brackish because mixing with seawater begins *inside* the estuary (as described in a later section). Figure 18.6*a* shows how surface water salinity is as high as 23 ppt at the mouth of the Hudson River estuary and increases rapidly to 31 ppt at 10 km offshore. Coastal oceanographers find it useful to trace these dilution contours because dissolved chemicals carried by the river fluid can often be assumed to follow the same dilution pathways as does salinity. For example, if Hudson River water were "red ink" instead of fresh water, one would find a gradual change of color, from red to shades of pink to loss of all color, in direct correlation with the increased salinity contours of Figure 18.6*a*. Biologists find this correlation especially useful in tracing the dispersion of trace element pollutants across a shelf region, especially because the measurement of salinity is easier or more accurate than direct measurement of the trace element concentrations.

The New York Bight into which the Hudson River empties is extensively studied for the impact of industrial and other wastes dumped from barges into offshore shelf waters (Figure 18.6*a*). (A "bight" is an open bay formed by a large-scale bend in the coastline.) Investigators rely on direct current measurements that yield mean flow as well as tidal excursion. The tidal excursion is large at the mouth of the Hudson estuary, about 8 km. Researchers also rely on analytical models in which the effects of tide currents, mean currents, wind-driven currents, and vertical and horizontal mixing are all simulated by means of the mathematical equations of fluid motion. Once the *equation* model is combined with a numerical description of the actual boundaries to the bay or estuary, it becomes a "simulator" that models the behavior of real fluids.

One of the early applications of such a model was to predict the "flushing times" of different sectors of

Table 18.1 Major rivers of the world, grouped according to the ocean basin into which each drains. (Compiled from various sources.)

Basin	River	Dissolved Load, million tons/yr	Delta Area, km²	Sediment Load, million tons/yr	Average Annual Discharge, m³/s
Atlantic	Amazon (Brazil)	200	100,000	363	180,000 (1)‡
	Congo			65	42,000 (2)
	Orinoco (Brazil)			87	28,000 (4)
	Rio de la Plata (Argentina)				19,500 (7)
	Mississippi	140	30,000	340	17,545 (8)
	St. Lawrence				10,400 (16)
	Rio Magdalena (Columbia)				8,000 (18)
	Olenek-Pyassina (USSR)				3,810 (35)
	Nile (via Mediterranean)	20	20,000	111	1,584
	Columbia	43		33	6,650 (21)
	Danube	38	4,000	82	6,450 (22)
	Niger (Nigeria)				5,700 (23)
	Sao Francisco (Brazil)				3,300 (31)
	Grij.-Usumacinta (Mexico)				3,265 (33)
Atlantic via Arctic*	Yenisei (USSR)				19,600 (6)
	Lena (USSR)		45,000		16,400 (9)
	Ob-Irtysh (USSR)				12,600 (13)
	MacKenzie-Peace (Canada)				7,500 (19)
	Yukon†	22		88	7,000 (20)
	Pechora (USSR)				4,060 (24)
	Kolyma				3,800 (28)
	Northern Dvina (USSR)				3,560 (30)
	Khatanga (USSR)				3,280 (32)
Pacific	Yangtze Kiang			499	35,000 (3)
	Mekong (Vietnam)			170	15,900 (10)
	Amur (USSR–China)				12,500 (14)
	Yellow (Hwang Ho; China)			1,887	16,750 (31)
	Xi Jian (China)	100		1,600	11,000 (15)
	Song-Koi (Vietnam)				3,900 (26)
	Fraser (Canada)				3,750 (29)
Indian	Brahmaputra (Bangladesh)			726	20,000 (5)
	Ganges (India)		80,000	1,455	15,000 (11)
	Irrawaddy (Burma)			299	14,000 (12)
	Godavari (India)				3,980 (25)
	Indus (Pakistan)			435	3,850 (27)
	Mahanadi (India)				2,940 (34)
	Zambesi (Mozambique)				2,500 (37)

*A subgroup of rivers discharging into the Atlantic discharge first into the Arctic basin.

†The Yukon is grouped with the Arctic drainage rivers because of proximity to the Bering Strait.

‡Numbers in parentheses order rivers according to annual discharge volume.

the shelf waters between Cape Cod and Chesapeake Bay (Figure 18.6b), one of which is the New York Bight sector (Ketchum and Keen, 1955). Flushing time is defined as the average time taken by a water parcel to move from a freshwater source (estuary) across the shelf to the limit of the open ocean (here taken as the shelf break or 200-m-depth contour). Using a combined river inflow of 5000 m³/s as an average freshwater input, these investigators ob-

tained an average flushing time of 500 tide cycles, about one and one-half years. These are useful guides in evaluating environmental impact.

By way of contrast to the Hudson River, note the extended plume of the Columbia River in Figure 18.7. A combination of considerable river flow plus a narrow continental shelf extends the 31-ppt salinity contour up to 180 km offshore, many times more than the 10-km extent of the same salinity contour

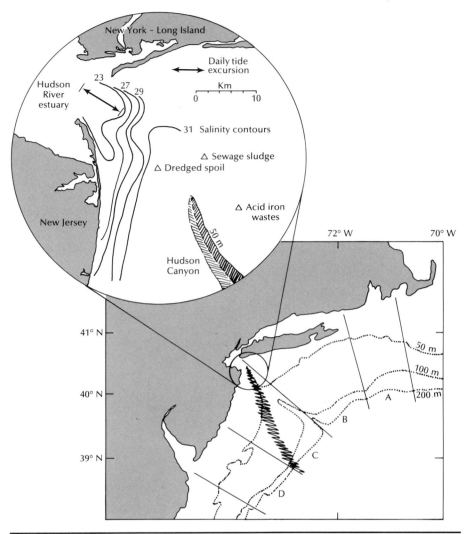

(a) Contours of surface water salinity off the mouth of the Hudson River estuary, New York. Salinity grades from a brackish 23 ppt inside the estuary to 31 ppt in a distance of 10 km. About the same distance is covered by the daily excursion of tide currents, which are the principal energy source for this mixing. Pollutants that are dissolved in the water follow the same dilution path as does the fresh water. In the past, solid wastes were dumped closer to the head of the Hudson Submarine Canyon.

(b) The zigzag feature models the movement of a parcel of river water across the shelf; each zigzag represents a tide cycle. Ketchum and Keen (1955) studied the "flushing" time of the region by dividing the shelf into segments and arrived at an average of 500 tide cycles needed to move river water out of the shelf zone into the open ocean. Because the tide here is semi-diurnal, 12.4 hours per cycle, the flushing time is therefore about *1 year* to cover a *net* displacement of about 220 km.

Figure 18.6 How freshwater moves from river mouth across the shelf zone to the open sea.

in Figure 18.6*a* for the New York Bight; flushing time for this West coast system is correspondingly much shorter.

The Flux of River-Borne Materials into Coastal Zones

Table 18.1 also lists eleven rivers that are major carriers of land erosion products, together moving almost 5 billion tons of suspended sediments to the sea each year. Judson (1968) estimates that the amount of river-borne soil sediment worldwide has increased from 9 billion tons/yr *before* the introduction of agriculture to 24 billion tons/yr today. It is interesting to compare this last number with the 240

million tons/yr estimate of fish production oceanwide (Table 1.2). These numbers are in the ratio of 100 tons/yr of sediment transport to the coastal zone to each ton of fish produced per year.

Although there are no definitive experiments to link directly the sediment load and the amount of fish produced, there are examples indicating clear correlations. Before the Aswan Dam was constructed on the Nile River in Upper Egypt, a healthy fishery for sardines existed off the river's delta. But the sardine catch began to diminish significantly within a few years after the dam began capturing silt; the delta itself began to be eroded away as the Nile no longer carried enough silt to sustain it against ocean erosion by waves and currents.

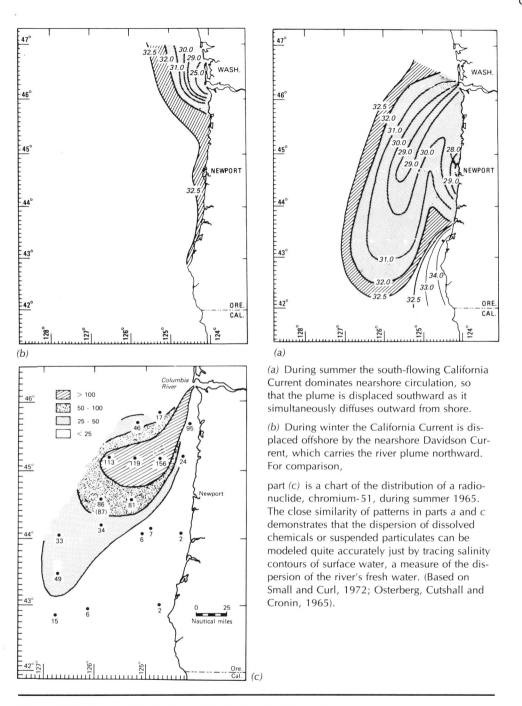

(a) During summer the south-flowing California Current dominates nearshore circulation, so that the plume is displaced southward as it simultaneously diffuses outward from shore.

(b) During winter the California Current is displaced offshore by the nearshore Davidson Current, which carries the river plume northward. For comparison,

part (c) is a chart of the distribution of a radionuclide, chromium-51, during summer 1965. The close similarity of patterns in parts a and c demonstrates that the dispersion of dissolved chemicals or suspended particulates can be modeled quite accurately just by tracing salinity contours of surface water, a measure of the dispersion of the river's fresh water. (Based on Small and Curl, 1972; Osterberg, Cutshall and Cronin, 1965).

Figure 18.7 Seasonal behavior of the Columbia River plume.

Lithogenous Material in Suspension. Lithogenous particles are inorganic and usually in the form of crystals. They are derived from the erosive weathering of rocks by chemical decomposition or mechanical breakup. Feldspar and quartz minerals make up the coarser fractions of suspended materi-

als in rivers such as the Columbia (Table 18.2). These are moved along the river bottom to eventual discharge at the river mouth, where they become a source of material for beaches along the shore. The finer fractions, such as the chloritic and micaceous materials tend to be very small in size and platey or

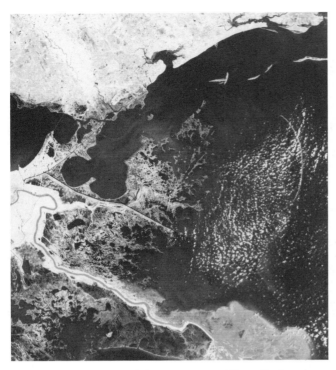

Satellite photograph of the Mississippi River discharging into the Gulf of Mexico. The plume is seen clearly in the lower right portion where the river gradually leaves its channel to mix its sediment-laden water with the clear seawater. Eddies caused by tides are outlined near the center of the photograph. Offshore top right are several barrier islands.

flat in shape. These remain in suspension much longer than the coarser sands and are thus deposited farther out to sea as well. Fine particles of clay minerals, such as the kaolin of Table 18.2, tend to *adsorb* other particles and dissolved ions of sea salts, so that the particles actually increase in size once the river water begins to mix actively with seawater. The par-

ticles flocculate into clusters and settle rapidly to the shelf bottom or within the delta being constructed by the river.

Flux of Dissolved Materials. Rivers carry dissolved materials in addition to their suspensoids (Table 18.1). For most of these rivers, the tonnage ratio of dissolved to suspended materials is about 1:3. Using the Columbia River again as an example, Table 18.3 lists the major chemical constituents and the tonnage of each that are discharged to the Northeast Pacific.

How can we evaluate this dissolved load in terms of its contribution to nutrient fertilization of coastal waters? One way would be to compare the amounts of *river-borne nutrient* with that delivered into a typical coastal upwelling region by *wind-forced upwelling*. Working with the nitrate numbers in Table 18.3, we calculate first that the Columbia River delivers 5100 g of nitrate per second into its coastal waters. From upwelling studies we know that the velocity at which deeper-layer waters are "pumped" *upward* in a wind-driven upwelling is about 0.01 cm/s, quite small. Given that the nitrate content of such deeper water is, say, 25 μmol/liter (see NH-105 data as an example, Table 8.1), over what area must vertical pumping at this rate take place to produce the same nitrate injection as does the river outflow? The answer is about 22.0 km^2, or equivalently about 5 km of coastline in an active, wind-driven upwelling region. River discharge by itself is therefore not an important source of phosphates and nitrates to support primary productivity in coastal waters, but the mixing energy it delivers to the process called *river-induced upwelling* is very important, as is shown in the discussion of mixing in estuaries given later in this chapter.

River-borne dissolved materials are unevenly dis-

Table 18.2 Particle characteristics of suspended lithogenous material in the Columbia River. (Based on Conomos and Gross, 1972.)

Lithogenous Component	Relative Size, microns of diameter	Morphology
Quartz	Large	Compact (equidimensional)
Felspar	All sizes	Compact
Lithic fragments (rock)	All sizes	Irregular
Chloritic	Small (0.15–0.40)	Platy
Kaolinitic	Small (0.12–0.20)	Platy
Micaceous (includes biotite and muscovite)	Generally large	Platy
Illitic	Small (0.05–0.08)	Platy

Table 18.3 Annual chemical input from the Columbia River into the Northeastern Pacific Ocean

Chemical Constituent	Annual Delivery, million metric tons/yr
Phosphate, PO_4	0.0095
Nitrate, NO_3	0.1610 (5100 g/s)
Silicate, $Si(OH)_4$	3.2640
Carbon dioxide, CO_2	
Alkalinity	22
Water, H_2O	6650

tributed in coastal zones by seasons. This is strikingly shown in Figures 18.7*a* and *b* which chart surface salinity off the mouth of the Columbia River during two seasons. If we assume that other dissolved materials are conserved as salt is, the patterns shown here hold also for their dispersion off the coast. During winter months the river plume hugs the shoreline to the north of the mouth. Spring is a transition month, and summer months show strong southward transport of the river plume (recall the results of our two oceanographic cruises in Chapter 8).

The Human Factor. Humans are increasingly impacting the coastal oceans. Figure 18.7*c* indicates, for example, the distribution of radioactive chromium-51 within the summer plume of Columbia River water. The atoms of chromiun were made radioactive by the neutron bombardment of the upper river water when it was used as a reactor coolant in the Hanford, Washington nuclear test facility. Although the contaminant distribution data given are for the summer of 1965, its distribution is very similar to that of freshwater dilution in the summer 1962 plume (Figure 18.7*b*). (For a more detailed discussion of radioactive wastes in coastal waters, see Chapter 19.)

Geological Aspects of Coastal Oceanography

Waves and currents of the coastal oceans shape the edge of the continent. In turn, the shape of a shoreline exerts a strong control over the nearshore wave regime and nearshore currents. We examine first the process of *wave refraction,* in which nearshore bottom topography controls how wave energy is distributed along the shore. Next we describe the phenomenon of *littoral transport,* which moves beach sands along the shoreline, then how the wave-induced *onshore–offshore* transport of sands seasonally changes the appearance and texture of a beach. Finally, the concept of a *budget* of sands and sediments into and out of a specified sector of a coast combines all these geological processes in the theory of "dynamic cells" of coastal sand transport.

Wave Refraction

Earlier we described how the speed of a surface wave decreases as it moves into continuously shallowing water enroute to a beach. We now look at this process in plane view to see its geological aspects.

Figure 18.8 shows a long, straight shoreline for which the offshore water depth contours are also straight and parallel to the shore. Assume that a train of uniformly crested surface waves approaches the straight shoreline at the angle α, as shown. This figure demonstrates how the wave crest lines are "bent" as they approach shore, so that the crests are almost parallel to the beach when they finally enter the surf zone. This is called wave refraction. It occurs because that part of a wave crest that first feels bottom friction, for example, at the 20 m-depth contour, is forced to reduce speed while the seaward portion of the same crest continues its approach at normal speed. Each crest line is *warped* so that it approaches the beach in an almost parallel way. An observer watching waves break on the beach seldom knows from which direction the wave train originally entered the coastal zone.

An incoming, uniform wave train distributes energy uniformly along the straight shore. This is illustrated in Figure 18.8 by constructing a pair of "ray lines" spaced one unit of crest line distance apart in open water, *x–x*, and noting that the *energy between these adjacent rays must be conserved as the train approaches shore.* This energy dissipates on shore between points *x'–x'.* Careful measurement will show that the distance *x'–x'* is slightly greater than *x–x*, in proportion to the angle at which the train arrives. Further, the energy dissipated in each adjoining beach segment of size *x'–x'* is the same because all parts of the wave train are equally *refracted.*

If the shoreline is not uniform, however, the energy in breaking waves tends to be concentrated upon headlands and dispersed within bays. The headland shore distance *x'–x'* in Figure 18.8 is much smaller than its counterpart *x–x* offshore. This means that wave energy *density* is concentrated on the headland in the ratio *(x–x)/(x'–x'),* a ratio greater than 1. In contrast, the distance *x'–x'* where a similar pair of rays intersects the shore in a bay is much

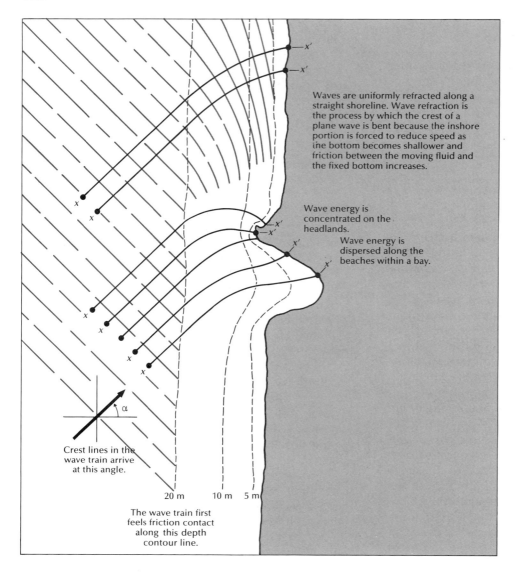

Figure 18.8 Relations between an approaching wave train and shorelines of different shapes. When shoreline topography is straight and uniform, wave crests are uniformly refracted and wave energy density is uniformly distributed along the beach. Wave energy density is much higher on headlands because refraction forced by bottom bathymetry causes a convergence of wave energy; conversely, bathymetry in coves and bays causes refraction that forces a divergence along wave crests, so energy density at the beach is much lower.

Waves are uniformly refracted along a straight shoreline. Wave refraction is the process by which the crest of a plane wave is bent because the inshore portion is forced to reduce speed as the bottom becomes shallower and friction between the moving fluid and the fixed bottom increases.

Wave energy is concentrated on the headlands.

Wave energy is dispersed along the beaches within a bay.

Crest lines in the wave train arrive at this angle.

20 m   10 m   5 m

The wave train first feels friction contact along this depth contour line.

larger than its *x–x* origin, so that energy density delivered inside the bay is sharply reduced.

If the headlands are eroded rapidly because of the wave energy concentrated there, and if coves and bays are either weakly eroded or tend to fill with deposits of sands, the net result should be *straight* shorelines, given enough time for these processes to work. Real shorelines are not so straight, but rather appear as a succession of rocky points or headlands separated by broad indentations of the shoreline. Bays and coves often contain low-elevation coastal valleys and river mouths (Figure 18.9). We are led to conclude that headlands erode slowly, in spite of wave energy concentration, and that erosion of the shoreline segments between such points has stabi-

lized. Moreover, if the balance of erosion and deposition is stable, we should be able to study the budgets of sediments into and out of segments of the coastal zone.

Budget of Coastal Sediments: The Concept of Coastal "Cells"

What are the sources of sands to a coastal sector? Clearly, rivers are the primary source at all times. The erosion of sea cliffs along the shore, particularly where the continental structure is subsiding, is an occasional source.

As coastal investigators acquired more and better data on coastal sediment movement, it became clear that balanced budgets would not work for *arbitrarily* chosen sectors of a coastline. Instead they found a definite "cellular" structure to sediment movement (Figure 18.9) in which rivers supply sediment, wave action causes a drift of sediment *along* the shore, and the cell terminates at either a rocky headland or a submarine canyon. In this cellular system of sediment *source*, *transport*, and *sink*, the ultimate sink has to be the abyssal floor of the adjacent ocean basin; an exception can be made for deposits in deltaic structures where the continuous loading of a coastal sector causes a subsidence in place, as in Figure 18.9.

River Transport. Table 18.1 lists the suspended loads carried by some major rivers. Similar data are

compiled for all riverine imports, large and small, when budgets are made for specific coastal sectors.

Shore Cliff Erosion. Shore cliff erosion is included in the coastal cell outline of Figure 18.9 because geological uplift of a region or a transgressing sea level can expose older shoreline to new wave erosion (see also Figure 5.9).

Littoral Transport. The refraction example assumes a straight shoreline and a uniform bathymetry offshore. Although not stated as such, the example also assumes that the shelf itself slopes away very slowly from the beach, so that the process of refraction is perfect and all incoming waves reach the surf zone parallel to the beach. But real shorelines are not per-

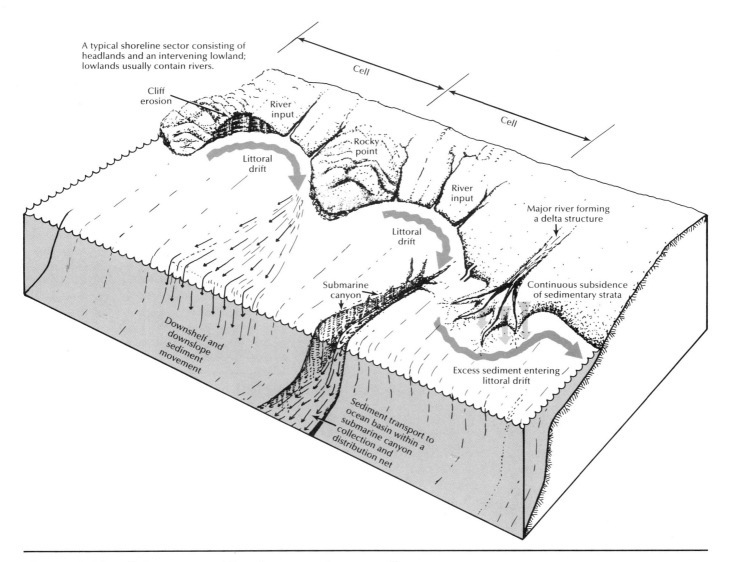

Figure 18.9 The cellular structure of littoral transport along a coastline.

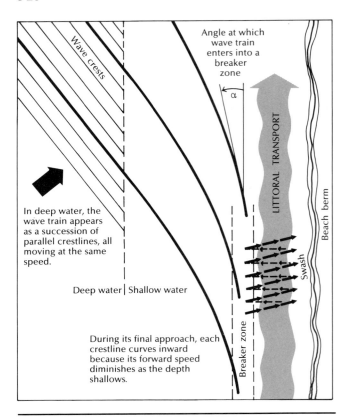

Figure 18.10 How littoral sand transport *along* a coastline is related to the angle at which the incoming wave trains approach the beach. When a wave enters the breaker zone at an angle, the accompanying swash moves up the beach face at the same angle, ∝, shifting sand laterally along the beach face. However, the return swash is grav-ity-controlled, and sand returns directly seaward. The re-sult is that each wave contributes a small, net displace-ment of sand in the alongshore direction.

fect; the rate at which the bottom shallows toward shore is too steep for perfect refraction to occur. The waves therefore enter the breaker zone at an oblique angle (Figure 18.10).

Because water particles at the crest of a breaking wave are moving in a direction perpendicular to the crest line, the resulting *swash* of water up the beach face is at a small angle to the beach face. But note that the *backwash* begins its movement back down the beach face in a direct *downhill* direction. As the backwash picks up speed, it also reentrains sands that dropped out of the swash phase, moving these sands downhill. Each wave cycle produces a net, small displacement of sands in a direction parallel to the beach. This is called *littoral transport* or littoral drift. The more oblique is the angle of arrival of

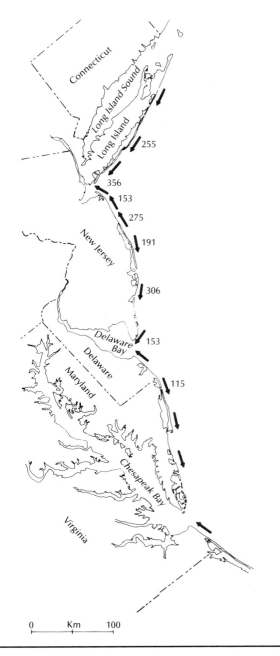

Figure 18.11 Littoral transport volumes for sectors of the Atlantic seaboard in thousands of cubic meters per year. Transports off the central New Jersey coast diverge: to the the north sands are moved toward the mouth of the Hudson River estuary; to the south sands move toward the entrance to Delaware Bay. (After Komar, 1976.)

waves in the surf zone, the more energetic is this littoral transport.

Littoral transport can change direction as the sea-sons change the direction from which the dominant wave trains approach a shoreline, but there is usually a net transport direction over the annual cycle. Beaches of Oregon and northern California have a net *southward* littoral transport.

Different sectors of the same shoreline may have littoral drifts in different directions, as shown in Figure 18.11 for a section of the New Jersey coast. It is possible that the same wave regime that produces the northwest transport along northern New Jersey also produces the southwest transport along Long Island, for these two shorelines would have a different orientation to the same incoming wave train.

It is instructive to compare the sizes of littoral transports in Figure 18.11 with the suspended sediment loads of major rivers (Table 18.1). The New Jersey transport numbers range from 115,000 to 365,000 m^3/yr. Using an average density of 3 for sand grains, we can compute the equivalent tonnage as ranging from about 350,000 to just over 1 million tons per year. In contrast, major rivers carry much larger amounts, up to more than 1000 million tons for the Ganges River. The reader may conclude from this comparison that only a part of the material injected into coastal waters actually participates in the littoral drift. The remainder can be deposited far enough offshore to be outside the possible manipulation of waves. Finer-size particles will remain suspended for longer periods of time, and also be de-

posited directly on the floor of the outer shelf (see also Figure 5.9).

Role of Submarine Canyons in the Coastal Cell. Recall from Chapter 5 that the submarine canyons that cut across the slope and shelf of a continent's margin are a major route by which coastal sediments are moved away from the margin onto the abyssal plain. There are at least two known mechanisms by which the downslope transport occurs. Scientists at the Scripps Institute of Oceanography have dived into the Scripps submarine canyon to find virtual "sandfalls," continuous streams of sand pouring into the canyon (Figure 18.12). The process constitutes a permanent loss of sand from the coastal cell. Alternately, sediment may accumulate and periodically move by turbidity currents.

Role of Headlands. Not all coastal cells lose sand and sediments via a submarine canyon. A rocky headland or promontory is a natural "block" to the continuity of littoral transport of beach sands (Figure 18.9). As the transport process encounters a headland and attempts to move sands out and around, it loses its identity in the deeper water and the entrained sand load settles out, usually so far onto the shelf proper that it will never reenter a coastal cell again. Thus rock headlands are another natural limit

Figure 18.12 An unusual photograph taken by divers in the San Lucas Canyon off Baja, California. Streams of sand are seen cascading down a rockfall in the canyon. Such transport can be considered as more or less steady, a continuous loss of sands from the littoral region to the ocean basin. (From Shepard, 1963.)

to the alongshore extent of a single coastal cell. These outer-shelf sediments may eventually encounter a canyon route to the ocean basin; they also accumulate along the slope itself and periodically slump away.

Cyclical Onshore–Offshore Sand Movement. A beach may lose sand to offshore transport during the winter season and regain it all during the following summer. Winter storm waves tend to be more energetic than summer wind waves. When the winter wave breaks, both the swash up the beach face and the backwash are very energetic. Beaches that appeared quite stable during summer months are now subjected to more turbulent energy in the swash–backwash, and so more beach sand is entrained in this up–down movement. The net result is a seasonal shift of sand away from the beach face and into bars located several hundred meters offshore (Figure 18.13*a*) and oriented parallel to the beach.

Summer waves are less energetic. Both the mass of water in the swash phase and the amount of turbulent energy it carries are less than in winter. Sand that is entrained by waves during their final entry into the breaker zone tends to settle out of the swash flow. As the swash water itself percolates into the porous sands of the beach and makes its way back to sea level, its sand load is filtered out onto the beach face and deposited there (Figure 18.13*b*). This is how beaches are reconstructed during the summer.

Coupling between Onshore–Offshore and Littoral Transports. Now we consider the seasonal onshore–offshore shift in beach sand in light of its relation to the littoral transport process. Each season sand *moves across the surf zone* between beach and offshore bar; while in the surf zone, it is also swept up in the alongshore transport prevailing at that time. If there is *net* alongshore flux, at least some of the sands are permanently shifted along the beach. In the typical coastal cell, alongshore sand movement is more or less continuous from river source to canyon sink; a specific beach locale will see as much new sand imported as is lost, so that the onshore–offshore seasonal shift is also continuous (although with different sand grains). But if the sand supply to the cell is interrupted, for example, when sediment is trapped in upstream hydroelectric or flood control dams, a local beach may lose sand in each onshore–offshore cycle, resulting in an incremental loss to the littoral-transport effect.

When a river discharges large amounts of sediment into the nearshore coast areas, and the tide range is relatively small but wind waves are vigorous, the alongshore drift builds special structures called barrier islands (Figure 18.14). An excellent example is the very long barrier island called Padre Island off the coast of Texas; it occurs as a narrow, low-lying strip of sand ranging for hundreds of kilometers. Between this barrier island and the inland marshes is a more or less continuous lagoon, varying in width from a few hundred meters to several kilometers, and connected to the open ocean via occasional narrow slots in the barrier island itself. Such islands are continually shifting position as sand is deposited at one end or edge and removed from the other end or edge. The barrier island whose position has stabilized often builds dunes or is covered with salt grasses.

Spits are similar deposits of sands but grow directly from a projecting point of land to enclose part of an inland estuary or lagoon (Figure 18.14). Generally, such spits grow outward in the direction of the *net* littoral drift.

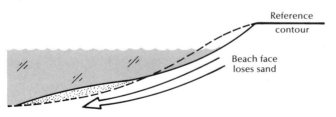

Sand stripped from the beach face is deposited offshore in bars oriented parallel to the shoreline and located seaward of the breaker zone.

(a) The more energetic winter storm waves will remove sands from the beach face and deposit them in offshore bars.

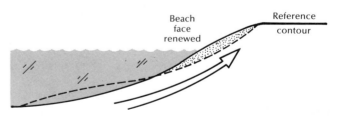

Offshore bars are reduced as waves move sands back toward the beach.

(b) Summer waves reverse the process and move sands back to the beach face.

Figure 18.13 The seasonal pattern of onshore-offshore movement of beach sands.

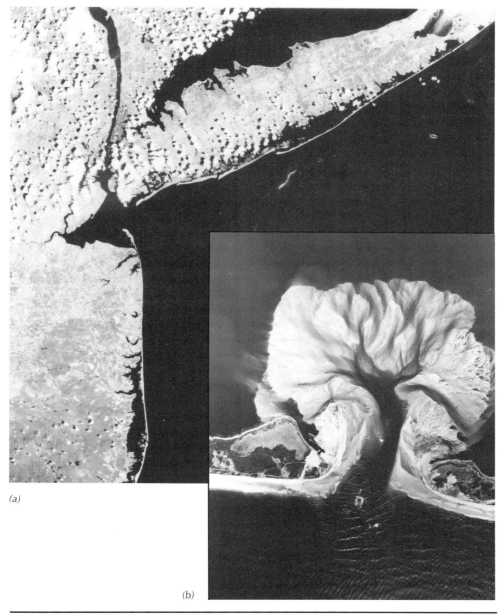

(a) A satellite view of barrier islands along Long Island south to New Jersey. Fire Island National Park is the long barrier strip off the south central portion of Long Island. Sandy Hook, New Jersey, the large sand spit deposited in Raritan Bay by waters from the south shore of the state, is a classic example of how coastal waves and currents deposit. The Long Island and New Jersey barriers and spits should be compared with the littoral drift diagram of Figure 18.11.

(b) An aerial view of the East Moriches gap in the Fire Island barrier, located in the upper right corner of view *a*. Littoral drift is from right to left, across the mouth of the gap. Tide currents surging into the lagoon carry and deposit some of the drift sand in the dramatic pattern shown here.

(a)

(b)

Figure 18.14 Examples of sand spits and barrier islands

Transport Convergence and Loss to Sand Dunes.
Where a cove has been carved out of a shoreline, the angle at which an incoming wave train enters it may be such that there is alongshore transport in the inland direction on *both* sides of the cove. Sands converge at the point of the cove farthest inland. Either the shoreline builds outward, leaving dunes of sand behind, or the prevailing winds may move some sands further inland as mobile dunes. Either way, the formation of dunes represents a sink to the coastal cell.

Biological Aspects of Coastal Oceanography

The coastal ecosystem is remarkably diverse in species because it is the interface between land and ocean, it is subject to extreme, sometimes violent changes in ambient conditions, and it is also the shallow zone within which the benthic food webs actively mix with the photic zone food webs. Stated another way, the coastal zone is extremely bountiful in terms of biological niches. In fact, if today a species diversity suddenly drops within a coastal sector, it is a strong indicator that some recent human activity, or natural disaster, has had a negative impact on the ecosystem.

It is not possible in these few pages to discuss all the reasons that the coastal zone is so biologically rich, even assuming that these were well known. Clearly, the flux of dissolved and suspended materials in river runoff is one nutrient source, but other factors are involved. We need also to examine coastal ecosystems from the view of energetics and to expand on the role of estuaries.

In Relation to Energetics

Earlier in this chapter we discussed how the energy in river runoff, together with the eddies generated by friction between ocean current and continent, moves material *horizontally* across the coastal boundary zone from land to sea. Equally important from the biological view are the strength and location of *vertical* mixing of shelf waters and the vertical overlap of distinct habitat layers. These concepts are illustrated in Figure 18.15.

Overlap of Photic and Benthic Habitat Layers. If we define a surface habitat layer as being the photic zone alone, it reaches to a depth of 100 m or less. If we base the definition on the behavior of large-size members of the photic zone community (for example, the myctophid fish of the deep scattering layer, Figure 3.2), the surface habitat layer reaches to 400-m thickness. A similar argument can assign a thickness to the benthic habitat layer. We assume that in the open ocean these habitat layers do not touch one another except through the one-way transfer of organic detritus from top to bottom. But in the shallower waters of continental margins these layers must eventually come into contact and overlap. Where that contact zone occurs (Figure 18.15), it defines in a systematic way the outer limit of coastal waters, or the working domain of the "coastalographer." In this view, the energetics of the fauna are the basis for vertical overlap.

Other criteria might be used to define the limits of a coastal zone. One could be the intersection between the shallowing continental slope and the maximum depth to which sunlight is biologically sensed. Such a choice is based on the maximum depth to

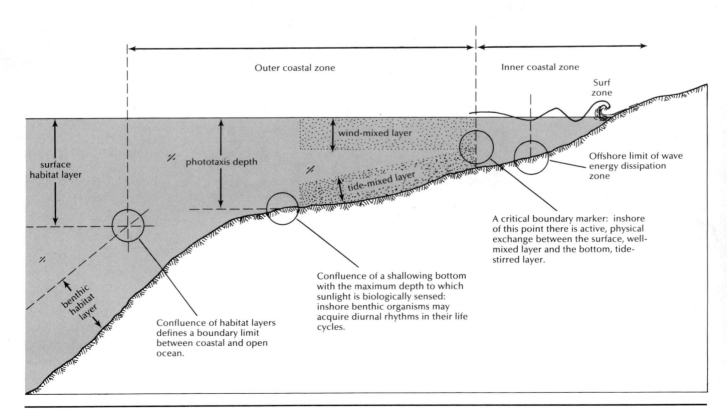

Figure 18.15 Several possible definitions of the boundaries of a coastal zone.

which benthic organisms could cue their behavioral rhythms to the diurnal light cycle; it would range between 500 and 700 m deep. If we choose instead the level at which light intensity falls to 1% of surface value (the rule of thumb measure of the depth to which viable photosynthesis occurs), overlap does not occur until about the 100-m contour, well shoreward of the shelf break (Figure 18.15) Based on these criteria, the seaward extent of coastalography is controlled by the energetics of solar radiation, or light transmission in water, or the process of photosynthesis itself.

Regardless of which overlap criterion is used, its significance is that the *shoreward* benthic community becomes an integral part of the coastal community and vice versa. Niches exist here that cannot exist offshore. For one example, refer to the discussion of the Dungeness crab harvest off Oregon (Table 12.1). Another example is the behavior of some pennate diatoms. They descend to the shallow floor of the continental shelf to spend the winter season in what is termed a "resting phase"; the following spring the cells resume active photosynthesis in the surface layer. We have also discussed, in Chapter 2, how benthic fauna have evolved with one or more surface-feeding larval stages in their life cycles, a supposed response to the high primary productivity in coastal surface waters.

Overlap of Dynamic Mixing Layers. During the late 1970s coastal oceanographers discovered a new type of zonation boundary in shelf waters. As shown in Figure 18.15, this zone marks the confluence between the surface well-mixed layer, in which mixing is dynamically forced by wind and wave energy, and a bottom boundary layer where the energy for vertical mixing comes from the friction of currents (mostly tide) against the shelf bottom. It is an important boundary because here begins a *physical exchange of water mass* between the two layers. Inshore of this confluence the water column is well mixed from top to bottom, but the offshore shelf water will retain a two-layer stratification.

The location of this dynamic mixing zone is usually marked by a sharp change in the temperature of the surface water, this because the bottom water is colder; any mixture with surface water has a lower temperature than the unmixed surface water offshore. This type of structure is called a "frontal zone." An excellent example of such a front was discovered by Schumacher and his colleagues (1979) on the Bering Sea Shelf along a line located about 140

km offshore and in 50 m of water (Figure 18.16a). Hydrographic sections across the front (Figure 18.16b), show how the front divides the Bering Shelf into an *outer coastal zone*, in which the surface and bottom layers form distinctly separate strata, and an *inner coastal zone*, in which the temperature and salinity characteristics are quite uniform from top to bottom, indicating vigorous vertical mixing.

Similar fronts have been found and studied elsewhere. Horizontal currents are usually parallel to the frontal zone, and in a direction that puts the well-mixed inshore water on the right-hand side looking downstream. We also find that these fronts are not very stable; they take on meandering turns and convolutions, and sometimes eddies break off in a process like that shown for the Gulf Stream eddies in Figure 10.11. Physicists and chemists note that such meanders and eddies are also mechanisms by which material flux passes across the boundary from lands to the open sea.

Biologists now recognize that this frontal zone is an important facet of the coastal ecosystem as well. There is increased biological production associated with the fronts. On the outer-shelf side the water column is density-stratified limiting primary productivity because it limits the rate of nutrient replenishment from below. On the inner-shelf side of the front, vigorous vertical mixing enhances primary production in the same way as wind-driven coastal upwelling. Fishermen and fishery specialists know that a strong correlation exists between larger catches of fish and proximity to the front itself.

Frontal Zones Correlated with Fish Populations. Recent studies show that these fronts in shallow coastal waters have key roles in controlling different stocks of Atlantic herring. This herring is rich in stocks, with dozens of them identified in the Northwest Atlantic alone (Iles and Sinclair, 1982). Populations of different stocks vary widely, from as little as hundreds of tons in one stock to millions in another.

Each stock has its own specific spawning time, but only recently have biologists learned that the larval stages of different stocks are also highly localized. The larvae on Georges Bank (Figure 18.17) belong to stock that is distinct from that on Nantucket Shoals or off southwest Nova Scotia. Further, the chart shows that *each stock's larvae are generally clustered within the thermal front* that surrounds each of the shoals. These thermal fronts mark the same type of coastal mixing boundary as described earlier for the Bering Shelf. They are formed by the same

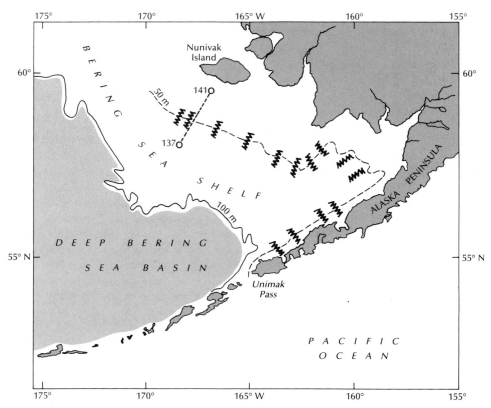

(a) A chart of the Bering Shelf showing the 50-m depth contour along which the front was found, marked here by zigzag lines.

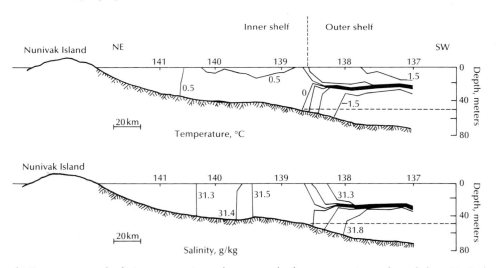

(b) Temperature and salinity cross sections taken across the front on a cruise track south from Nunivak Island. There is a sudden transition between a clearly stratified, two-layer water column offshore of the 50-m depth and the well-mixed water column inshore from this point.

Figure 18.16 Recent discoveries of frontal zones in shelf waters of the Bering Sea (Schumacher et al., 1979).

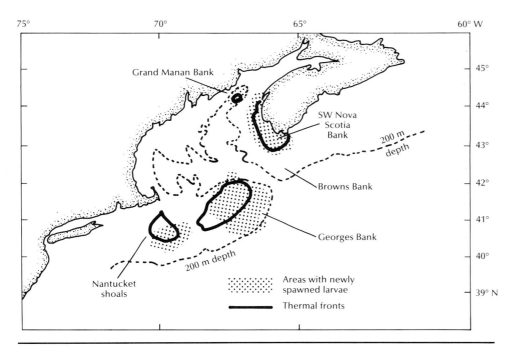

Figure 18.17 Correlation between the distribution of herring larvae shortly after spawning and the shoals of Nantucket, Georges Bank, Southwest Nova Scotia, and Grand Manan. Waters contained within the thermal fronts are generally well-mixed, top to bottom. (After Iles and Sinclair, 1982.)

mechanism of tide-stirred bottom layers mixing dynamically with wind-stirred surface layers. Biologists now conclude that different stocks of herring "home" to their specific frontal enclosures for spawning, but that the juveniles mix together on feeding grounds along with feeding adults. This new knowledge is useful in managing the fishery to avoid overfishing one specific stock. If a stock is eliminated, there is no evidence that a substitute herring stock would take its place over the specific shoals.

Estuaries and Wetlands

Marine science activities in estuaries and wetlands are even more specialized than coastal oceanography in general. There are several reasons for this.

1. Salinity changes rapidly over short distances in estuaries, in both horizontal and vertical directions. This means that animals must be able to move about freely to maintain acceptable salinity in their ambient fluid, or be able to adapt in osmotic pressure.

2. Water circulation patterns inside estuaries vary markedly with the *rate* of river outflow. There is strong seasonal variation in *where* in the estuary mixing of salt and fresh waters takes place.

3. Virtually all estuaries are now heavily altered by human actions. It is almost impossible to learn the true, original behavior of *any* estuarine ecosystem.

4. Many marine organisms that in their adult stage live mostly in offshore waters migrate into and out of estuaries to feed or to spawn; Atlantic menhaden and Pacific herring do this. Migration events are seasonal and are triggered by some characteristic seasonal change, for example, in the amount of freshwater runoff, in temperature, or in salinity (see Figure 12.7).

Water Circulation in Estuaries

Two factors govern the mixing of fresh water and seawater in estuaries: the *range* of the ocean tide at the estuary mouth and the *rate of river runoff* at the head of the estuary. Estuarine mixing ranges from the *well-mixed* type, in which tide range dominates, to the *salt-wedge* type, in which river runoff dominates.

The "Well-Mixed" Estuary. When the range of the ocean tide is large, or if the river runoff flux is low, or when both conditions hold at the same time, the result is an estuary in which the fluid is mixed rather well from top to bottom, but salinity varies in a regular pattern from a normal ocean value at the mouth of 30 ppt or so to 0 ppt at the estuary head. Figure 18.18a illustrates this for the Columbia River during the season when river flow is at its lowest and the Pacific tides at the mouth are large.

The energy for vertical mixing comes from the turbulence generated in the flood and ebb tide currents inside the estuary. The more rapid the tide current, the greater is the degree of turbulent mixing within the water.

We can also view this type of estuarine mixing in terms of the *amount* of fresh water supplied by the river during the same period (Figure 18.18b). The volume of water that the coastal ocean supplies during the cycle must occupy the volume between high- and low-tide levels, called the *tidal prism*. Here it is clear that the tidal prism is substantially greater in volume than the river runoff; flood and ebb currents

are therefore strong. Turbulence generated in these tide currents overwhelms the natural density stratification that would otherwise prevail when fresh water spreads outward over denser seawater. In a sense, we might say that the water column of the well-mixed estuary is dynamically destabilized from below.

In some extreme cases, the incoming flood tide enters as a completely turbulent, roiling wall of water that advances upstream as a *hydraulic wave*, called a "bore."

The Salt-Wedge Estuary. At the other extreme a large amount of river runoff dominates the mixing processes in the estuary. Tide range may be low or high, but the river water itself contains the most dynamic energy in the form of turbulence. As the fresh water meets and overrides salt water, its turbulent eddies "pick up" parcels of salt water and mix these upward into the overlying column of fresh water. This type of mixing, in which the *interface* between two different fluids *is eroded* in only one direction, is called '"entrainment." As we shall see, it has a

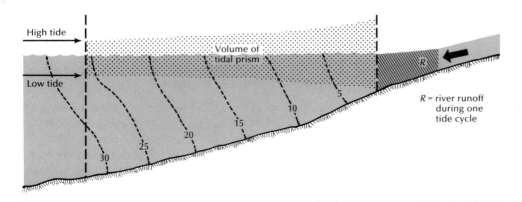

(a) The Columbia River Estuary conditions during the season of lowest river runoff, (in early winter), an example of a well-mixed estuary. Tide-forced mixing is considerable, and salinity values are quite uniform, top to bottom. (From Neal, 1972).

(b) When the range between high and low tide levels at the mouth of an estuary is large, the sea supplies a correspondingly large transport in and out on a daily tide cycle. If this volume change, called the tidal prism, is greater than the volume of river water entering the estuary during the same tide cycle, the mixing inside the estuary is dominated by the turbulent energy in these flood and ebb tide currents. Salinity becomes rather uniform from top to bottom at any one point in the estuary, as shown; further, seawater is uniformly diluted from its open-ocean value to zero salinity at the riverhead.

Figure 18.18 The well-mixed estuary, causes and conditions.

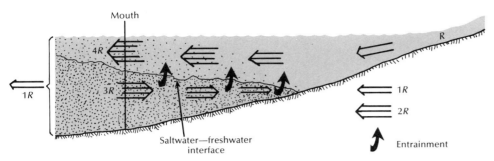

(a) Conditions under which the salt wedge develops. Vertical mixing is dominated by the large amount of turbulent kinetic energy within the river water, which entrains salt water upward to mix with fresh water. The mass of entrained salt water leaving the mouth of the estuary must be balanced by a like mass entering the estuary in its *salt wedge*. Upward entrainment of salt water is sometimes referred to as *river-induced* upwelling.

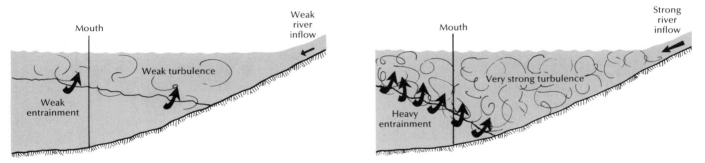

(b) The location of the salt wedge in relation to the mouth of the estuary depends on the relative strength of river inflow. Very high river runoff may force the salt wedge out of the estuary itself, into the adjacent coastal ocean. The Amazon River is an example.

Figure 18.19 The salt wedge estuary.

profound effect on the behavior of the coastal ecosystem.

In the entrainment process (Figure 18.19) R units of fresh water enter the estuary at its head; then a *net* flux of R units must exit the mouth in the same time period. But because the river water entrains salt water enroute, the total volume of brackish mixture exiting the mouth can be much larger than R units. For example, if $1R$ unit of river water entrains $3R$ units of seawater, a total of $4R$ units of brackish water must exit seaward to balance the estuary's freshwater budget. But the $3R$ units of seawater that exit in the brackish mixture must also be balanced by a corresponding inflow of $3R$ seawater units. This explains the peculiarity of the salt-wedge estuary, namely that seawater flows *inward* along the bottom to supply the continuous loss of seawater across the interface with turbulent river water. A "wedge" of seawater penetrates some distance upstream, hence the type name.

Where is the salt wedge located in the estuary? Is the nose of the wedge near the mouth or far upsteam? Answers depend on two factors, the amount of river runoff and the shape of the estuary itself. Runoff in the Amazon is so great that the salt wedge is literally pushed completely out of the mouth of the estuary; here oceanographers describe the extended "plume" of brackish waters in the offshore coastal ocean. For most rivers, however, the nose of the wedge is inside the estuary (Figure 18.19b). During the early summer months when runoff is greatest, the wedge is closer to the mouth of the estuary but is also very vigorous; the rate of entrainment and the flux of seawater into the upstream compensating flow are high. As the river runoff rate drops during the fall season, so also does the supply of turbulent energy for entrainment; the salt wedge then penetrates farther upstream.

The shape of the estuary exerts control over the type of circulation through its influence on the "speed" of river water, hence on its degree of turbulence. Rivers with narrow estuaries and high runoff force the wedge downstream. Wide rivers, even with large runoff, tend to be less turbulent so that the wedge is farther upstream.

A classic example of the salt-wedge estuary is at the mouth of the Mississippi River. Tide range is low in the Gulf of Mexico, 0.5 m or less, river runoff volume is high, and the river is quite wide. Its salt wedge has been found as far upstream as Baton Rouge, a distance of 200 km from its deltaic mouth.

The "Partially Mixed" Estuary. Circulation in most estuaries falls between the two extreme conditions just described. There is some degree of turbulent energy available for mixing in both the riverine layer and in the upstream flow of salt water along the bottom. Stated another way, surface layer salinity increases markedly within the estuary by the time the outflow reaches the mouth, whereas the bottom layer is steadily diluted in the upstream direction.

An excellent example of a partially mixed estuary is the Chesapeake Bay system. This estuary extends 300 km inland, receiving the outflow from a number of separate rivers along the way. Waters are brackish; the 5-ppt salinity contour in its surface waters occurs quite close to the head of the estuary during late summer, as indicated in Figure 18.20. Notice also how these surface salinity contours are oriented obliquely across the channel, giving the appearance that the *upstream salt water flows along its right-hand bank while the downstream brackish water flows along its right bank.* This is precisely the case; the turning effect of the earth's rotation deflects both the upstream and downstream flows to the right of flow direction.

Estuaries in Relation to Coastal Ecosystems

The United States has about 15,150 km^2 of estuarine waters. In 1970, together with the continental shelf waters, these yielded a total fish and shellfish harvest of 5,000,000 tons (Pruter, 1972). We now know that many of the shelf water species taken commercially are dependent on the estuarine waters during some part of their life cycles. McHugh (1966) estimates that 66% of commercial species on the Atlantic shelf are estuary-dependent, and there are even higher estimates for special areas. Using these numbers, we can calculate the *degree* of dependency: for each square kilometer of estuarine waters, the corresponding regional commercial catch is 200 tons/yr. Conversely, each square kilometer of estuarine water that is lost to landfill or pollution reduces annual commercial fish landings by 200 tons!

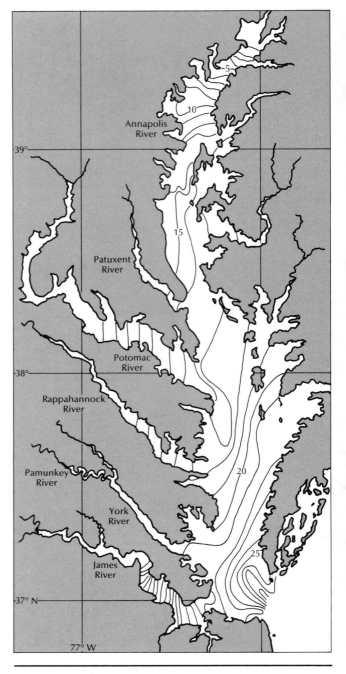

Figure 18.20 The Chesapeake Bay estuary is a fine example of a partially mixed estuary. The typical distribution of surface salinity in the system ranges from 28 ppt at the mouth to 1 ppt near the upper reaches. As can be seen, this bay system is complex, with many river inputs along both sides. The turning effect of earth rotation forces the inflow of the salt water against the right bank; notice how the 20-ppt contour extends obliquely up the right bank. (From McHugh, 1966.)

What is it about estuaries that makes their waters so important to the coastal ecosystem, and in what ways do the coastal fauna make use of them? Answers to both questions lie in two fundamental aspects of estuaries: the extremely high primary productivity in estuarine waters and their unique two-layered, dual-direction circulation system. A more germane question then is why is primary production so high in estuaries? The answer to this question also has two facets. First, the load of *dissolved* nutrients and particulate organic matter carried by the river supports algal production within the river water itself. Second, the turbulence-driven entrainment of seawater from the salt wedge is a type of *river-induced upwelling* that pumps nutrient-rich subsurface seawater upward into the photic zone.

River-Induced Upwelling: The Columbia River Case.

Entrainment along a salt-wedge interface produces the same biological result as does wind-driven upwelling along an open coast, namely enhanced primary productivity. The Columbia River estuary offers an excellent example. Figure 18.21*a* shows how entrainment "pumps" subsurface oceanic water into the estuarine area.

The river itself supplies the estuary with little free nitrate because algae growth upriver has depleted it. In contrast, as shown in Figure 18.21*b*, the nitrate content of the subsurface oceanic water is very high, more than 30 μg atoms/liter. The estuarine area into which the nitrate is entrained is potentially very productive, as documented in Figure 18.21*c*.

However, one important contrast needs to be drawn between the estuarine upwelling and coastal upwelling mechanisms. In the estuary, regardless of the intensity of the entrainment of seawater, the brackish mixture always remains less dense than underlying coastal ocean water, and so its nutrient load remains in the sunlit zone until exhausted. In contrast, wind-driven upwelling forces high-density subsurface fluid directly to the surface; a cessation of wind could allow this fluid to sink back down out of the photic zone before its nutrient load is fully used up. Further, winds are intermittent, whereas the turbulent river entrains day and night and continuously over the seasonal runoff cycle.

Estuarine productivity can be evaluated by comparing the highest value in the Columbia River example, 20 mg C/m^3(h) from Figure 18.21*c*, with the typical coastal upwelling value of 300 g C(m^2)(yr) from Table 1.2. To place these rates in comparable terms, assume a 6-month production season for the Columbia with 10 hours of production per day (daylight hours only) throughout a column 10 m thick; the total is 900 g C/(m^2)(yr) or 9000 kg C/(hectare)(yr), *three times* that for coastal upwelling. An instructive comparison can also be made with land crops. Good Iowa farmland produces 3000 kg of shelled corn per hectare per year; if fed to poultry, the corn produces about 1000 kg of chicken. In contrast, an estuarine conversion efficiency of 20% from plant to zooplankton, together with 20% conversion from zooplankton to commercial catch such as menhaden, would convert 9000 kg C/(hectare)(yr) into about 900 kg per hectare of harvestable fish meat, about the same amount as the chicken fed by an Iowa cornfield. But in addition estuaries yield shellfish commercially, so that the total yield exceeds that of Iowa farmland.

Estuaries as Nursery and Spawning Grounds.

With estuarine waters so highly productive, it is no accident that many coastal ocean species of fauna have evolved to "use" these rich zones during a part of or all their life cycles. Figure 18.22 categorizes the nekton found in estuarine waters into four types separated on the basis of the *degree* and *nature* of their dependency or use of the estuary. A few species of freshwater fish have been taken in brackish water, and very few are thought to be endemic to the estuary. By far the greatest variety of fish captured in inshore estuary waters, and certainly those that are most important commercially, are species that spend only a part of their life cycle in the estuary. These are classed into two types: the anadromous–catadromous species that pass through the brackish estuaries enroute from fresh to salt water or vice versa, and the open-ocean species that migrate into estuarine waters either to feed or to spawn.

The unique circulation within estuaries, in which salt water inflows in the salt wedge and exits via entrainment in riverine water, provides a mechanism by which the larvae and juveniles of many species can crop the high productivity of the estuarine zone for periods of time much longer than ordinary flushing times. The mechanism is illustrated in Figure 18.23 for both spawning and feeding species. Pacific herring migrate into estuaries up to the intertidal zones where they spawn, releasing heavy concentrations of sticky egg masses which attach to marsh grass such as *Zostera* (eelgrass). Atlantic smelt move farther upstream to lay eggs on the bottom in fast-running water zones; alewives similarly spawn bottom eggs but in the quieter waters of small rivers and upstream ponds. Striped bass spawn egg

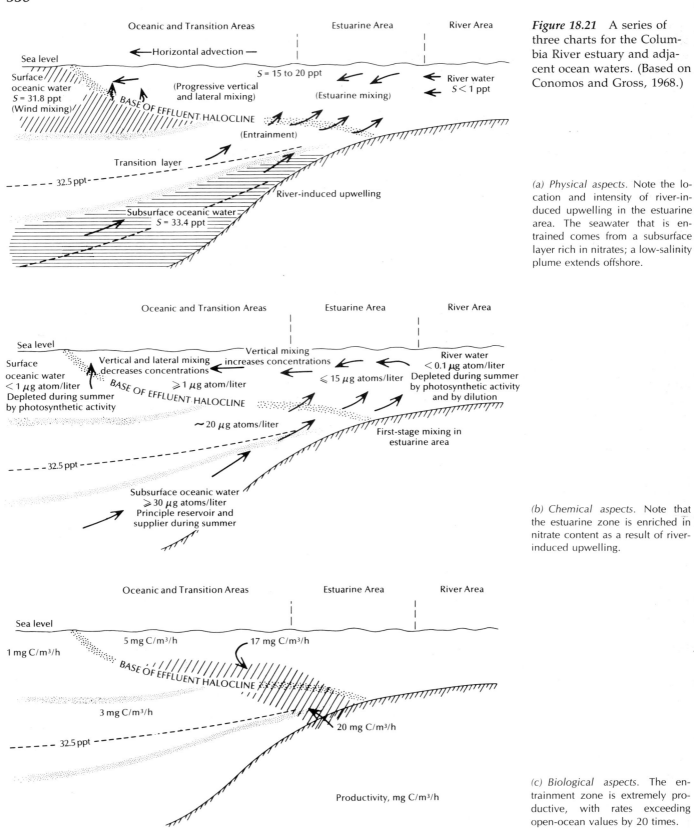

Figure 18.21 A series of three charts for the Columbia River estuary and adjacent ocean waters. (Based on Conomos and Gross, 1968.)

(a) *Physical aspects.* Note the location and intensity of river-induced upwelling in the estuarine area. The seawater that is entrained comes from a subsurface layer rich in nitrates; a low-salinity plume extends offshore.

(b) *Chemical aspects.* Note that the estuarine zone is enriched in nitrate content as a result of river-induced upwelling.

(c) *Biological aspects.* The entrainment zone is extremely productive, with rates exceeding open-ocean values by 20 times.

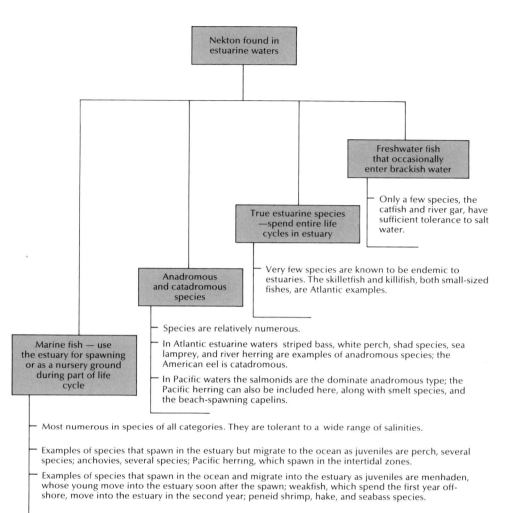

Figure 18.22 The categories of nekton that use the brackish waters of estuaries.

masses just above the saltwater line. Haedrich and Hall (1976) describe their behavior in this way.

The fish spawn in late May or early June above the salt water [Figure 18.23]. The semibuoyant eggs and larvae appear to spend some days or weeks very near the bottom of the river, drifting little as the bottom currents are not strong. By about two weeks of age the young bass are found throughout the water column and begin to drift downstream. As the fish become strong enough to swim feebly, they begin to migrate vertically, moving toward the bottom during the daytime and to the surface at night. Once within the influence of the estuarine salt wedge, the young fish are swept seaward during the night in the relatively fresh surface water, and upstream during the day when nearer the bottom. At about six weeks the fish are able to maintain their position by swimming against the current. Direct competition (among the species) is avoided by the use of different areas and times for spawning.

But regardless of the spawning pattern, inshore or offshore, the time at which many juvenile fishes reach the estuary closely coincides with periods of maximum food production, an apparent evolutionary strategy, so to speak.

Complexity of the River–Estuary–Coastal Ecosystem. By now the reader has acquired an introduction to the complexity of "coastal oceanography"—a rationale, if you will, of why the topic is separately treated in this text. But there is more to the subject that should be mentioned: effect of climatic variations, the physiological "cost–benefit" to the fish of exposure to salinity variation, and the human impact factor.

If the influx of salt water with its rich nitrate load into the estuarine zone becomes larger as the flux of river outflow (and turbulence) increases, multi-year

Adjacent ocean area | Estuarine area | Riverine area

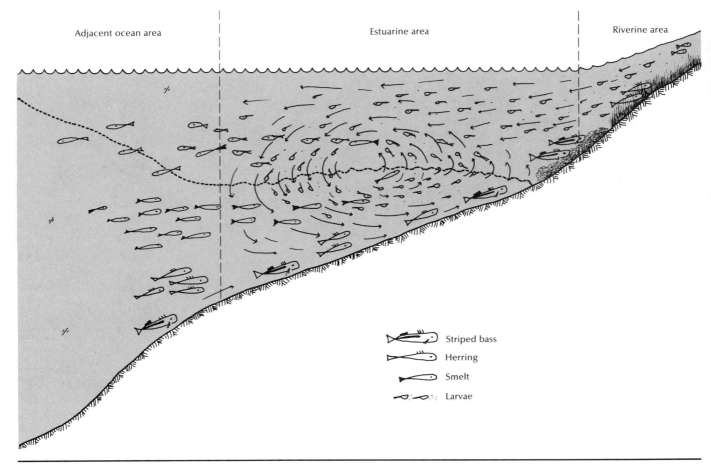

Striped bass
Herring
Smelt
Larvae

Figure 18.23 Small animals and zooplankton (not shown) can remain in the highly productive estuarine zone for extended periods of time by using the salt wedge as an upstream transport mechanism to compensate for downstream displacement while feeding in the riverine brackish water. Included in this scheme are the migrations by offshore species of nekton, such as herring, menhaden, striped bass, and smelts and the migrations of anadromous fishes, such as the salmonids, and catadromous species, such as the American eel.

variations in a region's hydrologic cycle may also trigger variations in the estuarine–coastal faunal community. Figure 18.24 shows the remarkable correlation between the *annual halibut catch* of Quebec province, Canada, and the *annual changes in the March freshwater discharge from the St. Lawrence River.* An increase in March runoff is reflected in an increase in halibut catch *10 years later!* The implication of such research is straightforward—the possibility of managing the halibut fishery using forecasts based on river runoff.

In the same sense that the halibut obtain a survival benefit in an increase of freshwater discharge, each species using the estuarine system also undergoes a physiological "stress" in doing so, in that it exposes itself to large osmotic pressure changes in the process, as well as to thermal shock should the temperature of shallow estuarine waters suddenly change faster than the organism can adapt. How does a researcher discover the numbers involved in this apparent cost–benefit interaction? What is the balance point to the organism between the energy cost of physiological adaptation and the availability of food in greater concentration?

Such questions have yet to be studied well, but we must do so. The need for answers is directly tied to the great diversity of conditions and species in estuaries. Many delicate balances exist between specialized ecological niches, the timing of the hydrologic cycle, and other factors like salinity, temperature, and turbidity. We already know that human intervention in the river–estuary–ocean system usually reduces species diversity. We usually alter one or more conditions, thereby eliminating the roles of

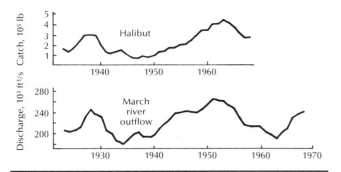

Figure 18.24 A comparison of *(a)* the annual catch of halibut fish of Quebec, Canada with *(b)* the annual rates of freshwater outflow from the St. Lawrence River during the month of March. These graphs have two intriguing aspects. (1) They are plotted with a *10-year* time difference, halibut catch lagging behind river flow. (2) The correlation is excellent, (better than 0.85), suggesting that production in the halibut fishery is greatly dependent on St. Lawrence River outflow during the month of March. (From Sutcliffe, 1973.)

species whose performance was delicately linked to them. Examples abound. The construction of upriver dams can alter the timing of the hydrologic pulse in the estuarine zone (recall Figure 12.7). The introduction of sewage wastes alters nutrient levels; (in the Chesapeake estuary populations of blue crab have noticeably increased in zones where large amounts of sewage are discharged).

Summary

On what basis is it justified to insert a minicourse on coastal oceanography inside a general oceanography text? There are many reasons; we summarize them under discipline categories.

1. *Physics.* Friction between continents and large-scale ocean motion is the main mechanism that offsets the torque by which the winds continuously "spin up" ocean currents. The process that mixes the friction force across the margin waters is turbulence. The reader can get no better *picture* of coastal turbulence than that shown in the satellite image in Figure 14.8; it is an eloquent demonstration of how materials from land can be moved to the open sea, including water.

2. *Chemistry.* The affinity that fine clay particles in suspension in rivers have for various chemicals makes coastal chemistry an essential part of environmental protection. Of course, chemistry has a much wider scope than just pollution. To determine the limits to chemical processes in the open sea, we must have accurate measures of the flux of all salts from land to sea, such as rates of sediment formation, dissolution of carbonates at depth, and many others.

3. *Geology.* Research into how sands and other coarse fractions move up and down the shore, along the shore, into and out of estuaries, into canyons and out to rises, or their incorporation into subsiding sediments of the margins themselves are all important coastal geology tasks.

4. *Biology.* It is in biology that the coastal ocean manifests a "zenith" in complexity. Ninety-nine percent of the seafood that we take from oceans comes from its margins; therefore, we must learn the complex cycles of all life forms in this region. Such cycles are closely tied to the physics of mixing over shelf and estuary. Thermal fronts, created where the well-mixed upper layer and the tidally mixed lower layer converge, are closely coupled to the behavior of large segments of coast communities. Biology is also closely tied to the bottom substrate, where different communities exist in mud floors and on sand or rock bottoms. The aspects of pollution studied by the chemists are eventually integrated into the understanding of biological communities as well.

5. A relatively new field of coastal oceanography is the study of estuaries. These are essential parts of coastal communities because they serve as nursery, feeding, reproduction, and transportation vehicles for so many of the species of coastal life, including *Homo sapiens*. Cycles of mixing, productivity, and sedimentation are complex results of the interaction between riverine flow and tide currents.

Study Questions

1. The first question to ask regarding the material of this chapter is why we should have to treat the coastal ocean *differently* from the open ocean? Because there are many parts to the answer, give only an outline of these several aspects.

2. Refer to Figure 11.1 on circulation in surface layers to find the major continental barriers to ocean currents. Where are they, and what change do these boundaries make on the impinging ocean current?

3. What is the impact on global weather patterns of

the fact that the major equatorward currents along the eastern shores of ocean basins move *cold* water?

4. The major mechanism removing clockwise torque from the ocean is the friction between a boundary current and the shoreline. True or False?

5. What is it about a tide current that makes it *respond* to the Coriolis force when water motion from ordinary wind waves does not?

6. What can you say about the *relative* amounts of energy dissipated by *tide currents* through friction with the continental shelf and of energy dissipated in the surf zone by breaking wind waves?

7. In what latitude zone do we find the major river outflows to the ocean basins?

8. Describe how a wave form changes as it enters the nearshore region and eventually breaks up in the surf zone.

9. To predict the distribution of a pollutant discharged into a river that empties into the sea, is it important to know whether the pollutant is dissolved in water rather than suspended? Explain.

10. How could you employ the data in Figure 18.7 to answer question 9?

11. *Refraction* is a physical phenomenon that determines how wave energy is diverted into bays and coves, thus explaining how these coast features are formed by erosion. True, false, or misleading?

12. Explain to the layperson why it is that waves coming *on*shore can force sands to be transported *along* shore.

13. What are the essential parts of a "coastal cell," the concept we use to explain the behavior of coastal erosion?

14. How do turbidity currents enter the erosion discussion of question 13?

15. Explain why it is that some beaches can be "stripped" of sand during winter months and yet be built back up to sandy conditions during the summer.

16. Is there a coupling between littoral transport and the onshore–offshore transport of sand in question 15?

17. How would you explain to a layperson that *real* but invisible boundaries separate the coastal from the open ocean?

18. Can you explain why a *front* like that in Figure 18.16 could influence how fishermen fish?

19. Write a short essay explaining how the *relative* strengths of (*a*) river runoff and (*b*) tide range govern whether an estuary is a well-mixed or salt-wedge type.

20. Write a short essay explaining how river-induced upwelling operates; include a definition of *entrainment*.

21. Biologists identify four categories of nekton that use the brackish waters of estuaries. List the four and give short definitions and an example of each.

22. Fishermen and fishery agencies seem always to be at odds over whether overfishing is depleting natural stocks. Using Figure 18.24 as a data base, take a position.

PART VII

SPECIAL TOPICS IN MARINE STUDIES

By which we acknowledge that the reader is now able and entitled to examine some of the most complex of issues or topics in marine sciences: the role of our own *societies in the marine world; the obscure but fascinating, yet vital oceans of polar regions; and the history of humankind's gaining a global view of earth's geology.*

CHAPTER 19   THE SEA AND MAN

The zone of our greatest impact on the ocean is along its shores; topics range from Law of the Sea to pollution to mining the sea floor and even to the taking of energy from the sea.

CHAPTER 20   THE POLAR OCEANS

A unique part of the ocean world, the polar oceans are solid and liquid, and the impact of sea ice plus the diminished solar heating makes their ecosystems very specialized.

CHAPTER 21   PLATE TECTONICS AND OCEAN BASINS

The history of oceanography is now very much a part of the revolution in global geology studies; only some two decades old, this revolution first began with the concept of continental drift, then proceeded to the theory of sea floor spreading, and is now consolidated into modern plate tectonic theory.

CHAPTER 19

THE SEA AND MAN

In the beginning of the book we learned that each of us has as *our share* about 1.40 acres of the earth's cultivable land and some 20 acres of its ocean. But it is presumptuous to believe that we *possess* such shares. We are "stewards of earth" for the short while that we move upon its surface, and it is from this perspective that this chapter is organized.

Strangely, as individuals we cannot possess the earth, but as a society within a nation-state we do possess ocean sectors. The first topic of this chapter is the *Law of the Sea*, an international convention that represents humankind's most recent effort to "part" the seas. This is necessary because the other topics of the chapter—living resources, nonliving resources, pollution, and energy—are all best presented within the framework of the rights and responsibilities of coastal nations, and these are described within the Law of the Sea Convention.

Background

Competition and fear provided the basis for the first *recorded* attempt by nations to "divide" the seas among them. In 1494 Spain and Portugal, then both world leaders in high-seas navigation, divided the oceans of the globe into "spheres of influence" for purposes of exploitation. Their Treaty of Tordesillas established the 30°W meridian as the divider between Spanish oceans and lands to the west and the Portuguese domain to the east; the treaty even had papal authority to back it. The English immediately rejected this "Iberian solution" and promoted the concept that certain resources of the sea, its fish, for example, could be claimed by a nation.

As the Dutch increased their own world trading into Java and Sumatra, they ran into problems with the Portuguese over oceanic dominion. Hugo Grotius, a lawyer for the Dutch East Indies Trading Company, published a paper in 1609 in which he stated that the seas were the communal property of all peoples. Called *Mare Liberum*, his opinion was that the fish of the sea are infinite in number and therefore belonged to no one. Obviously no nation could have an infinite number of anything. Moreover, the also infinite seas could not be damaged, and so did not need "protection." The English countered with a reply called *Mare Clausum* (1632) in which they claimed that seas contiguous with a nation's coast were indeed "territorial" since the nation's security depended on their defense. The 3-mile territorial strip came into existence as the *range* of a typical cannonball fired from shore batteries of that time.

The 3-mile limit became accepted international custom, and all other waters were designated as *high seas*. But in 1945, in large part because World War II was the first major war in which petroleum was a *strategic resource* and because the United States had become aware of the large resources of oil and gas in offshore waters, President Truman issued a unilateral claim to all mineral resources of the continental shelf out to the 200-m-depth contour (Proclamation 2667). A sister proclamation (2668) asserted the right of the United States to establish extended *fishery conservation* zones adjacent to our coasts. This was the first national move to "extend" the concept of national rights in contiguous waters. Other nations soon followed with declarations of their own. Several Latin American nations extended claims of sovereignty out to 200 nautical miles during the 1950s, including *exclusive* fishing rights. The Tuna Wars began, during which boats of United States fishing fleets were impounded and the United States forced to pay fines. During the 1960s this concept was expanded into a type of "patrimony" over seas to 200 nmi, in which the coastal state exercised exclusive jurisdiction over all economic affairs; it was the forerunner of the 200-nmi Exclusive Economic Zone (EEZ) principle that was embodied in the Law of the Sea (LOS) Convention of 1982.

The 1982 Law of the Sea Convention

The United Nations first called a General Conference on Law of the Sea (LOS) in 1958; it sought to adopt uniform conventions on four aspects of international sea law: (1) defining the *territorial sea*, (2) defining *high seas*, (3) determining what is a *continental shelf*, and (4) *fishing and conservation* of living resources. Some progress was made at both this conference and the one in 1960. But in 1967 the nation Malta injected a new factor: Who would have jurisdiction over the mining of deep-sea mineral beds? By 1970 the United Nations issued a "Declaration of Principles Governing the Sea-bed and Ocean Floor, and the Subsoil Thereof, beyond the Limits of National Jurisdiction" and called for a third LOS Conference. A vast battery of people from virtually all nations worked for the next 12 years generating the eventual 1982 LOS Treaty, now undergoing ratification. Two elements of the treaty especially impact on ocean-ography and are described in following sections.

The Exclusive Economic Zone

Quite possibly no other provision of the 1982 LOS Treaty has as great an impact as the provision that each coastal nation is *entitled* to a minimum 200-nmi exclusive economic zone (EEZ). Figure 19.1 shows that about 40% of all ocean area could be involved in EEZ claims, and Figure 19.2 shows the United States and its territories with the 200-nmi expansion in place. The United States has the largest EEZ of any of the world's nations, about 3.9 billion acres if the EEZ of its territories is included.

Geometry of the EEZ. What specifically is the EEZ? How is it defined? The 1982 convention specifies a number of distinct zones over which a coastal nation will exert a degree of control, the degree ranging from total to none. The first zone in Figure 19.3 is the *territorial sea*. It extends 12 nmi offshore and here the coastal nation exerts complete sovereignty except for the "right of innocent passage." The innocent passage aspect is rooted in old, customary maritime practice; it provides for passage through narrow straits, for example, even though such straits would be *closed* under simple geometry constraints of a 12-nmi territorial zone.

Adjacent to the territorial boundary is the *contiguous zone*. It extends control by a coastal state to an additional 12 nmi, within which the state may control contraband shipping or licensing for minerals

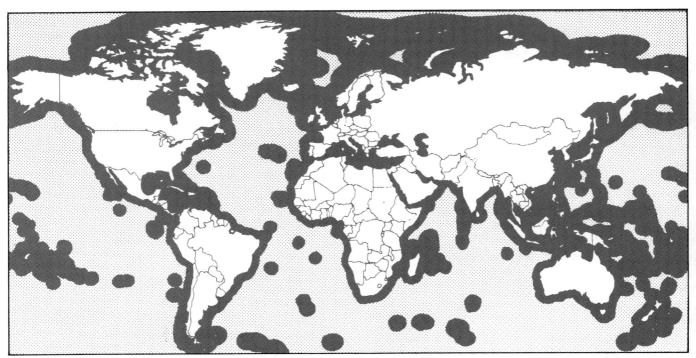

Figure 19.1 A chart of the world ocean claims if each coastal nation is entitled to a minimum 200-nmi Exclusive Economic Zone (EEZ). The reader is cautioned that the chart is not an equal-area projection; polar areas appear disproportionately large.

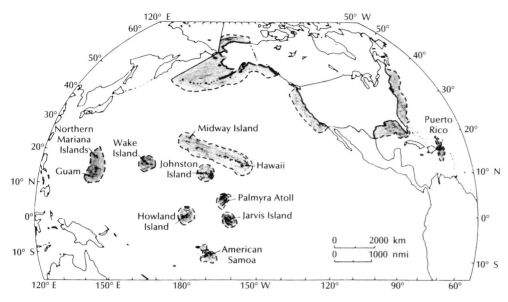

Figure 19.2 The United States Exclusive Economic Zone. In all, about 3.9 billion acres are covered, more than the 2.3 billion acres of land in the United States and its territories. (From Rowland and McGregor, 1984.)

exploitation (and within which no other nation may claim abandoned fish potential—more on this later).

The outermost boundary is the 200-nmi limit of the EEZ, a fixed limit that is measured from the same base line as the territorial sea. Within this zone the coastal state is entitled to all renewable resouces, indigenous or migratory, and to all seabed and subsoil resources.

A more complex definition is applied to the *continental shelf*. If the *real* shelf, as defined by its break to the continental slope at about 200-m depth, does not extend to the 200-nmi EEZ limit, a *jurisdictional shelf* is defined. This shelf loses all relation to bathymetry; it simply extends as far as the EEZ, even though the sea floor may by that point be either abyssal plain or even submarine trench. This stipulation provides especially for coastal states that have subduction zones, trenches, volcanic arcs, and unusually narrow shelves along their margins.

Where the continental shelf itself extends *beyond* 200 nmi, the LOS defines an *extended shelf*, which can stretch as far as 350 nmi offshore from the baseline (Figure 19.3*b*). However, the EEZ is not extended in a like manner.

The coastal state is entitled to all the resources within the 200-nmi limit and to the major part of the resources of the seabed and subsoil in the shelf part extending beyond 200 nmi. The LOS stipulates, however, that a part of the resource value taken from this extended shelf must be paid to the *Authority*, an agency of the United Nations, as described next.

Non-EEZ Ocean Remainder—The Area, the Authority, and The Enterprise

The 60% or so remainder of the oceans are called the *high seas* in the sense of transportation and fishing, or the *Area* when referring to other aspects, such as the mineral resources of the deep-sea floor. This Area is declared to be the common heritage of humankind. Its major known resource consists of the polymetallic nodules lying on the deep basin floor (see Table 5.2). The United Nations, anticipating that these mineral resources might soon be technically harvestable, set up the *International Seabed Authority* (ISA) as the agency in control of licensing and revenues. Mining of the deep-sea bed within the Area will require a contract from the ISA, and production will be controlled by the ISA to avoid undue stress on nations already mining similar minerals from land deposits.

The ISA also requires that the mining operation pay a part of the production value as a royalty. Whatever money remains after administration costs are paid is to be distributed among developing nations.

There are other facets of this Law of the Sea that are new and revolutionary in their own sense. The ISA can underwrite its own production arm, called the *Enterprise*. Funding at a scale large enough for the Enterprise to begin production is estimated at 1 billion dollars now (Oxman, 1983) to be raised in a combination of private and state loans. Uniquely, the Enterprise would come into existence only after

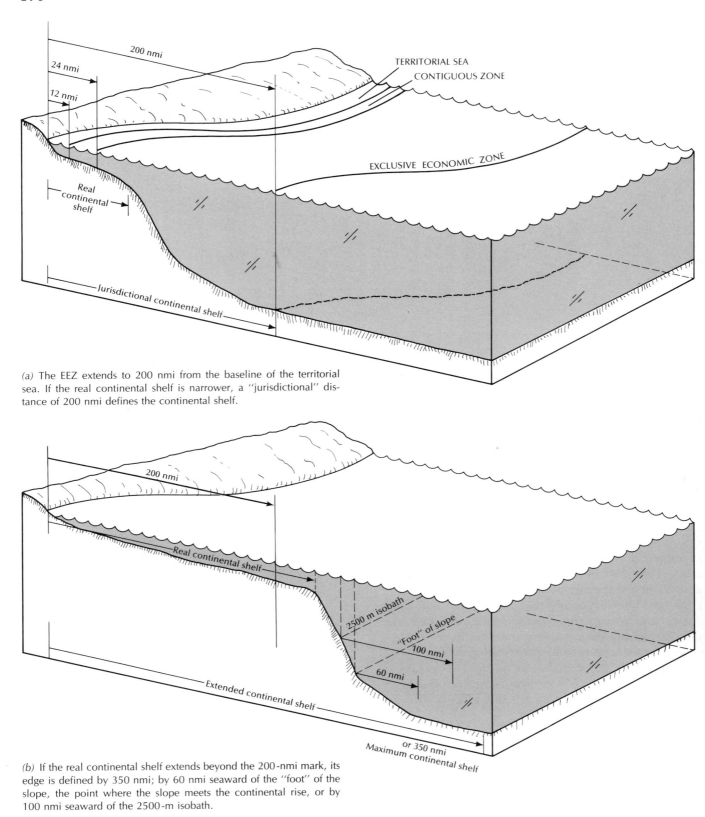

(a) The EEZ extends to 200 nmi from the baseline of the territorial sea. If the real continental shelf is narrower, a "jurisdictional" distance of 200 nmi defines the continental shelf.

(b) If the real continental shelf extends beyond the 200-nmi mark, its edge is defined by 350 nmi; by 60 nmi seaward of the "foot" of the slope, the point where the slope meets the continental rise, or by 100 nmi seaward of the 2500-m isobath.

Figure 19.3 Definitions of the Exclusive Economic Zone and of the Continental Shelf.

pioneer production by private companies has been proved profitable. Such pioneer companies would be required not only to share technology with the Enterprise but to agree as well to sell to it their own operations should the ISA require it.

The ISA is to be governed by a 36-member council whose makeup is of interest to understanding how the new LOS will work. Membership must include four of the nations that consume the largest amounts of the metals to be mined, together with four members from the largest land-based producers of these same type minerals and four from states that have invested the most heavily in mining of the seabed; developing countries would hold the remaining seats.

1983—The Reagan Proclamation

The United States has refused to ratify the LOS Convention (1985). Instead, President Reagan in 1983 issued a unilateral claim to a 200-nmi EEZ for the United States, which is almost identical to the EEZ provision of the 1982 LOS. In effect, the United States position is to accept the EEZ as customary international law, but it rejects entirely the idea of an international authority over the seabed resources.

Rights and Responsibilities of Coastal States

The 1982 LOS Convention states the *rights* of a coastal state:

- Exclusive sovereign control over exploration for and development of all living and nonliving natural resources of water, bed, and subsoil.

- Exclusive control of other economic activities such as producing energy, installing artifical islands, or mariculture farming.

- The right to be informed of and participate in marine science projects or to withhold consent for research under specific conditions.

- The right to control all dumping of all wastes.

- The right to board, inspect, impose fines on, and arrest ships suspected of violating international standards of safety or pollution.

The *responsibilities* of the coastal state are the following:

- To ensure the conservation of living resources in the waters of the EEZ.

- To promote the optimum utilization of these resources by (1) determining harvest capacity and (2)

granting permits under reasonable conditions to foreign vessels to fish for the surplus, if any exists, beyond its own harvest.

- To participate in preexisting fishery conventions for the protection of endangered species and for the management of species that migrate across EEZ boundaries.

- To submit to arbitration boundary disputes that are not settled by the parties concerned, in order that other rights and responsibilities can be exercised effectively by all parties concerned.

All *states*, whether coastal or landlocked, *have rights* within the EEZ of any coastal state:

- All states have the right to navigate the high seas, fly over them, lay submarine cables and pipelines, and carry out the operations associated with these activities, such as maintaining pipelines.

- Ships of visiting states, called flag states because of the tradition of sovereignty of ships at sea, must observe international rules.

All states have the following *responsibility:*

- To ensure that its own rights are exercised with due regard for the rights of others.

The 1982 LOS provides that unresolved disputes between states over the LOS must be submitted to binding arbitration at the request of either state. Exceptions are made when there are earlier international agreements on borders.

Commentary on Rules and Concepts of the 1982 LOS

Never before in history have so many of the world's nations attempted such a far-reaching convention. It is akin to being a World Constitution. But even more striking are several included features.

- The control of scientific research in the EEZ by the coastal state.

- The concept that a coastal state must *give up* "unused" renewable resources.

- The establishment of a "world enterprise."

The coastal state is responsible for and has rights to conduct all science research within its EEZ. This is *a major new concept.* Heretofore, foreign vessel research was conducted by "customary" procedure anywhere *except* within the narrow territorial zone of a coastal state. Under the 1982 LOS, *all* research activity within the EEZ is subject to the consent of the coastal state. A foreign state may apply to the coastal state for permission to conduct research, and the convention stipulates that the coastal state must either allow or refuse consent within 4 months of the request. Not replying is taken as affirmative consent. But permission can be denied if the coastal state claims: (1) the research is of direct consequence to the exploration of natural resources, hence infringes on proprietary rights; or (2) the research activity will itself do "harm" to the environment; or (3) the request for consent contains inaccurate information; or (4) the request was filed without allowing an adequate lead time of 6 months; or (5) the request fails to provide adequate opportunity for the granting coastal state to *participate fully in the research activity;* or (6) the request fails to recognize the coastal state's control over *which* results of scientific research can or cannot be published.

These controls could be helpful or detrimental to progress in marine science. On the one hand, the coastal state can benefit from investment by foreign nations in expensive marine research in its waters; it receives new data and its own research advances. On the other hand, the new permit procedures could bog down, and delays might reduce the ben-

efit-cost basis on which much work in science is funded today.

The idea that a coastal nation can be required to allow another nation to harvest "unused fish resources" is a controversial point. It has many ramifications. For one, a coastal nation with a *high* living standard, (and *high* wages), will not be able to compete internationally when fish prices are low and may therefore elect *not* to fish a resource within its waters. Should another nation whose seamen work for lower wages be allowed to "take" the resource?

Whether or not the 1982 LOS Treaty is ratified by the 60 nations needed to place it into effect, the *negotiations* over the 1960–1982 period have already established new definitions of "customary maritime international rules." Moreover, almost all nations of the world have participated. The idea of the open sea as a "common heritage" is now indelibly stamped upon the future.

Fishing the Living Resources

A dramatic way to begin the discussion of fisheries is to demonstrate with Table 19.1 that the *per capita*

Table 19.1 World per capita production of basic commodities, 1950–1981, with peak year underlined. (From Brown and Shaw, 1982.)

Year	Wood, m³	Fish, kg	Beef, kg	Grain, kg	Oil, barrels
1950	—	8.4	—	251	1.5
1955	—	10.5	—	264	2.0
1960	—	13.2	9.3	285	2.5
1961	0.66	14.0	9.6	273	2.7
1962	0.65	14.9	9.8	288	2.8
1963	0.66	14.7	10.7	282	3.0
1964	0.67	16.1	10.1	292	3.1
1965	0.66	16.0	9.9	284	3.3
1966	0.66	16.8	10.2	304	3.5
1967	0.64	17.4	10.4	303	3.7
1968	0.64	18.0	10.7	313	4.0
1969	0.64	17.4	10.7	311	4.2
1970	0.64	18.5	10.6	309	4.5
1971	0.64	18.3	10.4	330	4.7
1972	0.63	16.8	10.6	314	4.9
1973	0.64	16.8	10.5	332	5.3
1974	0.63	17.7	11.0	317	5.2
1975	0.61	17.2	11.3	316	4.8
1976	0.62	17.7	11.6	337	5.1
1977	0.61	17.3	11.5	330	5.2
1978	0.61	17.3	11.4	351	5.2
1979	0.60	16.9	10.9	331	5.3
1980	0.60	16.1	10.5	324	4.9
1981 (preliminary)	—	—	10.4	331	4.5

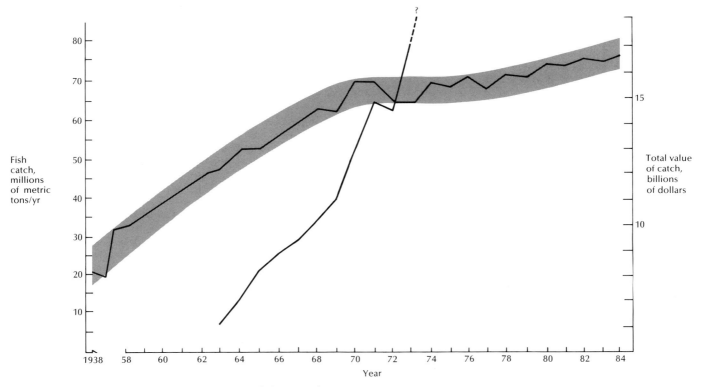

(a) The shaded line shows the total world fish catch by year, the other its market value. Although the total catch has held steady since about 1970, the value has climbed steadily.

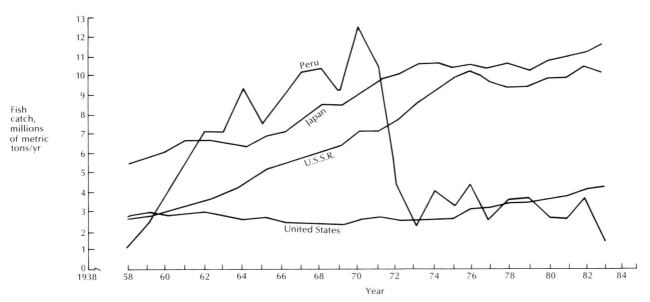

(b) The fish catch of leading fishing nations for the same time span. The large rise in the Peru catch was caused by the overexpansion of fishboat capacity during the 1960s, and its later decline by the cata-strophic El Niño of 1971–1972. Together with overfishing, El Niño decimated the anchoveta resource of Peru. (Compiled from FAO Yearbook statistics, 1968–1983.)

Figure 19.4   Fish catches.

catch of fish worldwide peaked in 1970. To understand this better, look first at the world trend in the tonnage of annual fish catch in Figure 19.4a. The fish catch increased rather uniformly from 1958 to 1970 but has leveled off since then at about 70 million tons per year. But world population continues to rise, so that per capita fish catch has already peaked and is now falling. The reader is urged to compare this level of about 70 million tons of fish catch with the Ryther estimates of Table 1.2. If indeed more fish can be caught per year on a sustained basis, as Ryther claims, why is it not happening? The answer to this question has many facets.

First, there is the matter of price. The value of the total catch has been climbing sharply (Figure 19.4a). Although the total *catch doubled* during the decade 1963–1973, the total *value of the catch tripled!* Fish meat has become more expensive. This has led to more consumption by rich nations and less by developing nations, lesser amounts of seafood consumed within nations that need the foreign exchange provided by seafood exports, and excess fishing pressure on the few species with high market value, leading to overfishing and declining harvest. Meanwhile, too little is invested in nonexportable fishing and in broadening the fishing base for both food *and* export value.

Second, there is the matter of which fish will sell and which will not, a factor related more to personal taste and to custom and tradition. Table 19.2 lists the four nations that are leading consumers of each of several selected fish or shellfish. The Japanese lead all others in eating groundfish (which include flounder, sole, cod, haddock, and especially hake), about 30 kg per person; we in the United States eat only 4.5 kg each. Japan also leads in overall consumption of salmon, 36% of all salmon eaten, even though Canadians eat more salmon per person. In Europe, the Danes win all records for per capita consumption of *one* type of fish, taking over 100 kg of sardines and herring per person per year. But this figure includes a large amount of fish meal derived from these fish but fed to chickens and swine, which are then consumed either internally or exported for cash.

There are many more aspects that complicate the fishery picture. For example, it has been said that *one* driving force for the creation of the 1982 LOS Convention was the repeated postwar failures of the Russian wheat harvest to satisfy that nation's needs. Russian studies during the 1950s showed that, calorie for calorie, they could produce a food calorie in fish meat from the world oceans for one-third of the man-hours needed to produce a food calorie of beef. Is this the reason for the subsequent buildup of their open-ocean fishing fleet? By the early 1970s the Russians captured about 11% of the world total of open-

Table 19.2 The consumption of selected fishes by the four major consumer nations, 1973. (Compiled from FAO Yearbook statistics.)

Species or Type	Country	Total Consumption million kg	Percentage of World Total	Per Capita, kg
Groundfish: Soles, flounder, hake, cod, etc.	USSR	3564	30	14
	Japan	3214	27	30
	United States	938	8	5
	United Kingdom	700	6	13
Tuna	United States	627	37	3
	Japan	381	22	3.5
	South Korea	121	7	3.5
	France	57	3	1
Salmon	Japan	169	36	1.5
	USSR	81	17	0.3
	United States	79	17	0.4
	Canada	59	12	2.6
Sardine, herring	USSR	843	16	3.4
	Denmark	514	10	103
	Japan	481	9	4.5
	South Africa	396	8	16.7
Shrimp	United States	318	29	1.5
	India	167	15	0.3
	Thailand	90	8	2.3
	Japan	89	8	0.8

ocean fish with a fishing fleet exceeding in tonnage that of all other fishing fleets *combined* (Wertenbaken, 1983). The capture by the Russian fleet increased faster than that by all other countries except Peru. (Figure 19.4*b*).

Peru has been a special case. Its fishery is based heavily on the anchoveta and grew at a fantastic rate during the 1950s and 1960s. Each year the goverment subsidized the expansion of the fishing fleet. Aircraft were used to locate fish schools, and the purse seine technique for taking the schools was very efficient. Each year more tonnage was taken, processed into fish meal, and either converted into local chickens or exported as animal feed, principally to Denmark and Holland. In 1970 Peru took a record 12 million tons. But in the winter of 1971–1972 a warm, strong ocean current pushed into the coast waters of Peru, scattering the anchoveta schools and reducing their spawning efficiency (see the discussion of El Niño in Chapter 10). At the same time the fishing pressure reached an all-time high. The combination proved catastrophic to the ecosystem, so much so that in 1972 Peru harvested less than 1 million tons of anchovetas (The total Peru tonnage in Figure 19.4*b* is supplemented by other fish such as the jack mackerel.) Today Peru harvests less fish meal than its Chilean neighbor to the south. The anchovetas never returned to their former, dominant role in that offshore ecosystem.

Where Are All the Fish?

Figure 19.5 divides the world oceans into regions with the potential catch for each region given by the relative area of its circle. The numbers refer to different types of fish and shellfish and the shaded portions of the circles indicate how much of the estimated potential was caught in 1968. For example, the large circle in the western South Atlantic represents an estimate of 12 million tons total *potential catch;* the smaller partial circles inside indicate the relative degree of exploitation of the potential catch, by types. About 25% of the potential for groundfish, the category marked 1, was harvested in 1968. The potential for pelagic fish, the category marked 2, is about 40% or 4.8 million tons but was almost totally unfished in 1968. Other unexploited resources are in the Indian Ocean, the central Pacific Ocean, and the Southern Ocean.

The regions of Figure 19.5 are too coarse to show that the greatest fraction of fish production occurs in the coastal oceans (see Table 1.2). This is better shown in Figure 19.6. Notice that only tuna and

whale species are potential fishery stocks in the open ocean. Most of the fisheries are coastal.

Managing Fish Stocks That Migrate

One of the most difficult fishery tasks is to manage migrating fishing stocks. Nowhere is the problem better illustrated than along the western coast of Africa, where many small coastal nations divide up the 200-nmi EEZ into pieces much, much smaller than the ranges of migration of many of the indigenous fishes. Figure 19.7 shows the approximate distances *along* the shoreline over which a few of the species are known to migrate; the heavy lines are national boundaries.

The problems of managing such species and harvesting them equally are enormous. The entire life cycle of *each* stock must be known very well, first, and then the interactions among species within the local food chains must also be learned. If the small states of Africa are to join the research, they will need many trained marine ecologists. This interaction of humans and the sea is destined to be long in coming.

Fish Stock Patterns versus Human Demography

Another difficulty in managing fisheries for human food is that the heaviest concentrations of economic fisheries are in global regions where human populations are the lowest. Look at Figure 19.5 again and note that the circles of largest potential catch are in the subpolar and polar regions. Conversely, potentials are smallest within the temperate latitude belt 40°N to 40°S, where the major populations of humans occur. One exception to this pattern is the large tropical fishery potential in the extensive shallow seas and islands of Indonesia, the fifth-largest nation with over 140 million people. Another large tropical fishery potential, shown in the West Indian Ocean, is attributed to the intense seasonal upwelling along the coast of Somalia and the Arabian Peninsula, but both these regions are high desert areas and very sparsely populated (see Figure 11.1 for the Somalia Current, number 30).

The Antarctic Fishery—A Special Case

The most striking example of the inverse pattern between human population centers and economic fishery regions is in the Southern Ocean. There are no indigenous peoples living on Antarctica, and only a few hundred thousand live south of the 45°S paral-

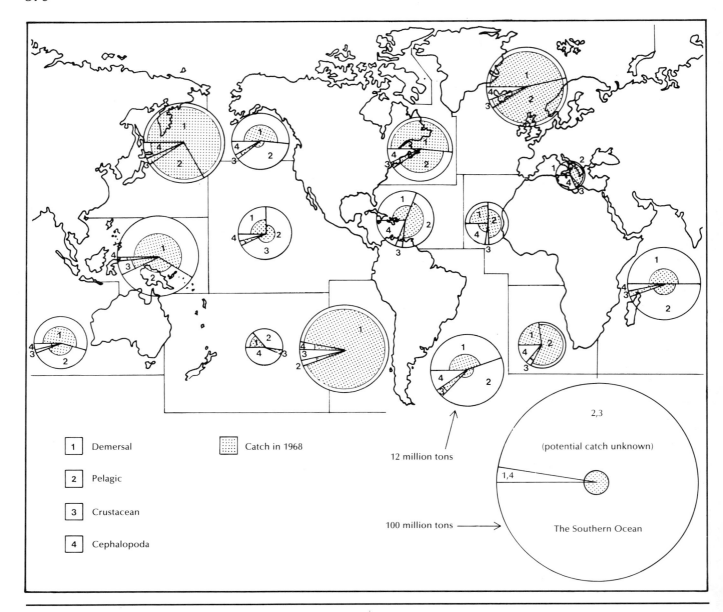

Figure 19.5 The distribution of the potential fish catch within certain regions of the world oceans. Numbers refer to types of fish. The diameter of the outer circle is proportional to the *potential* catch; the diameters of the shaded, partial circles are a measure of the actual harvest in the region in 1968. (Compiled from FAO Yearbook statistics, 1968.)

lel. But Figure 19.5 shows that at least 100 million tons per year can be harvested from this region, mostly in the form of the crustacean "krill," *Euphausia superba*. (See Chapter 20 for a detailed discussion of the krill fishery.)

The Aquaculture Resource

Aquaculture, the farming and husbandry of freshwater and marine organisms, is an ancient practice. The Romans cultured oysters and the Chinese were growing carp fish in farm ponds before that. Currently, aquaculture provides about 9% of the world's water-derived protein (Bell, 1978). Pillay (1973) estimates that about 3.7 million tons of finfish are produced yearly in 42 countries. When the production of mollusks (1 million tons) and seaweed (⅓ million tons) are added, the world production reaches at least 5 million tons per year.

Table 19.3 contains data on how and where fin-

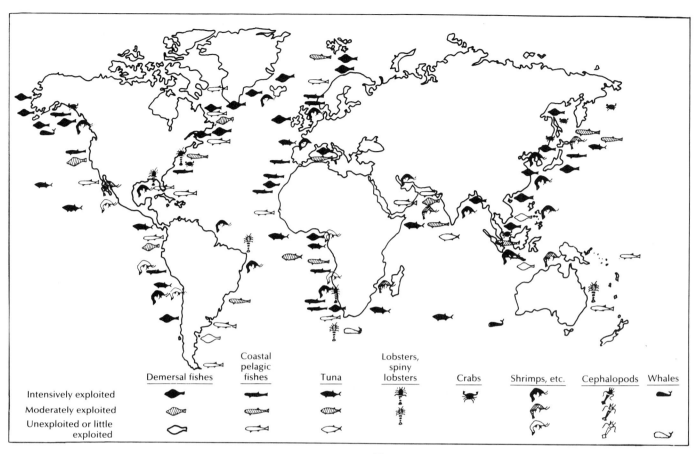

Figure 19.6 Distribution over the world oceans of the major fishery species and their degree of exploitation. (Compiled from FAO Yearbook statistics, 1968.)

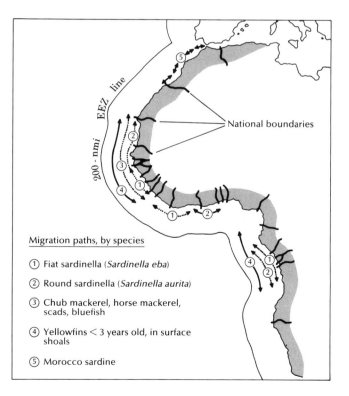

Migration paths, by species

① Fiat sardinella (*Sardinella eba*)

② Round sardinella (*Sardinella aurita*)

③ Chub mackerel, horse mackerel, scads, bluefish

④ Yellowfins < 3 years old, in surface shoals

⑤ Morocco sardine

Figure 19.7 The extended 200-nmi exclusive economic zone of the western coast of Africa, together with the land boundaries of coastal nations and the migration paths of selected fish species. (Based on Gulland, 1972.)

Hauling aboard a mid-water trawl full of hake. (Courtesy William G. Pearcy, Oregon State University.)

fish and other types of water-related organisms were cultured in 1970. China cultured about 33% of its total annual catch of finfish, mostly in species such as carp and milk fish. Of the nations grouped in part (a) of Table 19.3, which like China culture a large percentage of their total fish supply, the first four—China, India, Sri Lanka, and Indonesia—all have abundant supplies of fresh water and low-lying coastal plains on which to construct culture ponds. Israel, on the other hand, is forced to recycle virtually all its fresh water at least one time, but it still derives a large fraction of its total fish catch through culture. And Hungary, a land-locked nation, albeit with the Danube River flowing through, supplies most of its finfish consumption from aquaculture.

On the other hand, some of the world's greatest open-ocean fishing nations do not engage in extensive finfish culture. Japan is the outstanding example (Table 19.3b), but notice also that it cultures instead much more seaweed than any other nation. Norway, in spite of its abundant fjords and freshwater lakes, does not yet engage in aquaculture to any significant extent.

Table 19.3c lists some nations that have diverse culture production, from fish to mollusks. Spain produces a large quantity of mussels, much of it exported. The aquaculture production of the United States is greatest in oysters, matching Japan's seaweed production both in volume and in the fact that the product is internally consumed for the most part. In short, what a nation produces depends on many factors, including eating habits, climate, and economic needs.

Potential Role of Aquaculture

In 1950 the United States imported 25% of its total supply of fish products; by 1972 this reliance on imports had jumped to over 65% (Bell, 1978). Viewed in these terms, the need and potential for aquaculture products are both great. In the past most cultured products have been in the higher-priced, higher-demand species such as oysters, salmon, and catfish. As the demand for fish products increases with population increase, so will the demand for less exotic species such as mullet, bass, bream, sunfish, crappie, and carp.

There is one very important point to make in assessing the role of today's aquaculture. By and large it is devoted to the production of luxury food, and because the output product price is high, the feed used in present-day culture farms is fish meal itself. Stated another way, we are producing exotic, high-priced fish using trash, low-priced fish. We can do this as long as trash fish are available, either in our own fishery or imported from abroad. But the demand for fish protein is going up all over the world, and many of the regions where such fish are available are also regions of rising human population density, so that local demand may eventually consume all available fish protein.

The point is that the United States may soon see a sharp rise in local demand for lower-valued aquaculture products. Many of the fish species best adapted to culture in ponds, carp and tilapia for example, are also vegetarians. Feeding costs are then for fertilizers that can be dissolved in pond water to support algae production, which in turn supports fish growth. We need to devote more research to improving productivity within such system and to developing the market schemes by which cultured, nonexotic fish can be accepted as "good" table fare

Table 19.3 Comparison of fish produced by a country's aquaculture with marine fish captured by the same country, 1970. (Compiled from FAO Yearbook statistics and Pillay, 1973.)

Nation	Total Fish Catch, metric tons	Fish Production by Aquaculture, metric tons	Finfish, % of total	Other Aquaculture Products, metric tons			
				Shrimps and Prawns	Oysters	Mollusks	Seaweeds
(a) Six Nations in Which Aquaculture Produces a Very High Percentage of Total Catch							
China	6,880,000	2,240,000	32.6				
India	1,845,000	480,000	26.0				
Indonesia	1,249,700	266,300	11.3				
Sri Lanka	87,700	15,000	17.1				
Israel	28,200	10,200	36.0				
Hungary	26,600	19,700	76.0				
(b) Six Nations in Which Aquaculture Production Is a Small Part of Total Catch							
Japan	9,994,500	85,000	0.9	1,800	194,000	10,800	357,000
Denmark	1,400,900	11,000	0.8				
Burma	442,700	1,494	0.3				
Norway	3,074,900	600	0.02				
Nigeria	155,800	127	0.1				
South Korea	1,073,700	40	0.003		45,700	53,600	16,000
(c) Six Nations in Which Aquaculture Production Is Diversified							
Philippines	1,049,700	94,570	9.0	2,500			
Spain	1,498,700	50	0.003	1,800		109,700 (mussels)	
Italy	391,000	18,000	4.6			13,700 (mussels)	
Mexico	402,500	9,026	2.2	43,500			
United States	2,766,800	40,200	1.5		350,500		
Malaysia	390,000	25,648	6.6	250		28,600 (mussels)	

The "Salmon Ranch" Concept—Open-Ocean Mariculture

Open-ocean mariculture is a very special type of mariculture. In the "ranching" approach, the industry carefully controls the spawning of larvae and larval growth up to the juvenile state. At this point the juvenile is "imprinted" to the local, natural stream to which the industry wishes it *to return* after growth to the adult stage, and then it is released to the open ocean. The presumption is that the juvenile will follow its ancestral patterns of roaming the open seas for food, storing energy in its body flesh for the eventual return to home rivers to spawn and complete the cycle. The process works only with species that have evolved with an open-ocean growth phase as part of the life cycle and that are guaranteed to return to the point of origin for harvesting.

This "ocean ranching" industry has grown rapidly during the 1970s and is expected to double again during the 1980s to a peak production of *6 billion* juvenile releases per year, worldwide. Ranch salmon already contribute more than 20% of the world supply. Japan leads the world in this industry, with the

USSR close behind; together, these two nations release 75% of the "roaming" salmon. Figure 19.8 gives a perspective on the Japanese industry. Not only are the numbers of released juveniles increasing exponentially, but the return rate is keeping pace as well. In 1979 some 800,000,000 juveniles were released from the Japanese hatcheries on Hokkaido Island and about 15,000,000 returned, yielding a return-to-release ratio of about 1 out of 50.

Mining the Sea Floor

Our economic and military strength is founded on minerals. At today's rates, each citizen will require over a lifetime at least 0.5 ton of lead, 0.5 ton of zinc, 2 tons of aluminum, and 45 tons of iron and steel (Pendley, 1983). For some strategic minerals our dependence on imports ranges between 80 and 100%. We import 100% of the cobalt required for the critical hardening of steels and in vital parts of jet engines. Nickel is mined in the United States but more than 80% is imported. Almost 100% of the manganese we use comes from South Africa and South America,

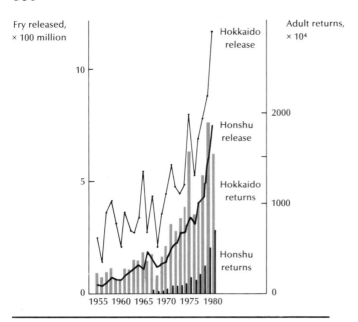

Figure 19.8 Graphs of the release of juvenile salmon by the Japanese salmon-ranching industry, with comparative graphs of the rate of return. (From Japan Fisheries Agency Annual Report, 1980.)

and this metal is fundamental to the multiple alloys of steel and iron supplied to our factories. Virtually all chromium is imported from Africa and the Soviet Union.

What can be taken from or under the ocean floor that is of use in this technical age? Where is it? How did it get there? Can it be mined? What is its value today? Answers are organized into four different headings: (1) basins as collectors of sediment and the petroleum factor, (2) placer deposits in the nearshore sediments, (3) ferromanganese and phosphorite crusts, and (4) the polymetallic sulfides in rift muds.

Basins and Petroleum and the EEZ

Petroleum deposits seem to be concentrated along the margins between oceanic basins and continents; this is true both of present-day continental margins and of older margins, which may now be in the interior of continents, along what were shores of ancient seas. The deposits formed wherever basins existed to capture organic material as well as other sediments eroded from continents. Figure 21.12 shows how a shallow basin might have formed behind a reef during the extensive transgressions of shallow seas in the Mesozoic to Triassic ages, 65 to 225 Ma. And Figure 21.13 shows how other basins

can form where continents collide with ocean crustal plates, and how ocean floor sediments, organic and inorganic, can be scraped off the oceanic crust to form layered structures along the margin of a continent.

Within the EEZ of the United States are many such basins (Figure 19.9). No one knows how much oil and gas exists within them. Some estimates place the offshore potential at about one-third of the total world reserves. Offshore deposits have been the focal point of remarkable developments in the technology of drilling and production during the past two decades (Figure 19.10).

Sand and Gravel Resources and Placer Deposits

The worldwide offshore production of sand and gravel is greater in volume and value than that of any other nonfuel mineral. Nearly 1 billion tons were produced in 1980 at a value of about $3 billion (USGS Survey Circular 929). Land deposits currently supply most of the demand in the United States, but already coastal cities are finding that offshore mining can deliver sand and gravel at lower cost than overland delivery. Construction of offshore gravel platforms for drilling wells in Alaskan waters requires large amounts of gravel. Fortunately, huge deposits of sand and gravel sediments exist, many located near large industrial user sites. Figure 19.11 shows locations around the continental United States and Alaska.

Beach placer deposits have been known for centuries; descriptions of the West coast of North America written in the mid-sixteenth century mentioned the *black sands*. The first United States commercial mining of beach placers was at Nome, Alaska for gold. Beaches of Southeast Asia have been processed for centuries for tin. More recently, there has been renewed interest in United States offshore placer deposits of rutile and ilmenite, both major source minerals for titanium metal. These are heavy metals and crystals of these minerals have collected over geologic time at the mouths of rivers. Because of their high density, the crystals drop out of suspension as soon as the river flow enters the sea and loses its turbulent momentum. Over geologic time, the beach sands have been worked and reworked, with the lighter sands washed away to leave the heavy minerals in a concentrated deposit. Figure 9.11 also shows the locations of major placer deposits around the United States coastline. On the Pacific continental shelf the sands are estimated to carry almost 1% of heavy minerals by volume, on

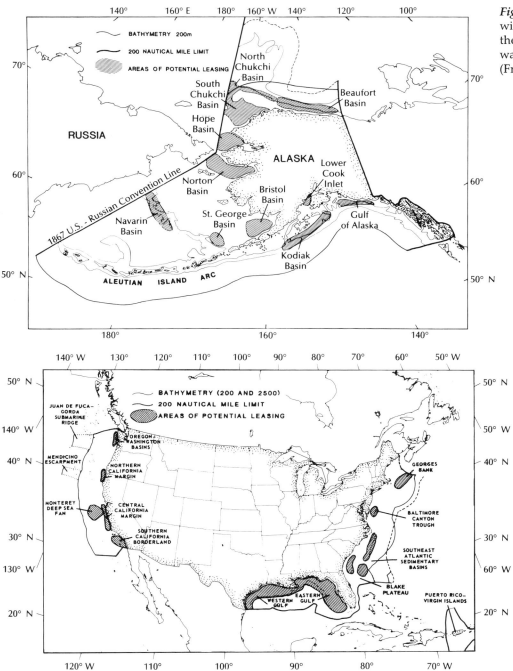

Figure 19.9 Locations of basins with oil and gas potential within the EEZ of the United States (Hawaii and territories not shown). (From Peck, 1983.)

the average. Alaskan placers may be especially rich in both heavy minerals and gold and platinum and could become the first *offshore* United States placer resources to be developed. The demand for titanium in our country is great, and we currently meet over 40% of it with imports. Domestic production so far has been from relict beach deposits in Florida and Georgia.

Ferromanganes Nodules and Co-crusts, Phosphorites, and Brine Pools

The decade 1975–1985 has been the "era of enlightenment" regarding the shape, composition and geological history of the sea floor. Historically, however, the first evidence of sea floor mineralization surfaced during the *Challenger* voyage, 1873–1876, in the form

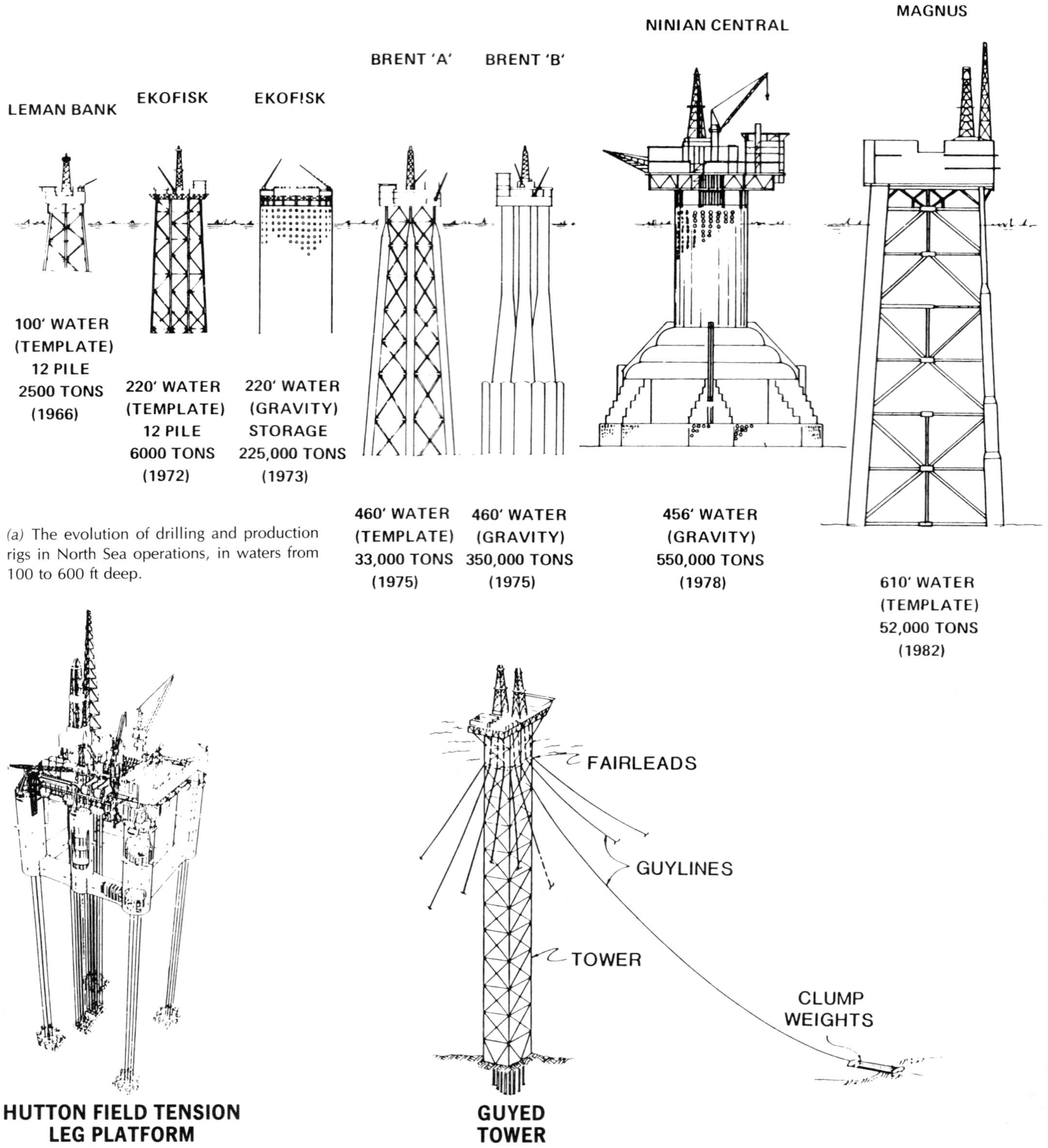

LEMAN BANK

EKOFISK

EKOFISK

BRENT 'A'   BRENT 'B'

NINIAN CENTRAL

MAGNUS

100' WATER
(TEMPLATE)
12 PILE
2500 TONS
(1966)

220' WATER
(TEMPLATE)
12 PILE
6000 TONS
(1972)

220' WATER
(GRAVITY)
STORAGE
225,000 TONS
(1973)

460' WATER
(TEMPLATE)
33,000 TONS
(1975)

460' WATER
(GRAVITY)
350,000 TONS
(1975)

456' WATER
(GRAVITY)
550,000 TONS
(1978)

610' WATER
(TEMPLATE)
52,000 TONS
(1982)

(a) The evolution of drilling and production rigs in North Sea operations, in waters from 100 to 600 ft deep.

FAIRLEADS

GUYLINES

TOWER

CLUMP
WEIGHTS

**HUTTON FIELD TENSION
LEG PLATFORM**

**GUYED
TOWER**

(b) Some ideas on how the depth of operation can be extended in the future.

Figure 19.10 Offshore petroleum production aspects (From Crooke and Otteman, 1983.)

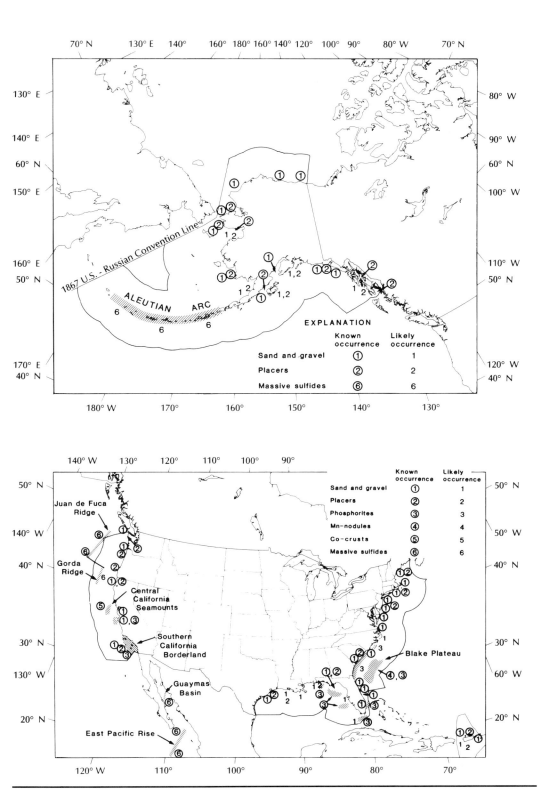

Figure 19.11 Locations of hard-mineral resources in the EEZ of the United States. (From Ballard and Bischoff, 1983.)

of ferromanganese nodules taken in deep-sea dredge hauls. Phosphorite and other crusts and brine pools are recent discoveries.

Ferromanganese Nodules. Highly mineralized nodules ranging in size from marbles to cobbles are widely distributed over the sea floor. They consist mainly of ferrous and manganese oxides (see Table 5.2). Little is known of how the nodule deposits actually form, nor of the rate at which they form. Found lying on the surface of deep-sea muds (Figure 19.12), these deposits suggest that their formation rates are more rapid than the rate at which muds form but this is not the case (see Table 5.1). The richest deposits occur in the eastern equatorial Pacific between latitudes 5 and 15°N (Figure 19.13).

Serious study of *how* to mine such deposits off the sea floor in waters 4000 m deep began during the late 1950s as the extent of the deposits became known and the projections for United States needs in strategic minerals became clearer. Such studies continue today. Various techniques have been proposed, from remote-controlled tractors that "rake up" nodules into containers to giant "suction cleaners" that would be pulled across the ocean floor. But the reality of deep-sea mining appears to be fading into the future rather than coming into clearer focus. The reason for this is the more recent discovery of ferromanganese "crust" deposits in shallower waters and of the possibly richer polymetallic sulfide deposits in rift zones.

Ferromanganese Co-Crusts. Ferromanganese encrustations were discovered only in the last decade. These crusts form over basaltic substrate and so are associated with undersea volcanic mounts. The crusts have a mean thickness of about 2 cm and occur in the depth range from 1000 to about 2500 m. The metals contained are similar in type to those of the deep-sea nodules (see Table 5.2), except that the cobalt concentration is much higher, and for this reason these encrustations are called *co-crusts*. Cobalt content ranges from about 0.4% for the deeper crusts to 1.2% at seamount depths shallower than 2500 m, and is concentrated in the outermost 0.5 cm of the crust, regardless of its thickness.

The importance of ferromanganese co-crusts was recognized only about 1981, after data on cobalt content as well as the distribution of the co-crusts had accumulated. Figure 19.13 shows that extensive areas of such crusts occur within the EEZ of the Hawaiian Islands and other islands under United States jurisdiction in the Pacific. The most favorable areas

Figure 19.12 (a) Manganese nodules on the Pacific Ocean floor, and (b) emptying a dredge of nodules. (NOAA.)

are where the sea floor is at least 20 million years old and preferably over 80 million years. Moreover, the best deposits occur where other sediments are prevented from accumulating; hence the importance of strong currents to keep the tops and sides of seamounts free of biogenous oozes. For this reason we see the encrustation on basaltic boulders as in Figure 19.14a. Co-crusts often appear together with phosphorite deposits, in which case the combined mineralization appears as a kind of "pavement" over

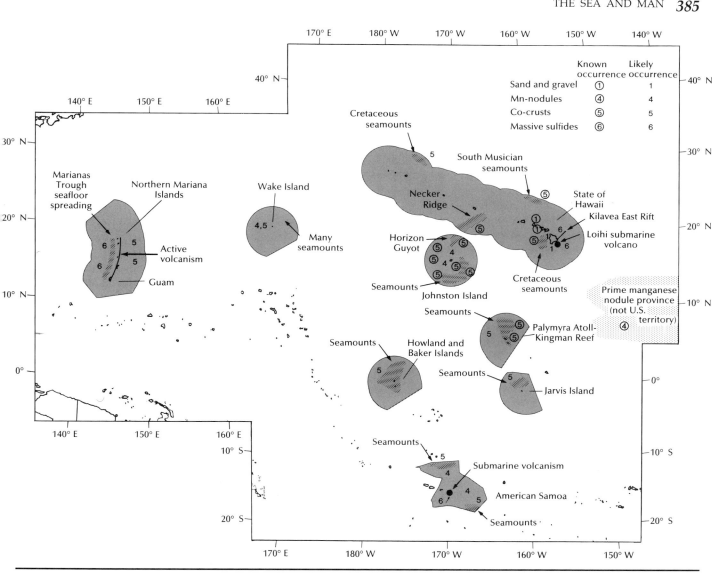

Figure 19.13 Deposits of hard minerals within the EEZ of the United States, Hawaii, and Pacific island territories. (From Ballard and Bischoff, 1983.)

the bottom. The photograph of Figure 19.14*b* is from a station over the Blake Plateau off Florida and shows such pavementlike crust mixed with nodules at about 800-m depth, which is quite shallow.

The monetary value of the cobalt makes the co-crust deposits considerably more valuable than the deep-sea nodules. At 1983 prices, a single seamount could yield over $1 billion worth of ore, and there are over 100 seamounts within the United States EEZ in the Pacific (USGS Survey Circular 929). These factors, together with the shallower depths of co-crust deposits compared to those deep-sea nodules, will probably dictate that initial offshore mining by the United States will be for these co-crusts.

Phosphorite Sands, Crusts, and Nodules. Phosphorite deposits occur in many places within the United States EEZ, but the most extensive and economically attractive deposits are on the Blake Plateau off Florida and off California's coast (Figure 19.11). *Phosphorite* is a term applied to a sedimentary deposit comprised mostly of phosphate minerals such as *apatite* and fluorapatite.

It appears that the phosphates are deposited wherever there is extensive upwelling, as in coastal waters. Many of the richer deposits occur at depths between 30 and 300 m or are associated with the shelf break. This association with upwelling zones follows directly from the fact of high primary pro-

Figure 19.14 (*a*) A typical ferro-manganese encrustation of basalt boulders dredged from the Horizon Seamount in 1983 by the United States Geological Survey vessel *Lee*. See Figure 19.13 for the location of the seamount.

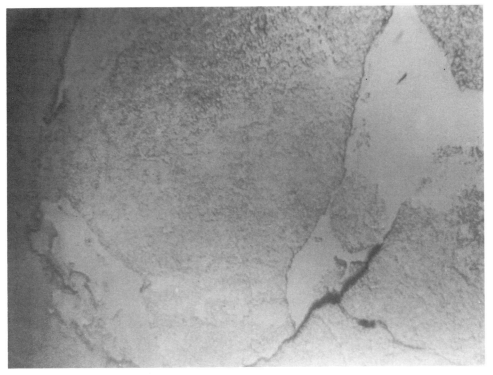

(*b*) Ferromanganese-phosphorite pavement on the Blake Plateau off Florida, in about 800 m of water. The field of view is about 5 m across. (Photos courtesy Frank T. Manheim, USGS.)

duction in nutrient-rich upwelled waters; marine plants fix the dissolved phosphates into organic compounds, and these in turn accumulate in high concentration in the organic sediments. It is believed that a low level of oxygen content within these decomposing organic layers is a necessary condition for the deposition of phosphorite. Deposition also seems to require *low* rates of deposition of *other* types of sediments such as biogenic materials or clastic sediments eroded from continents. On the Blake Plateau, from Florida to the Carolinas, phosphates are deposited because strong currents of the

Gulf Stream prevent other sediments from forming. Off California, the deposits tend to occur on ridge tops, flanks of seamounts, or other such areas where extraneous sediments do not accumulate. In some areas the deposits appear as uncemented sand grains and in others as cemented crusts.

The demand for phosphate for fertilizer continues to rise worldwide. The United States presently exports phosphate taken from strip mines in Florida; these are (uplifted, ancient shallow-water deposits). But as land-use restrictions increase, with correspondingly increased costs of production, the commercial exploitation of the extensive shallow marine deposits on our continental shelves will become a reality.

Polymetallic Sulfides

Even more recent are the discoveries of metallic sulfide deposits associated with the hydrothermal activity in the mid-ocean rift zones. (Chapter 21 describes the rifting process at mid-ocean ridges.) The first discovery of a submarine "hot spring," more formally called a hydrothermal vent, was on the Galapagos Rift in 1979 when scientists first descended to a rift zone in the deep submersible vehicle ALVIN. Since then additional discoveries have been made (Figure 19.11) on the East Pacific Rise at 21°N, the Juan de Fuca Ridge at 48°N, and the Gorda Ridge at about 42°N. The latter discovery is within the United States EEZ and is thus very important.

Geologists understand the mineralization process for these sulfide deposits quite well because they had earlier studied similar ancient deposits now mined on land. The copper mines of Cyprus are an example. The mineralization typically includes iron, manganese, zinc, lead, and silver. Figure 19.15 indicates how the deposits are formed at the East Pacific Rise (Scott, 1983). Cold seawater percolates down into the fractured, newly solidified oceanic crust and is then heated and forced to convect upward as its density is lowered through heating. Exposed to new and still hot rock, heated seawater leaches metal ions from the rock and also gives up some ions to the rock. When the metal-rich hot solution is vented into the cold seawater overlying the rift zone, precipitation occurs almost immediately; the area around a vent becomes a deposit of sulfides of these various metals.

Whether such deposits will be economical to mine depends on many factors. Where the rate of sea floor spreading is low, zones will have poor mineralization because the hydrothermal activity is weak.

Both the East Pacific Rise and the Galapagos Rift, zones examined to date, have medium spreading rates, about 5 cm/yr, and appear to have a potential for mining. In addition, magma chambers that feed different rift areas have different compositions, so that their resulting mineralizations will be different, some rich in silver, others rich in zinc, and so on. For example, some samples taken from the Juan de Fuca Ridge assayed as follows: zinc, from 29.7% to 59.2%, value from $237 to $473 per ton; silver, from 124 to 290 ppm, value from $30 to $72 per ton (McLain, 1983). This is indeed rich ore. But the deposits occur in small areas immediately around the ventholes. Although a single vent deposit may be several meters thick, unless there are numerous such deposits in a given region, mining may not be

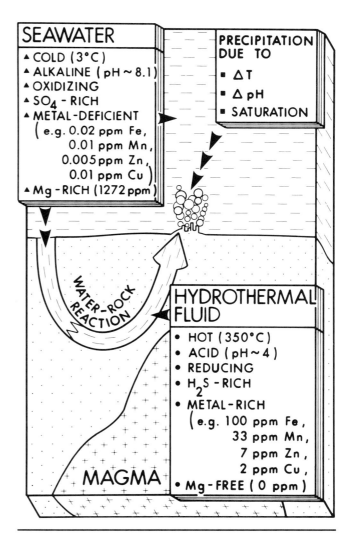

Figure 19.15 A model of how the metal sulfide deposits are formed. (From Scott, 1983.)

economical. One scheme for mining would use a standard clamshell bucket (Figure 19.16), but with precision navigation and propulsion equipment aboard the mining ship. A clamshell device equipped with a visual monitor and propulsion could be quite selective in its "grab." The photograph of a hydrothermal vent area, Figure 19.17, indicates that accurate placement would be essential to "grab" the minerals and avoid the rocks.

The metallic sulfides are not immediately consolidated after precipitation from the hot brine solution. They exist initally as "muds" in the vicinity of active vents and collect in depressions. A clamshell device would not recover such deposits, but some type of submersible pump might be used instead. Such pumps have been suggested for "mining" the brine solutions found at the bottom of the Red Sea.

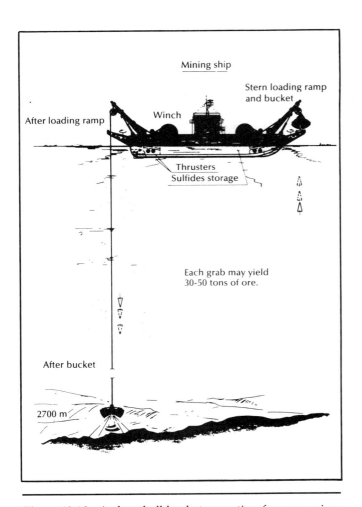

Figure 19.16 A clamshell bucket operation for recovering consolidated crusts of metal sulfides. (From Amann, 1983.)

Figure 19.17 A polymetallic sulfide ledge near a submarine vent on the Gorda Ridge; see Figure 19.11 for the location of this ridge. Snails on this ledge are about 2-cm size and are part of the vent community living at depths of over 2000 m in this sector of the Pacific Ocean. (Courtesy David Kadko, Oregon State University.)

The Impact of Modern Urban Society on Oceans

Humans have always impacted coastal oceans. Today we find the oldest evidence of human life and activity on the shores of an ancient African lake, in what is now called the Olduvai Gorge; the evidence is several millions of years old. Our kind always impacts the coastal seas because we literally live *on* them. But it is modern urban society that has transformed the impact from one of dependency on seas for food to a suite of *other types of dependency*, of which the major is the *use of oceans as waste dumps*.

How do modern humans treat the oceans differently than their predecessors? Although there are many more of us, we still converge to live in narrow strips along the coastal ocean (Figure 19.18*a*). About 65% of all humans live within 500 km of the shore-lines. Coastal human density is growing not only with the population at large but also at the expense of rural people density. In the case of the United States, this coastal area in which the majority of our people live is almost doubled by the addition of the EEZ area to 200 nmi (360 km); the Great Lakes area is considered here as the equivalent of ocean coast-line (Figure 19.18*b*).

The major difference today is that *we move things into and out of our coastal zones*, a result of the fact that the coastal zones cannot both house us *and* feed us. Worldwide, we now use 100 million tons of fertilizers to sustain agriculture. The phosphate fertilizer used in the United States comes mainly from

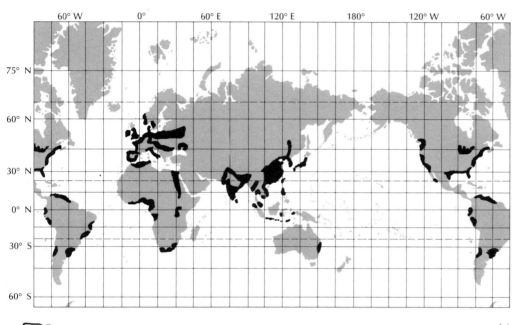

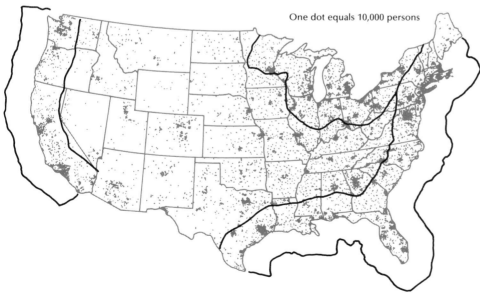

(a) About 65% of humans live within 500 km of a major water body. Dark areas indicate a population density greater than 75 persons/km^2.

(b) The United States population distribution from the 1980 census; the lines show the 200-nmi EEZ ocean boundary and the 500-km inshore distance.

One dot equals 10,000 persons

Figure 19.18 How human populations are distributed.

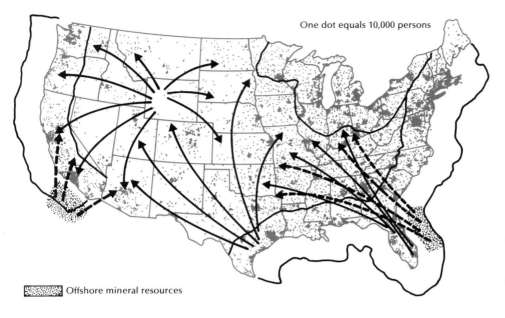

One dot equals 10,000 persons

Offshore mineral resources

(a) The initial transport of phosphate and nitrate is from mining areas into the Great Plains for cereal production and to the subtropical belt, Florida across to California, for vegetable and fruit production.

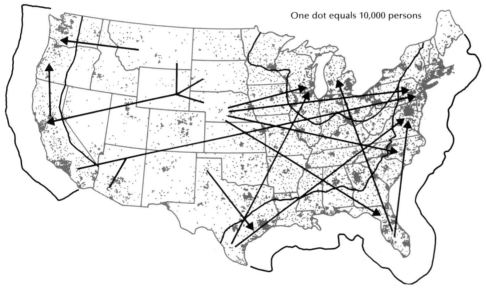

One dot equals 10,000 persons

(b) The majority of foodstuffs produced are shipped to the coastal zones for consumption and eventual release to the local environment.

Figure 19.19 How agricultural chemicals become concentrated in coastal zones.

two regions, and both are ancient shallow-sea deposits of phosphorite (Figure 19.19*a*). Phosphate is strip-mined from uplifted, ancient seabeds in Florida and then distributed widely over the interior of the United States as well as exported. Ancient shallow-water deposits of phosphate rock in Wyoming, are mined and distributed to the high-output agricultural areas of California, Arizona, and the wheat-growing regions of the Central Plains. As the Wyoming beds become exhausted or the Florida beds im-

pacted by environmental restrictions, the industry will turn to the shallow beds off the California and Florida shores (see Figure 19.11).

Nitrate fertilizer today is manufactured for the most part from natural gas; the petrochemical plants of Texas and Louisiana send much nitrate into the interior. It too is widely distributed to the food crop producing areas.

In turn, we transport huge amounts of foodstuffs and other raw products from the interior of the

United States into the densely populated coastal zones. For example, each food item we eat has been moved *an average of 1300 miles* before it is set upon the dinner table. Figure 19.19*b* shows the cereals of the great plains moving into the coastal zones for consumption by the two-thirds (180 million) of us who live there. When one coastal zone moves material *out*, the foodstuff usually travels directly to another coastal zone: for example, produce from the coastal plains of Texas and California and Florida is sent to the Northeast coastal strip.

The average quantity of foodstuff consumed per person in the United States is about 700 kg/yr. Thus the 180 million coastal people require 126 million kg, or 126 million tons, of food alone delivered to the coastal zones each year. Now remember that the phosphates and nitrates in our food do not enter a permanent sink in our bodies, for most is passed on immediately and all returns to earth eventually. The point is clear: much of the materials we bring into our coastal zones is either new stuff from the interior or a recycling of material that we had sent into the interior for further processing.

So the real difference between modern society and that of our ancestors is the *collection of enormous amounts of material from far distances and their transformation within our coastal living sectors into products and waste*. The phosphates and nitrates that were not returned in the form of food and fiber will, for the most part, be recycled back to the coastal zone by the hydrologic erosion processes. And coastal oceans, because they occupy the lowest elevations, are the eventual sink for our waste products.

Fishery products are also collected into our urbanized coastal zone, where they are either consumed or processed for transshipment to the interior of continents. Most of the refuse of fish, for example, remains within the zone, either as waste or for further processing into fish meal. Petroleum is a *large import* to the coastal zone, where most of it is consumed as fuel or as raw material for manufacture. Out of the total world oil production in 1969 of 1820 million tons, 1180 million tons entered coastal zones by tanker transport. And even if the petroleum is locally produced, it still represents an import in the sense of being extracted from depths of several thousands of meters and released at ground level, directly into our human coastal habitat. When seabed mining becomes a reality, it too will extract material that is loosely distributed over the sea floor and will concentrate the ore into the coastal zone for processing and consumption.

The point is made! Ocean pollution is concentrated within coastal zones and its impact will increase. How serious is it, what are the pollutants, and how does coastal oceanography fit into the picture? We will deal with these questions only in a general way, recognizing that whatever spills to the seas can be *good* or *bad*, sometimes both, and almost always a question of *degree*. Further, we must realize that we know almost nothing about long-term effects of toxic materials, particularly at low levels of exposure.

Specific Pollutants: An Interaction Matrix

In general, the way to evaluate whether a pollutant has an *impact* is to develop an *interaction matrix* that outlines where, how, and how much interaction can occur between a pollutant and the environment. Table 19.4 is one such matrix, and the following sections discuss some of its intersections.

Table 19.4 A matrix of specific pollutants matched against the general environmental concern or impact on the coastal zone environment

Specific Pollutant	Concentration and/or rate of buildup	Rate of mixing and transport from the region	Rate of biodegradation	One-of-a-kind event, or the amount of future addition	Food chain concentration factor	Effect of the pollutant on vital life processes	Specific biology and chemistry of pollutant interaction with the environment
Trace metals	A				A	A	
Plant nutrients	B	B					
Organics, petroleum			C	C			
DDT, DDE, PCB, etc.			D		D	D	
Solid wastes	E					E	
Radioactivity		F					F
Pathogens		G				G	
Thermal (heat)	H	H				H	

Trace Metals (Matrix Points A)

Trace elements are defined as those occurring in natural concentrations of 1 ppm or less. An earlier discussion in Chapter 7 pointed out that many of these have distributions in the vertical water column that are similar to, and correlated with, plant nutrients. We know that many marine organisms concentrate trace metal ions; for example, Table 19.5 shows that shellfish flesh may contain lead at concentrations thousands of times greater than that in open-ocean waters (Brooks and Rumsby, 1965). Therefore, a first approach to determining whether the increases in the concentrations of the heavy metals that we generate are harmful is to compare open-ocean concentrations with the enrichment that the organisms themselves perform. This comparison easily suggests that human-introduced increases in heavy metals are of little consequence. The fact that the natural distribution of these trace elements closely resembles that for nutrients suggests instead that organic life requires them to be present.

But there are cases of severe harm caused by excessive concentrations of one or more of these heavy metals. Coastal shellfish banks near the mouth of the Quintero River in Chile virtually disappeared after a copper smelter began discharging waste water into the river. In Minamata, Japan, during the 1950s, hundreds of citizens contracted mercury poisoning by eating fish and shellfish from Minamata Bay. The mercury entered the bay in the waste waters from local industry. Survivors retained permanent body damage, in visual impairment, incoherent speech, unsteady weight gain, and hearing deficiency.

Marine plants can also be impacted. Experimental studies show that the photosynthesis process in the giant kelp *Macrocystis* off California coasts is 50% deactivated after a 4-day exposure to mercury at 0.05 ppm, copper at 0.1 ppm, nickel at 2.0 ppm, and zinc at 10 ppm (Merlini, 1971).

What is surprising is how much pollution reaches the ocean via the atmosphere (Table 19.5). Each year 350,000 tons of dry-cleaning solvents evaporate into the air, along with 1,000,000 tons of gasoline (314,000,000 gallons!) (Hood and McRoy, 1971). The 10% of lead produced by mining that moves to the oceans by means of air comes from burning leaded gasoline. (As of now, leaded gasoline is to be phased out of production in the United States).

As for the heavy metals carried by rivers much of their amounts never reach the open ocean. Instead, the metal ions appear to become complexed with, or attached to mineral particles in the suspended sediments, and so are deposited either in the estuaries or on nearshore sands. This suggests that river contamination is quickly transferred to the estuarine or nearshore biosphere, especially to benthic habitats (see also Chapter 5).

Plant Nutrients (Matrix Points B)

There are two paths through which human-generated phosphates and nitrates enter the ocean: through the disposal of domestic wastes and via the water runoff from agricultural watersheds where phosphate and nitrate fertilizers are applied. Both of these souces are implied in the distribution charts of Figure 19.19. Phosphates and nitrates may be directly applied to land crops inside the coastal zone or imported from the fertilized interior lands in the forms of food and fiber or carried as dissolved nutrients by river runoff.

Nutrient additions to coastal waters have the greatest impact when they cause specific plants to grow excessively, to the point that species diversity is reduced as favored plants overwhelm other species. In freshwater lakes and rivers this impact is called "eutrophication." Wherever it occurs, the dynamics of the entire food chain can be altered. In

Table 19.5 World natural and mining production of heavy metals, their principal routes into the oceans and the heavy metal enrichment ratios found in selected marine organisms. (Based on data compiled from various sources)

Metal	Natural Production, 10^3 tons	1970 Mining Production, 10^3 tons	Transport by Rivers to Oceans, 10^3 tons	Atmospheric Washout, 10^3 tons	Concentration in Open Ocean	Enrichment Ratio in Upper 200 m of Open Ocean	Enrichment Ratio in Marine Shellfish		
							Scallop	Oyster	Mussel
Lead	180	3,000	100	300	0.00003	11 to 1	5,300	3,300	4,000
Copper	375	6,000	250	200	0.002	2 to 1	3,000	13,700	3,000
Cadmium	—	10	0.5	10	0.0001	4 to 1	2,260,000	318,000	100,000
Nickel	300	500	10	30	0.002	1.04 to 1	12,000	4,000	14,000
Zinc	370	5,000	700	—	0.01	1.25 to 1	28,000	110,000	9,100

THE SEA AND MAN **393**

fact, one of the procedures that biologists use first when studying a potential impact is to measure whether species diversity in the food chain has been reduced, that is, whether some species have been eliminated. Another measure of impact is a sudden change in abundance of one species over another.

That the problem can be severe is shown in Figure 19.20. Sewage disposal outlets distributed over the highly populated coasts of Florida have a combined output of about 500 million gal/day (about 1.6 million m³/day, or about 1/100 of the Nile River flow). Using the average concentrations of total

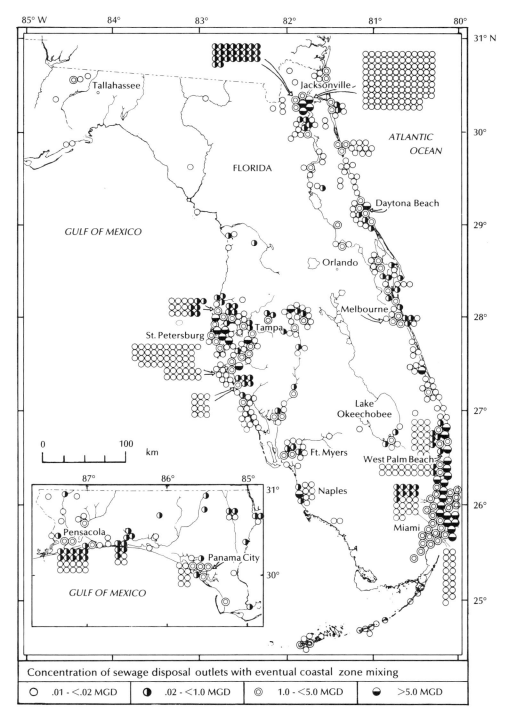

Figure 19.20 Distribution of sewage outfalls to both coastal ocean and coastal rivers in Florida. (From Haddad, Manheim, and Marr, 1976.)

Concentration of sewage disposal outlets with eventual coastal zone mixing

| O .01 - <.02 MGD | ◑ .02 - <1.0 MGD | ◎ 1.0 - <5.0 MGD | ◖ >5.0 MGD |

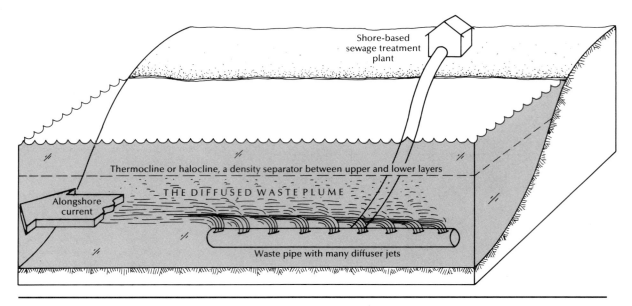

Figure 19.21 The diffusion of liquid sewage waste into coastal oceans. The diffuser pipe is installed below the principal density interface, and the waste is jetted into seawater at high velocity so that it will entrain (mix with) large volumes. The mixture is thus adjusted to a density insufficient to break through the interface into the surface layer.

phosphates and nitrates in household sewage as a basis (Foyn, 1971), we compute that about 100 tons of these plant nutrients are injected every day into the coastal zone of Florida. To put this amount into perspective, it is a *daily* equivalent to the total fertilizer used *annually* on about 1000 acres of farmland. Over a year's time, this equivalent land fertilization would grow to 365,000 acres, or about 600 square miles. Again, the reader should bear in mind that these are *new* nutrients, imported into the coastal zone as discussed earlier.

Domestic sewage that is piped directly to sea is usually ejected from a series of nozzles called diffusers (Figure 19.21). The purpose is to force enough mechanical mixing between the waste fluid, which is almost always less dense than seawater, and the surrounding seawater itself, so that the mixture eventually becomes denser than the overlying, warm ocean surface water. The plume of waste water–seawater mixture will then rise no higher than the density barrier that occurs at the thermocline. This way, no sewage material reaches the surface, and coastal currents eventually transport the material away from shore.

Petroleum (Matrix Points C)

A 1971 report by the National Academy of Sciences summarized how petroleum is involved with the

marine environment. The reader can extract several perspectives from a study of Table 19.6.

First, *natural* seepage of oil into marine waters is of the same magnitude as that lost by seepage from offshore producing wells. Second, the amount of oil injected into the seas by *operations* of tankers and other ships is *five times* as large, *each year*, as that lost in the Torrey Canyon ship accident and the drill-well blowout at Santa Barbara, California. Third, the largest amount of oil injection by direct human activity stems from the day-to-day industrial and commercial activities within the coastal zone (0.8 million ton/yr) and deliberate dumping (0.5 million ton/yr). These are the cumulative effects of many small injections: dumping waste oils into sewage, losses during transport and handling, washing down service stations, and many other domestic and industrial activities. But note that the largest single injection of petroleum pollutants occurs through the vaporization of volatiles of petroleum and its derivatives, an indirect path from human society to the oceans.

Oil is washed into the sea during tanker operations because ballast water is pumped into and out of the ship's tanks, inevitably emulsifying oil with water and pumping it overboard. The shipping industry is modifying its procedures and installing oil-capture equipment to reduce such contamination.

Much research and development has focused on techniques to reduce the severity of local oil contam-

Table 19.6 How petroleum is involved with the marine environment. (Based on National Academy of Sciences Report 1973.)

	Millions of Tons per Year
World oil production (1969)	1820
Oil transport by tankers (1969)	1180
Oil that is injected into the marine environment directly through human activities	
Offshore oil production (seepage from wells)	0.1
Tanker ship and other ship operations	1.0
Refinery, industrial, and automotive wastes	0.8
Deliberate dumping	0.5
Accidental spills: Torrey Canyon tanker (0.117), Santa Barbara offshore (0.011), plus others	<u>0.2</u>
Total impact directly related to human activity	2.6
Natural seepage into the marine environment	0.1
Atmospheric input from continents through vaporization of petroleum products	9.0*

*Based on assumed 10% return to ocean of all petroleum hydrocarbons entering the atmosphere (SCEP Report, 1970).

ination, such as during a tanker accident. Spills that occur in bays and harbors can now be contained by deploying a curtain of plastic material around the spill (Figure 19.22). Spills that occur in open waters, and thus subject to strong wave and wind forces, rapidly spread out. The Santa Barbara spill generated a slick that covered about 800 square miles, but it was dissipated in about 2 weeks, mostly by microbial decomposition. The role of bacteria in oxidizing crude oil is very important, so much so that serious research is now in progress to develop a "super-bacteria," one that could be released onto an oil spill and make "short shrift of it indeed." Most recently, it has been learned that emulsifying oil with water is an important first step in promoting bacterial decomposition of the oil. So it seems that the best way to disperse and remove an oil slick at sea is to marshal as many small boats as are available to pass back and forth through the slick, their propellers generating the emulsion that supports rapid bacteria production.

Figure 19.22 Photograph of a typical oil containment curtain installed within a harbor. (Henri Bureau/SYGMA.)

Solid Wastes (Matrix Points E)

Solid wastes cover a wide assortment of waste materials, from construction debris to the sludge solids from sewage treatment plants. In fact, the first major United States pollution control law was the Rivers and Harbors Act of 1899; it was formulated in part because dumped construction debris was clogging harbor channels in the New York area.

The dumping of sewage sludge has increased significantly during the past few decades. Sewage sludge is the mix of liquids and solids produced during sewage treatment, and its disposal has been a problem of growing concern. We know now that many harmful elements attach to particles, as when DDT and PCB adsorb to silts suspended in fluids or when heavy metal ions attach to clay particles. Thus sewage sludge is really a concentration of toxic elements. Field research shows that sewage sludge dumped in shallow continental shelf waters has a long-lasting impact on the surrounding benthic community. In some cases the residual organic material in the dumped sludge may demand and consume too much of the local supply of oxygen dissolved in the seawater, creating respiratory problems in local marine life.

Another category of solid wastes are the gravels, sands, and muds dredged from harbor bottoms and barged out to dumps on the continental shelf. At first thought such material seems harmless, of no potential threat to marine life in the area where it is dumped. However, toxic materials from sewage and industrial wastes, so often introduced into inland waters, tend to *adsorb* to particles of muds and silts, and so are readily concentrated in the very materials that need to be dredged. Dumping dredge spoils in another area simply moves the pollutant a bit farther offshore. Dredging and dredge spoil dumping also have a direct effect on the benthic organisms that are mucus-type filter feeders. The suspended silts turbulently released near the bottom can have disastrous effects on local populations of clams, for example.

Today, all aspects of ocean dumping in the United States are regulated under the 1972 Marine Protection, Research and Sanctuaries Act (or Ocean Dumping Act). The London Dumping Convention of 1976 is an international program through which information about dumping practices is exchanged between member nations.

Radioactivity (Matrix Points F)

Radioactivity is an extremely complex problem. Here we shall touch only on two aspects: first, the discharge of *low-level* radioactive nuclides from many dispersed sources, with the corresponding question of its long-term impact; and, second, the question of deep-sea disposal of *high-level* radioactivity, such as that encountered in *spent* reactor fuel and in spent reactors themselves.

Figure 19.23 explains radioactivity as a process by which an initially unstable atom spontaneously gives up some internal energy in order to arrive eventually at a stable atomic structure. The loss of energy proceeds through a specific chain of discrete events, some of which are shown for the initial, unstable atom of uranium-238. The number that accompanies the name of the atom (238 here) is the sum of the number of protons and neutrons in its nucleus. As the atom loses protons and neutrons during disintegration, this sum number decreases; in effect, the material remaining in the atom is then a *different* element. Each element that occupies a step in the decay sequence, even if for a very short time, is called a *radionuclide*; each has a unique half-life, defined as the period of time over which exactly one-half of an initial number of atoms will have spontaneously degenerated to the next step down the sequence.

Figure 19.23 also explains that the discrete energy losses during decay are of three types: *alpha particles*, *beta particles*, and *gamma radiation*. Alphas are the most massive; each is actually the core of a two-proton–two-neutron helium atom. Their large size means that alphas penetrate only a short distance before being stopped (this sheet of paper stops most alphas). But because of their massive size, *they disrupt a large number of atoms per molecule in a short distance*—this is the meaning of "ionizing radioactivity." If the radionuclide disintegrates while inside the human body, for example, the ejected alpha will ionize thousands of molecules of tissue inside a small volume, and this is what makes *alpha radiation* more harmful to humans than either of the other radiation forms. Beta particles are single electrons. They have strong ionizing power but spread their damage over a much larger volume, for they can penetrate deeply into material before losing their energy. Gamma rays are the most penetrating of all radiation forms; lead shields are needed to protect against them.

How to Quantify Ionizing Radiation? One measure of radioactivity is to specify its number of disintegrations per second, called becquerels; 37 billion of these are called a *curie*. We can now distinguish between low-level and high-level waste material. *Low-level* waste material emits less than 370 becquerels per gram of mass and *high-level*-waste material emits

WHAT IS RADIOACTIVITY?

The atoms of some elements are *unstable*. Their atomic structure changes with time, and with each change they lose some internal energy. It is this ejected energy that damages living tissue.

This example begins with an unstable element uranium-238. Each 4,500,000,000 years exactly half the existing ^{238}U atoms decay one step down the chain by losing energy. Energy is lost in three ways.

1. The ejection of alpha particles.   Each alpha particle takes with it two protons and two neutrons, leaving 234 of these in the remaining nucleus, which is then classified as a different element, thorium-234. The half-life of ^{234}Th is only 24.1 days.

2. The ejection of beta particles from the ^{234}Th nucleus. Beta particles are high-energy electrons, very small in mass but ejected at very high speeds. The loss of one beta particle from ^{234}Th converts the remainder to protactinium-234. Its number of neutrons and protons is the same as for ^{234}Th but not for very long. Within 1.14 minutes, half of all protactinium atoms eject a beta particle to become a more stable uranium-234.

3. The emission of gamma rays.   When the nuclei are required to undergo structural changes, as they do when they eject particles, the events are usually accompanied by the emission of packets of energy in the form of electromagnetic waves, otherwise called photons, radiation, or just radioactivity, hence the origin of the name.

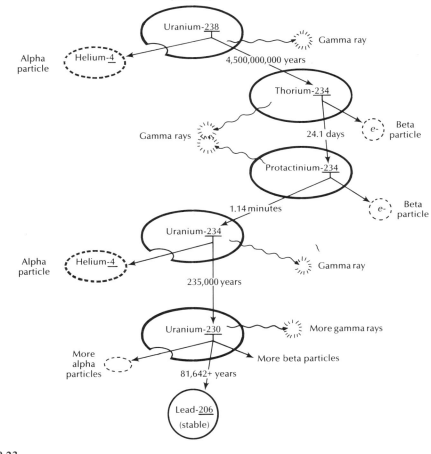

Figure 19.23

over 37 billion becquerels of alpha particles (1 curie) or 3700 billion of either beta particles or gamma rays.

Radiation Dose for Humans. Radiation *dose* measures the amount of biological harm that the human body can tolerate. The lethal dose that causes death to 50% of those exposed (LD-50) is one such mea-sure. It is set at 5 sieverts per year. For beta and gamma exposure, each sievert unit means an energy absorption of 1 joule/kg of body weight. For alpha particles, which are more lethal, a sievert is set at from 0.1 to 0.05 joule/kg. Annually, we each absorb about 0.0013 sievert from natural causes, 30% of which comes from the natural radionuclides carbon-

14 and potassium-40 inside our bodies; we get an additional 0.01 from medical and dental care. Altogether, we receive about 0.05% of the LD-50 each year (Park et al., 1983).

Radiation Dose for Marine Organisms. Much research had been done on the effects of ionizing radiation on specific marine organisms. Experimental studies are made using high dose rates and short exposure time, and results are reported as LD-50 numbers for exposure to specific radionuclides. We now know that it takes about 100 times more exposure to kill a protozoan, for example radiolarians and bacteria, than is needed to kill a human, which suggests that the more complex the organism, the more susceptible it is to radiation damage. Other studies show that the radiation dose needed to kill adult fish can be 30 times *more* than the LD-50 dose for the embryo stage of the same fish. This suggests that radiation has its most damaging effect on reproduction, and that we must not base a strategy for radioactive waste disposal on the response of adult organisms!

Field data on exposure to low-level waste products from nuclear power plants are now available from a number of years of discharge of such wastes from the Windscale power plant in Great Britain. Table 19.7 compares measured exposures of both pelagic and benthic species from the Irish Sea near the Windscale outfall to exposures of similar organisms in the natural environment. Some interesting points can be drawn from these data. First the largest amount of exposure occurs in the food chain phytoplankton–zooplankton–pelagic fish for *natural* radioactivity. This is a straightforward example of how

trace materials accumulate to the higher trophic levels. The natural radionuclides here are probably carbon-14 and potassium-40. Strangely, accumulation of Windscale artifical radionuclides occurs from phytoplankton to zooplankton, but not to the pelagic fish. Second, mollusks receive most of their exposure from the sediment radioactivity, suggesting that the Windscale *zooplankton are gathering into their fecal pellets radionuclides that then pass directly to the benthic realm.*

Radioactive Waste Control Strategy: Maximum Permissible Concentration or Critical Pathway? A pressing demand on marine science today is to discover precisely the maximum permissible concentration (MPC) of ionizing radiation that can be safely tolerated. In this respect, regulations for discharge of radioactive contaminants are designed for *human safety*, not for animals. Fortunately for marine animals, it appears that if we set acceptable levels for human safety, they are automatically protected because of their higher tolerance levels. But unfortunately we cannot accurately determine MPC numbers for extended exposure to low-level radiation by extrapolating from LD-50 studies done at high dose rates for short periods. This is the critical problem and defines the critical need.

A control strategy depending on manipulation of the *critical pathway* has been applied in some locations (Table 19.8). A critical pathway is the means by which radiation contamination is the most rapidly and intensively passed on to humans. The strategy is to identify the many food and recreation links between humans and the local coastal ecosystem, and to set nuclear waste discharge rates so that the most

Table 19.7 Examples of natural and nuclear waste radiation dose rates* taken by marine pelagic and benthic species in coastal waters. (From Park et al., 1983.)

Radiation Source	Phytoplankton[†]	Zooplankton	Pelagic Fish	Mollusks
Natural background ‡				
Cosmic radiation	40	40	40	40
Internal radioactivity	200–600	200–30,000	100,000–800,000	100,000
Water radioactivity	30	20	10	10
Sediment radioactivity	—	—	—	100–2,000
Waste disposal from the Windscale, U.K., Nuclear Plant				
Internal radioactivity	20,000–200,000	50,000–600,000	40–100	1000–5000
Water radioactivity	20–300	20–300	1–200	4–100
Sediment radioactivity	—	—	—	3000–300,000

*Units are microsieverts per year equivalents.

[†]Phytoplankton, zooplankton, and pelagic fish were sampled at 20-m depth and mollusks on the seabed.

[‡]Natural background data were taken in waters free of artificial radionuclides.

Table 19.8   Some critical pathways and the individuals affected by reactor, chemical reprocessing, and research radionuclides identified at selected coastal sites. (From Templeton, 1980.)

Site	Critical Pathway	Principal Exposed Group
Hanford, Washington (plutonium production reactors)	Oyster flesh	General public on coast
Windscale, U.K. (chemical reprocessing)	Fish flesh	Local fishermen
	Porphyra (seaweed)	General public (recreation)
	Estuarine sediment	Local fishermen
Bradwell, U.K. (power reactors)	Oyster flesh	Oyster fishermen and families
Dungeness, U.K. (power reactors)	Fish flesh	Local fishermen and families, bait diggers
	Beach sediment	
Dounreay, U.K. (chemical reprocessing)	Detritus on fishing gear	Local fishermen
La Hague, France (chemical reprocessing)	Fish flesh	Local fishermen
	Seaweed	
Trombay, India (research and development)	Fish flesh	Local fishermen
Tokaimura, Japan (chemical reprocessing)	Fish	Local fishermen
	Shellfish	

direct link to human beings will result in lower radiation exposure than the maximum allowed annual dose.

Disposal of High-Level Radioactive Wastes. There is widespread and growing concern about how to dispose of high-level radioactive wastes. At this time, a primary source of this waste is in the spent fuel rods from nuclear power reactors. Table 19.9 lists the major radionuclides contained in the spent fuel rods and how exposure to them is the most likely to damage the human body. It is an interesting list. Note that the most intense radiation comes from the group 1 radionuclides with a short half-life, fewer than 5 years. These should be, and are, simply stored in temporary holding tanks for about 10 years, during which their activity will rapidly decay so that eventually they can be disposed of as low-level wastes. This is not to imply that no danger exists. In fact, the cesium-144 radionuclide is extremely dangerous because it easily enters into biological reactions. However, none of the radionuclides in this group is an alpha-particle emitter; these radionuclides emit only beta and gamma rays.

The second group has members with half-lives up to about 100 years and are still dangerous 100 years after disposal. Some members are also alpha-particle emitters, particularly the plutonium radionuclides; it is this characteristic that makes plutonium so dangerous to living tissue.

The radionuclides with the longest half-lives are in the plutonium series, with that of plutonium-239 the longest at 24,100 years. However, their activity level is the lowest of all listed radionuclides.

What can be done with this stuff? The possibilities are few.

1. We can continue to store the wastes in protected land areas for the present. Although securing wastes and guarding them for public safety is a major cost of such storage, there is one advantage that overwhelms others. It is rooted in the idea that these radionuclides are today's wastes only because we do not know of a use for them. Continued storage for 100 years into the future, assuming technology advances at a rate similar to that of the past century, would assure that the radionuclide *resources* remain at hand. The plan avoids decisions today that might be irreversible, or stupid, or both.

2. We can prospect for the most quiescent piece of the sea floor and then bury the wastes in canisters deep within the sediment. By the time sea floor spreading will have carried the material to a distant subduction zone the required number of half-life decay periods will have elapsed even for the plutonium series. As of this writing, this scheme is the one most actively being explored.

Pathogens (Matrix Points G)

Studies are needed to document the incidence, duration, and number of types of virus found in coastal waters. We may also need to examine the interaction between human viruses and marine organisms, and

ionuclides in spent fuel rods from nuclear power plants, their radioactivity energy and the body organs
cted. (From Park et al., 1983.)

Radionuclide	Half-life, yr	Type of Radiation*	Critical Organ	Radiation in density Years after Disposal[†]				
				0	10	100	1000	10,000
Group 1								
^{144}Ce	0.778	β,γ	Bone, Gi tract[‡]	78,000	11	—	—	—
^{106}Ru	1.01	β,γ	GI tract	41,000	41	—	—	—
^{134}Cs	2.062	β,γ	Whole body	8,900	310	—	—	—
^{147}Pm	2.6234	β,γ	GI tract, bone	4,100	290	—	—	—
^{125}Sb	2.7	β,γ	Kidney	480	36	—	—	—
^{60}Co	5.271	β,γ	GI tract	330	89	—	—	—
Group 2								
^{85}Kr	10.7	β,γ	Lung, whole body	400	220	0.67	—	—
^{3}H	12.33	β	Whole body	28	16	0.11	—	—
^{241}Pu	14.4	α,β,γ	Liver, lymph nodes	4,800	2,700	24	—	—
^{244}Cm	18.11	α,γ	Bone	72	49	1.4	—	—
^{90}Sr	28.8	β,γ	Bone	5,700	4,400	490	—	—
^{137}Cs	30.17	β,γ	Whole body	7,700	6,100	760	—	—
^{238}Pu	87.74	α,γ	Liver, lymph nodes	85	78	38	—	—
Group 3								
^{240}Pu	6570	α,γ	Bone, red marrow	18	18	18	16	6.3
^{239}Pu	24,100	α,γ	Bone, red marrow	12	12	12	12	8.9

*α is an alpha particle, β a beta particle, and γ a gamma ray.
[†]Radiation intensity units are becquerels x10^{12}.
[‡]Gastrointestinal tract.

how these pathogens can be transported through food organisms.

Thermal Buildup (Matrix Points H)

In recent years, as the total generation capacity of a single power plant has increased to 1000 megawatt (MW) the plant's need for cooling water can easily exceed the cooling capacity of small, local rivers. Then an oceanic plant site is selected for its much higher cooling capacity. What are the main criteria to consider in evaluating the impact that such a plant might have on the local coastal community?

First, recall that metabolic activity rates increase with temperature, so the warm-water effluent from a generating plant can have positive values. For example, such water can be further processed through an aquaculture facility where growth rates and profits can be enhanced. Along shorelines where the bottom steepens quickly, the cold-water intake might be positioned below the thermocline so that the warm-water discharged later is not only warm,

and density stable on top, but also nutrient-rich, creating a type of artificial upwelling.

On the other hand, alongshore currents are always somewhat turbulent, so that the warm-water effluent is transported seaward not along a specific channel but rather in a disorderly way. Then biological communities in the area can be alternately bathed with cool and then warm waters. For large temperature differences, such exposure can be traumatic to some marine organisms. In summation, each potential siting of a thermal effluent source along a shore must be studied as an individual case.

Energy from the Sea

After the 1973 oil embargo the United States began a serious search for alternate ways to gain energy for industry and commerce. It was inevitable that the oceans be studied. Table 19.10 is a simple summary of the *types* of energy sources present in the oceans, the amount of energy in each type, and rough estimates of the dates at which the technical problems of extraction might be solved as well as when production might begin. We examine briefly the advantages and disadvantages of some of these and describe the thermal-gradient technique in more detail.

Waves and Wind Power

The main arguments for developing wave energy extraction technology are, first, that wind waves are the *cumulative* result of winds blowing over large ocean areas, a storage of diffuse wind power, so it makes sense to tap into this "reservoir"; and, second, that waves deliver their accumulated energy directly to the shores of densely populated coastal zones where energy is in demand. On the other hand, wave and wind power have distinct disadvantages. First, the nearshore flora and fauna depend on wind and wave energy to mix water (see Chapter 18), so that extraction of such energy near shore would have direct impact on this community. Second, in order to achieve economy of scale, both wind and wave power machines must be installed over wide areas, and so can produce unacceptable aesthetic changes.

Currents and Tides

Popular technology magazines often publish artists' conceptions of huge submerged turbines anchored in the Gulf Stream, busily turning electrical generators by extracting energy from the Stream itself. Those of us who have worked at sea, *or* paid the bill for others to do so, know these to be "spinning" dreams. Nevertheless, I have visited remote islands where strong tide currents might be tapped using such submerged turbine generators. An example is the tide channel between Guardian Angel Island and the east shore of Baja, California. Called *Sale Si Puedes* in Spanish, "(Leave If You Can!,)" the channel has tide currents exceeding 4 knots. A local com-

munity would do well to trade off the high cost of importing oil for a suitable offshore turbine generator unit.

As for "harnessing" the total tide displacement in enclosed bays, the French built the first large-scale successful unit on their Rance River in 1966. Here the *pre-dam tide range* exceeded 8 m. A dam containing 24 turbine generators closes off an inlet of 22 km^2; the plants produce over one-half billion kilowatt-hours per year, tapping about 18% of the total energy available. Expressed in relative costs, Rance Tide power is produced *at about twice the cost* of conventional river dam–turbine installations. There are few suitable locations for this type of plant (see Figure 18.3 for some of the best potential sites). The most thoroughly studied North American site is at Passamaquoddy Bay on the United States–Nova Scotia border. The Bay has an average pre-dam tide range of 5.5 m, a potential maximum average power delivery of 1800 MW, and has been studied a number of times since the 1930s. It is still under study.

Let us put the potential of tide power into perspective. If fully developed worldwide, tide power would still account for only 1% of the available total hydroelectric power (Hubbert, 1969). In turn, hydroelectric power worldwide supplies only 6 to 7% of energy consumption. Viewed in this light, tide power is not an appreciable part of the global energy picture.

Artificial Upwelling, Bioconversion, and Osmotic Pressure

The late John D. Isaacs of Scripps Institute of Oceanography once proposed that nuclear power reactors be installed on the ocean floor so that the heat generated would warm the surrounding seawater enough to force the water to convect upward to the surface. The nutrient-rich water of this *artificial upwelling* would support heavy plant growth. If the plants were seaweeds, for example, they could be

Table 19.10 Types of oceanic energy sources, with estimates of amounts of power available, dates by which the technology needed for extraction might be achieved, and dates by which exploitation might begin. (From Constans, 1979.)

Oceanic Energy Source	Theoretical Estimate of Power Available	Date Technology Developed	Date Extraction Might Begin
Thermal gradients	$40,000 \times 10^6$ MW	1990	2010
Salinity gradients	1400×10^6 MW	2000	2050
Bioconversion	10×10^6 MW	1990	2000
Waves and tides	6×10^6 MW	1980	1990
Currents and winds	25×10^6 MW	1990	2000

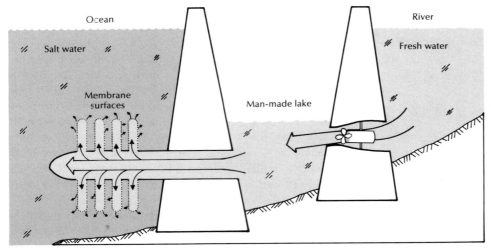

Figure 19.24 How electricity might be generated using the difference in salinity of seawater and fresh water. Fresh water from the man-made lake will diffuse across membrane surfaces into seawater on the other side; the osmotic pressure is enough to draw the lake level down as much as 200 m below sea and river levels. Hundreds of square kilometers of membrane surface are needed for a plant to achieve economy of scale.

processed to yield a variety of foodstuffs and chemicals and the residue burned for additional energy recovery.

Isaacs also proposed that the difference in salinity between seawater and fresh water represents a large potential as an energy source. The same osmotic regulation problem described for marine organisms in Table 7.4 is also an energy source that could build large hydraulic pressure heads, as illustrated in Figure 19.24. The key element in this process is the *membrane* that would "pump" fresh water into seawater without itself rupturing under the pressure head or clogging up with debris. Table 19.10 indicates that an enormous amount of energy is available in this potential resource, and that it is renewable directly with the hydrologic cycle of evaporation from seas and precipitation and river runoff from lands. But the estimated date that exploitation might begin is the year 2050, at best, just a guess!

Thermal Energy in the Oceans: The "Possible" Conversion

Of all the potential energy sources listed in Table 19.10, the largest by far exists in the temperature difference between upper-layer warm waters and deep-layer cold waters. The *physics* of this tremendous energy reservoir is easy to explain (Figure 19.25). Using the deep-water–shallow-water temperature difference of 15°C, and the specific heat of water of 1 cal/(g)(°C) (Table 6.1), we calculate that the total heat energy "stored" in *each cubic meter* of upper-layer water relative to lower-layer temperature *is 15 million calories*, or about 60 million joules. If released in one second, the power generated would be 60 million joules/s, or *60 megawatts*. Let us put this into another

perspective: extracting *all* the energy (theoretically) available from 20 m³/s of this seawater is equivalent to the generating capacity of a modern 1200 MW nuclear power plant! How easy it is to appreciate the enormous theoretical estimate of thermal-difference energy listed in Table 19.10!

But the reality of *Ocean Thermal Energy Conversion* (OTEC) is something else. The problem of "getting" useful energy is that we cannot devise an OTEC machine with better than about 2 to 3% efficiency when the difference between inlet and outlet fluid temperatures is only the 20°C or so we can get from the ocean. This means that for an OTEC plant to operate with the same 1200 MW capacity as the nuclear plant described earlier, it must pump *1000 cubic meters of seawater per second!* To put this into perspective, note from Table 18.1 *that the average flow rate of the Nile River is 1583m³/s*. Indeed, there are *real* limits to the OTEC business.

An OTEC plant must move huge quantities of water. Figure 19.26 is a sketch of one version of such a plant. Whatever the plant design, the central element is the long and wide-diameter pipe that must reach to depths of many hundreds of meters to suck up huge volumes of cold, deep water. The cold water is pumped to the surface where it forms the "cold" end of a fluid cycle; warm surface seawater pumped into the plant at the same rate forms the "hot" end of the fluid cycle. A working fluid, usually ammonia, is caused to circulate between the two end conditions by being evaporated at the hot end, creating a gas under pressure, and condensed at the cold end, creating a vacuum. The circulating gas transfers energy to a turbine that then turns the generator itself.

Though it may appear that the task of pumping

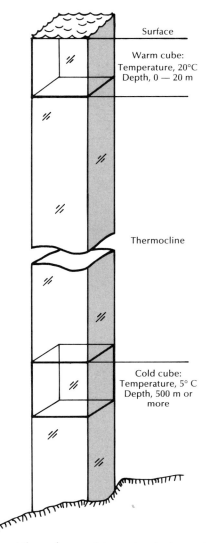

The *warm* cube at the surface contains water at a temperature 15°C *higher* than that of water in the *cold* cube located below the permanent thermocline. Therefore, each gram of surface water contains 15 cal *more* heat energy than that in the colder deep water. And, with each cubic meter containing 1,000,000 g, 15,000,000 cal of *heat energy difference* is potentially extractable from each cubic meter of cold water pumped to the surface.

Figure 19.25 The difference in heat energy content per cubic meter of surface water at a temperature of 20°C and of deep water at 5°C.

so much water, not to mention the task of installing and maintaining such a huge pipe would eliminate the OTEC as a profitable way to obtain energy from the sea, there are good reasons for going ahead with its development. For one, the large ocean areas where the temperature difference between the surface and 1000-m depth is 20°C or more cover the en-

tire central oceanic belt (Figure 19.27). Within this belt are hundreds of centers of population where an OTEC could reduce markedly the dependence on expensive imported fossil fuels.[1] Fresh water can be one of the by-products from an OTEC, so that thousands of Pacific islands now uninhabited could be populated.

But there is another by-product of the OTEC that is even more enticing for development. The cold, deep water is nutrient-rich—it is "upwelled water" in very large quantities. Mariculture on a large scale has great potential as a *co-industry* with the OTEC. Last but not least, the water discharged from the OTEC is still quite cold compared to air temperatures in subtropics and tropics. A final useful by-product is cooling water for other industries along a coast.

Summary

With the proclamation by President Reagan in 1983, the United States made claim to a 200-nmi exclusive economic zone in which we have all rights to renewable resources in the fluid realm and all nonrenewable resources of the bed and subbed. At the same time, the United States assumes the responsiblities of managing all renewable resources within this zone, and this means an agreement to conduct all the marine science research needed to know the marine biosphere with all its food chains, to acknowledge the fishery conventions needed to manage migratory species, and to protect endangered marine species.

The fishery outlook worldwide has been considerably altered by the adoption of the EEZ concept. The major distant-water fishing nations (Japan, Russia, Korea, and Spain) now must enter into joint-venture fishing enterprises with coastal nations to gain access to their resources. Nations that formerly never fished beyond a few kilometers from shore may now begin actively to develop their own EEZ resources and to conduct corresponding marine research. Looming over all this is the prospect of an Antarctic fishery in krill that portends to double the present annual fish catch. Aquaculture products will

[1]By April 1986 the price of crude oil had plummeted to one-third of its cost at the time I wrote the first draft. But make no mistake: the thermal difference between upper and lower ocean layers will still be here, and probably tapped heavily, long after our followers have torched the last drop of recoverable oil.

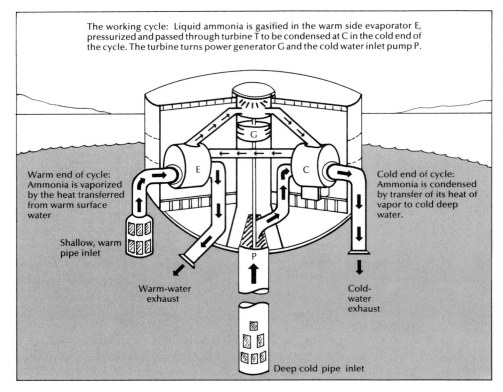

The working cycle: Liquid ammonia is gasified in the warm side evaporator E, pressurized and passed through turbine T to be condensed at C in the cold end of the cycle. The turbine turns power generator G and the cold water inlet pump P.

Warm end of cycle: Ammonia is vaporized by the heat transferred from warm surface water

Shallow, warm pipe inlet

Warm-water exhaust

Cold end of cycle: Ammonia is condensed by transfer of its heat of vapor to cold deep water.

Cold-water exhaust

Deep cold pipe inlet

Figure 19.26 A possible design for an Ocean Thermal Energy Conversion (OTEC) plant. This type of converter plant must be installed in water deep enough that the cold-water inlet pipe can pick up water at least 15°C colder than surface water at the same location.

continue to increase in importance worldwide, and the concept of "ranching" the open "ocean range" is being rapidly developed in the line of salmon fishery.

Mining the sea floor has already become a worldwide preoccupation of nations, most of which are territorially lacking in one or more of the strategic metals such as cobalt, manganese, or copper. Just a scant 15 years ago we were facing a bleak future in minerals self-sufficiency as our land mines of iron, nickel, and precious metals were being exhausted of high-grade ore. Today, the new discoveries of ferromanganese nodules on the Blake Plateau off Florida, crusts on our many Pacific seamounts, and polymetallic sulfide deposits on rift zones—all within our 200-nmi EEZ—have once again placed our industry on a confident base of raw materials.

No problem has been more serious than that of contamination of our coastal waters. It is not petroleum, either in transport or in offshore production, that poses the major problems. Instead, toxic chemicals are our most serious problem, made worse in the marine sphere by the tendency of many marine organisms to "horde" trace elements. Today many trace elements are toxic in coastal waters. We must dedicate our energies to the resolution of this problem in a major way, now.

Part of our interaction with the sea comes from the potential of taking energy from the sea. Of the many potential prospects, only one "looks good" today—the Ocean Thermal Energy Conversion—and this only when developed in conjunction with aquaculture based on the upwelled water and use of its cooling effects in subtropical latitudes.

Study Questions

Law of the Sea

1. In all, how much new "territory" does the United States gain with its exclusive economic zone (EEZ)?

2. What does the EEZ *entitle* to the coastal state? (In this chapter "state" means "nation.")

3. In all, how much does the "free ocean" shrink after the EEZs of all coastal states are taken into account?

4. Briefly, explain the rationale behind the United

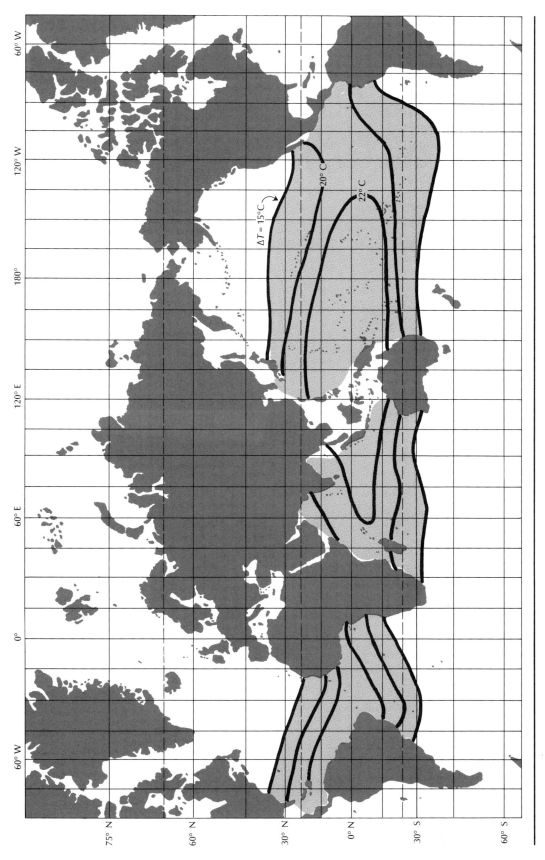

Figure 19.27 Distribution of oceanic regions where the temperature difference, surface to 1000 m, is adequate to operate an OTEC plant. (Based on data from the National Oceanographic Data Center.)

States refusal to ratify the Law of the Sea Convention.

5. What are the *rights* and the *responsibilities* of a coastal state regarding marine science expeditions?

6. Do you agree with the LOS Treaty "requirement" that a coastal state either *harvest* its fish resources itself or *license* other states to do so?

7. In a very real sense, the LOS now demands that *every* coastal state have a high level of expertise in the marine sciences. True or false? Do you agree with this aspect?

Fishing the Living Resources

8. Which nations have been big winners in the fish business? What is the rationale for the story that the creation of the LOS was a consequence of a 1950s Russian fishery study?

9. Where are major fisheries yet to be developed?

10. Simply stated, what is the major impediment to the ability of a coastal state to manage the fisheries within its own EEZ?

11. Which nation produces the highest percentage of its total fish catch using aquaculture?

12. About what percentage of United States fish consumption is imported? Do you think that imports are mostly of frozen meats for human consumption?

13. What is the difference between fish ranching and fish mariculture? Between mariculture and aquaculture?

Mining the Sea Floor

14. What minerals are likely to be extracted from the ocean floor over the next 50 years, and for what applications?

15. Nodules, co-crusts, and sulfide deposits each have a distinct location or scheme of formation. Explain. Which type is deposited the most rapidly?

16. Of what commercial use are the phosphorite deposits? Are these deposits usually found in warm, clear ocean water?

17. Where and how are the co-crust deposits formed? What difficulties do you foresee in exploiting these?

18. Where and how are the polymetallic sulfides formed? What difficulties do you foresee in exploiting these?

19. Are the polymetallic sulfides found only near active hydrothermal submarine vents or are they present over widespread zones of the ocean floor? If yes, why can't we mine these deposits anywhere, and particularly near the continental margin so as to reduce transportation costs?

The Impact of Human Societies on the Ocean

20. Because most of us live within 500 km of the seacoast, virtually everything we do and throw away somehow impacts the coastal ocean. But what is the argument that even the things we do in the interior of the United States, far removed from the coast, also impact the coastal zone?

21. The matrix of Table 19.4 summarizes the environmental impact of specific pollutants on a small set of very generalized concerns. Take just *one* of these seven general headings and develop it to more specific concerns, impacts, and so on. For example, the general heading "Effect of the pollutant on vital life processes" can immediately be subdivided into "Good effects" and "Bad effects."

22. Define what is meant by "Enrichment ratio in marine shellfish," a column head in Table 19.5.

23. Comment on the types and sources of pollutants that reach the ocean via atmosphere transport.

24. In designing an acceptable sewage outfall system to be located offshore from coastal city, what factors about seawater circulation—vertical, horizontal, or both—must be taken into account?

25. How do you explain that tanker ship and other ship operations are the single largest source of oil injected into the marine environment, as listed in Table 19.6, but the accidental tanker spill contributes only about one-fifth as much?

26. How do accidental spill plus industrial wastes plus ship operations together compare with the atmospheric input of petroleum products to the ocean?

27. Has the use of DDT been reduced worldwide since the publication of *The Silent Spring* by Rachel Carson?

28. Why are radionuclides potentially dangerous to life?

29. Which of the three forms of *radiation* are potentially more dangerous to human life? Why?

30. What is the argument against basing a *radiation exposure* limit on the dose rate required to harm the adult?

31. Describe the philosophy of radioactive waste disposal that seeks to determine the *maximum permissible concentration* or the *critical pathway.*

32. Using Table 19.9 as a guide, which radionuclides in spent fuel rods from nuclear power plants are the most "dangerous"?

33. Write an essay on the use of the deep-sea bed as a disposal area for radioactive wastes.

Energy from the Sea

34. Compare the three "sources" of energy from the sea that supply the greatest total amount.

35. Why is it, then, that the first commercial power plant to extract energy from the sea uses the *tide* energy, the least powerful category of Table 19.10?

36. Our classroom is typically 10 m wide by 10 m long by 3 m high, containing $300 m^3$ of volume. How many classrooms like this would have to be filled and emptied in one second, half with 5°C water pumped from 1000-m depth off Puerto Rico and the other half with 25°C surface water, before the available difference in heat content of the two waters would equal the 1000 MW of power produced by a modern nuclear power plant?

CHAPTER 20

THE POLAR OCEANS

The oceans of the pole regions, north and south, hold a special fascination over us. To a great measure, I think, this is because the expeditions that opened the polar oceans are so recent in history that we can almost touch and identify with the persons who "crossed" them first.

To the marine scientist, the fascination of polar exploration lies also in the new questions that can be asked. How does the presence of a solid water cap over an ocean change the interaction between ocean and atmosphere? How do fishes and marine mammals adapt when the saltwater temperature is lower than the freezing point of body fluids? Why are the waters under pack ice in Antarctica teeming with life when ocean water under similar pack ices in the Arctic is almost barren? Why are polar regions so arid, semideserts in a frozen state?

It is a complex region. It is said that the Eskimo language has more than 60 different words to describe different types of ice. Evidently, ice in its natural state has many distinct, perhaps subtle characteristics; certainly ice formed from seawater differs greatly from freshwater ice. Such differences have substantial impact on the behavior of oceanic ecosystems, as we shall see.

Which Are the Polar Oceans?

At the outset, we need a definition of what is meant by polar oceans. What are they, and are all polar oceans the same?

Polar Ocean North

The map projection in Figure 20.1 shows the region of planet earth around the North Pole. The Arctic Circle at latitude 66.5°N is usually taken as the southernmost boundary of the so-called "Arctic Ocean," but there is no specific ocean region having that name. Instead, we find a succession of *seas* about the outer periphery of the central polar basin: beginning east of Greenland, the Greenland Sea, the Norwegian Sea, the Barents Sea, the Kara Sea, the Laptev Sea, the East Siberian Sea, the Chukchi Sea, and the Beaufort Sea. All these seas are confined to shallow waters less than about 500 m deep, with the exceptions of the Beaufort and Norwegian seas.

In the north the term Polar Ocean is reserved for that area around the Pole itself. Its boundaries coincide more or less with the 80°N circle, and with the edges of the deep Arctic Basin system, the exception being the part of the basin system called the Canada Basin, which forms the floor of the Beaufort Sea.

Figure 20.2 is an interesting view of the overall Arctic Basin. The physiography identifies several smaller basin areas separated by submarine floor features such as the Nansen and Alpha cordilleras and the Lomonosov Ridge. The Nansen Cordillera is actually an extension into the Arctic Basin system of the North Atlantic mid-ocean ridge, a rift zone of active sea floor spreading.

In this chapter the term Arctic Ocean is used to encompass both the Polar Ocean and the several seas that surround it. About one-half of the total earth area north of the Arctic Circle is oceanic, some 9 million km^2; for comparison, the Mediterranean and Black seas make up about 3 million km^2 together.

Polar Ocean South

In contrast, almost all of the area within the Antarctic Circle at 66.5°S is filled with the continent Antarctica itself. The few oceanic areas that exist south of this latitude are called seas, as in the Arctic. In Figure 20.3 find the Ross Sea, the Amundsen Sea, the Bellingshausen Sea, and the Weddell Sea.

There is no Antarctic Ocean per se. Oceanic waters that surround the continent and its seas are sim-

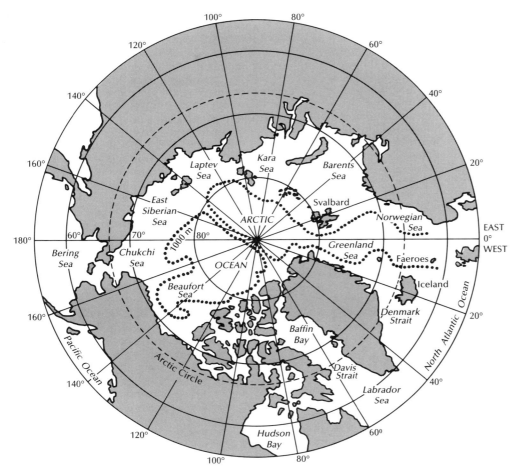

Figure 20.1 The Arctic Ocean with its peripheral seas. The dotted line is the 1000-m depth contour. Notice that the Arctic basin itself covers the area around the Pole to roughly the 80°N latitude. Notice also the extensive shallow seas along the Siberian and Norwegian coasts. The only deep connection between the Arctic basin and other ocean basins is the relatively narrow strait between Greenland and Svalbard (Spitabergen).

ply called the "Southern Ocean." But unlike the Arctic Ocean, where most of the boundaries are against land masses, the northern boundary of this ocean is a special feature of the ocean itself, representing a sharp change in the temperature of surface water. This feature is called the Antarctic Convergence, and it is distributed around the southern region as shown in Figure 20.3. It is also called the Polar Front, because it marks the transition from cold Subantarctic Surface Water to the much warmer surface waters of the South Atlantic, South Pacific, and Indian Ocean Central Water masses. The reader is referred also to Figure 9.6, where the front is marked on the Atlantic Ocean cross section by the letter P (for Polar Front). In still another context, the reader can refer to Figure 11.1, which shows the major surface currents in the world oceans; here the position of the front coincides with the northernmost boundary of the Antarctic Circumpolar Current.

The major features of the Southern Ocean, then, are that it is bounded by the Antarctic Convergence and that it contains the Circumpolar Current. In

terms of area, it is about four times larger than the Arctic Ocean, some 34 million km².

Why and How Do Polar Oceans Differ from Other Oceans?

Oceans at high latitudes are much different from oceans at mid and low latitudes from both physical and biological views. We look first at the physical aspects and in a later section the biology and ecosystem aspects.

Differences in Geography and Geology

Apart from their geographical uniqueness in being located at a pole (Arctic) or surrounding a polar continent (Antarctic), there is little that differentiates the shape and bathymetry of polar basins. The Arctic Basin (Figure 20.2) is typical with its broad continental shelves, slopes, and rises, and the usual abyssal plains. The basin of the Southern Ocean is also typ-

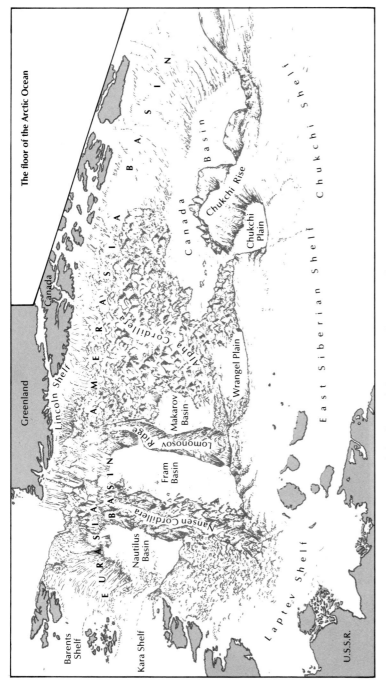

Figure 20.2 A physiographic diagram of the floor of the Arctic Ocean (Beal, 1966). This aspect emphasizes the width of the continental shelf off the Siberian coast, as wide as 1000 km in places. Also shown are the Lomonosov Ridge and the Nansen and Alpha cordilleras. The Lomonosov differs from the other two in that it is not a spreading rift; rather, it is thought to be a remnant of continental material, much like the Chukchi Rise.

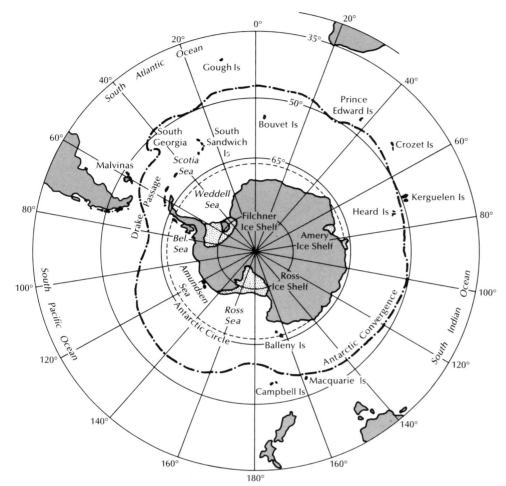

Figure 20.3 The Southern Ocean around Antarctica is bounded on the north by the Antarctic Convergence, a well-defined frontal zone created where warm central surface waters meet the cold polar surface waters. The front is also the northern boundary of the Antarctic Circumpolar Current. Although shown here as a smooth contour, the front is in reality highly convoluted and variable in location.

ical, with one notable exception. The margins of the Antarctic continent have continental shelves that are much deeper than most, with depths at the shelf break averaging 600 m in contrast to a world average of 135 m. Reasons for this difference are varied. One hypothesis holds that during epochs when sea level was much lower (see Figure 4.12*d*) icebergs scoured off much shelf material; another holds that the entire continent, including shelf margins, is depressed by the weight of ice.

Differences in How Temperature Change Alters Seawater Density

Polar ocean surface waters are at near-freezing temperature the year-round, partly because of the negligible seasonal heating by solar rays and partly because of the extensive ice pack. Because the thermal expansion coefficient of water is very small at low temperatures (see Appendix 2), the density of cold water is quite insensitive to changes in temperatures. The reader is referred back to Figure 9.1 for a graphic explanation. Look at the two different water types noted by points *A*, typical of arctic surface water, and *B*, typical of a subtropical ocean surface water. Type *A* water is at −0.6°C and 30.7-ppt salinity. In the vicinity of point *A* the contours of constant density are almost vertical, meaning that density changes little with temperature. For example, near point *A* a change of 3.8°C along the heavy vertical line involves a density change of 0.2 sigma-*t* units; around point *B* a similar density change occurs for a shift of only 0.8°C in temperature.

This means that whenever polar ocean behavior is governed by changes in density, salinity is the controlling factor. Notice also in Figure 9.1 that the change in density with salinity is almost the same for points *A* and *B*, from which we conclude that density is almost a direct function of the concentration of dissolved salts (in Appendix 2, this function is described by beta, β), regardless of whether the water is cold or warm.

Differences in Density Stratification from the Flux of Fresh Water

As discussed in earlier chapters, the salinity of surface waters at low and mid-latitudes is controlled by the flux of fresh water across the air–sea interface, that is, by evaporation E and precipitation P. *But in polar oceans this relation does not hold, especially in the Arctic Ocean.* This point is central to the need for this special chapter on polar oceans.

First, we examine the global distribution of water flux between the atmosphere and the earth's surface, oceans plus lands. Figure 20.4*a* plots the quantity $P - E$, precipitation minus evaporation, for the entire globe by increments of 5° bands of latitude. A strong net rainfall rate occurs within the 5 to 10°N band; compare this with the description of the ITCZ zone shown in an earlier chart (Figure 7.4) of surface salinity in the world oceans. Notice also the strong net evaporation bands in latitudes 20 to 30° north and south; these are the zones of Central Waters of high surface salinity. Then, near the Arctic and Antarctic circles, in latitudes 50 to 60° north and south, the system is again dominated by net precipitation. In between these zones of net P and net E, respectively, there must be a zone where precipitation and evaporation balance out, where the net $P - E$ is zero. Such a null zone for air–sea water exchange occurs at about the 40° latitudes both north and south.[1]

We next compare the $P - E$ distribution just given with a graph of how surface salinity changes across the entire planet (Figure 20.4*b*) along a route from the North Pole across the Arctic Ocean and down the middle of the North and South Atlantic oceans all the way to the Antarctic continent. In the central zones between the Arctic and Antarctic circles, salinity is *inversely* proportional to $P - E$; where there is net precipitation, surface salinity is lower, and vice versa. Important exceptions to this inverse rule occur on the poleward sides of the two polar circle lat-

itudes. Within the Arctic Ocean, surface salinity is almost totally independent of $P - E$.

One reason that $P - E$ does not control surface salinity in polar oceans is simply because there is so little precipitation! Both polar zones are deserts; the average rain, snow, and sleet combined are no more than 100 mm/yr—about 2.5 inches, the equivalent of one afternoon shower in Costa Rica!

Why then is surface salinity so low in the Arctic Ocean? The explanation is twofold: first, there is a large runoff of river water from the Siberian and Canadian rivers; and, second, the Arctic Basin is sharply limited in communication with other oceans. Only between Greenland and Spitsbergen is there a deep channel (Figure 20.1), and this connects with the North Atlantic. Thus, because Arctic Ocean water is limited in the rate at which it can mix with adjacent ocean bodies, the high rate of river inflow is able to keep the surface layer salinity quite low. Consequently, a large density difference exists between the lighter, fresher surface layer and the deeper, more saline water that enters the basin from the Atlantic. This density stratification governs the flow patterns within the Arctic Ocean and controls the thickness of the pack ice cover, hence the albedo and solar absorption—in short, the role of the Arctic Ocean in modifying earth's climate. More on this later.

The Mixing Role Played by the Antarctic Circumpolar Current

In contrast to the Arctic Ocean, the surface salinity of the Southern Ocean stays relatively constant (Figure 20.4*b*), even though the net precipitation is about two times greater within the Antarctic Circle. Moreover, the runoff of fresh water from the Antarctic continent is almost as great as the flow of rivers into the Arctic Ocean; the fact that this flow is in the form of ice shelves and icebergs does not alter its impact, for the ice eventually melts at the surface. Why then is not the Southern Ocean surface salinity low, and hence the ocean column strongly stratified in density, like that of the Arctic? The answer is that there is substantially greater mixing between the Southern Ocean and the surrounding bodies of water.

One of the mixing mechanisms responsible is the Antarctic Circumpolar Current. Driven by the strong westerly winds in latitudes 45 to 60°S, this current flows around the globe touching each of the major ocean basins, the Atlantic, the Pacific, and the Indian (see Figure 11.1). Not only does this current

---

[1]The reader is encouraged to compare this discussion with the data shown in Figure 6.4. The 40° latitude is also the approximate zone where incoming solar radiation and outgoing back radiation balance (Figure 6.4*a*); further, it also has the highest horizontal flux of heat transported poleward (Figure 6.4*b*). All this makes sense: the net evaporation on the equatorward side of 40° latitude must be transported poleward to sustain the net precipitation.

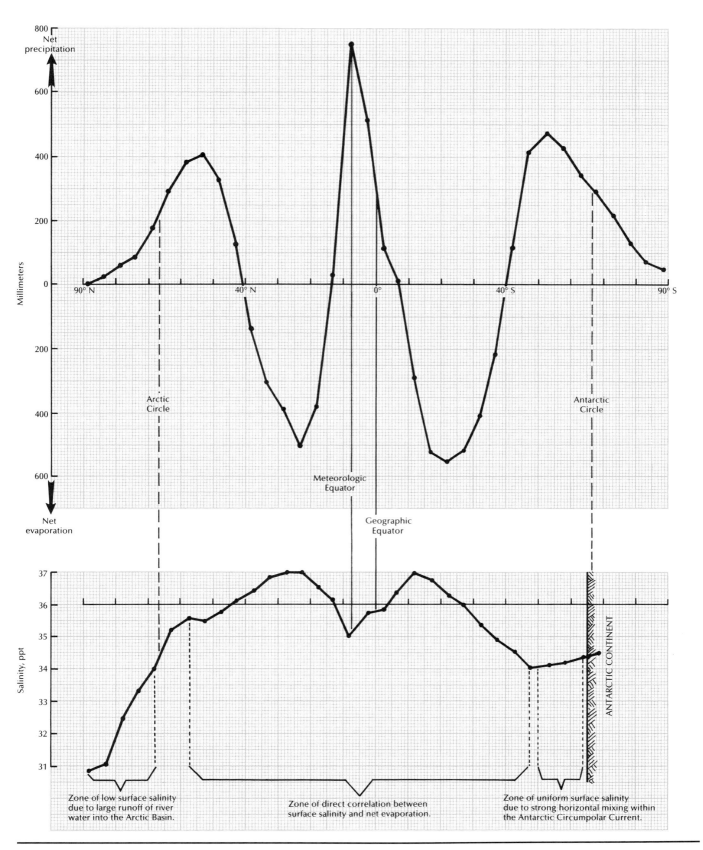

Figure 20.4 The close correlation between net precipitation or evaporation and the salinity of surface waters, plotted across latitudes from North to South poles. (a) With the exception of a narrow zone of excess rainfall at the equator, the latitudes poleward of 40°N and 40°S are zones of net precipitation; zones equatorward have net evaporation (b) Zones of highest surface salinity correlate with zones of excess evaporation.

mix waters laterally into and out of the Southern Ocean, but it does so at all levels from the surface to the very bottom. Figure 11.3 gives evidence that this current flows at deep as well as shallow levels and must therefore be a coupling mechanism between surface and deep circulation patterns. In Figure 9.4 Antarctic Circumpolar Water is shown to be the "hub" to which all major ocean water masses are connected. It is no idle metaphor that we call this the "great communicator current."

Exclusion of Salt from Crystallizing Seawater: THE GIANT ANTARCTIC SEA ICE MIXING MACHINE.

The most intriguing of mixing mechanisms stems from the freezing of seawater itself. Ice is a crystalline form of water. When formed from fresh water it is translucent, almost glasslike, with a faint blue coloration. But ice formed from seawater is very much different: it is a mass of platelets and small crystals joined together in a somewhat random pattern, with entrapped pockets of liquid seawater that is of much higher salinity than the parent seawater, almost brinelike. With time, these brine pockets migrate downward within the mass and eventually exit out the bottom, leaving a much more uniform crystalline formation behind. In fact, one-year-old sea ice when melted makes highly potable water, but freshly frozen sea ice still retains about 5 ppt of sea salts. Important to our discussion here is the fate of the other 25 to 30 ppt of salt that is immediately *excluded* as seawater crystallizes.

The series of sketches in Figure 20.5 shows what can happen. Initially, the polar ocean loses heat to the atmosphere and a mixed layer forms at the surface with its temperature just at the freezing point. Further heat loss then causes ice to form (Figure 20.5b) and the excluded brine to be convected down in plumes that mix with the remaining surface layer fluid; density of this remaining fluid increases as more and more salt is added from more and more ice being formed. Eventually (Figure 20.5c) the remaining surface fluid can become dense enough to break through the stabilizing layer into the deeper water below. This is precisely what happens in the waters around Antarctica during the formation of Antarctic Bottom Water. This water mass flows away toward the north.

Continuity now requires that the fluid leaving the Antarctic region via Bottom Water be replaced (Figure 20.5d). The compensating flow occurs at mid-depth layers and brings in waters that originate in the downwelling zones off Greenland, flow south as North Atlantic Deep Water (see Figure 9.6), and surface in the Southern Ocean hundreds of years later as a nutrient-rich upwelling. It is this upwelling that is responsible for the great biological productivity of the Southern Ocean.

But there is one additional effect. When the sea ice of Figure 20.5e melts in the following season, the surface layer is diluted, just as though rainfall or river runoff had taken place. This less dense, cold water is also forced to flow northward, away from the Antarctic region. The continuity argument requires this export to be replenished by an inflow, again of water at the intermediate levels, that is, North Atlantic Deep Water. In this way the sea ice effect produces an upwelling both during the winter when it forms and during the summer when it melts.

The description of freezing and melting processes just given is the basis for the term "ANTARCTIC SEA ICE MIXING MACHINE"; but why is it also described as "giant" in size? The answer is that the annual formation and melting of sea ice involves giant sectors of the planet earth. Let us illustrate with the charts in Figure 20.6. These are polar stereographic projections showing the extent of sea ice cover over the Southern Ocean for the austral summer (February 23, 1984) and winter (August 23, 1984). The numbers labeled on different areas of these charts specify how "close" the ice pack is; for example, the numbers 5–7 mean that between 50 and 70% of that area is covered with pack ice. In winter the great majority of the sea ice is between 90 and 100% "close," that is, it almost completely covers the sea. In summer, the ice edge retreats poleward and the pack ice remaining is quite "open."

To put the summer-to-winter sea ice change into perspective, ice pack cover in winter is about 8% of the total area of the Southern Hemisphere, whereas the summer cover reduces to about 1%. In terms of area, this is a change of some 18,000,000 km², a very large area. To draw an analogy, picture the North Atlantic between the American and the Euro-African continents, from the tip of Florida at 25°N to the tip of Newfoundland at 45°N, alternately covered with an ice pack 1 m thick during winter and completely icefree in summer. The potential impact of such a North Atlantic feature on our Northern Hemisphere winter is obviously great. It has no less import in controlling weather and climate of the Southern Hemisphere, and it is a "giant" effect in the forcing of large-scale ocean circulation as described earlier. Hence the metaphor "Giant Sea Ice Mixing Machine."

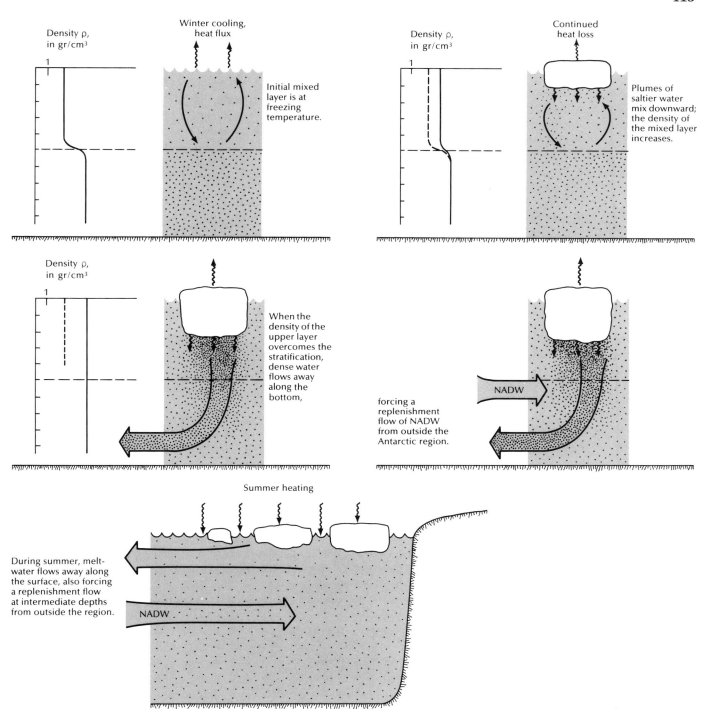

Figure 20.5 How vertical circulation, forced by sea ice forming and melting around Antarctica, replenishes surface water with upwelled water from intermediate depths. The source is the North Atlantic Deep Water (NADW).

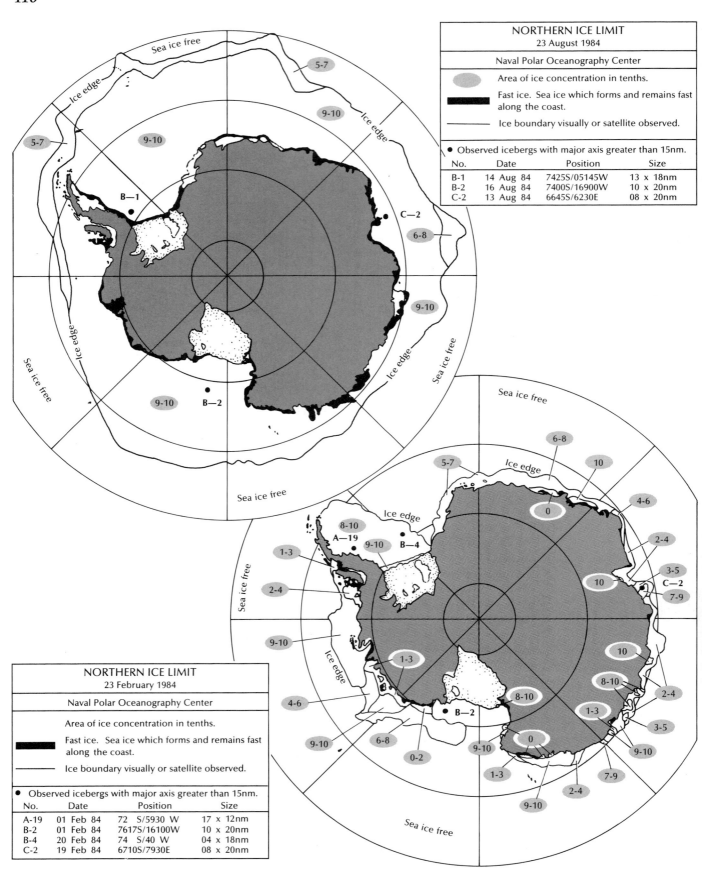

NORTHERN ICE LIMIT
23 August 1984

Naval Polar Oceanography Center

Area of ice concentration in tenths.

Fast ice. Sea ice which forms and remains fast along the coast.

Ice boundary visually or satellite observed.

● Observed icebergs with major axis greater than 15nm.

No.	Date	Position	Size
B-1	14 Aug 84	7425S/05145W	13 x 18nm
B-2	16 Aug 84	7400S/16900W	10 x 20nm
C-2	13 Aug 84	6645S/6230E	08 x 20nm

NORTHERN ICE LIMIT
23 February 1984

Naval Polar Oceanography Center

Area of ice concentration in tenths.

Fast ice. Sea ice which forms and remains fast along the coast.

Ice boundary visually or satellite observed.

● Observed icebergs with major axis greater than 15nm.

No.	Date	Position	Size
A-19	01 Feb 84	72 S/5930 W	17 x 12nm
B-2	01 Feb 84	7617S/16100W	10 x 20nm
B-4	20 Feb 84	74 S/40 W	04 x 18nm
C-2	19 Feb 84	6710S/7930E	08 x 20nm

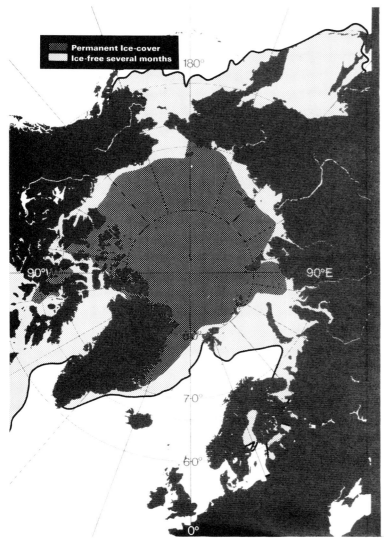

Figure 20.7 The areas of the Arctic Pole covered by pack ice in summer (August) and winter (February). Much of the continental margins become ice-free during at last part of the summer. The high-latitude zone north and east of Norway, which is ice-free throughout the year is the route taken by the warm surface waters of the North Atlantic Drift Current entering the Arctic Basin. (From Rey, 1982.)

Does the sea ice of the Arctic produce similar driving forces? Broadly speaking, the answer is yes, but it must be a qualified answer. First, the annual change in ice-covered sea surface is much less, perhaps one-fourth as great as that of the Antarctic.

Second, the Arctic retains a perennial ice pack over its central basin waters. Figure 20.7 shows both of these aspects in the Arctic. Third, the waters that do undergo seasonal cycles of freeze-up and melting are for the most part shallow.

Figure 20.6 The northermost limits of the pack ice in the Southern Ocean during late summer (February) and late winter (August). (From NOAA, 1984.)

August 23, 1984. Sea ice is at its greatest seasonal coverage, and the pack ice is everywhere quite tight, with less than 10% open water. No ships can break through to coastal bases; with air traffic slowed by severe winter storms, this is the season of "wintering over" for crews at Antarctic bases. Notice that Iceberg C-2 has moved about 800 km westward along the coast since its sighting in February. Much fast ice surrounds the Continent.

February 23, 1984. The area covered by pack ice is now at its seasonal minimum. The sea route into the United States base at McMurdo Sound is open to ship traffic without need for an icebreaker. In the Weddell Sea, however, prevailing surface winds and currents cause the relatively tight pack ice to jam against the Antarctic Peninsula, making this the least-explored ocean sector in the world. Very little fast ice remains around the continent.

Biological and Ecosystem Aspects of Polar Oceans

The previous section presented a number of reasons why the Arctic and Southern oceans are different in behavior from other oceans, and very clearly different from each other. Both types of differences are essential elements in a discussion of life in polar oceans, and the physical processes involved set the stage, so to speak, for a study of the biological and ecosystem aspects of polar oceans. Before beginning such studies, let us review these differences and processes.

1. Polar oceanography is more recent and much less developed than marine studies of temperate and tropical oceans. In large measure this is due to their remote geography and to the high cost of exploration at high latitudes. Our data base is meager for both polar oceans. Arctic data are sparse because the perennial ice pack over the central basin excludes all research ships except icebreakers. Antarctic data are sparse because of its remote location and high logistics costs. Moreover, to obtain data in winter means exposing a ship to the dangers of being frozen into the pack ice and crushed by ice floes as winter storms churn the ice field.

2. Because the temperature effect on water density is negligible in colder water compared to the salinity effect, the processes that control the flux of fresh water into polar oceans are also responsible for their circulation, particularly in the Arctic Ocean.

3. The unique exclusion of dissolved salt ions from the crystalline structure of solid water is the physical basis for the tremendous influence of seasonal sea ice freeze-up and melting on ocean circulation, particularly in the Southern Ocean.

4. The Arctic and Southern oceans are completely different in their communication with other oceans: the Arctic is isolated, with deep interconnection only with the Atlantic and then only in one strait between Spitsbergen and Greenland; the Southern Ocean with its Circumpolar Current mixes with all major oceans and at all levels from the surface to the deep abyssal basins.

5. The Arctic and Southern oceans are importantly different in vertical stratification: the Arctic is strongly stratified, with surface waters limited in nutrient replenishment from deeper levels and hence with low primary production; the Southern Ocean is weakly stratified and has massive upwelling, virtually unlimited nutrients in the photic zones, and high primary production.

Based on these differences, two major aspects of life in polar oceans are the relative absence of speciation in the Arctic and the tremendous biological productivity of the Southern Ocean.

Low Speciation in the Arctic Ocean

One of the most curious and intriguing facets of life in polar oceans is the undeniably small number of species of both plants and animals in contrast to the much greater species diversity in temperate and tropical seas. Nowhere is this more striking than in the Arctic Ocean. Excerpts from Dunbar (1982) summarize the Arctic situation very well.

> The paucity of fauna and flora of the Arctic is well known. As examples, there are only perhaps 25 or 30 fish species that belong strictly to Arctic Water (of thousands of species oceanwide); of marine planktonic copepods (1500 species oceanwide), only about 40 are known in the Arctic Ocean; of known species of chaetognaths (over 30 oceanwide) only two are found in the

Oceanography through a hole in the ice! Much of what we know about the Arctic Ocean is gained by lowering instruments through holes made in ice floes. (Courtesy R. Baumann, Oregon State University.)

Arctic. And among some 85 species of euphausids known, four are found in sub-Arctic water (e.g., the Barents Sea) and *none* in the Arctic (Polar) water.

Among the four groups of benthos of the continental shelf (Crustacea, Echinodermata, Mollusca and polychaete worms), the more Arctic the environment, the smaller the number of species. The conclusion must be that low diversity of Arctic fauna is apparent everywhere.

The fact is that until we understand the *Arctic* ecosystem well enough to explain the speciation aspect, we really do not understand *any* ecosystem well enough. A number of factors have been proposed as "reasons why" there are so few species in the Arctic.

Low Level of Primary Production. The low values for primary production measured in the Central Arctic Ocean contrast sharply with the higher productivity of some subarctic seas, for example, the Bering Sea and the Southern Ocean. For example, Dunbar (1982) lists the primary production of Arctic Ocean Central Water at 0.6 g C/(m²)(yr); contrast this with numbers like 0.63 g C/(m²)(*day*) reported for the central Bering Sea (Hood, 1974), 1 g C/(m²)(*day*) in the Weddell Sea of Antarctica, or the value of 300 g C/(m²)/(*year*) listed in Table 1.2 for upwelling areas in general.

I have underscored the time units *year* and *DAY* in these production data for a reason. The fact is that scientists simply are not able to get into the Bering Sea or the Weddell Sea of the Antarctic in wintertime, so we cannot estimate a *year* average. In contrast, scientific stations have been set up on ice floes in the Central Arctic and manned year-round; cruises in upwelling regions have also been studied year-round.

One cause of such low production in the Arctic has to be the strong density stratification that inhibits mixing with deeper waters and thus subdues nutrient replenishment of the surface photic zone. And, as discussed earlier, heavy river runoff into the basin, limited mixing with adjacent oceans, and the process of sea ice freeze-up and melting, which forces salts downward in the water column, are largely responsible for this strong stratification.

Another cause is the restriction of light by the ice cover itself. A photograph of pack ice (Figure 20.8) shows the white, snow-covered ice to be a strong reflector of solar radiation. Open water where the sun's rays are mostly absorbed is almost black, but such openings in the wind-driven, moving pack ice account for only about 10% of the surface area in the perennial ice cover. The light intensity in water just beneath the floating ice is a small fraction of that

striking the surface; perhaps less than 1% of surface light reaches below ice that has a snow cover, with values rising to a few percent as snow melts off. Thus ice cover severely restricts the light available for photosynthesis in the waters below.

Still another aspect of primary production in ice-covered water is the timing of the phytoplankton bloom. In mid-latitudes the spring bloom begins in late March, the time of year when the length of day is rapidly increasing. But in high latitudes the sun's rays have a low grazing angle. This plus the ice cover delay the bloom. Figure 20.9 shows how these factors come together in the nearshore fringes of the Beaufort Sea; the phytoplankton bloom begins only in July, after the ice cover has substantially melted away.

Still another feature of plant growth in ice-covered seas (Figure 20.9) is the intriguing way algae have evolved to grow within the ice itself. The word *epontic*, meaning out of water, has been assigned to these species, although strictly speaking they still

Figure 20.8 Aerial photograph of the perennial ice pack of the Arctic. Ice floes often take on a diamond shape. The open gaps, called leads, here average about 1 km in length. (Courtesy Victor T. Neal, Oregon State University.)

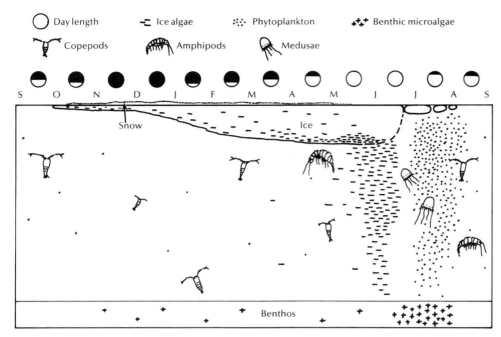

Figure 20.9 The annual cycles of ice algae, phytoplankton, and benthic microalgae in the near-shore area of the Beaufort Sea. At this latitude, about 70°N, the months November and December have no direct sunlight, and May, June and July have continuous sunlight 24 hours each day. Toward the end of May the snow cover melts off the ice, allowing a sharp increase in the light available to the epontic algae, which then begin a bloom. During June the pack ice begins to melt thinner and break up, allowing more light into the fluid ocean. Phytoplankton begin to bloom, but their production time is extremely short. The shrimplike organisms shown here are amphipods; *Calanus* copepods are also shown. (From Horner and Schrader, 1982.)

live in water, even though inside the ice. These algae produce a bloom that begins in April and reaches a maximum coinciding with the meltoff of snow cover, about the first of June. As the ice melts, these algae are released into the water. Epontic algae communities contain primarily pennate-shaped diatoms and microflagellates, the latter about 6 micrometers or so in size.

Epontic algae may account for as much as 25% of primary productivity. But how do they grow in ice, and how can they contribute to grazer and higher animal production if encased in ice? Answers are still being developed, but we do know that the underside of pack ice is not a solid surface but rather a spongy region of random platelets of ice intermixed with fluid. These epontic algae live in the interstices of the ice–fluid interface. There they exchange gases and receive nutrients from the liquid water just as do other phytoplankton. Some are entrapped along with brine pockets during rapid freeze-up and may live this way for extended periods of time.

Amphipods are the dominant grazers of epontic algae. They resemble euphausids in form (Figure 20.9), with sizes ranging up to a few centimeters. I have sat at the edge of a hole cut from the ice, with an electric light bulb lowered into the water below the ice to illuminate its underside, and watched as amphipods skittered along the underside. There are more species of these animals than of planktonic copepods in Arctic waters, which suggests the impor-

tance of ice algae in the Arctic ecosystem. Higher-order animals, fish that prey on the amphipods, as well as some birds, are part of the system. The underside of the ice becomes a kind of "upside-down" substrate, somewhat solid, on which a special community has evolved.

Very Short Productive Season. From Figure 20.9 we conclude that the highly productive phase of the annual cycle is quite short in the Arctic, not like the cycle pictured in Figure 12.4 for temperate oceans. This must have a pronounced influence on the life cycles of the grazers, for there may not be enough time to grow to adult spawning stage in one season, and the timing of larval stages with available food can be a problem. The lifetime of the larger copepods in the Arctic, for example, is at least one year, and is probably an evolved feature of the ecosystem so that spawning at the beginning of a second year coincides with a new phytoplankton bloom.

One net effect of a prolonged life cycle in grazer species is that more energy is stored within the ecosystem; therefore the standing biomass of herbivores is relatively high. It is partly for this reason that the subarctic seas like the Bering are so phenomenally rich in fish production. The ice pack melts off sooner in the season at subarctic latitudes, and primary production is somewhat longer than in the Central Arctic, but it nevertheless pulses extremely strongly. Because the Bering Sea is weak in stratification, vertical mixing supplies large quantities of nutrients to the photic zone. As a result we find large commercial tonnages of a few fish species; halibut, pollock, and turbot are examples. In the Atlantic, the cod and herring fisheries in the North Sea and the Greenland and Labrador seas are also rich. These latitude regions have few fish species but large biomasses in their populations, conditions ideal for commercial fishing. In contrast, we would describe a tropical ocean ecosystem as one in which physical conditions are very uniform year-round, life cycles are shorter, energy passes through the food webs rapidly, and standing biomass is thus reduced and commercial fisheries are limited.

We can carry these arguments one step further. Arctic copepods have evolved with a long life cycle in order to "time" spawning periods to the short season of plant blooms, but it would also be to their advantage to increase the number of eggs and larvae produced per spawn. The tendency of arctic copepods and amphipods to have large sizes is evidence that they do this, for the number of eggs an individual can produce increases exponentially with its size. The antarctic *Euphausia superba* reaches adult sizes of about 5 cm, although it is still primarily a herbivore. It then follows that the number of species in the evolving arctic community would remain low because of the strong competition among the various larvae for limited plant resource.

Low-Temperature Water. We know that warm-water organisms slow down their activities when exposed to colder water. But it would be an error to extrapolate this tendency to animals in the near-freezing waters of the polar oceans. It must be just as important that a fish be able to "move fast" to capture prey in cold water as in warm. And the evolutionary advantages of being swift are so overwhelmingly important to species survival that the principle must also apply in polar oceans. We also know that different enzymes used in metabolic processes work at different rates, so that an organism evolved within the Arctic ecosystem could have the same metabolic rate as one evolved at low latitudes simply through choice of enzyme. In summary, the rigors of the arctic climate is not a useful explanation either for meager speciation or for low primary production.

But the rigors of the arctic climate do have an impact on its community structure. Among the top canivores are many species of warm-blooded animals, the seals, otters, whales, birds, and bears. The fact that temperature-regulating organisms are so

prevalent in the polar oceans is itself a fascinating aspect; more on this later.

"Age" of the Arctic Ocean Ecosystem. Because of the extensive shallow seas of the Arctic, and the shallowness of the Bering Strait, around 50 m, the lowered sea level of the Pleistocene period, about 150 m lower some 40,000 years ago (see Figure 4.12d), made dry lands of huge areas of the Arctic Basin. As for the circulation of the Arctic Ocean, it was cut off from the Pacific as well as from the warm Gulf Stream water that today transports heat into the Barents Sea. Topographically, the Arctic was converted to a deep ocean basin with virtually no shelf area and, like the present Mediterranean Sea, with only one narrow communication to another ocean.

Such topographic and circulation changes, together with temperature change, must have had a drastic impact on the organization of living communities in the Arctic Ocean. This is the basis for the argument that the low speciation of the present Arctic Ocean reflects its ecological "youth," that there has simply not been sufficient time for the ecosystem to fill and develop its "niche" potential. Proponents of this argument point to the higher diversity of species in the Antarctic as supporting evidence, because the already deep 600-m waters of its continental shelves would be little affected by a sea level reduction of only 150 m; hence there would be little impact on its living communities. This topic is still under study.

The Sampling Aspect. Some scientists rationalize the present seemingly meager speciation in Arctic Ocean flora and fauna as primarily the result of insufficient sampling. More species *are* being discovered as more research is carried out, and clearly the extent of sampling of the communities of the Arctic is far less than that of, say, the North Atlantic communities between Woods Hole, Massachusetts and Bermuda. This then is an open argument, to be resolved only over time. Whatever the outcome, it is undeniably true that the Arctic Ocean ecosystem is unique.

High Production of the Southern Ocean

Just as the Central Arctic Ocean is noted for its low production, the Southern Ocean is noted for its very high production. There are two facets of its ecosystem that warrant special mention: its unique food chains and the unique physical factors that support

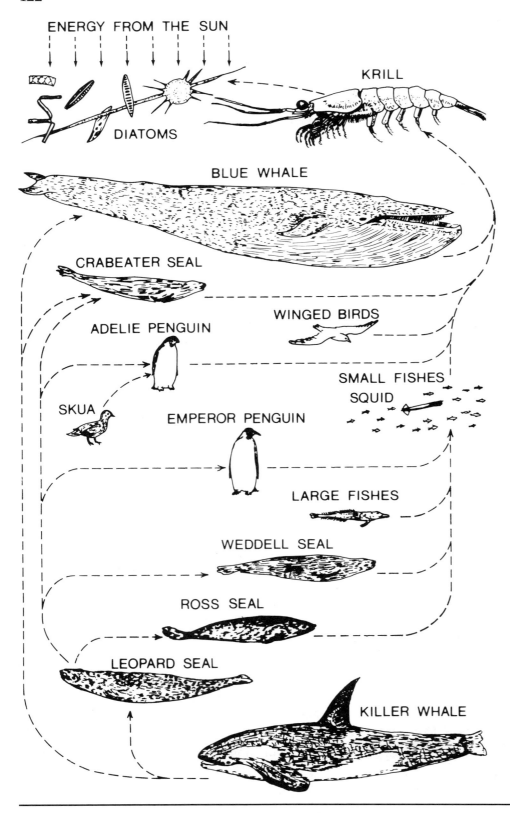

Figure 20.10 The major food linkages in the Southern Ocean. (From Murphy, 1962.)

them. Figure 20.10 shows in simplified form the major linkages in the food chains. Notice that the number of trophic steps between the primary producer and a top carnivore is small, only *two* for the blue whale, the crabeater seal, the Adélie penguin, and most winged birds. The other seals, the emperor penguin, and the large fishes are but three levels above the primary producers. To a large extent, textbook classification of the Antarctic as "rich in marine resources" stems from the fact that short, direct food chains allow the ecosystem to store a large biomass in high trophic levels. Because the animals in these levels also have high commercial value and are highly visible, the richness here is more apparent than in other oceans.

Primary Productivity. Actual plant production rates may not be much different from those of other oceans. Various numbers have been published, but it is important to understand that all represent limited sampling of that distant ocean. Comparatively, sampling of primary productivity in Arctic waters has been much more extensive, and there are virtually *no* samples of primary production in Antarctic waters during the winter season. Values such as 100 g C/(m^3)(yr) have been estimated and are equivalent to those in Table 1.2 for coastal ocean waters in general.

There are two factors to keep in mind when comparing Southern Ocean productivity with that in other waters. First, nutrient measurements over ex-

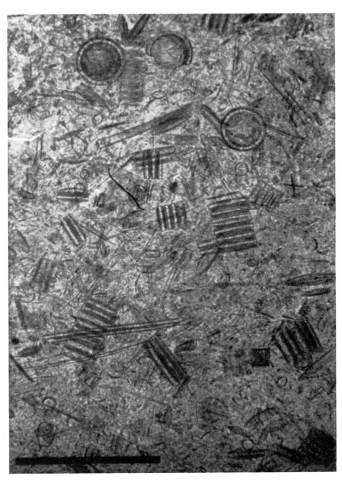

(a) Photomicrograph of the setae on the legs of *Euphasia superba*, showing the close spacing, about 7 microns, by which the animal captures its food.

(b) Photomicrograph of the stomach contents of one *E. superba*, showing its predominant diet of diatoms. Krill have hard plates inside the stomach with which they crush and grind the siliceous shells of these phytoplankton.

Figure 20.11 Feeding devices and stomach contents of Antarctic krill. (From Nemoto, 1968.)

tensive areas during the spring and summer bloom periods show that nitrate is usually present in excess quantities; stated another way, nutrient depletion is not a limiting factor in these waters. Compare this condition with that shown in Figure 8.6 where the nitrate content associated with Oregon upwelling water is high but further offshore nitrate depletion is virtually complete by midsummer. It is the massive upwelling in the Southern Ocean that maintains such high nutrient concentrations throughout the year. But unlike the Oregon case, the light available for photosynthesis at latitudes of 55 to 70°S is substantially less. Another factor is the heavy grazing pressure exerted by krill, whereby new production is rapidly removed.

The Phenomenon of Krill. Dominant over all other herbivorous plankton of the Southern Ocean is the species *Euphausia superba* (Figure 20.10). Adults range to 5 cm in length; their large size subjects them to direct predation by many top-level carnivores, from birds to large seals and of course baleen whales. At the same time, *E. superba* is equipped with capture mechanisms that enable it to feed on microscopic diatoms. Figure 20.11a is a photomicrograph of the spines and filtering setae of its sixth leg, and Figure 20.11b shows the stomach contents of a specimen. In both photographs the heavy bar measures 100 microns of distance. The setae form a mesh with average spacing of about 7 microns, which is more than adequate to capture diatoms in the 10- to 80-micron size shown in Figure 20.11b.

Within the Antarctic ecosystem, *E. superba* bridges at least one and perhaps two trophic levels normally present in central and low-latitude waters. *Copepods are the trophic level conspicuously absent in Antarctica.* Chaetognaths, which predate copepods in temperate waters and which have adult sizes also of about 5 cm, are not important here. Further, both copepods and chaetognaths are selective predators, each singling out and capturing individual prey.

In contrast, *E. superba* filters water rather indiscriminately, taking whatever particles its setae "net" captures. It is because of this feeding behavior that these animals can congregate in huge swarms and literally "sweep" the ocean clean of its flora and small fauna. Such swarms are found only within the upper 200 m, and mostly in the upper 50 m. They can be extremely extensive. During fieldwork in 1982, scientists from the University of Washington mapped out one "superswarm" that measured 5 miles in width and 11 miles in length. It contained

an estimated 2.5 million tons of krill at densities of from 2000 to 5000 individuals per cubic meter of ocean. A common size for a swarm would be 50 × 600 m, and its depth would be less than 50 m; fishermen locate swarms using acoustic devices. The baleen whale, a natural predator, also locates swarms using its evolved acoustic sensory organs.

The distribution of swarms is not uniform. Figure 20.12 shows the densest concentrations of mature krill eastward from the tip of the Antarctic Peninsula, sandwiched between the edge of the summer ice pack to the south and the Polar Front to the north. Less dense populations exist around the continent, and to a large extent some krill live everywhere south of the Front. But note that the figure shows summertime distribution; we simply do not have wintertime data. The densest summer concentration of krill is in the Atlantic sector of the Southern Ocean, which conforms with our previous discussion of massive upwelling of nutrient-rich North Atlantic Deep Water into the photic zone.

Much of what we know about krill has been gleaned from over 200 years of hunting its major predators, the baleen whales. In fact, the best estimates that we have of the actual production of krill come from the numbers of whales found in the early days of whaling, together with a modern analysis of stomach contents and estimated daily intake per whale. For example, a mature blue whale may take up to 3 tons of krill per day during its feeding season of about 4 months. On the basis of this, estimates of the tonnage of krill taken originally by whales alone range from 40 to 270 million tons. To estimate the total production throughout the Southern Ocean, we must add the tonnage of krill taken by winged birds, penguins, and seals. It is a prodigious number, but one not at all reliably estimated yet.

Krill Predation by Whales. With so much food in the form of small, uniform individuals that swarm together so densely, it is not surprising that the baleen whale evolved with a unique harvesting technique. Within the mouth cavity of these large animals are rows of long, platelike structures called baleen. Each baleen plate is a composite of many hairlike strands that are solidified except along one edge where the hairs separate into individual strands; might be compared to a very long paint brush with bristles along the side of the handle, not the end. In Figure 20.10 the baleen plates hang vertically down from the roof of the mouth cavity of the blue whale. The plates are closely oriented side by

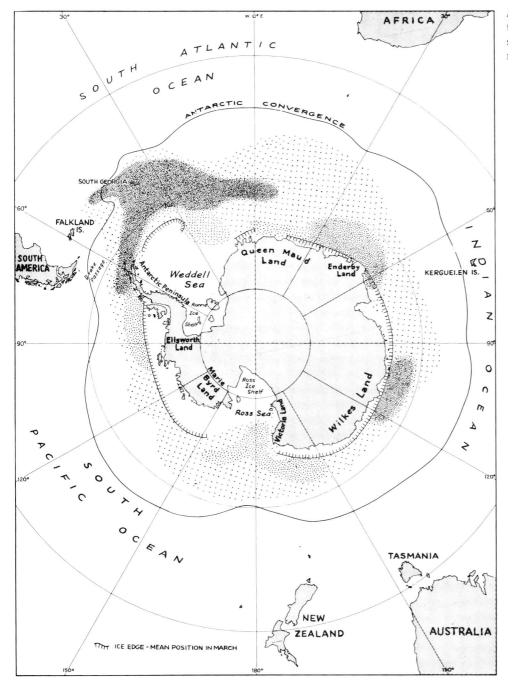

Figure 20.12 Distribution of mature krill *(Euphausia superba)* in the southern Ocean during summer months. (From Potter, 1969.)

side and mesh their hair edges to form a matlike surface that lines the interior of the mouth.

To feed, the whale first locates the krill swarm with its acoustic sensor. It then enters the swarm and ingests huge quantities of krill-laden water into its mouth cavity. The longitudinal striations evident

along the underside of its jaw (Figure 20.10) are folds in the skin; they allow the mouth cavity to expand during the ingestion process, literally "drawing" the water–food mixture into the mouth. By pushing its huge tongue forward, the whale then forces the mixture out of the mouth but through the matted hair surface, trapping the krill. Its final action is to swallow the capture.

During years of whaling and simultaneous study of other life forms in the Southern Ocean, whalers learned that the blue whale preferred to feed on one-year-old krill, which tended to congregate inside the icepack regions. Fin whales, on the other hand, would prey on swarms of two-year-old krill, generally found in open water outside the pack ice limits. There are also close relationships between the krill and the phytoplankton they feed on. Their preferred plankton species is the diatom *Fragilariopsis antarctica*; investigators have found that whales are often sighted in waters containing these diatoms, but almost no sightings are made where other species of diatoms are dominant.

Krill Predation by Man. In recent years, several nations have begun fishing the krill itself. Because whale stocks are now so low that whaling is either uneconomical or banned, and because the world demand for fish meal is steadily rising, krill fishing may become an important part of the ocean harvest.

For example, assume that the annual sustained harvest by whales during past centuries was 150 million tons per year, a middle-range value of estimates described in the previous section. This is roughly twice the size of today's total catch of all other marine animals. About 15% of a krill is convertible into fish meal, or about 22 million tons per year. At a New York price of $400 per ton, about twice the value of soybean meal, the potential market value of the krill resource is about $8 billion annually. If a major part of this protein harvest was further refined into a food item salable to humans, its resource value could be doubled. Krill meat has been processed into a paste form and sold as "shrimp butter" for use in salads, stuffed eggs, and the like.

But there are limits to a potential krill bonanza. First, it is a small animal, and this means towing a fine-mesh net to capture it, with the attendant high fuel costs. It may well happen that we can never take the krill as efficiently as does the whale. A mature blue whale can cruise with an energy equivalent of about 10 horsepower and take over 3 tons of krill a day. If the cost of diesel fuel rises substantially, we would have to revert to whaling as the only economical method of harvesting krill. Currently, world attention is focused on a restoration of the whale populations. Experts estimate that at least 50 more years are needed to restore whales to the point that whaling can again be sustained and profitable, and at that point some 100 to 200 billion tons of krill must

be reserved for whale sustenance. In short, we may have difficulty saving the whale and eating krill too.

High-Trophic-Level Carnivores in Polar Oceans— The Role of Pack Ice

The Southern Ocean food cycle (Figure 20.10) and the arctic and subarctic food webs, (Figure 20.13) have one factor in common that makes them differ markedly from food webs in lower latitude oceans: the higher carnivore levels are filled mostly by warm-blooded animals, many of them large— whales, seals, walrus, sea lions, and both flying and nonflying birds. Except for the whales, these animals have one common behavior trait—all make use of the broken pack ice as a solid substitute for terra firma. To the seals, walrus, and birds, the floating ice pack is a place to rest, to hide from predators, to reproduce young, and to hitch rides across large stretches of ocean. As discussed earlier, the ice zone is a fertile area for primarily production and production of krill and associated fauna such as fish.

In the Antarctic. The truly astounding visual impact of the Southern Ocean ecosystem is found in areas of loose pack ice. Here I saw flying birds in uncountable numbers, thousands of kilometers from shore, and groups of penguins resting on ice floes. Other ice floes will have crabeater seals, which prey on krill directly (Figure 20.10), or leopard seals, which prey on the penguins, or Weddell seals, which prey on larger fish. Surfacing in the open areas of the ice pack are the blue whales. Also in the fringes of the ice pack are the killer whales, which capture small seals and penguins or hunt in packs for the larger whale species. (I once sat spellbound in Chile listening to Pablo Macaya, one of the last whalers, recount how he saw a pack of orcas wound a large whale, wait while it bled to death, and then consume it.)

In the Arctic. In the subarctic seas such as the Bering, the role of the pack ice in supporting high-trophic-level carnivores is clearly demonstrated. For example, walrus and bearded seals are benthic feeders. With the pack ice to rest upon between hunting excursions, these animals forage over the entire breadth of the Bering Sea Shelf, some 500 km wide and over 700 km long. They tend to congregate along the edges of the broken pack ice. Figure 20.13*b* shows a general food web for Bering Sea bearded seals and walrus.

The walrus preys mainly on benthic fauna, of which clams and polychaete worms are the domi-

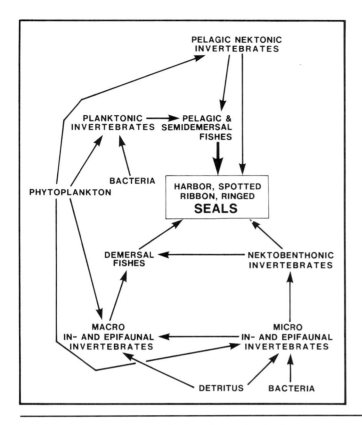

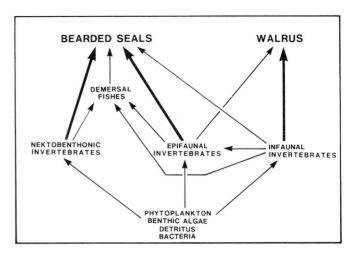

Figure 20.13 Generalized food webs (a) for harbor, spotted, ribbon, and ringed seals and *(b)* for bearded seals and walrus in the subarctic Bering Sea. (From Lowry and Frost, 1981.)

nant forms (see Table 20.1). This beast has two handsome ivory tusks for digging these clams; an adult walrus may dig several bushels of clams each day. Using recent estimates of 200,000 walrus in the Bearing Sea area alone, we conclude that the production of clams and infaunal worms must be prodigious indeed.[2]

The bearded seal feeds primarily on the crab, shrimp, snail, and octopus, all benthic epifauna, as well as demersal fishes such as flatfish (sole and flounder) and sculpin. It also makes use of the pack ice as both resting place and transporter. Other seal

species are also found in the ice pack: the ringed seal, harp seal, and harbor seal are primarily pelagic feeders, depending on fish of intermediate size such as the capelin. Populations of all seals in the Bering and Chukchi seas are estimated at over 1,000,000.

The (white) beluga and gray whales also feed seasonally in the Bering Sea. Both are baleen whales, but differ in their preferred prey. The beluga feeds on semidemersal and pelagic fishes, whereas the gray consumes shrimps and amphipods, which are epifaunal crustaceans. The "hair" of the baleen in a gray whale is much coarser in texture than that of the baleen whale that filters the smaller, planktonic forms such as krill.

Climate and Ice Cover on the Ocean

So far in this chapter we have studied the polar oceans much more than the polar atmosphere. But polar explorers have long observed that changes in the area covered by pack ice are strongly correlated with many aspects of climate. By Climate we mean

---

[2]It is. And the production cycle is somewhat unusual in that it is only partly dependent on the detritus that arrives at the bottom from phytoplankton production overhead. The Yukon and Kuskoquim rivers have extensive marshlands in which the eelgrass *Zostera* is produced at phenomenal rates (see Table 1.1 for production in marshlands). During the spring ice breakup, riverborne ice floes tear up these shallow beds and deposit large quantities of eelgrass on the coastal shelf. Here bacteria break down the material into detritus, which is then captured by filter feeders such as the clam (see Figure 13.4).

Table 20.1 The major species within the six types of marine animals commonly taken as prey by seals and walrus in the Bering Sea. (From Fay, 1981.)

Prey Type	Major Species
Pelagic and semidemersal fishes	Walleye pollock, saffron cod, arctic cod, Pacific cod, capelins, rainbow smelts, herring
Demersal fishes	Eelpout, sculpins, flatfish, sand launces
Pelagic nektonic invertebrates	Euphausids, hyperiid amphipods
Nektobenthonic invertebrates	Mysids, shrimps, gammarid amphipods, octopuses
Epifaunal invertebrates	Crabs, snails
Infaunal invertebrates	Clams, polychaete worms, echiuroid worms, priapulids

the principal conditions of a region's atmosphere: for example, mean air temperature, mean annual rainfall, and mean annual cloudiness and winds, but also the extremes of these conditions throughout the annual cycle. For this reason our study of polar oceans must include their atmospheric connection.

Connections

In general, the presence of an ice cover sharply reduces the rate at which heat and moisture pass from ocean to atmosphere. Ice, and particularly ice with a snow cover, is a relatively good insulator. The onset of the freezing season in the Arctic begins in the fall as the amount of solar radiation rapidly falls, approaching the darkness of the arctic night. But open-ocean water continues to lose heat by long-wave radiation, so that the surface water cools toward its freezing point. During this process of cooling, the surface layer is driven into vertical convection motions that reach to a depth dependent on the density stratification. If winds and waves are present, these assist in the vertical mixing (recall Figure 8.3 for an example of the winter well-mixed layer in a non-freezing ocean). Eventually, the mixed surface layer is cooled to its freezing point and ice begins to form at the surface. The ice grows rapidly at first; formation of 10 cm in the first day is not unusual. But the rate of new ice added each day diminishes exponentially, so that usually only about 100 to 150 cm of ice form during the entire season of freezing, 100 days or more, even though the overlying atmosphere may be tens of degrees below freezing temperature. Ice is a poor conductor of heat energy. Although the surface of the ice cover continues to lose heat by radiation to outer space as the ice thickens, the fluid ocean below supplies less of the surface heat loss, and more is supplied by the atmosphere itself, which grows increasingly colder and dryer.

This is a dramatic behavior of nature, a rapid seal-ing-off of the exchanges between ocean and atmosphere that normally occur across the fluid–fluid interface. The reader will recall the earlier discussion of air–sea interactions I and II (Chapter 6), in which the transfer of sensible heat and latent heat of vaporization are key elements in the heat budget of the fluid ocean. In the ice-covered ocean, however, these transfer processes are effectively stopped. The net result is that the temperature *difference* between the tropical and polar atmosphere is sharply increased. This leads to more vigorous circulation in the global wind patterns, which in turn means that the atmosphere carries greater heat flux to polar regions until an equilibrium is reached. The point is clear: if seawater did not impose this "cap" upon itself by changing to solid crystal form, global wind circulation would be less intense and temperatures of both polar air and sea would be higher as the ocean assumed more of the task of poleward transfer of heat.

Arctic Ice and Atmosphere Connections: Northern Hemisphere

Figure 20.14 is a simplified chart of the pattern of westerly winds in the Northern Hemisphere. First locate the positions of the two high-pressure cells over the central areas of the North Atlantic and North Pacific ocean basins; these are the same high-pressure cells shown in (Figures 10.5 and 10.7).

Poleward of these central highs, and weaving wavelike between alternating high- and low-pressure influence, is the core of upper-level winds known as the *jet stream*. Normally, we find about four complete waves in this circumpolar belt. Moreover, the wave pattern tends to be stationary, tied to the locations of the central high-pressure cells. But just as the positions of these highs migrate during the summer–winter cycle (see Figure 10.5), so also does the jet stream pattern shift in a west–east cycle.

This jet stream exists because of the large temperature difference between the dry cold polar high-pressure zone and the warm central high-pressure zone. The polar high exists because the perennial arctic ice pack insulates the ocean from the overlying air.

The relationship can be summarized as follows. In summer, the ice cover reflects over 60% of incident sunlight; this prevents the ocean from warming but causes the overlying air to warm faster than it would were the ocean ice-free and thus absorbing most of the incoming solar radiation itself. In winter, the ice cover prevents the ocean from supplying heat to the overlying air, and the air temperature drops to very low levels as back radiation cools the ice surface. Therefore, the atmosphere over an ice-covered ocean cools more intensely during winter and warms up more in summer than does an atmosphere over open water.

Arctic Ice and Climate Changes

The direct consequences of all these effects of arctic ice cover are the short-term weather and long-term

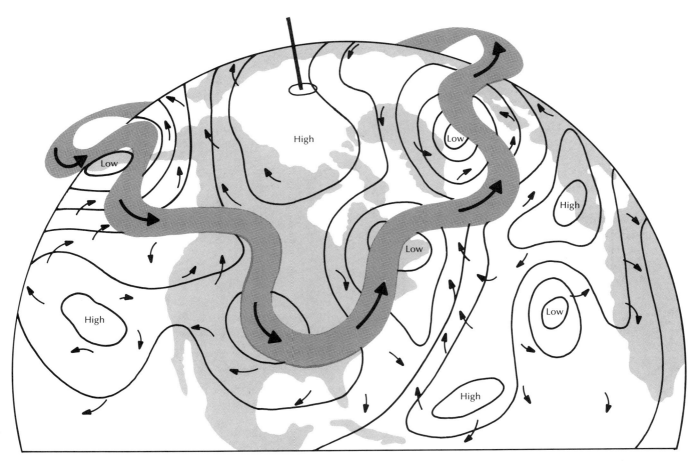

The North Central Pacific and Atlantic oceans each have the characteristic high-pressure cell with anticyclonic winds; the region over the perennial pack ice of the Arctic Polar Ocean is also a high-pressure cell. In general, the jet stream shifts southward during winter as the polar ice pack area expands, and wind intensifies. During summer the pack ice limit recedes toward the pole, and the jet stream shifts northward and decreases in intensity. An east–west displacement of the crests and troughs of waves in the jet stream may accompany the north–south shifting. When the jet is moving south, as shown here over the central United States, the air is cold and dry. Conversely, the jet moving northward along the eastern seaboard of the United States will be considerably warmer and moister, having entrained moisture from the regions around the Gulf of Mexico. This alternating dry, equatorward and wet, poleward behavior of the atmosphere is the major mechanism for a *net* poleward flux of heat and moisture.

Figure 20.14 The upper-level jet stream in the atmosphere of the Northern Hemisphere.

climate changes in the Northern Hemisphere. The air temperature difference from pole to tropics is larger in winter, so that the intensity of circulation in the atmosphere increases in winter; there are more storms, and the tracks of storms move southward just as the edge of the pack ice moves southward in winter. When circulation intensifies, the pattern of wavelike crests and troughs in the jet stream (Figure 20.14) shifts eastward. The eastern part of the United States receives higher-than-average exposure to the cold arctic air masses sweeping down off the Canadian shield; at the same time the Pacific Northwest gets more storms from the southwest and more rainfall.

There are longer climatic trends in which arctic ice cover and climate changes correlate well. Accurate records of arctic conditions have been kept for over 100 years. Russian scientists report that between 1890 and 1940 the mean ice thickness in the Soviet Arctic decreased by one-third (from 365 to 218 cm) and that the area covered by pack ice decreased by 10%. During this same period, the whole global circulation was intensifying; westerly winds in the North Atlantic increased some 20%, cyclones took more northerly paths, and the Gulf Stream became stronger and passed farther north, contributing to a 5°C warming of waters in the Norwegian Sea. It is worth noting that on time scales of 50 years the oceanic community composition easily adapts. Figure 20.15 shows how the northern limits of the European herring fishery shifted northward over the period 1900 to 1940. Since 1940 this trend has reversed.

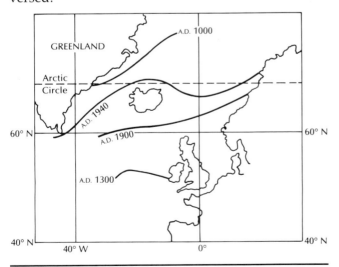

Figure 20.15 The southern limits of the European herring fishery at four dates during the past thousand years. (From Fell, 1975.)

On a still longer time scale, we know from the Viking migrations to Iceland around 950 A.D. that from this time until about 1200 there was hardly any summer pack ice around the island. Figure 20.15 shows how the herring fishery in 1000 extended even north of Iceland, but by 1300 it had moved substantially southward to the latitude of the British Isles. Concomitantly, a Greenland colony was founded in the tenth century and flourished for some 300 years before being abondoned around 1410 A.D.

Summary

Polar oceans are such special places, and in so many ways, that a summary in paragraph form would be too lengthy. An enumeration is better.

1. The dynamic behavior of polar oceans is controlled by processes that alter salinity. In the Arctic this control is exerted by rivers flowing off the great Siberian and Canadian watersheds. In Antarctica, the process is the freezing and melting of sea ice, mechanisms responsible for the GIANT ANTARCTIC SEA ICE MIXING MACHINE.

2. A very large factor that accounts for the differences in dynamics between the Southern and Arctic oceans is that the Southern ocean is completely open to vigorous mixing with all others through the Antarctic Circumpolar Current. The Arctic is highly restricted from communication with other oceans; the Bering Strait is very shallow, as is the Barents Sea north of Norway, and the only deep flow is with the North Atlantic.

3. Following the closure of the land bridge between the American continents, the overall circulation pattern of the world ocean changed. Only during the past 4 million years (roughly, the same span as covered by *Australopithecus*) has the North Atlantic become saltier than the Pacific, the Arctic region been glaciated repeatedly, and Antarctica been covered with its several kilometers of ice. Could a massive open canal across Panama restore the world to a warmer climate overall?

4. The arctic biological system is unusal in its meager speciation. Possible explanations are low primary production; a short productive season, which in turn controls reproduction cycles; low-temperature water; the immaturity of the biological system because the basin itself is young; and the possibility that we simply have not sampled it adequately.

5. The antarctic biological system is unusual in its high production coupled with an exceptionally short food chain between microscopic diatoms and macroscopic whales. The main reason is the massive upwelling, whereby surface waters are not nutrient-limited.

6. An unusual feature of both oceans is the dominance of warm-blooded organisms, many of them large, at the top of the food chain. The first reason is that the larger the size of the animal, the larger is its volume to surface ratio, hence the less loss of energy to "staying warm." And the rate at which warm bodies can deliver energy for the *chase* is much higher than that for cold-blooded animals. Second, when cold-blooded animals have to develop their own antifreeze to prevent tissue from freezing at the depressed temperatures of polar water, the problem of energy production in muscles becomes acute. These animals simply could not compete in polar waters. Third, the extensive ice floes make the tasks of the warm-blooded animal much easier.

7. Last, there is the special effect that ice-covered oceans have on world climate. Because the whiteness of ice increases the albedo of polar oceans, they have little ability to absorb solar radiation. Thus ice cover is responsible for large equatorial–polar temperature differences in the atmosphere, which in turn creates the jet stream determining climate in mid-latitude zones.

Study Questions

1. There are at least two major differences between the Arctic Ocean and the Southern Ocean. Describe them.

2. The similarity between these two polar oceans lies in the identification of their marginal seas, always the waters that lie between the "ocean" and a continental boundary. Name and locate on a suitable chart the various marginal seas and state if each is a shallow or deep sea.

3. An even more interesting comparison is that of polar oceans with other oceans. In what ways are polar oceans *different?*

4. No course in oceanography would be "successful" without the student's learning:

  (*a*) What is it that constitutes the GIANT ANT-ARCTIC SEA ICE MIXING MACHINE?

  (*b*) What important role in ocean mixing is performed by these sea ice processes?

5. Similarly, it would be negligent if the course did not require the student to place the annual sea ice progession–regression into perspective. How much earth *area* is involved in the annual variation in sea ice? Make a Northern Hemisphere area contrast that would dramatize the importance of annual changes in sea ice coverage in the Southern Ocean.

6. One of the more intriguing problems currently being studied for Arctic Ocean waters is the meager speciation in these waters. Develop several reasons to explain this speciation aspect.

7. Perhaps the most intriguing of all questions is why the Southern Ocean is so productive. This question has different facets:

  (*a*) Why is the Southern Ocean so productive at the primary trophic level?

  (*b*) Why has the Southern Ocean ecosystem developed such a short food chain between primary productivity and the blue whale?

8. If the blue whale has been reduced in population to the point of being endangered, has any other organism or organisms replaced the blue as a major predator?

9. What do you think of the proposal to rear salmonids in South American rivers and estuaries for release into the AACP and its swarms of *Euphausia superba?*

10. Discuss the krill of the Southern Ocean from the view of the in potential value in present-day markets.

11. Pack ice modifies the sea surface in an extremely important way—it provides substrate and cover. Discuss the significance relative to the population of walrus in the Bering Sea.

12. What is the basis of the food chain of the Bering Sea in which the walrus plays such an important role?

13. At the risk of overplaying the term "important process," what is the role of the perennial arctic sea ice in the overall question of what triggers an ice age?

14. For years Russia had studied the possibility of

diverting river water from its eventual exit into the Arctic Basin to irrigate the Central Asian regions around and east of the Caspian Sea.[1] How might this affect the global climate?

15. Justify the statement "There is about as much river runoff into the Antarctic Ocean as there is into the Arctic Ocean."

[3]In April 1986 the Kremlin announced that this river project has been *canceled*.

16. Write a short essay explaining why the primary production in arctic waters is *(a) delayed* farther into summer and *(b)* more *peaked* than that in temperate waters?

17. If cold temperatures decrease the rate of metabolism, why do we find so many warm-blooded species in the polar seas?

18. Name the principal predators of krill in Antarctica.

CHAPTER 21

PLATE TECTONICS AND OCEAN BASINS

The most fascinating revolution in the development of natural science since the time of Darwin is being written in our own lifetime. We are witnessing the crystallization of a new theory on the structure of the earth itself. Called plate tectonics theory, it describes a geological evolution—how the present distribution of continents and ocean basins has evolved from a much different earth over geologic time.

Just as fascinating as the new geology itself is the story of the *evolution of our understanding* this new global geology. Marine geology and geophysics have played a key part in this recent history, so that much of the research by which the new theories are being shaped is taking place at oceanographic institutions.

In this chapter we build a historical perspective of how the oceans have changed during past geological epochs.

Connections

The first perceptions that the continents as we know them today could have once been joined together were gleaned in the explosion of maritime exploration begun in the fifteenth century. Portuguese sailors explored and charted the west coast of Africa, a task completed in 1497–1498 by Vasco da Gama when he rounded the Cape of Good Hope and opened the Indian Ocean to Western civilization. Meanwhile, in 1492, Columbus touched the New World for the first time since the landing by Lief Ericsson in 1001 on the shores of Labrador. John and Sebastian Cabot followed quickly with voyages into the Northwest Atlantic (1497, 1498). Ferdinand Magellan touched the northern coast of Brazil in 1519 and proceeded to chart the coast of South America down to the strait that bears his name today. By the mid-sixteenth century Gerhardus Mercator and others were busily producing new maps that by then included the eastern and western shores of both the North and South Atlantic Ocean basins, and thus the stage was set for the first observations that the *shape* of the Atlantic Ocean suggests that its east–west boundaries would "fit" together remarkably well.

Indeed, references to the "correspondence in shapes" of Atlantic Ocean coastlines appear as early as 1620 in comments by Sir Francis Bacon in his *Novum Organum.* Others noted these complimentary-shaped coasts and deduced that the Old and New Worlds became separated by the Great Flood. But the early charts from which such deductions came were grossly inaccurate, because mariners had no reliable way *to reckon longitude.* Although the Flemish astronomer Gemma Frisius first proposed in 1530 the use of a time-keeping device for reckoning longitude at sea (Lewis, 1982), almost 250 more years passed before such devices were produced.

Latitude at sea was easily measured, and quite accurately; (you need know only the day of the year and the maximum elevation angle of the sun above the horizon during that day). But for longitude you must *know the precise difference in time* between a local "high noon" and that at some selected reference longitude (Greenwich, England), and this requires a reliable chronometer. John Harrison developed and successfully tested such a timepiece in 1761. Acceptance quickly followed: Captain James Cook carried three chronometers on his historic circumnavigation exploration of Antartica in 1772–1775. From then on new maps were quite accurate, and so the correspondence of shapes across the Atlantic attracted more and more attention, not only of geographers and geologists, but also of botanists and zoologists.

At about the same time that accurate *solar* time

was first given to mariners, a different kind of time was conceived—this was the continuous, immense time scale on which could be marked the geologic events of the earth's history. James Hutton (1726–1797), an Edinburgh, Scotland physician who also studied geological formations, developed the concept that geologic processes today are the same as those of the past. This fundamental principal of geology is called *uniformitarianism*. William Smith (1769–1839) developed the idea that different geological strata originated during the same period if each carried the *same assemblage of fossils*. The work of these two men, together with the writings of Charles Lyell (1797–1875), the father of modern geology, provided the genius of Charles Darwin with precisely the "tool" he needed, a "framework of geologic time epochs" upon which to hang his theory of natural selection of species (Faul, 1978).

From that point forward science expanded exponentially and across all disciplines. Of particular value was the development of paleontology, the science dealing with fossils of past geological periods, for it gave us independent evidence of connections between sedimentary rock structures that today are in widely separated locations.

The stage was set by the end of the nineteenth century: accurate maps of the world existed; fossil sequences had been collected from many regions of the world; ancient climates were being studied, based in large part on the fossil record; and geology was a widely developed science because of industrial demands for all types of mined minerals. It was a time for a new synthesizer to gather together the evidence on which the concept of *continental drift* would be based, and eventually accepted some 50 years later. It was time for Alfred Wegener.

Wegener (1880–1930) was not the first scientist to be intrigued enough by the "fit" of continent shapes to search for evidence that they had been linked together in a large, ancient land mass. But he was without doubt the first to put together an unparalleled suite of clues and evidence to support the existence of such a land mass, gathered from all disciplines of science known in his day. He proposed that the breakup of the ancient land mass, called Pangea, began about 200 million years ago. Since then the separated continents have moved across the face of the earth to their present positions (Wegener, 1929). Today we know that continents are still "drifting."

It is interesting to note how Wegener began the scholarly work that was to last his lifetime. In a letter to his wife in January 1911, he wrote:

My room neighbor Dr. Take has received for Christmas the big Andree *Handatlas*. We admired the magnificent maps for a long time. Presently an idea came to me. Look at the world map again please: doesn't the east coast of South America fit precisely with the west coast of Africa, as if they had been connected formerly? It agrees even better when one considers the bathymetric chart of the Atlantic Ocean and compares not the present continental coastling, but the margin of the continental slope in the deep sea. I must pursue the idea.
(Drake, 1976).

Wegener died at what was perhaps a time of peak intensity in the polemic "put-down" of his writings (see, for example, Schuchert, 1928). His thesis would not be "accepted" for another 30 years or more. Even in 1963, when I took my first course in geological oceanography, the professor refused to be drawn into a discussion of the Wegener hypothesis. But by the end of the 1960s this phase of revolution in earth science was over; today, the details of modern plate tectonics are being consolidated by a host of investigators.

The Main Evidence

Let us first list the "evidence" by which the fact of continental drift eventually became widely accepted; then the evidence for sea floor spreading, which was soon recognized as the mechanism through which such drift was possible; and finally the evidence allowing their consolidation in a theory of plate tectonics. The reader is urged to read rapidly through the list to get a sense of both its chronology as well as its interdisciplinary aspects. Later sections expand each of these facets.

Evidence Supporting Continental Drift

1. The continents seem to "fit" together, much as the pieces of a jigsaw puzzle, as if once they comprised a contiguous supercontinent called Pangea.

2. By charting the ancient climate of different parts of continents, we can show that fitting these continents together produces not only a "shape" fit but also logical contiguous zones of glaciation or tropics of Pangea; this science is called paleoclimatology.

3. When rocks solidify from a molten state they retain a permanent record of the earth's magnetic field at that time. After charting this "remanent magnetism" from many ancient rock formations, scientists were forced to conclude that either the magnetic poles had drifted over the earth surface during geologic time, or the continents on which the rocks solidified had themselves drifted.

4. Similarity in assemblages of fossils in coal beds and sedimentary rocks from continents on *both* sides of the Atlantic, north and south, suggested that these stratified beds were formed under similar conditions and possibly in the *same region,* and that they were later separated as their respective continents moved apart; this science is called paleobiogeography.

5. Absolute dating of the *times of formation* of geologic features like mountain chains, obtained by measuring the radioactive decay of materials originally locked into such features, produced evidence that rock formations found on widely separated continents today were formed at the *same* time, hence probably in the same region.

Evidence Supporting Sea Floor Spreading

In spite of the accumulated evidence, the theory of continental drift was rejected by many because there was *no plausible source of energy* for such massive movements in the earth's crust. The next phase in the unfolding of plate tectonic theory produced evidence that the floor of *each* ocean basin was itself in motion, and that movement was directed *away from* the mid-ocean ridge zones:

1. Nowhere had expeditions dredged up rocks from the ocean floor that were older than about 150,000,000 years. How had the old basin crust disappeared?

2. The discovery of patterns of anomalous magnetism within the basin crust suggested that the basin crust is moving away from the center lines of the mid-ocean ridges, *a crucial piece of evidence.* For the first time a self-consistent explanation existed not only for the mid-ocean ridges but also for the fracture zones across the ridgelines and for the earthquake epicenters along both the ridges and the fracture zones; this science is called paleomagnetism.

3. If the ridgelines were a "source" of new oceanic crust, the material coming from them would have to originate from deep within the lithosphere and would therefore be much *hotter* than the older basin crust far removed from the ridgelines. Scientists designed devices to measure the upward heat flow from the ocean floor and used them to prove the hypothesis.

4. It then followed that if the slablike oceanic crust cooled as it moved away from its origin in the

ridgeline, it should also shrink in volume. This would cause the basin floor to appear at greater depths the farther it was from the spreading source, a result readily confirmed by numerous depth measurements. This shrinking basin floor idea was confirmed in two other ways: basin terrain becomes *smoother* with distance from the ridge, an expected result because sediment fills the depressions in the floor and accumulates thickness with distance from the ridge; and second, the oldest volcanoes in strings, such as the Hawaiian chain, appear to sink with time.

5. The development of the chemistry of radioactivity led to theories on the radioactive decay of materials deep within the mantle of the earth. *This was the "plausible" source of energy so long awaited.* The heat from radio active decay could drive convection cells deep within the mantle. This cell motion could then be the mechanism by which solidified crust is "dragged" away from the ridgelines, and thus explain sea floor spreading.

Evidence for Plate Tectonics

There remained the problem of explaining the disappearance of oceanic crust older than about 150,000,000 years, along with a host of questions on *how* basin crust material actually moves. The two decades from 1955 to 1975 saw a virtual explosion of seismic and magnetic exploration of the basin crust, together with a large-scale project to drill cores from the sediments of all the major basins. These data yield the evidence that the crust of the earth, both continental sial and oceanic sima, consists of a few large entities called plates, perhaps 14 in number; that these are in relative motion across the plastic mantle of the earth's lithosphere; and that the boundaries where different plates are in contact are also zones of strong seismic activity.

1. By 1960 the accumulated data on earthquake foci revealed patterns of stress in the earth's crust, but located deep within the crust. These data were interpreted as evidence that wherever the spreading basin crust collided with a segment of crust carrying a continent, the basin crust was forcibly underthrust beneath the continent. The process is called *subduction,* and it was used to explain vanishing ancient basin crusts. The subducted segments of crust remelted as they eventually reentered the hot, plastic mantle material.

2. If the subduction process with all its ramifications is typical of a collision between a basin crust

CASE IN POINT

THE MEXICO CITY EARTHQUAKE, 1985

The 1985 Mexico City earthquake disaster originated with a tectonic plate adjustment along a fracture, part of the subduction process where Pacific Ocean basin crust moves beneath the continent.

Serg Dorantes/Sygma.

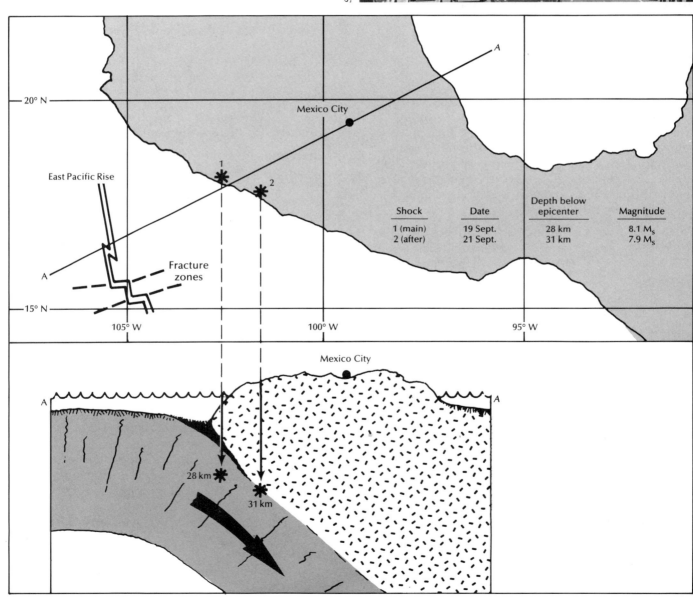

Shock	Date	Depth below epicenter	Magnitude
1 (main)	19 Sept.	28 km	8.1 M_s
2 (after)	21 Sept.	31 km	7.9 M_s

plate and a continental plate, a collision between two continental plates should cause extensive uplifting and lateral thrust, since neither plate could be underthrust. The Himalayan Mountains are such an upthrusting, created by the collision between the Indian continent and the Asian continent.

3. Strings of volcanoes such as the Hawaiian chain are evidence that the basin crust is linearly moved over a still deeper "hot spot" within the mantle. If the hot spot activity is sporadic, the episodic eruption of magma *through* the overlying crustal plate would create separate volcanic islands.

4. All the evidence that supported the intermediate theories of *continental drift* and *sea floor spreading* are also part of the foundation for plate tectonic theory.

Continental Drift

The "Fitting" of Continental Shapes

Reproduced in Figure 21.1*a* is a chart from Wegener's 1929 book *The Origin of Continents and Oceans.* All the major continents of today were joined in a supercontinent, Pangea. The geologic date of Wegener's chart is about 280 Ma (million years ago) the "shapes" outlined with heavy borders are not the precise shapes of continents today. With a few exceptions, however, Wegner's construction is still valid today. For example, he places India as already connected with the greater Asian continents. Modern tectonics specialists believe that India was then a part of the southern subgroup called Gondwanaland, composed of South America, Africa, India, Australia, and Antarctica, and that an open seaway called Tethys separated Asia from Gondwanaland.[1]

By the time of the Lower Quaternary, some 2 to 3 Ma, the continents had separated into a pattern much like that of the present earth (Figure 21.1*c*). The Antarctic was already astraddle the South Pole, and the North Atlantic, although not quite as wide nor perhaps as deep as today, already had all its principal features.

Paleoclimate, Sea Level, and Ocean Currents during the Drift

It is interesting that one of the early studies by Wegener was of ancient climate and how changes in climate might be related to continental drift. Koppen and Wegener (1924) assembled data on the location of land glaciers during the Carboniferous period shown in Figure 21.1*a* and found these to be centered in Gondwanaland, near what is today the tip of South Africa. This location is marked in Figure 21.1*b* as the probable location of the South Pole at that time; the stippled circle represents the earth's surface within 30° of latitude of this pole. Most of the glaciation traces *(E)* of that epoch are contained within this 30° circle, just as the glaciation of today is around either the South Pole or the North Pole. Wegener combined these data with the known locations of coal beds in the Northern Hemisphere today—Appalachia, Newfoundland, Ireland, Scotland, Poland, and Central Asia—and deduced the probable location of the equator during the Carboniferous "coal-forming" period, shown in Figure 21.1*b* as the heavy curve following along the locations of coal beds marked *K*. With the equator and South Pole in place, the North Pole of that epoch was located in the North Pacific, as shown.

Assuming that the scenario just given is correct, it follows that the strong glaciation around the South Pole (because its entire 30° sector is on land) would have locked in much water as ice. A direct consequence would be the *lowering* of sea level.

Yet the geological evidence is that ice cover was comparatively thin around the South Pole of that time, much less than the very thick (3 Km) ice observed when the geographic pole is surrounded by land, as Antarctica is today. At the same time, there was extensive *transgression* of shallow seas over the continents, as shown by the stippled areas in Figure 21.1*a*. In fact, the transgression reached its peak about 75 Ma when about *40% of all land* was flooded (Van Andel, 1977). It was extensive in time also; earlier, during the Carboniferous epoch of 280 Ma, a shallow sea, the Tethys, was continuous from the eastern seaboard of Asia to the western coasts of the Americas.

Is it correct then to assume that the extremely "low" sea level of 75 Ma was the direct result of polar ice cover at the time? Certainly in very recent times we have shown the direct link between water melted off the polar ice caps and the rise in sea level (see Figure 4.12). And before the most recent ice advance, called glaciation, there were a number of advances and retreats (interglacials) that reach back 10 to 15 Ma but were *not* accompanied by extensive transgressions like the Tethys. Even today, when the ice cap is very thick and sea level must therefore be at a low, the melting of all the ice on Greenland and Antarctica would elevate the sea surface only

---

[1]The concept of Tethys as a separating sea is also manifest in the use of "Laurasia" as the name for the part of Pangea located in the Northern Hemisphere and "Gondwanaland" for the remainder located in the Southern Hemisphere.

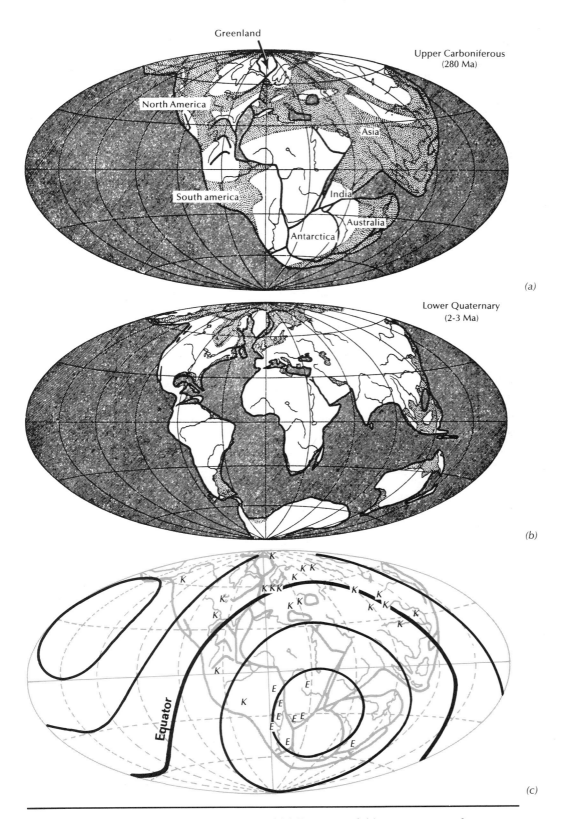

Greenland

Upper Carboniferous
(280 Ma)

North America

Asia

South america

India

Australia

Antarctica

(a)

Lower Quaternary
(2-3 Ma)

(b)

Equator

(c)

Figure 21.1 Wegener's (1929) construction of *(a)* Pangea and *(c)* recent geography;
(b) Koppen and Wegener's (1924) construction of the latitude occupied by Pangea during
the Carboniferous period.

438

about 60 m. This is *not nearly enough* to reproduce the vast inundation of continental area by the Tethys (Figure 21.1*a*). In short, geophysicists had to find an alternate explanation for the Tethys phenomenon. The explanation was found, and is discussed in detail in the next section as evidence supporting the theory of sea floor spreading.

The fact that the earth of the Carboniferous period had weak polar ice caps and extensive inland flooding of shallow lands does not explain the mildness of its climate. Such factors were the *result* of a mild and uniform climate. We now believe that the system of *ocean currents* in that epoch could have created the uniform and moderate climate. Figure 21.2 shows one interpretation of how the single continent Pangea could have influenced ocean circulation and

hence the distribution of heat from equatorial to high latitudes. Overall, the circulation patterns are simple, and the inferred heat distribution is much more direct than that of today's complex oceanic circulation. Major currents carried warm waters directly into both polar regions. Perhaps more importantly, surface waters remained exposed to polar cooling for much *less* time than exposed to equatorial heating, from which we would deduce that the ocean surface was much warmer during the Carboniferous period. It is also likely that the negligible cooling in polar latitudes could not duplicate the massive sinking of surface water (because cooling increases density) in today's oceans. Therefore, even the deeper layers of the ancient oceans would have been warmer by today's measures.

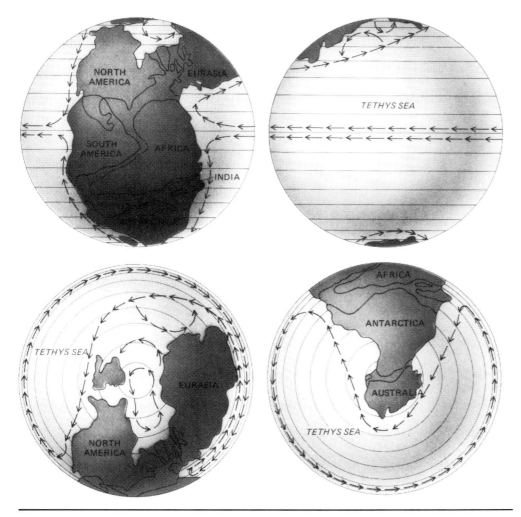

Figure 21.2 A simple model of the main circulation patterns in the oceans during the geologic period of Pangea, before about 200 Ma. The two sets of views are *(a, b)* from points above the equator, and *(c, d)* from points above the geographic poles. (From Van Andell, 1977.)

Remanent Magnetism in Rocks—Pole Drift or Continent Drift?

First, recall that the positions of the poles and equator in the previous discussion are based on evidence of ancient climates. There is still another way to discover *where* the poles were in ancient times, and that is to study the clues "locked" into rocks when they are first formed. When a molten lava pours out onto the surface of the earth and begins to cool, some of its iron oxide molecules "line up" with the lines of the earth's magnetic field because the iron oxide molecules themselves have a "magnetic polarity," like tiny compasses. After solidification, the rock retains this permanent memory, or remanent magnetism, of where the magnetic poles were located at that time.

Figure 21.3 shows that the earth's magnetic field lines are almost vertical near the magnetic poles, but almost horizontal at the magnetic equator. Therefore, the "dip" in the remanent magnetism signal yields both the direction to the magnetic pole at the time of solidification as well as the "latitude" at the place where the lava spewed forth. Now if we assume that the magnetic pole and the geographical pole are coincident, or at least nearly so, the magnetic "latitude" can also be interpreted as the *geographic* latitude at which the lava poured out.

Geophysicists have made great use of this remanent magnetism data in rocks, but only after the technique was combined with a "dating" method that could also tell the *age* of the rock, that is, how long ago it solidified. By carefully collecting rocks of *different* ages but from the same place, for example,

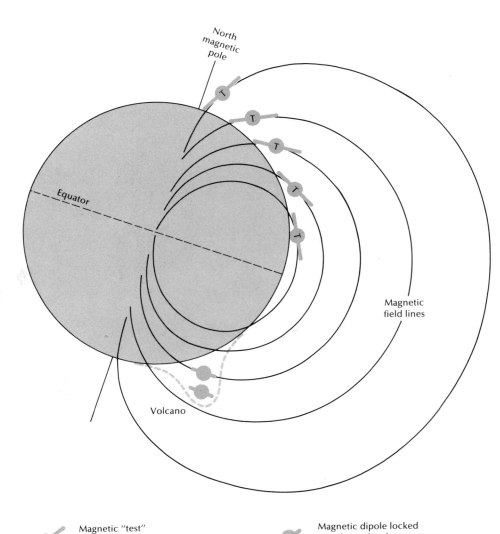

Figure 21.3 A cross section of the earth with its family of curves representing magnetic field lines. Near the magnetic poles the magnetic field lines are almost vertical relative to the earth's surface, as demonstrated by the orientation of the test magnetic dipoles. At the equator the earth's magnetic field is almost parallel to the surface. When a molten lava flow is extruded and cooled and then solidifies some of the magnetized molecules within the lava are oriented to the local magnetic field and frozen in place. The location of the magnetic pole at this time is preserved in the rock, even though the magnetic pole may later shift to some other position on earth.

the same volcano, they discovered that the remanent magnetic signals often *changed* with time in the sense of pointing to successively *different* locations for the magnetic poles. This finding brought forth the term "pole wandering." If we assume that a continent does not itself move, the only plausible explanation for the changing location of the magnetic pole relative to the continent is that the pole shifted position. This was initially a working hypothesis, because there was no absolute explanation of why a magnetic pole exists in the first place; a shift in its location could not be denied out of hand. An example of such pole wandering is shown in Figure 21.4*a*.

But Figure 21.4*a* also shows that the meandering path of the north magnetic pole as determined from South American rocks differs greatly from that obtained using rocks from Africa. The periods associated with these apparent paths are noted in millions of years ago (Ma). Notice that the shapes of the two paths are much the same from 400 to about 250–300 Ma (Tarling and Tarling, 1975) but become different

for more recent geologic epochs. In fact, if the two continents are translated in position and rotated a bit, their respective pole meander paths are brought into very close coincidence for the time frame 400 to 250 Ma, BP, as demonstrated in Figure 21.4*b*. This was a stunning result, first seen during the 1950s. It was a substantial justification for the Wegener hypothesis that the continents have drifted apart since about 200 Ma. Similar analyses were made for other continent groups, with similar results.

Similarities in Fossil Assemblages—Land Bridges or Continental Drift?

Sedimentary rocks, including beds of coal, contain fossil traces of organisms that lived at the time the rocks were formed or the organic matter was laid down for eventual transformation into coal. Paleontologists can now place rocks into a chronological order based on their fossil content, from the present time to as far back as about 570 Ma. This period

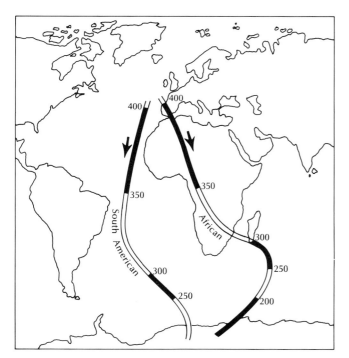

(a) From rocks gathered on the continents of Africa and South America indicate that the position of the *north* magnetic pole for each continent "wanders" over geologic time. Notice that the shapes and curves of the two tracks are remarkable similar for the time span 400 to about 280 Ma.

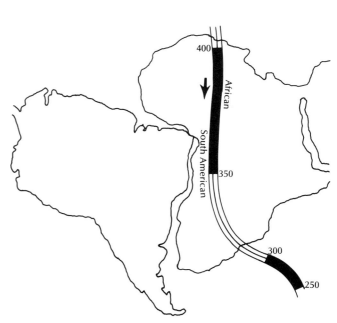

(b) If now the continents with their respective magnetic pole tracks are rotated to the positions shown, a best fit based on the shapes of their coastlines, their pole wander tracks coincide over the time frame 400 to about 280 Ma. The conclusion is that the continents were "together" during this period.

Figure 21.4 The "connection" between the migration of the *north* magnetic pole during geologic time and the theory of continental drift. (From Tarling and Tarling, 1975.)

marked the beginning of invertebrate animals that constructed shells able to leave a permanent record in compacted sediment. As the fossil record was being constructed from samples around the world, it became evident that a very *specific combination of fossils* could be associated *only* with a very specific point in geologic times, that is, coherent with the evolution of species themselves, and that rocks with the same *fossil assemblage* must have the same *age* regardless of which continent carried the rocks. This was a "dating" technique invaluable to geologists as well as biologists.

By the early 1900s it became clear that such identical fossil assemblages occur in various "pairs" of continents: South America and Africa; India and Australia; Africa and Australia and Antarctica. In the absence of other hypotheses on how such *living communities* could have existed simultaneously on continents thousands of kilometers apart, the biologists and zoologists invented the *land bridge* thesis. It proposed that land bridges connected each of the pairs of continents having like communities at some point in past time, but that the land bridge had subsequently sunk beneath the sea. If then the separated parts of an original community became distinctly diversified, it was simply interpreted as evidence of evolution, or of the natural selection of species as nature's way of changing living communities.

Wegener's thesis struck directly at the heart of the land bridge argument in that it offered alternative explanations for the same biological data. Stated another way, most of the biological arguments that had been made in support of land bridge connections were equally suitable arguments for continental drift. Stated still another way, the actual boundaries along which the original Pangean mass divided into separate continents were *not* also biogeographical boundaries. Desert regions were not broken away from tropical jungles, for example; instead, deserts were broken in two, jungles were separated, and mountain ranges were divided. Strangely, Wegener received little support from biologists, even though the geophysical argument *against* the land bridge concept, based on different densities of sial and sima, was well known by the time of his 1929 work.

Absolute Dating by Radioactivity—The Drift Theory Is Confirmed

Most rocks at the time they are formed contain traces of radioactive elements such as uranium, thorium, rubidium, or potassium. Since the discovery of radioactivity by Henri Becquerel in 1896, the detailed pathways by which these elements decay into elements of different forms have been worked out. Each step in the decay process has a characteristic

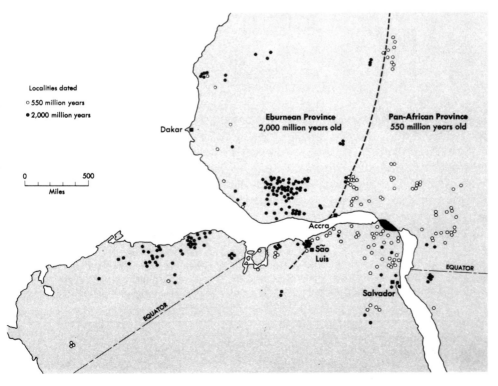

Figure 21.5 How a boundary between rock formations of two distinctly different ages was mapped, using radiometric age measurements. When the South American and African continents are placed into "shape fit" orientation, the rock boundary line can be traced directly from one continent to the other. This supports the theory that the continents were together at least until 550 Ma, the age of the more recent of the two rock types. (From Eicher, 1968.)

half-life (see Figure 19.23). The decay of uranium-235 into lead-207, with a 700,000-year half-life, is one such process used extensively in geophysics to date the age of rock (find the ratio of atoms of this pair, then multiply by 700,000). Many other ratio pairs can be used the same way, for example, the decay of potassium-40 to argon-40, with a half-life of 11,850 years. Rock ages measured this way are called *radiometric ages.*

One of the classic examples of how radiometric ages have confirmed continental drift theory was the discovery of continuity in a rock age boundary between Western Africa and South America (Figure 21.5). The boundary was first mapped in Africa between rocks of 2000 million years in the Eburnean Province and those only 550 million years old in the Pan-African Province. Later, geologists mapping the Altantic seaboard of Brazil discovered rock formations of the same ages and types as those of West Africa. More importantly, the boundary between the rock formations occurred with precisely the correct orientation to be a continuation of the African boundary *if these two continents were moved together* as shown, that is, into the relative positions each would have had in the ancient Pangean supercontinent.

Thus by the mid-1960s science had accumulated an overwhelming suite of evidences that indeed the continents of the present had their origins in the breakup of a larger and older continent structure. But widespread acceptance of the drift theory at this time was possible only because of the development of a new global geology theory during the decade 1955 to 1965. Parallel to the definitive studies on drift geography was the new theory of sea floor spreading. It was to provide the first plausible explanation of "how" the continents could have been forced to separate during the geologic epoch from 200 Ma to date. But even more important, the new theory supplied the *rates* at which the drift had occurred and proved that the drift is still happening today.

Sea Floor Spreading

Evidence Provided by the Ages of Ocean Crust

Nowhere from the basin floor have we found rock material that is older than about 150 million years. The first systematic sampling of material from the basin floor was done during the *Challenger* voyage (1873–1876). These early data were sparse, for at best, the crew of the *Challenger* was able to scrape only an occasional rock off the basin floor.

One of the most productive procedures for exploring the basin floor was developed during the latter half of the 1960s. The *Glomar Challenger,* is a vessel outfitted specially for drilling cores from sediments on ocean floors covered by deep waters. Leg 6 of the Deep-Sea Drilling Program used this vessel to extract a series of cores from the floor of the Western Pacific at the locations shown in Figure 21.6. Later, the core samples were dated using a combination of radiometric dates and the known dates of various layers in the sediments. The result is this sequence of geologic epoch "lines" to the west of Hawaii. When combined with previous drill core data, the entire Pacific basin was contoured with time lines, beginning with rocks of the Pleistocine age at the crest of the East Pacific Rise itself, the mid-ocean ridge of the Pacific. The oldest exceed the 136 million years that separate the Cretaceous from the Jurassic geologic periods. The sequence terminates at the Bonin trench east of Japan and the arcuate line of the Marianas Trench. Data like these provided very strong evidence that new basin crust is introduced into the basin along the mid-ocean ridges and eventually disappears into the trenches.

Patterns of Anomalies in the Earth's Magnetism— Critical Evidence

Gravity and magnetism have long held a special fascination to earth scientists. Because of the technological advances during World War II geophysicists were able in the 1950s to begin a systematic mapping of the earth's gravity and magnetic fields. For magnetic studies at sea, a sensitive device called a magnetometer is towed behind a ship at some distance (to avoid the warping of the earth's field by the ship's iron hull). It records the "strength" of the local magnetic field. By combining magnetic tow data from a series of closely spaced tracks, scientists were soon able to chart the strength of the earth's magnetic field over large areas of the basins, including the mid-ocean ridges.

Subsequent charts plotted from these magnetic data showed peculiar, often *striped* "bands" in which the magnetic strength alternated between higher-than-average and lower-than-average values. These higher and lower differences were called positive and negative *anomalies* in field strength. Their bands seemed to be oriented parallel to the mid-ocean ridgelines. In Figure 21.7 the bands are parallel to the Juan de Fuca ridge crest of the Northeast Pacific. Although similar patterns in relation to mid-ocean ridges in other oceans were being found, no suitable

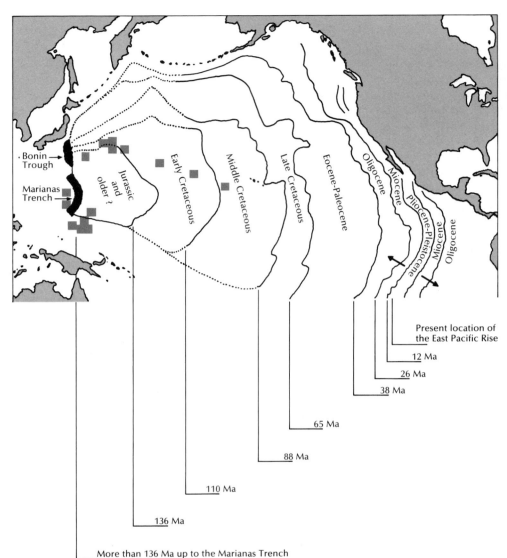

Figure 21.6 The age sequence of the oceanic crust of the Pacific Ocean, beginning with the most recent rocks of the Pleistocene, from 2 to 3 Ma, in the vicinity of the East Pacific Rise. Samples taken farther from the Rise are progressively older; the oldest appear to be over 136 Ma. The sequence stops at the arcuate line coincident with the Marianas Trench. (From Fischer et al., 1970).

explanation of why these patterns occur had yet been made.

Meanwhile, work proceeded on paleomagnetism "locked in" to young lava flows on land, together with the age of the young rocks. It was soon evident that a new phenomenon had been discovered—at various times in the past *the earth's magnetic poles had reversed positions;* north became the south magnetic pole, and then later the poles reversed again, and again and again. A "time" scale was then measured for the sequence of magnetic flip-flops using the radiometric potassium-to-argon decay ratio, with the results shown in Figure 21.8a. In this graph, earth periods during which the magnetic poles were like those of today's earth, with magnetic north nearby

geographic north, are called "normal"; reversed polarity has opposite locations.

It remained for two English geophysicists, Matthews and Vine, to combine the evidence offered by the age of floor material with that of the reversals of the earth's magnetic field. They reasoned that if new, molten lava was extruded toward the surface along the mid-ocean ridges and cooled, it would take on the magnetic polarity of the earth's field at this time, and that the earth field combined with the field of magnetism in the newly cooled lava would produce a total field strength *higher than average.* But later, after the earth's own field had reversed, the net field strength over this section of crust would be *lower than average* because the "locked in" magnetism

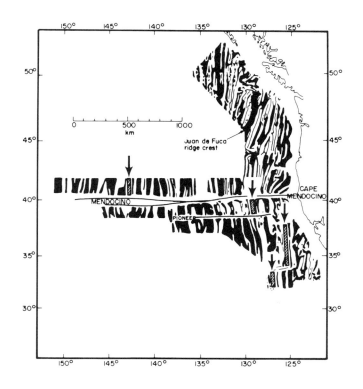

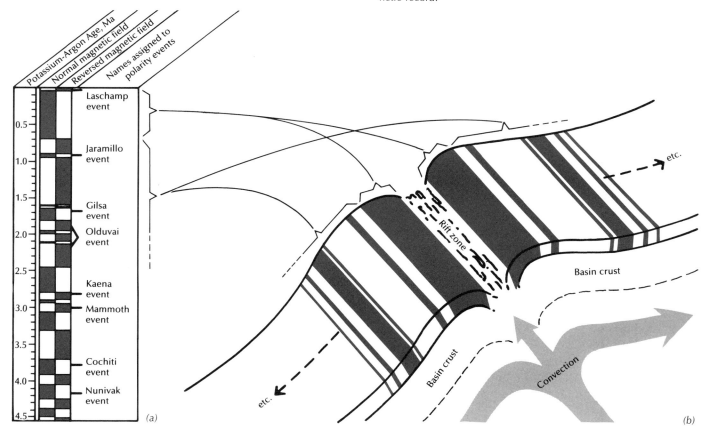

Figure 21.7 Distribution of magnetic anomalies in the Western Pacific. Notice that the pattern of stripes lies parallel to the crest of the mid-ocean ridge in the region of the Juan de Fuca Ridge. Arrows show the displacement of a given magnetic anomaly line across a major fracture, the Mendocino. (From Raff and Mason, 1961).

Figure 21.8 Geomagnetic reversals, key evidence for seafloor spreading.

(a) A time scale for geomagnetic reversals. Columns show when the magnetic field was *normal,* like that of today, or *reversed,* and the geologic period during which the reversals happened. Clusters of events have specific names. (After Cox, 1969.)

(b) Sections of basin crust containing the normal and reversed magnetic record.

of the rock itself would be opposite to that of the reversed field. They reasoned further that if new lava was continually extruded into the ridge zones, it would push the previously solidified, and magnetized, material away from the ridgeline. As Van Andel (1977) put it: "They saw the basin floor as a gigantic tape recorder, recording the history of the earth's magnetic field on a tape newly formed in the rift and moving out with the seafloor." The concept is diagramed in Figure 21.8*b*.

One final step was needed. The *rate* of sea floor spreading could now be calculated from the measured distance along the sea floor between successive magnetic events and the interval of time between their occurrences. Two immediate uses were made of these rate numbers. First, the ages of material gathered in core samples from the basin floor were matched with the ages inferred from the magnetic anomalies study. Second, distinct *dislocations in the striped magnetic patterns* were found at intervals along the ridgelines. These were interpreted as *fracture zones*, areas where separate pieces of the basin floor were being pushed away from the ridgeline at different rates (see Figure 4.1). One of the most prominent of such fracture zones occurs directly to the west of Cape Mendocino, California.

The consolidation of all these different views and data provided science with the first plausible confirmation that the basin floors had been *in motion* during past geologic times. But more work needed to be done before the theory of sea floor spreading itself was confirmed. There was still the enormous problem of *"the"* energy source capable of forcing all such motion.

Radioactive Energy Sources within the Earth— The Heat Flow Evidence

For centuries we knew that earth temperatures increased with depth, an observation made in many existing deep mines, and that the rate of temperature rise with depth was fairly uniform. For lack of better evidence, nineteenth-century scientists such as Lord Kelvin concluded that the earth was simply cooling off from an initial, much warmer state. This view of earth severely limited the science of geology because the *time scale* of a simple cooling earth was just not long enough to fit the geological processes picture. It also limited the Darwinian concept of evolution.

When Becquerel discovered that uranium emits "rays" that activate photographic plates, even in total darkness, and Marie Curie discovered radium in

1903, science was on its way to solving the limitation of a cooling earth. In 1906 R. J. Strutt calculated that the heat generated by radioactive minerals in the earth's crust could itself explain the present-day heat flow from the earth's interior. Further, such heat would be generated continuously, so that the present-day processes are in "steady-state" equilibrium, regardless of when earth began. These advances in physical chemistry gave a big boost to the Darwinian theory because they allowed "geologic time" to be extrapolated much farther into the past than possible before. But even more, they identified *"the"* energy source that could support the theories of sea floor spreading *and* continental drift.

The heat released in the earth's mantle by radioactive decay could generate the plumes of magma that seemed to be extruding new crust in the mid-ocean rift zones. It then followed that the ocean waters above the warm ridges would receive heat at a faster rate than waters elsewhere. During the 1960s scientists made many measurements of heat flow upward through the sediment layers covering the ocean floor. They measured up to 5 μcal/(cm^2)(s) of heat flow above ridges compared to only 1 μcal/(cm^2)(s) or so for basin floors far removed from ridges, and so confirmed the hypothesis in principle. However, theory had called for a much higher heat flow from ridge zones; this "deficiency" was a problem that remained unsolved until a decade later when scientists in submersible vessels discovered the "hydrothermal vents" of the rift zones (see also Chapter 19).

The Earthquake Pictures

During the mid-1950s following the development and testing of hydrogen nuclear explosives, governments supported the installation and monitoring of large numbers of seismograph stations all over the world. One of the very useful results from the network of seismic stations was an accumulation of data on the locations of earthquake epicenters, by the thousands, and with high accuracy. The 1961–1967 worldwide record of earthquakes with foci at depths of 100 km or less below the surface is shown in Figure 21.9*a*. A striking feature is the concentration of seismic events around the rim of the Pacific Ocean basin, a feature that fits the old description of the Pacific as an ocean "ringed with fire."

Of equal significance in Figure 21.9*a* are the vast areas of all ocean basins where there are few if any earthquakes. They all take place in narrow zones that coincide precisely with the mid-ocean ridgelines

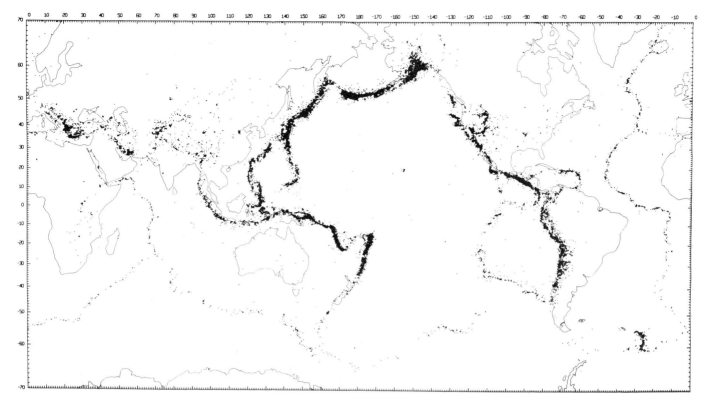

(a) Locations of shallow earthquake epicenters over the time span 1961–1967. The most frequent earthquakes occur at depths of less than 100 km. (From Barazangi and Dorman, 1969.

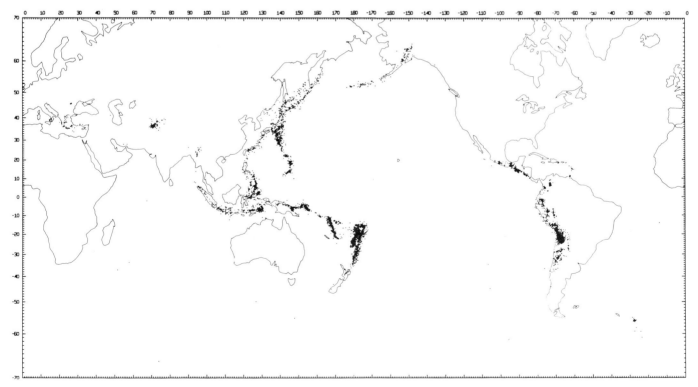

(b) Distribution of epicenters for severe earthquakes, of magnitudes greater than 7.5 on the (Richter scale), over the time span 1897– 1974. Severe earthquakes generally occur deep in the earth's crust, at depths ranging from 100 to 700 km. (From NOAA.)

Figure 21.9 Zones of earthquake epicenters mark active boundaries of the earth's crust.

and their associated fracture zones. The only *relative* movement within a basin crust itself, once formed, is along the fracture zones and then only for short distances outward from the ridge itself. We conclude that the basin crust away from its source zone tends to move as an *integral unit*. This is a vital element in the construction of a theory of global plate tectonics.

So it was that Wegener's speculations came to fruition. He had proposed that continents drift apart, a hypothesis that won at last widespread acceptance. But he had also implied that continents were forced to move *through* the ocean floor, much as a plow through earth sod. Sea floor spreading, once accepted, put an end to the controversy over *how* continents could move over the lithosphere of earth—it was the lithosphere itself that moved. At the same time, sea floor spreading introduced a new problem, that of explaining the disappearance of older basin crust. The solution to this problem was the route by which scientists arrived at the most recent, some say the final, chapter in the evolution of global geology—plate tectonics theory.

Plate Tectonics

As the evidence grew in favor of the theories of continental drift and the associated sea floor spreading,

there grew also an anxiety for a unified theory that could explain the total global geological and geophysical picture. The new earthquake data provided other interesting evidence. For example, Figure 21.9*b* shows the distribution of very *strong* seismic events, those that register over 7.5 on the Richter scale. There are almost no strong earthquakes in the vicinity of mid-ocean ridges. Instead, a comparison of Figure 21.9*b* with the locations of major trenches in Figure 4.7 yields a strong coincidence between strong earthquake foci and submarine trenches. Earthquakes located along the ridges and fracture zones were both weaker and shallower.

Formation of Large "Plate" Blocks of Basin Crust

H. H. Hess, in a classic paper (1962), noted that studies of basin crust using the propagation of sound waves had revealed the basin crust to be *thin, uniform,* and *stratified* in layers. Further, the same type of structure could be found in the crusts of all major ocean basins. The uniform layering could be interpreted as evidence that a crust moved as a unit rather than being broken into pieces or folded upon itself. Further research produced a model of how the new sea floor is formed at the ridges (Figure 21.10). Central to this picture is a plume of rising, hot plastic material of the asthenosphere, the plastic part of

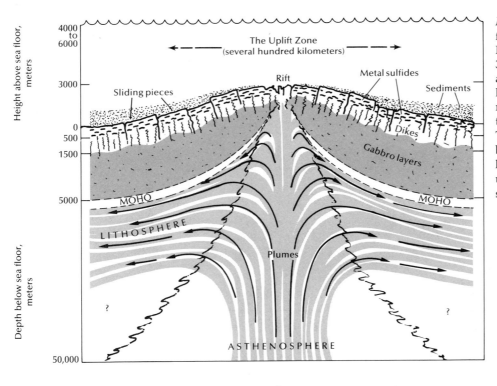

Figure 21.10 One model of the formation of a mid-ocean ridge. Ridge tops are generally about 3000 m below the sea surface over an uplift region that extends for hundreds of kilometers across the ridgeline. The lithosphere appears to move horizontally as a unit. The MOHO is a boundary zone between the crystallized mineral layers above, the gabbro, and an upper mantle layer of unknown structure.

the earth's mantle. As the uppermost regions of the plume cool and solidify, pieces tear apart and slide away off the sides of the uplift. More magma intrudes upward into these narrow, elongated rifts. Some of this magma is extruded outward to make contact with the overlying seawater; this part is rapidly cooled to form the characteristic pillow lava so often seen on the surfaces of mid-ocean ridges. Although this rapidly solidified lava is only a few hundreds of meters thick, in it is locked the earth's magnetic field effect forming the magnetic anomaly patterns. Below the pillow lava layer is the layer formed by repeated rifting and dike filling, typically up to 1500 thick; it moves laterally and downward away from the ridgeline.

Moving with the upper skin is the solidified, much thicker part of the stratified plate. Composed mainly of a coarse-grained basaltic mineral called gabbro—the coarse graining comes from slower cooling—this layer of the stratified crust is usually about 5000 m thick. The total thickness of the solidified lithosphere to this point then is about 7000 m. Below this level is a transition zone of variable thickness, and below this lies the upper mantle itself. The bottom of the upper mantle, some 50 km deep, marks the lowest level of the solidified crustal plate.

The transition zone is itself an interesting phenomenon. Its discovery was made possible by the development of seismic sounding. An array of sensitive sound wave detectors (called seismometers) is set up first, and then an explosion is detonated some distance away from the array. Sound wave energy travels through the earth to be picked up by the sensitive detectors. Some energy is reflected directly from the boundaries between layers of different type materials, for example, between the intrusive dike layer and the gabbro of Figure 20.10. In this way the thicknesses of such layers are determined. In contrast, some wave energy propagates *along* a bed for some distance before being scattered back toward the receiving array; this gives the investigator clues to the type of minerals and crystalline structure of the layers because each type transmits sound waves at a characteristic speed. In the transition zone just described, sound pressure waves move at speeds of more than 8.1 km/s below the transition, in upper mantle material, and at only 6.2 to 6.7 km/s above the transition. This transition is called the Mohorovičic[2] discontinuity; it is thought to mark a phase change in mantle material. During the 1960s a

special project, called Mohole, was begun with the objective of drilling a core hole down into the upper mantle. It was never carried out, but it did generate such great interest that a different program, the Deep-Sea Drilling Project, was begun and continues to this day the extremely fruitful work of extracting cores from the sea floor.

Hess also proposed that wherever the moving plates converged, there also one of the plates was drawn downward under the other. The downward thrust would explain the formation of trenches, for example, and the eventual reentry of the down-moving plate into the asthenosphere would complete the mass balance needed to offset crust formation at the spreading centers. The state was set for the *consolidation of the revolution* in global geology.

A Theory for Interaction of Plate Boundaries

In principle, because the entire earth is "encrusted," the large plates can interact in only three ways: they can collide (converge), they can slide past one another (transform slip), or they can move apart (diverge). Sea floor spreading had become an accepted explanation for *divergent* boundary motion. And the existence of trenches in the sea floor seemed to be directly connected with the *convergent* type. In 1968, Morgan pointed out that a theory already existed that would explain how plates should interact. The famous mathematician Leonhard Euler (1707–1783) had proved that relative motion between two rigid plates on a sphere could be described in terms of a rotation of both plates about a common axis.

This proved to be a very useful turning point. Geologists began to develop the rotation data for existing boundaries; the emerging view was that of six major plates (Figure 21.11) rotating slowly relative to one another. At any specific junction between plates the relative motion is convergent, divergent, or sliding, or a combination of sliding and one of the other two. Another type of classification is on the basis of whether the boundary is *active*, usually meaning seismically active with periodic relative movement. These classifications are summarized in Table 21.1, along with examples that are also located on the plates diagramed in Figure 21.11.

Passive Margins—Continent and Basin Crust

Perhaps the best-studied example of the passive margin is that of the junction between the eastern seaboard of the United States and the western North Atlantic sea floor. Figure 21.12 shows a geological

[2]It was discovered in 1908 by a Serbian geophysicist, Andrija Mohorovičić.

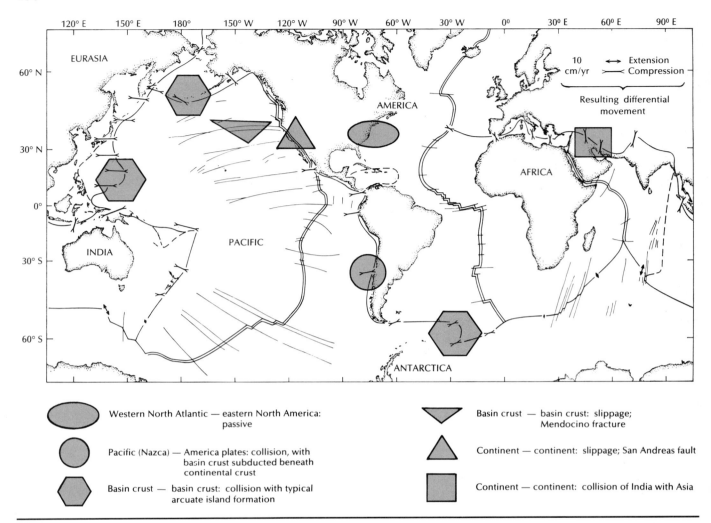

Figure 21.11 The boundaries of the six major crustal plates. Examples of each of the major plate boundary classes described in Table 21.1 are given. The line with arrowheads facing in opposite directions (←→) denotes a divergent boundary, typical of all spreading centers along mid-ocean ridges. The line with arrowheads facing each other (→←) means a convergent boundary. In both the distance between arrowheads is proportional to relative plate speed. (From Le Picheon, 1968.)

cross section of this region from the oceanic ridge and rift zones (as detailed in Figure 21.10) into the North American continent itself. It illustrates two of the more interesting aspects of plate tectonics: first, that the basin crust continues to slope downward away from the spreading center, and second, how sediments accumulate on the spreading basin crust.

The Slope of Basin Crust. The slope of basin crust is very evident in the cross section of Figure 21.12. Near the rift zone the cooled and solidified crust of the rising plumes easily tears apart for two reasons: The slope of the uplifted sea floor is steepest here,

so that the cooled crust is put under tension by the tendency to slide downhill; and the solidified crust is thinnest here and more easily torn apart. This is the root cause of the numerous *dikes* created as additional magma intrudes into the rifts and *cements* the pieces into a solid. With distance from the central rift zone, cooling and solidification reach deeper into the asthenosphere and the crust thickens, but the thicknesses of the topmost layers remain the same. Eventually, even this thickening process ceases as the deep magma is increasingly insulated from the cooling ocean waters above.

Magma shrinks in volume as it cools and solidi-

Table 21.1 A classification of plate boundaries based on the type of relative movement between adjacent plates and whether the boundary is active or passive; examples of each are listed.

Type of Relative Motion	Description	Seismically Active or Passive	Example
None	No relative movement between plates	Passive	Western North Atlantic with Eastern Seaboard of North America; see Fig. 21.11, Fig. 21.12
Divergent	Plates move apart, generally in opposite directions	Active (shallow earthquakes)	Virtually all mid-ocean ridges; see Fig. 21.10
Convergent	(1) Basin crust thrust under continental crust. (continental subduction zone)	Active (deep earthquakes)	Chilean Coast: Pacific (Nazca) plate and South America; see Fig. 21.11 for location.
	(2) Basin crust thrust under another basin crust (island-arc subduction zone)	Active (deep earthquakes)	Scotia Arc; Aleutian Arc; Marianas Arc; see Fig. 21.11 for locations.
	(3) Collision between continental crusts	Active	India-Asia; the Himalayan Mtns.; see Fig. 21.11 for location.
Transform	(1) Basin crust slipping by adjacent basin crust	Active (shallow earthqukes)	Mendocino Escarpment; see Fig. 21.11 for location, and Fig. 21.7 for magnitude of slip.
	(2) Continental crust slipping by other continental crust	Active (shallow earthquakes)	San Andreas fault, California; see Fig. 21.11 for location.

fies. It then follows that one of two things will happen: either the denser solid crust will *overturn* into the plastic, less dense material of the asthenosphere below, or it will *float* ever deeper into it. Because of the structural integrity of the solid crust, it does not break into pieces and "sink" as would a broken ship. Nevertheless, wherever a rift does appear in the crust, the tremendous weight of this overlying crust forces magma upward, and often these form undersea mounts, as shown in Figure 21.12. Seamounts can initially protrude above the sea surface, then be flattened by erosion, and later subside below the surface because of the continued deepening of the parent crust with distance from the spreading center. Figure 4.4 shows that the floor of the Pacific

basin has many more seamounts than does the Atlantic.

Apparently, basin crust subsides until it reaches a depth of about 6000 m below sea level some 75 to 100 million years after formation (Van Andel, 1977). In Figure 21.12 this is about the depth of the top of the crust where it contacts the American continent. Several thousands of meters of sediment overlie the crust at this point.

Sediments Accumulate over the Basin Crust. Figure 21.10 shows that the first sediments to accumulate on newly formed basin crust are metal sulfide "muds." These are metallic compounds that are soluble in *hot* seawater, among them are sulfides of

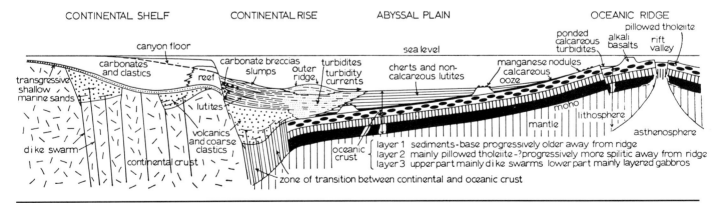

Figure 21.12 A cross section showing the relation between the continental crust, oceanic crust, and sediment layers for a passive margin such as the boundary between the North Atlantic ocean basin and the American continent. (From Dewey and Bird, 1970.)

iron, manganese, copper, nickel, lead, and cobalt (see Chapter 19; Figure 19.15). Although there is reason to believe that the entire ocean basin has this mud sediment, only near the ridges is it uncovered enough to be mined.

As the crust moves farther from the rift zone it accumulates *biogenous* sediment, chiefly in the form of calcareous shells of marine organisms. Still farther away from the rift zone, and deeper within the water column, the crust no longer accumulates calcareous sediments because the shells tend to dissolve in the cold waters below about 4000 m. This phenomenon, called carbonate compensation at depth, prevents additional accumulation of such sediment (see Chapter 5). However, calcareous sediments already laid down may remain and be gradually covered by lutites, the reddish clay typical of sediments in deeper parts of ocean basins.

Closer to the continental margin, the nature of sediment suddenly changes from the rather uniform muds to a more disordered mixture of different types of material, from mud to particles known to be eroded from continents. These are *turbidites,* deposits generated by turbidity currents that periodically sweep down the continental slopes and submarine canyons to carry assorted sediments out onto the abyssal floor, the continental rise.

Passive Junction Between Continent and Basin Crust. As seen in Figure 21.12, the junction between continent and basin crust is quite complex, so much so that the term "transition zone" is used as a broad generic description. Much coarse sediment eroded from the continent is collected here. The continent is itself rifted along its perimeter. But the most important aspects of the junction are (1) that it is seismically quite inactive; (2) that the broken, rifted, sometimes folded junction has been the instrument for collecting organic sediments, later converted by nature to petroleum; and (3) that both the continental shelf and the coast itself tend to be broad, with extensive coastal plains and intricate, highly productive estuarine systems.

Active Margins—Basin Crust Subducted under Continental Crust

Figure 21.13 shows a generalized cross section of a subduction zone in which basin crust is drawn downward under a continental block. There are many variations. But several salient features are common to all.

Downthrust and Earthquake Epicenters. The basin crust, being thinner and denser than continental crust, is either pushed or drawn downward to underthrust the continent. Shallow earthquake foci are often measured in the region where the basin crust is subject to sharpest *bending.* At least part of the seismic activity, then, results from the episodic tearing and faulting of the crust in this way.

At the time of downthrust, the basin crust is quite old and quite cool. As it moves downward it gathers heat from the surroundings, but it probably enters the plastic asthenosphere while still relatively consolidated. The friction between this and the continental crust down to its deepest part at, say, 100 km, is one source of *deep earthquake foci* (see also Figure 21.9).

Trenches and Prisms of Accretion. As seen in Figure 21.13, the downthrusted basin crust leaves a void in the sea floor as it curves downward—this is the origin of the deep trenches found along much of the periphery of the Pacific Ocean. Depending on many factors, such trenches may or may not be substantially filled with sediment. In fact, there are two principal sources from which sediments could arrive in a trench. The first and most obvious is the simple *offscraping* of sediment from the top layers of the basin crust as it slides and scapes its way under the continent. If the relative movement between the plates is episodic, its usual behavior, we find these sediments *accreting* to the edge of the continent in wedge-shaped pieces, in the *prism of accretion.*

Sometimes the accreted wedges build up above sea level. Then the basin created landward of this ridge may fill with the sediments eroded from the continent. Eventually such filled basins can become coastal plains, coastal highlands, or part of the continental shelf, depending on the rate of accretion versus the rate of erosion and other factors.

Deep Melting and Andesitic Volcanoes. Once the downthrust basin crust passes completely below the base of the continent and into the asthenosphere itself, it can be melted and the molten lava can again rise to the surface. This is the source of the volcanic activity by which mountain chains are created inland from the zone of collision, for example, the Andes Mountains of South America and the mountains of Central America. What is significant about these volcanoes is that the mineralization of their extruded lava is distinctly different from that of the lava extruded along the mid-ocean ridges. The continental

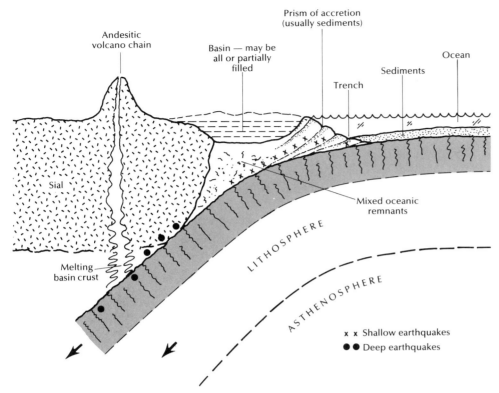

Figure 21.13   A generalized cross section of a convergence between a basin crustal plate and a continent. The basin crust is subducted, accompanied by deep seismic activity and its partial remelting to supply molten material for the buildup of volcanic mountain chains. Blocks of sediment are often scraped off the converging ocean floor and "stacked" into a formation which may or may not rise above general sea level. Between the volcanic chain and the actual coast we may find a coastal plain, a series of coastal mountain chains, or an unfilled basin. The trench usually marks the outermost end of the system.

volcanoes are andesitic, that is, significantly richer in silica-based rocks (50 to 65%) than are the basalts of seamounts and ridges (50% or less). This andesitic feature of volcanic material extruded near subduction zones has been known and studied for at least a century. The discovery of submarine vents along mid-ocean rifts provided a partial explanation of why oceanic basalt might have a different mineralization. Not only are metals selectively removed from the cooling magma as sulfides, but some ions are also selectively removed *from* the percolating seawater to form pockets of concentrated sulfides within the crust blocks. Among these are iron, silver, gold, copper, and lead sulfides. As old basin crust is subducted, Figure 21.13, some of the biogenous sediment cover, often silica rich, is also carried along. When the old crust–sediment mixture is liquefied,

the new magma that results will be richer in silica, hence andesitic; as it extrudes to form mountain chains like the Andes, the sulfide concentrations become the mineral ores that we mine for copper, gold, and so on.

Active Margins—Basin Crust Subducted under Other Basin Crust

For over a century geologists have noted that "arcs" of islands are far too numerous and similar in shape to be the result of random processes. A "first" explanation came in 1968 when Frank pointed out that the geometry of a thin, rigid shell over a sphere places a constraint on how the shell can be distorted. If the shell of a Ping-Pong ball is pushed in, the "fold" will take on a characteristic *arc* shape. Figure 21.11 identifies three of the more prominent island arcs: the Aleutian Island chain, the Marianas Islands, and the Scotia arc in the Southern Ocean.

The island arc structure has been of special interest, not only because it occurs so often but because it occurs away from the direct influence of continents. Moreover, the type of volcanic material forming the islands of the arcs is very similar to that making up large continents, so that a good understanding of these island arcs can provide new insight into the origin of continents. Figure 21.14 gives a cross section of the Aleutian arc that is constructed from data taken in three different ways.

1. First, the geophysicists used sound waves to probe the layering of the upper few kilometers of crust. These showed that the volcanoes themselves

are supported by a thickened section of the same material that makes up the oceanic crust; the density is 2.9 tons/m^3, (tcm). Other information from soundings showed that the density of basin crust below the Moho level was 3.4 tcm, slightly denser than that of the asthenosphere at 3.35 tcm; this supported the idea that the basin crust, once started into a downward plunge, would continue to "fall" into the mantle by virtue of its greater density.

2. Next, the investigators plotted the locations of earthquake epicenters. These indicate a swarm of seismic activity at about 60-km depth, between the trench and the volcanic island arc. Also shown are the deep epicenters, up to 250 km below sea level.

3. Just beyond the earthquake swarm the investigators found a vertical zone in which density was

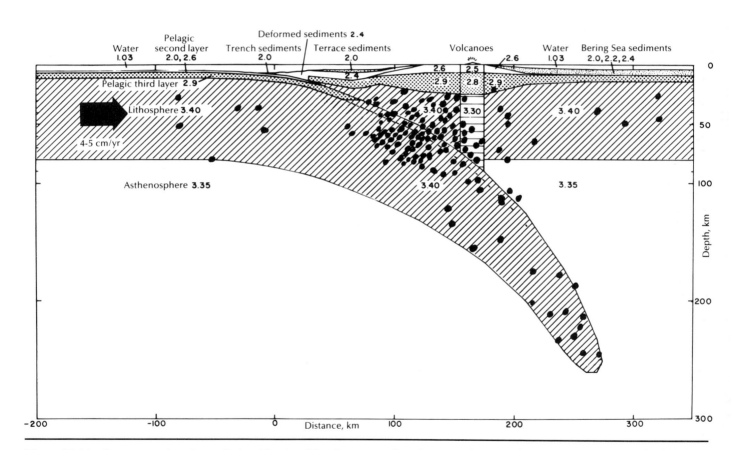

Figure 21.14 A cross section through the Aleutian Island arc, south to north. The North Pacific oceanic crust is approaching the Bering Sea crustal plate at a relative speed of about 4 to 5 cm/yr. Numbers placed on the section are the densities of the various layers, in tons per cubic meter (tcm). The material in the crustal plate is denser (3.40 tcm) than that of the asthenosphere below (3.35 tcm), so

that the crustal material, once started downward, is expected to continue sinking deep into the earth's mantle. The downward progression of the crust is marked by numerous earthquake foci, both shallow and deep. Seismograph measurements have revealed the location of the volcanic "pipe" of slightly less dense material which is the feeder channel to the volcanoes. (From Grow, 1973.)

slightly less than that of the surrounding material at the same level, indicating that lightweight magma is being pushed upward and is feeding into the volcanoes themselves.

4. On the opposite side of the volcanic arc from the trench is another ocean basin, that of the Bering Sea. Its only distinguishing feature is that it carries a thicker layer of sediment.

Active Margins—Basin Crust Slipping by Basin Crust

Margins where two basin crust slip by each other are active in innumerable places in ocean basins, wherever a fracture zone exists. The reader can refer to Figure 21.7 as an example of the very large slippage between the crust north of the Mendocino fracture zone and the crust just to the south, about 1150 km.

Active Margins—Slippage Between Two Continental Plate Margins

The classic North American example of slippage between two continental plates is that along the San Andreas fault in California. Here two continent blocks are each moving toward the northwest, but the seaward block moves about 6 cm/yr faster, causing an occasional, shallow earthquake. Just to the south is the special rift zone between Baja, California and mainland Mexico; these continental blocks are believed to be moving apart.

Paleoceanography: Messages in the Fossil Record

To this point, discussion of plate tectonics has focused on the "solid" earth crust. We mentioned the "liquid" oceans briefly when examining why the climate of the Carboniferous period was so moderate over so much of the earth. From the oceanographer's view, the revolution of plate tectonics is just as exciting as it is to the geologist and geophysicist. The same fossil records taken from cores of deep-ocean sediments by geologists and used in the construction of global geology are also clues to the evolution of life forms in the marine ecosystem. The reader should note that the ecosystem we study in today's ocean is a *successful* one—we do not find *living evidence* of unsuccessful experiments by Mother Nature. But the fossil record is "full" of glimpses of past ecosystems that failed. Paleoceanography, then, is a study of the life forms of ancient seas, how and when they evolved and disappeared, and the conditions of the ancient ocean water itself. It is a study intimately linked with plate tectonics and so is best described here in this chapter.

All readers of science are aware of the theory that the earliest forms of life evolved as simple molecular structures in the shallow primordial seas, and that complexity in life forms has increased steadily since then. Now that plate tectonic theory has shown how continents in the past have alternately come together and then dispersed again, we can ask how, if at all, does the fossil record reflect these global land changes.

To help answer the question, Valentine and Moores (1970) studied how the number of families of fossil species has changed over geologic time. They find a marked decrease in the number beginning at the time when the Pangean supercontinent was "sutured" together about 250 Ma, shown as event 2 in Figure 21.15. (Geologists use this term to describe the welding together of separate blocks of continents that collide during plate movements.) The next event in time is the opening of the Tethys Seaway about 200 Ma, which divided Pangea. The number of fossil families begins a new rise at this same time, and continues to rise with the later formation of present-day continents.

Geologists believe that before Pangea was formed there existed an earlier group of separated continents. Paleontologists also find that the earlier periods had larger numbers of families of life forms than when Pangea existed. The lowest number of families seems to be correlated with an even more ancient supercontinent of about 600 Ma.

In short, there is a definite connection between a diversity of life forms in the fossil record and the fragmentation of land crust and the dispersal of crustal pieces over spherical earth. Scientists are not yet sure why this connection exists, but the simplest explanation is that as land masses split up, some into polar regions, others into subtropical deserts, the number of "niches," the jobs that organisms can perform within the ecosystem, also increased, and so life forms evolved to fill these niches.

For example, the total length of coastlines, the number of estuaries, and other coastal features increased rapidly as the supercontinent broke up. And since each continent covered a latitude range, so that each had a corresponding *range* of climates for its coastal features, each continent ecosystem could

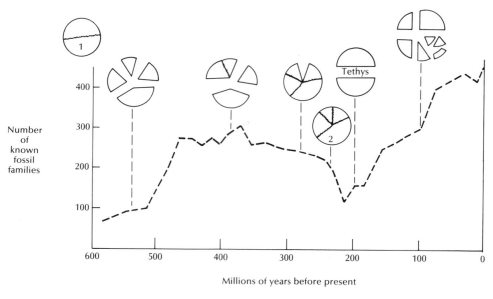

Figure 21.15 The number of families of organisms found in the fossil records has varied over geologic time. Also indicated are the estimated epochs during which the continents were combined into supercontinents, the first at over 600 Ma, and the second, Pangea, at about 250 Ma. The epochs of supercontinents appear strongly correlated with a minimum number of fossil families. This suggests that these were epochs of minimum diversification among living organisms. In between the epochs of supercontinent, continental pieces were distributed much as today. In these periods diversification seems to have been maximum. (From Valentine and Moores, 1970.)

evolve with a distinct suite of life forms. Ocean currents would also shift with the changing positions of continents over the globe. Because oceans exert strong control over climate, such shifts could both create new habitats as well as destroy old ones. In fact, the fossil record and supposed earlier positions of continents can be used to reconstruct long-ago aspects of climate.

Study Questions

1. It appears that the foundation for a theory of plate tectonics could not be built until two aspects of "time" were developed first; both were necessary and both were keystones. Discuss.

2. The list of five principal arguments supporting the theory of continental drift given in the beginning of the chapter is not in chronological order. In your opinion which came first? Would any of these pieces of evidence taken singly have convinced you of the correctness of the theory?

3. The phenomenon of radioactivity provides key support to the hypothesis of sea floor spreading in at least *two* distinct ways. Explain.

4. The development of which *technology* was necessary before oceanographers could map the island arc–trench configurations that turn out to be key elements of crust subduction and hence of plate tectonics theory?

5. Koppen and Wegener in 1924 deduced the location of the equator at the end of the Carboniferous period (280 Ma) using what argument? Were their data reinforced by the then known locations of land glaciers?

6. One of the critical arguments that the theory of plate tectonics had to satisfy was: "If all the polar ice that exists today was melted, the rise in sea level would not come close to covering as much *land* as was actually covered during the peak transgression period about 75 Ma." Stated another way, earth surface temperature changes alone cannot explain why some 40% of all land was covered by shallow seas in times past. How did sea floor spreading theory help resolve this problem?

7. One of the basic procedures used by scientists is to examine *all* possible explanations for an observed phenomenon. How was Wegener's drift theory an alternate explanation to the biologists' "land bridge" theory?

8. To interpret the "age" of the ocean floor on the basis of the age of a rock dredged up from the surface of the ocean basin would be to ignore alternate explanations for how the rock arrived at the place where it is found. Explain.

PLATE TECTONICS AND OCEAN BASINS **457**

9. When marine geologists learned that the ages of islands in the Hawaiian chain increased in the same direction as they apparently "sink" below sea surface, what conclusion did they draw in support of sea floor spreading?

10. Once the sea floor spreading theory was accepted, scientists reasoned that the heat flow from the earth up into ocean waters should be *much* higher over the ridges than elsewhere. When later measured, the actual heat flow, though higher, was much less than expected. We now know of another process that extracts heat from the hot rising magma. Explain.

11. Which picture of earthquake epicenters would tell us of seismic activity along spreading ridges—that for shallow earthquakes or that for deep epicenters?

12. A theory that requires the crust of the earth to consist of a number of large "plates," each of which moves as an integral unit, must also explain *how* these plates interact along their boundaries or margins. How, for example, does the observation that nowhere is the ocean basin older than about 200 million years "fit" this theory of large crust plates drifting over the earth's surface?

13. In the vicinity of the spreading centers are found the "hot springs," and the sea floor around the typical submarine geyser is covered with *muds* consisting of sulfides of various metals. Explain briefly why these deposits form.

14. A study of Figure 21.13 will yield clues why the margins of present-day continents are likely places to find petroleum deposits. Can you give the connecting argument that explains this?

15. It was for a while a mystery why the early island volcanoes that lie in "arcs" are richer in silica than island volcanoes that do not. How does the theory of plate tectonics explain this silica enrichment?

16. Discuss briefly how the theory of sea floor spreading accounts for such large-scale formations as the Mendocino Escarpment.

17. Briefly state the argument why geologic periods of *super*continents like Pangea are also periods of low numbers of families of organisms.

18. What was the Tethys Seaway?

CARTOGRAPHY: THE ART AND SCIENCE OF MAPPING EARTH FEATURES

Medieval travelers on land were always able to express distance and direction in terms of the *time* needed to walk west with the sun or east against it, together with additional time to walk north with one's shadow (Northern Hemisphere) or south against it, in order to get from one point to another. Early sailors also navigated "by the sun," but because of the variances in winds and ocean currents, "time" was of much less value in gauging distance across the ocean.

Therefore, early navigators sailed their ships first to a desired *latitude*, or sun angle above the horizon, and then headed east or west as the destination required, knowing that sooner or later they must make the desired landfall. James Clavel, in his epic novel *Shogun*, demonstrated how Portugese navigators of the sixteenth century first reached Japan using just such a scheme, sailing first northward into the North Pacific to latitude about 35°N and then turning west. Latitude is obtained at sea by calculating the complement of the angle that the sun makes above the horizon at local noon, an angle easily measured. One device for measuring the angle is the astrolabe, an instrument developed in antiquity by astronomer–sailors who needed a handy reference to the changes in sun angle across the seasons. Made usually of bronze, the device processes the local high-noon angle, or angle of known stars above the horizon, through a set of dials that accounts for the seasonal march of the sun's path across the earth's sky.

But a reliable, easily used method for finding longitude at sea was simply not available to early navigators. Gemma Frisius in 1530 was the first known astronomer to propose the use of a *time-keeping* device for this purpose. If a navigator sets his "clock" to the local high noon of a known point on earth, called a reference meridian, and sails away *knowing* that the timepiece keeps accurate time over periods of several weeks at least, he can *confidently* convert the difference in hours between a faraway local high noon and 12:00 noon at the reference meridian into an equivalent difference in longitude, at the rate of 15° longitude difference per hour of solar-time difference. In spite of this known need for an accurate timepiece, it was not until about 1760 that the first reliable "ship's clock" was produced by John Harrison, an English inventor.

Galileo Galilei, the Italian scientist, had proposed in 1582 that the ordinary pendulum was sufficiently accurate to "regulate" clockwork to the precision prescribed by Frisius. But the gravity-controlled pendulum simply does not work well aboard a pitching, rolling, heaving ship at sea. The first part of Harrison's invention was to replace the gravity pendulum with a wheel oscillating against a spiral spring. In effect, a wheel is nothing more than a series of stick pendulums arranged radially into a circle about a common suspension point, so that the effect of a ship's motion on one "spoke pendulum" is exactly compensated for by its opposite effect on the opposing spoke pendulum of the wheel. The other part of Harrison's invention, and the most important, was to solve the remaining problem that the oscillating period of the pendulum changed as the ambient temperature changed, a consequence of the thermal-expansion property of the iron spring itself. A clockwork gained or lost time according to how temperature changed. By combining brass and iron, metals of different expansion rates with temperature, Harrison produced a pendulum that self-compensated for temperature effects. In 1764, on a test voyage from Plymouth, England, to Jamaica that took 7 weeks, Harrison's Number Four chronometer was only *38.4* seconds fast, allowing a corresponding error in longitude of less than 20 kilometers distance! James Cook took three such chronometers on his famous circumpolar voyage around Antarctica from 1772 to 1775.

Development of Mapmaking and the Search for the Magnetic Poles

Of course, mapmaking as an art and science developed long before Harrison's solution to longitude measure. Early maps clearly show the great distortions that resulted as artists translated sailors' accounts of voyages onto paper charts. One major cause of the distortion was the fact that the earth's magnetic field is itself strongly "warped." This warp, called *declination* when known for a specific point on earth, is defined as the angle of difference between the north-pointing needle of a magnetic compass and the direction to the true geographic pole. The warp was known to vary over the face of earth, but without accurate measures of longitude the necessary correction was just as poorly defined as was longitude itself. In fact, then, the solution to longitude measure available in Harrison's chronometer immediately opened the next phase of high-seas navigation—the search to pinpoint accurately the positions of the two magnetic poles, north and south, and to map the magnetic field warp over the face of oceans in between. Cook performed magnetic

reconnaissance during his circumnavigation of Antarctica, but it was 1909 before an English expedition reached the south magnetic pole, located near the edge of that continent at about 140°E longitude. The search for the north magnetic pole during the nineteenth century was beset by tragedy, notably the loss of the entire 129-man crew of the 1845 Franklin expedition. This pole is located now on Bathurst Island in the Canadian Arctic Archipelago.

Point Designation: The Standard Grid of Latitude and Longitude

The need for accurate mapping was apparent to all maritime nations, and led to the adoption of standards for designating lines of latitude and longitude. Figure A1.1 shows the standards now used. Latitude lines are easily perceived by the viewer: they are the

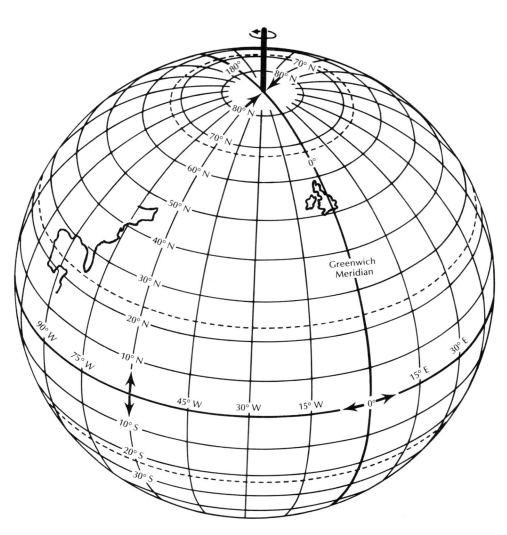

Figure A1.1 The sphere as it is sectioned into longitudes, *meridians*, and latitudes, *parallels*. Each meridian is the intersection between the surface of the sphere and a plane that slices through both pole points and the earth's center. Drawing successive meridians at 15° intervals around the sphere marks off a total of 24 solar hours of time at 1-hour intervals. Latitudes are a series of concentric circles so spaced that the *distance along the earth's surface* between successive latitudes has the *same scale*, 60 nmi/degree of latitude, all over the globe. A universal scheme for communicating distance across the earth's surface, then, is to express distance in equivalent degrees of latitude.

intersections of the spherical earth's surface with planes sliced through the earth, each parallel to the plane containing the equator. This equatorial plane divides the earth's surface into two equal hemispheres, north and south, each one centered around its own geographical pole located on the spin axis of earth.

Notice that latitude lines are not arbitrarily located. Rather, they are set so that the distance along the earth's surface between adjacent latitudes, measured in the north–south direction, is always the same regardless of where over the sphere such distance might be measured. Numerically, we divide the earth's circumference into latitude quadrants, each containing 90 degrees (°); their numbers increase consecutively away from the reference equator, 0°. Whenever designating a latitude, then, the user must also include a symbol N or S to identify the specific hemisphere. For example, latitude 50°N passes nearby the southern tip of England (Figure A1.1). With the earth's circumference divided into four quadrants of 90° each, for a total of 360°, and with each degree subdivided into 60 minutes of arc, we obtain a total of 21,600 minutes of arc along the earth's circumference. This number is designated as the total number of *nautical miles* around the earth; the distance between two adjacent latitudes that differ in number by *1* minute of arc is thus the definition of a nautical mile. Mariners also set the standard for the *rate* at which a ship moves over the sea as being one nautical mile of distance covered in one hour, called the *knot*. So the latitude scale becomes the reference not only for distance but for speed as well.

Longitude lines are also easily visualized. They are the intersections of the spherical earth's surface with planes that slice through the center of earth and its two pole points. Unlike the latitudes, which have the unique equator, no one longitude is naturally different from all the rest. Faced with the need for a reference longitude, the English mariners chose the one that passed through Greenwich, England (Figure A1.1). Henceforth all longitude lines located *east* of Greenwich would carry the E letter in addition to a number, and all longitude lines *west* of Greenwich a W. Numerically, the total number of longitude "degrees" was also set at the value 360° of arc, as for latitudes, with each degree subdivided into 60 minutes, again as for latitude. But the distance across the earth's surface between adjacent longitudes changes continuously as the lines converge toward the poles. The longitude scale has no value for measuring either distance in miles or speed

in knots; its sole geographical value is for pinpointing location.

Because the earth spins and produces the diurnal cycle of day and night, we can assign "time intervals" to space differences between longitudes. In Figure A1.1, for example, the longitudes are drawn at intervals of 15° of arc, a spatial measure. With 24 such intervals around the earth, each also corresponds to one *hour* of solar *time*. It is important to emphasize that the original choice of *24* as the number of hours in a day was an arbitrary one. The practice arose as a convenience to the customs of commerce and other societal functions. It relates back to sundials, which date from at least one millennium B.C. The half-circle sweep of a shadow cast by the sundial was divided into "useful" parts. One logical scheme would divide the semicircle into 3 parts using the radius of the circle as a chord measure; further subdivision would produce 6 units per daylight period, then 12, then 24, then 48, and so on. The choice of 12 units must reflect what was considered "convenient" for a basic unit, an hour. Hours varied in length from season to season until the thirteenth century when the concept of equal hours was introduced. With the advent of mechanical clocks in the fourteenth century, equal hours came into general use.

Simulating the Spherical Earth's Surface on Flat Paper

A scheme for *simulating* the earth's surface on a flat sheet of paper is also needed for mapping. There are two basic methods of transferring the features of spherical earth onto flat paper: (1) mathematically converting points and lines in spherical space into equivalent points and lines in two-dimension space, then constructing the new map; and (2) projecting the features and latitude–longitude grid of the sphere onto a paper sheet and tracing these. Sometimes both techniques are used in constructing a single map.

Projections: How Made and Why

There are several types of projections, but we will discuss only two, the cylindrical and the gnomonic.

Cylindrical. Start with a transparent globe on which are inked the grid of latitude and longitude lines plus the outlines of all continental features. Next, surround the globe with a cylinder of paper, the paper touching the globe along its entire circumference. Third, project the globe's features onto the cylinder of paper by installing a point source of light within the transparent globe at its center. Then trace the projected images over the cylinder surface. This completes the projection itself; now to produce a flat sheet map cut the cylinder along its length and unfold it. Figure A1.2a illustrates the resultant map of the American continents, which is but a part of the map of the entire world registered on the complete cylinder.

Notice that both the meridians and parallels are traced as straight lines, and that these line sets intersect as perpendiculars (orthogonals). But notice also that the paper distance between successive latitude lines changes drastically from the equator toward the poles; the paper distance from the equator to 40°N, involving 40° of latitude difference, is less that the paper distance from 60 to 70°N, a difference of only 10° of latitude. Area projection is therefore highly distorted, and the distortion increases with distance away from the equator.

In 1569 Gerardus Mercator was the first to apply a particular modification to the cylindrical projection to produce a chart with special appeal to ocean navigators. This modification applied a formula that yields a unique relationship between the *paper* scales of longitude and latitude. It makes these scales equal wherever the grid lines cross. Figure A1.2b shows this modification that now bears the name Mercator projection, even though it is not a true projection. Relative to the projection of Figure A1.1a, its paper

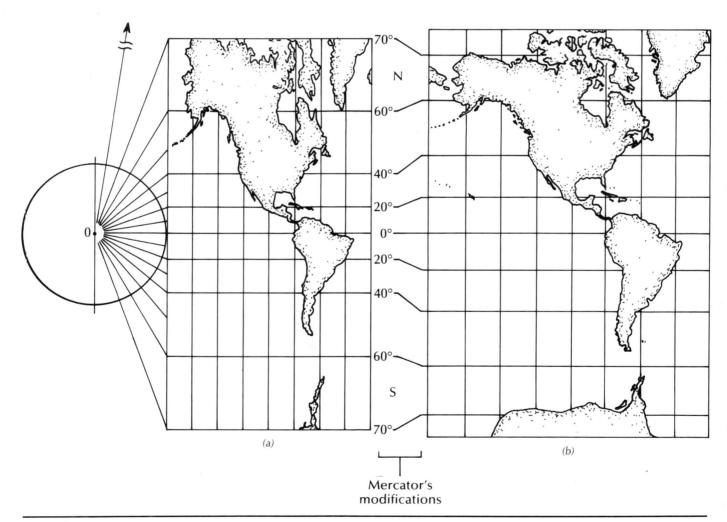

(a)

(b)

Mercator's modifications

Figure A1.2 A comparison of (a) a direct optical projection of the globe onto a cylinder and (b) the modified chart produced by Gerardus Mercator in 1569, now known simply as the Mercator.

scale of latitude expands from the equator to about 60° and contracts at higher latitudes. The Mercator chart has two unique features.

1. A straight line drawn over his projection has the same compass bearing everywhere along the line. This means that ocean navigators can draw a single line connecting their starting and end ports, measure the "heading" they must make good in order to achieve the voyage, and then simply keep this heading on the compass throughout the voyage. However, as is shown later, such a constant heading does not yield the shortest sailing time between the two ports.

2. Because the local scales of distance in both latitude and longitude directions are the same, navigators can quickly determine a paper distance in terms of "equivalent" degrees of latitude and, using the conversion formula of 60 nmi per degree of latitude, find the accurate distance between ports. They are then ready to estimate sailing time, schedules, provisions needed, and so on.

Gnomonic. The gnomonic projection is also an optical projection from the center of the globe. But for this projection the paper upon which the global features are traced is a flat sheet, made tangent to the globe at a single point only. Figure A1.3*a* illustrates how the latitude–longitude grid would project upon a plane tangent to the earth at a point on the equator. All meridians trace out as straight lines on the paper plane, because they all lie in planes passing through the light source that casts their shadows. However, the paper intervals between adjacent longitude lines increase with distance away from the tangent point, an effect similar to that suffered by latitude lines in the cylindrical projection of Figure A1.2*a*. In this gnomonic projection latitude lines trace out hyperbolic curves on the paper and so cannot intersect meridians at right angles, as they do in the cylindrical projection.

Figure A1.3*b* shows another gnomonic projection, this time with the paper tracing plane made tangent to the globe at the north pole, where all meridians converge to a point. Again, the meridians project as straight lines radiating from the pole. The latitudes project as perfect circles, but with paper intervals increasing with distance from the pole. These projections are universally used for mapping the polar regions; examples are seen in Chapter 20.

All gnomonic projections, made with the projection light placed at the center of the globe, have one very special feature: a straight line drawn on the

gnomonic map represents the shortest distance between the two points it connects. This is so because the shortest distance along the surface of a globe is a line that is also the intersection of the earth's surface with a plane slicing through the center of earth. The line is called a *great-circle route*.

Constant Bearing versus Great Circle—Mercator versus Gnomonic Projections: Which Map To Use and Why?

In laying out a cruise plan to gather oceanographic data systematically, we usually choose the Mercator projection. There are several reasons for the choice. First, it is an "easy" map on which the scientist can communicate with the ship's captain and crew. A proposed ship's track is usually laid out in straight lines. The scientist determines ahead of time the desired, *optimum* intervals between sampling stations and lays these points off along the track, using the scale of local latitude marks as distance markers. The captain and chief scientist together make plans for alternate stations should bad weather or other problem force a change of station plans, and such changes are easily calculated on the Mercator projection. Second, because most oceanographic work involves rather detailed sampling within a localized area, the Mercator feature of equal local scales of east–west versus north–south paper distances is a great advantage in the design of areal sampling plans. Finally, the fact that a straight line means a constant ship heading, or compass bearing, makes for ease in navigating. (The reader is urged to review Figure 13.5.)

Scientific cruises that cover long distances are usually laid out with shortest total sailing time a priority, not ease in navigating or sampling. During extended voyages there is always the inherent danger of running short on provisions, fuel, medicines, and so on. Emphasis is then placed on the shortest route of sailing, the great-circle route. On the Mercator projection the great-circle route can be misleading; it is a curved path and hence requires constant changes in ship's heading as the cruise progresses. Figure A1.4*a* shows that a cruise path laid out as a straight line from Sydney, Australia, to Marrakech, Morocco, through the Panama Canal actually requires a traverse of 21,980 km. In contrast, the great-circle route between these two end ports covers a distance of 20,900 km, over a thousand kilometers less (but only if it were possible to sail "through" Washington, D.C.!).

On a gnomonic projection the same cruise route

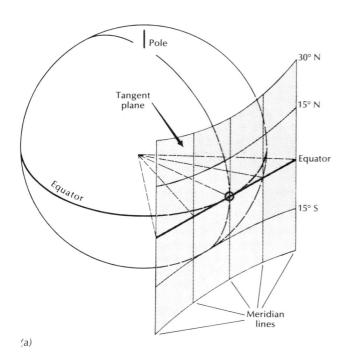

(a)

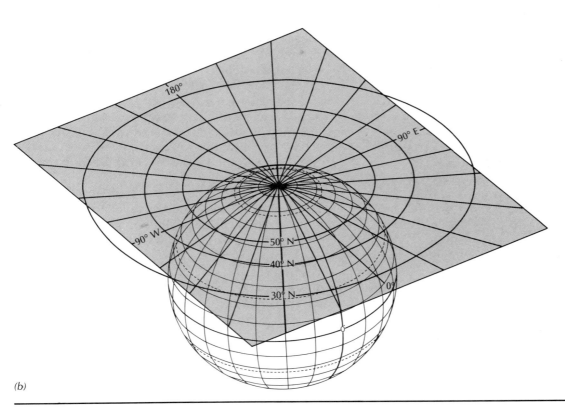

(b)

Figure A1.3 Two gnomonic projections, in which the globe features are projected upon a plane that is tangent to the globe at (a) the equator, and (b) the north pole.

464

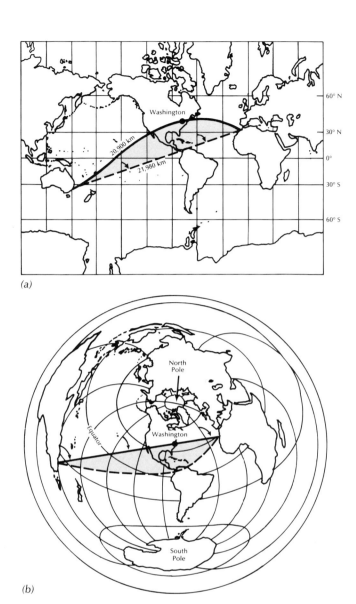

(a)

(b)

Figure A1.4 A comparison of the traces of great-circle routes on Mercator and gnomonic projections. *(a)* On the Mercator the great circle is a curved trace; *(b)* on the gnomonic the great circle is a straight line.

is a curved track passing through the Panama Canal, but it requires continuous heading changes, as indicated in Figure A1.4*b*. In general, gnomonic charts are used in aircraft navigation because of the ex-

treme importance of fuel management. Unlike oceanographic cruises, air flights do not make frequent stops for station sampling.

Equal-Area Charts

For special purposes oceanographers choose still another type of map to present as faithfully as possible the ocean area in true scale and in true shape. The Goode Homolosine base chart is an excellent choice. Figures 5.10 and 5.11 are examples; because the intent of these figures is to compare *areas* of the basin floors covered by distinctly different types of sediments, the choice of an equal-area chart is logical. Notice, however, that the equal-area advantage is had at the expense of a complex method of separating pieces of the spherical earth's surface, each then treated as a separate entity. Because these are ocean basin charts, the separations are made within the continents, wherever possible, and this creates the strangeness of the chart to the novice. The Goode Homolosine and other types of equal-area—shape-retention charts are not projections; they are mathematical designs.

Cartography in the Era of Satellite Navigation and Sensing

Today every large oceanographic research vessel, as well as all commercial vessels, navigates by measuring its positions in relation to the very accurately known positions of satellites. These position data are available day or night, without reliance on stars or sun or horizon. The era of mapmaking as a necessary technological skill in support of maritime traffic is over.

Every major oceanographic vessel also has sophisticated computers aboard, with software capable of generating, on demand and at sea, base charts of various kinds. Further, these computers also take ship position data and plot position and track directly upon the base charts, entering in addition whatever sampling data the scientists wish to include. The ocean basins are rapidly yielding up their secrets to humankind.

THE EQUATION OF STATE
FOR SEAWATER DENSITY

A picture is said to be worth a thousand words. In the sciences, an *equation* has equal power; it sums mathematically a multitude of words into a single picture. We use just such an equation to describe how the density of seawater varies as a function of three independent controlling variables, temperature, salinity, and pressure. Called the "equation of state," it is written as

$$\rho_{S,T} = \rho_{\text{reference}} (1 - \underbrace{\alpha\,\Delta T} + \underbrace{\beta\,\Delta S} + \underbrace{\gamma\,\Delta p})$$

modifier terms

whereby the density of seawater under any condition is found by evaluating a set of variable terms that modify a known density called the *reference*. Let us dissect the equation term by term.

$\rho_{S,T}$ is the unknown density of a seawater sample. It is computed once we know the salinity S and the temperature T of the sample taken at pressure p (depth). These *in situ* variables define the values of the three modifier terms; applying them, we discover how the density of the *in situ* sample deviates from that of the reference density.

$\rho_{\text{reference}}$ is generally taken as the density of water at salinity $S = 35$ ppt, temperature $T = 0°C$, and pressure $P = 0$, the latter meaning the pressure at the level of the sea surface, which is actually one atmosphere. Its value is 1.02812 g/cm^3.

$\alpha\Delta T$ is a product that modifies the reference density for the *effect* of temperature: its two factors have these definitions.

  $-\alpha$ describes precisely how seawater *decreases* in density as temperature *rises*; hence the minus sign assigned to it. It is called the *coefficient of thermal expansion*.

  ΔT is the difference between the temperature of the sample *in situ* and that of the reference water at 0°C.

The product term $\alpha\Delta T$, when subtracted from the reference value of 1.02812 g/cm^3, corrects for the reduction in density caused by the sample's being warmer than the reference.

$\beta\Delta S$ is a product that modifies the reference density for the *effect* of dissolved salts; its two factors have these definitions.

  β describes precisely how the density of seawater *increases* as the amount of dissolved salts increases. It is called the *coefficient of saline contraction*.

  ΔS is the difference between the salinity of the sample and that of the reference at 35 ppt.

The product term $\beta\Delta S$ corrects density for the saltiness of the sample in relation to the 35-ppt salinity of the reference.

$\gamma\Delta p$ is a product that corrects density for the *effect* of pressure changes in relation to pressure at sea level. Its two factors have the following definitions.

  γ describes precisely how the volume of a sample changes as it is compressed. It is called the *compressibility coefficient*.

  Δp measures the departure of ambient pressure at the sample's depth from pressure at sea surface, $p = 0$.

The product term $\gamma\Delta p$ corrects for the change in density caused by the sample's being subject to compression in the interior of the ocean.

Application

In practice, we find that most of the changes in seawater density in the vertical direction occur in the uppermost 1000 m. In this range the contribution of the pressure-modifying term, $\gamma\Delta p$, is very small compared to that of the terms controlled by temperature and salinity changes. We neglect the effect of compressibility unless our studies are specifically directed at the movement of deep waters, where compression is considerable and the changes in T

:

and S are much smaller than in the surface layer.

By neglecting pressure effects, we can then map the density of seawater onto a domain governed only by temperature T and salinity S. This is the T–S nomograph described in Chapter 9. It is a graphical representation of the equation of state (Figure A2.1).

When we evaluate the behavior of the coefficients α and β, we discover that these are *not* constant coefficients. In particular, α varies in a highly non-linear way as the temperature itself changes. For example, compare the points marked A, B, and C in Figure A2.1. These points are separated by equal increments of 10°C along the temperature scale. For each we find the amount of temperature change that alone could cause a density shift of 0.0004 g/cm³. At point A the change is 5.8°C, or 3.8°C *more* than the 2.0°C change at point B. But point B requires only 0.4°C more than point C. A *linear* relation in the thermal behavior of density would require that a

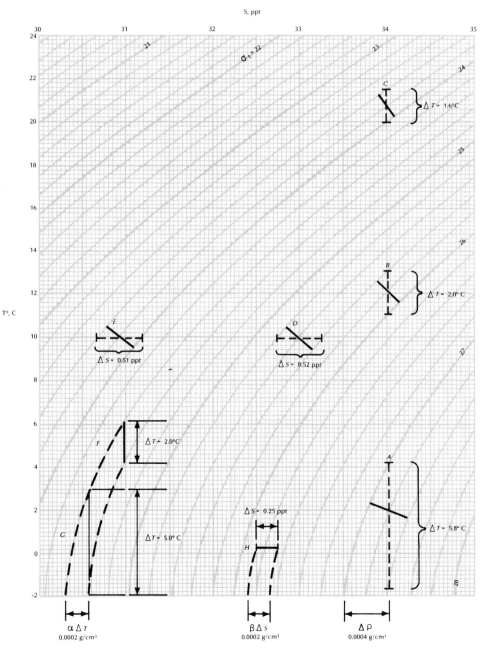

Figure A2.1 The T–S nomograph, a graphical representation of the equation of state for seawater density.

given density change be coupled to the same degree change from A to B as from B to C. It is this *nonlinear* behavior in α that forces such a strong curvature in the density contours of the T–S nomograph. The β coefficient, on the other hand, is almost linear, as seen in the constancy of the salinity change ΔS coupled to similar density changes in seawater samples of different salinities, at points D and E.

The nonlinear behavior of water density as a function of temperature has a specific application in the physical oceanography of polar oceans. At point F we show that a *change* of 0.0002 g/cm^3 in the $\alpha\Delta T$ term when ambient water is around 5°C temperature is correlated with a temperature *change*, ΔT, of only 2.0°C. But when ambient water is around 0°C, as at point G, the same change in density requires a ΔT of 5.0°C. The conclusion is that in polar seas, where pack ice keeps the surface water temperature near its freezing point, temperature changes have almost no effect on density changes. Here the physics of currents and vertical convection mixing are controlled almost entirely by changes in the salinity of seawater. For example, a ΔS of only 0.25 ppt forces the same density change of 0.0002 g/cm^3 (point H), as does the 5°C temperature change of point G, for seawater whose temperature is near 0°C.

REFERENCES

CHAPTER 1

Broecker, W. S., T. Takahashi, H. J. Simpson, and T. H. Peng. Fate of fossil fuel carbon dioxide and the global carbon budget. *Science* 206:409–418, 1979.

Fishing Information, 1978, No. 9. National Marine Fisheries Service, NOAA.

Ryther, J. H. Photosynthesis and fish production in the sea. *Science* 166:72–76, 1969.

Whittaker, R. H. and G. E. Likens. The biosphere and man. In *Primary Productivity of the Biosphere,* Eds. H. Leith and R. H. Whittaker. Springer-Verlag, New York, 1975: 306 pp.

Woodwell, G. M. The carbon dioxide question. *Scientific American* 238(1): 34–44, 1978.

Woodwell, G. M. and E. V. Pecan. Carbon and the biosphere. Atomic Energy Commission Symposium Series 30:CONF-720510, 1973.

CHAPTER 2

Beebe, W. *The Arcturus Adventure.* G. T. Putnam Sons, New York, 1926: 439 pp.

Fell, B. *Introduction to Marine Biology.* Harper and Row, New York, 1975: 356 pp.

CHAPTER 3

Chung, Y. and H. Craig. ^{226}Ra in the Pacific Ocean. *Earth and Planetary Sci. Ltrs.* 49:267–292, 1980.

Craig, H. and K. K. Turekian. The GEOSECS program: 1976–1979. *Earth and Planetary Sci. Ltrs.* 49:263–265, 1980.

Fell, B. *Introduction to Marine Biology.* Harper and Row, New York, 1975: 356 pp.

GEOSECS Atlas, Vols. 2, 4, and 6. U.S. Government Printing Office, 1981.

Hammond, A. L. Use of radioactive trace elements in the study of ocean motion. *Science* 195(14):165, 1977.

Heezen, B. C., M. Tharp, and M. Ewing. The floors of the ocean. I. North Atlantic. Geological Society of America Special Paper 65; 1959: 122 pp.

Sciremammano, F. and R. D. Pillsbury. Spatial scales of temperature and flow in the Drake Passage. *J. Geophys. Res.* 85(C7):4015–4028, 1980.

Sumich, J. L. *An Introduction to the Biology of Marine Life.* William C. Brown, Dubuque, Iowa, 1980 (2nd ed.): 375 pp.

CHAPTER 4

Barazangi, M. and J. Dorman. World seismicity maps compiled from ESSA, Coast and Geodetic Survey, Epicenter Data 1961–1967. *Bull. Seismol. Soc. Amer.* 59:369, 1969.

Emiliani, C. The temperature decrease of surface seawater in high latitudes and of abyssal-hadal water in open oceanic basins during the past 75 million years. *Deep-Sea Res.* 8:144–147, 1961.

Heezen, B. C. and M. Tharp. The Floor Of The Oceans. (Copyright Marie Tharp.)

Keigwin, L. Isotopic paleoceanography of the Caribbean and East Pacific: Role of Panama uplift in late Neogene time. *Science* 217:350–353, 1982.

Kennett, J. P. Paleoceanographic and biogeographic evolution of the Southern Ocean during the Cenozoic, and Cenozoic microfossil datums. *Palaeogeography, Palaeoclimatology, Palaeoecology* 31:123–152, 1980.

Kennett, J. P. Cenozoic evolution of Antarctic glaciation, the Circum-Antarctic Ocean, and their impact on global paleooceanography. *J. Geophys. Res.* 82(27):3843–3860, 1977.

Komar, P. D. *Beach Processes and Sedimentation.* Prentice–Hall, Englewood Cliffs, NJ, 1976: 429 pp.

Luyendyk, B. P., D. Forsyth, and J. D. Phillips. Experimental approach to the paleocirculation of the oceanic surface waters. *Geol. Soc. Amer. Bull.* 83:2649–2664, 1972.

Scorpio expedition, 1967. Data from National Oceanographic Data Center, Washington, D.C.

Shepard, F. P. *Submarine Geology.* Harper and Row, New York, 1963 (2nd ed.): 557 pp.

Sverdrup, H. U., M. W. Johnson, and R. H. Fleming. *The Oceans.* Prentice–Hall, Englewood Cliffs, NJ., 1942: 1087 pp.

Van Andel, T. H. *Science at Sea: Tales of an Old Ocean.* Freeman, San Francisco, 1977: 186 pp.

Wegener, A. *Die Entstehung der Kontinente und Ozeane.* Vieweg and Sohn, Braunschweig, Germany, 2nd ed. 1920; 3rd ed. 1922, translated into English in 1924; 4th ed. 1924, revised by A. Wegener and translated into English in 1929.

Weyl, P. K. The role of the oceans in climatic change: A theory of the ice ages. *Meteorol. Monograph* 8:37–62, 1968.

Wüst, G. *The Stratosphere Of The Atlantic Ocean.* Scientific results of the German Atlantic expedition of the Research Vessel *Meteor* 1925–1927, Vol. 6, Section 1. Berlin and Leipzig, 1935. English translation edited by W. J. Emery for the National Science Foundation, 1978: 111 pp.

CHAPTER 5

Berger, W. H. Deep-sea sedimentation. In *The Geology of Continental Margins*, Eds. C. A. Burk and C. L. Drake. Springer-Verlag, New York, 213–241, 1974.

Bullard, E. C., J. E. Everett, and A. G. Smith. The fit of continents around the Atlantic. In *A Symposium on Continental Drift*, Eds. P. M. S. Blackett, E. C. Bullard, and S. K. Runcorn. The Royal Society, London, 1965: 323 pp.

Darwin, C. *Voyage of the Beagle*. Bantam Books, New York, 1972: 439 pp.

Drake, C. L. and C. A. Burk. Geological significance of continental margins. In *The Geology of Continental Margins*, Eds. C. A. Burk and C. L. Drake. Springer-Verlag, New York: 3–10, 1974.

Emery, K. O. The continental shelves. In *Oceanography: Readings from Scientific American*, Ed. J. R. Moore. Freeman, San Francisco: 143–154, 1971.

Heezen, B. C. and M. Tharp. Physiographic Diagram of the North Atlantic. Geologic Society of America, copyright Bruce C. Heezen and Marie Tharp, 1968.

Hopkins, T. S. On the circulation over the continental shelf off Washington. Ph.D. thesis, University of Washington, Seattle, 1971: 204 pp.

Komar, P. D., R. H. Neudeck, and L. D. Kulm. Observations and significance of deep-water oscillatory ripple marks on the Oregon Continental Shelf. In *Shelf Sediment Transport*, Eds. D. J. P. Swift, D. B. Duane, and O. H. Pilkey. Dowden, Hutchinson, and Ross, Stroudsburg, Pa. 601–620, 1972.

McIntyre, A., N. G. Kipp, A. W. H. Be, T. Crowley, T. Kellog, J. V. Gardner, W. Prell, and W. F. Ruddiman. Glacial North Atlantic 18,000 years ago: A CLIMAP reconstruction. *Geol. Soc. Amer. Memoir* 145:43–76, 1976.

Meade, R. H. Sources and sinks of suspended matter on continental shelves. In *Shelf Sediment Transport*, Eds. D. J. P. Swift, D. B. Duane, and O. H. Pilkey. Dowden, Hutchinson, and Ross, Stroudsburg, Pa.: 249–262, 1972.

Shepard, F. P. *Submarine Geology*, Harper and Row, New York, 1963: 557 pp.

Swift, D. J. P. Continental shelf sedimentation. In *The Geology of Continental Margins*, Eds. C. A. Burk and C. L. Drake. Springer-Verlag, New York: 117–135, 1974.

CHAPTER 6

Harvey, J. G. *Atmosphere and Ocean*. Artemis Press, Sedgwick Park, Horsham, Sussex, 1976: 143 pp.

Michener, J. *The Covenant*. Fawcett, New York, 1982: 877 pp.

Sverdrup, H. U., M. W. Johnson, and R. H. Fleming. *The Oceans*. Prentice-Hall, Englewood Cliffs, N.J., 1942: 1087 pp.

Trump, C. L., S. J. Neshyba, and W. V. Burt. Effects of mesoscale atmospheric convection cells on the waters of the East China Sea. *Boundary-Layer Meteorology* 24:15–34, 1982.

Von Arx, W. S. *An Introduction to Physical Oceanography*. Addison-Wesley, Reading, Mass., 1962: 422 pp.

Weyl, P. K. *Oceanography: An Introduction to the Marine Environment*. Wiley, New York, 1970: 535 pp.

CHAPTER 7

Barth, T. F. W. *Theoretical Petrology*. Wiley, New York, 1952: 416 pp.

Dittmar, W. Report on researches into the composition of ocean water collected by H.M.S. *Challenger*. *Challenger Repts. Phys. and Chem.* 1:1–251, 1884.

Goldberg, E. D. and G. O. S. Arrhenius, *Geochim. et cosmoch. Acta*, 13:153, 1958.

Gross, M. G. *Oceanography: A View of the Earth*. Prentice-Hall, Englewood Cliffs, N.J., 1977 (2nd ed.): 497 pp.

McClellan, H. J. *Elements of Physical Oceanography*. Pergamon Press, Elmsford, N.Y., 1965: 151 pp.

Montgomery, R. B. Water characteristics of Atlantic Ocean and of World Ocean. *Deep-Sea Res.* 5:134–148, 1958.

Quinby-Hunt, M. S. and K. K. Turekian. Distribution of elements in sea water. EOS, *Trans. Amer. Geophysical Union*, Vol. 66, 5 April 1983.

Shepard, F. P. *Submarine Geology*. Harper and Row, New York, 1963: 557 pp.

Sverdrup, H. V., M. W. Johnson and R. H. Fleming. *The Oceans* Prentice-Hall, Englewood Cliffs, N.J., 1942: 1087 pp.

CHAPTER 8

Barstow, D., W. Gilbert, and B. Wyatt. Hydrographic data from Oregon waters 1968. Oregon State University, Dept. of Oceanography, Report 36, November 1969.

CHAPTER 9

Archives, National Oceanographic Data Center, Washington, D.C.

Clarke, M. R. Structure and proportions of the spermaceti organ of the sperm whale. *J. Marine Biol. Assn. United Kingdom*, 58:1–17, 1978.

Johnson, R. E. Antarctic Intermediate Water in the South Pacific. Ph.D. thesis, Oregon State University, Corvallis, 1972.

Wüst, G. *The Stratosphere of the Atlantic Ocean*. Scientific results of the German Atlantic expedition of the Research Vessel *Meteor* 1925–1927, Vol. 6, Section 1. Berlin and Leipzig, 1935. English translation edited by W. J. Emery for the National Science Foundation, 1978: 111 pp.

CHAPTER 10

Michener, J. *Centennial*. Random House, New York, 1974.

Neshyba, S. Upwelling by icebergs. *Nature* 267, 5611:507–508, 1977.

Quinn, W. H. The false El Niño and recent related climatic changes in the southeast Pacific. In *Proceedings of the Fourth Annual Climate Diagnostics Workshop*, Institute of Environmental Studies, University of Wisconsin, Madison: 233–244, 1979.

Richardson, P. L. Gulf Stream ring trajectories. *J. Phys. Ocn.* 10(1):90–104, 1980.

CHAPTER 11

Knauss, J. A. *Introduction to Physical Oceanography*. Prentice-Hall, Englewood Cliffs, N.J., 1978: 338 pp.

Stommel, H. *The Gulf Stream*. University of California Press, Berkeley, 1960: 202 pp.

Stommel, H. The abyssal circulation. *Deep-Sea Res.* 5:80–82, 1958.

Stommel, H. Westward intensification of wind-driven ocean currents. *Trans. Amer. Geophys. Union*, 29:202–206, 1948.

CHAPTER 12

Barber, R. T. Seminar given at Oregon State University, 1978.

Brown, K. and R. Berry. Pounds and value of commercially caught fish and shellfish in Oregon. Oregon Fish and Game Commission Report, 1979.

Hardy, A. *The Open Sea*, Vol. 2. Houghton Mifflin, Boston, 1965: 322 pp.

Hurd, L. E., M. V. Mellinger, L. L. Wolfe, and S. J. McNaughton. Stability and diversity of three trophic levels in terrestrial successional ecosystems. *Science* 173:1134–1136, 1971.

Parsons, T. R. and M. Takahashi. *Biological Oceanographic Processes*. Pergamon Press, Elmsford, N.Y., 1973: 186 pp.

CHAPTER 13

Barnes, R. D. *Invertebrate Zoology*. W. B. Saunders, Philadelphia, 1963: 623 pp.

Fell, B. *Introduction to Marine Biology*. Harper and Row, New York, 1975: 356 pp.

Fleming, R. H. The control of diatom population by grazing. Conseil Perm. Internat. p.l'Explor. de la Mer, *J. Conseil* 14(2):210–227, 1939.

Fryxel, G. Personal communication, 1983.

Johnson, M. W. and E. Brinton. Biological species, water masses and currents. In *The Sea*, Vol. 2, Ed. M. Hill. Wiley-Interscience, New York: 381–414, 1963.

Koehl, M. A. R. and J. R. Stickler. Copepod feeding currents: Food capture at low Reynolds number. *Limn. and Ocn.* 26(6):1062–1073, 1981.

McGowan, J. A. Oceanic biogeography of the Pacific. In *The Micropaleontology of Oceans*, Eds. B. Funnell and W. Riedel. Cambridge University Press, Cambridge, England: 3–74, 1971.

Metcalf, M. M. The Salpidae: A taxonomic study. United States Natural Museum Bulletin 100 (2, 2):5–193, 1919.

Miller, C. B. Personal communication.

Uribe, E., S. Neshyba, and T. Fonseca. Phytoplankton community composition across the West Wind Drift off South America. *Deep-Sea Res.* 29(10A):1229–1243, 1982.

CHAPTER 14

Beebe, W. *The Arcturus Adventure*. G. T. Putnam, New York, 1926: 439 pp.

Espinoza, F. R., S. Neshyba, and M. Zhao. Surface water motion off Chile revealed in satellite images of surface chlorophyll and temperature. In *Proceedings of the International Conference of Marine Resources of the Pacific*, Ed. P. M. Arana. Catholic University of Valparaiso, Chile: 41–57, 1983.

Hardy, A. *The Open Sea*, Vol. 2. Houghton Mifflin, Boston, 1965: 322 pp.

Jerlov, N. G. *Marine Optics*. Elsevier, New York, 1976: 231 pp.

Miller, C. B. Personal communication, 1983.

Neshyba, S. Pulsed light stimulation of marine bioluminescence in situ. *Limn. and Ocn.* 12(2):222–235, 1967.

Pak, H., G. F. Beardsley, Jr., and R. L. Smith. An optical and hydrographic study of a temperature inversion off Oregon during upwelling. *J. Geophys. Res.* 75(3):629–636, 1970.

Sumich, J. L. *An Introduction to the Biology of Marine Life*. William C. Brown, Dubuque, Iowa, 1980 (2nd ed): 375 pp.

Weyl, P. K. *Oceanography: An Introduction to the Marine Environment*. Wiley New York, 1970: 535 pp.

CHAPTER 15

Alvariño, A. Distributional atlas of chaetognatha in the California Current Region. *CALCOFI Report*, Vol. 3, 1965.

Bakun, A. and R. H. Parrish. Environmental inputs to fishery population models for Eastern Boundary Current regions. In *Workshop on the Effects of Environmental Variation on the Survival of Larval Pelagic Fishes*, Ed. G. D. Sharp. IOC Workshop Report 28, UNESCO: 67–104, 1980.

Barnes, R. D. *Invertebrate Zoology*, Saunders College Press, Philadelphia, 1980 (4th ed.): 1089 pp.

Berner, L. D. Distributional atlas of the Thaliacea in the California Current region. *CALCOFI Report*, Vol. 8, 1967.

Brady, H. B. Report of the Foraminifera dredged by HMS *Challenger* during the years 1873–1876. In *Reports of the Scientific Results of the Voyage of the HMS Challenger*, Vol. 9, 1884.

Brinton, E. Distributional atlas of Euphausiacea (Crustacea) in the California Current region, Part I. *CALCOFI Report*, Vol. 5, 1967.

California Cooperative Oceanic Fisheries Investigation

(CALCOFI) Atlas Series, Vols. 1–30. California Dept. of Fish and Game, NOAA and Scripps Institute of Oceanography.

Campbell, A. S. Tintinnids. In *Treatise on Invertebrate Paleontology*, Part d, 3, Ed. R. C. Moore. Geological Society of America, Boulder, Colo., 1954.

Dodge, J. D. *Marine Dinoflagellates of the British Isles*. Her Majesty's Stationary Office, London, 1982: 303 pp.

Geisbrecht, W. Systematik und Faunistik der pelagischen Copepoden des Golfes von Neapel unde der Angrenzenden Meeres—Abschnitte. *Fauna und Flora Neapel* 19 (Atlas), 1982.

Hart, J. L. *Pacific Fishes of Canada*. Bulletin of Fishery Research Board of Canada 180, 1973.

Horn, M. H. Diversity and ecological roles of noncommercial fishes. *CALCOFI Report*, Vol. 21:37–47, 1980.

Mayer, A. G. *Medusae of the World*. Carnegie Institute of Washington, Washington, D.C., 1910.

McGowan, J. A. The thecosomata and gymnosomata of California. *The Veliger*, 3(suppl.):103–135, 1968.

McGowan, J. A. and C. B. Miller. Larval fish and zooplankton community structure. *CALCOFI Report*, Vol. 21, 1980.

Morton, J. E. The biology of *Limacina retroversa*. *J. Marine Biol. Assn. United Kingdom* 33:297–312, 1954.

Reid, F. Personal communication, 1986

Schewiakoff, W. Die Acantharia des Golfes von Neapel. *Fauna uud Flora Neapel* 37 (Atlas), 1926.

Sleigh, M. A. *The Biology of Protozoa*. Elsevier, New York, 1973: 315 pp.

Steinbeck, John. *Cannery Row*. Viking Press, New York, 1945.

Sumich, J. L. *An Introduction to the Biology of Marine Life*, William C. Brown, Dubuque, Iowa, 1980 (2nd ed.): 375 pp.

Whedon, W. F. and C. A. Kofoid. Dinoflagellata of the San Francisco region. I. On the skeletal morphology of two new species, *Gonyaulax catenella* and *G. acatenella*. *Univ. Calif. Publ. Zool.* 41:25–34, 1936.

CHAPTER 16

Bolin, R. L. and Abbott, D. P. Studies of the marine climate and phytoplankton of the central coastal area of California, 1954–1960. *CALCOFI Report*, Vol. 19:23–45, 1963.

Collette, B. B. and F. H. Talbot. Activity patterns of coral reef fishes with emphasis on nocturnal-diurnal changeover. *Bull. Nat. Hist. Mus. Los Angeles County*, 14:98–124, 1972.

Darwin, C. *Voyage of the Beagle*. Bantam Books, New York, 1972: 439 pp.

Gonor, J. J. Personal communications, Marine Science Center, Oregon State University, 1985.

Hixon, M. A. and W. N. Brostoff. Damselfish as keystone species in reverse: Intermediate disturbance and diversity of reef algae. *Science* 220:511–513, 1983.

Moyle, P. and J. Cech, Jr. *Fishes—An Introduction to Ichthyology*. Prentice-Hall, Englewood Cliffs, N.J., 1982: 593 pp.

Nybakken, J. W. *Marine Biology: An Ecological Approach*. Harper and Row, New York, 1982: 446 pp.

Oxenford, H. Personal communications, Bellairs Research Institute, St. James, Barbados, West Indies, 1985.

Randall, J. E. *Caribbean Reef Fishes* T. F. H. Publications, N.J., 1968: 350 pp.

Smith, C. L. and J. C. Tyler. Space resource sharing in a coral reef fish community. *Bull. Nat. Hist. Mus. Los Angeles County* 14:125–170, 1972.

Weibe, W. J., R. E. Johannes, and K. L. Webb. Nitrogen fixation in a coral reef community. *Science* 188: 257–259, 1975.

CHAPTER 17

Bowden, K. F. *Physical Oceanography of Coastal Waters*. Halsted Press, New York, 1983: 288 pp.

Chelton, D. B., K. J. Hussey and M. E. Parke. Global satellite measurements of water vapor, wind speed and wave height. *Nature* 294(5841):10–16, 1981.

Cornish, V. *Ocean Waves and Kindred Phenomena*. Cambridge University Press, Cambridge, England, 1934.

Phillips, O. M. *The Dynamics of the Upper Ocean*. Cambridge University Press, Cambridge, England, 1966: 261 pp.

Reid, R. O., (Personal communication), Texas A&M University, Dept. of Oceanography and Meteorology, 1964.

Smith, R. L. Poleward propagating perturbations in currents and sea level along the Peru coast. *J. Geophys. Res.* 83:6083–6092, 1978.

Weigel, E. P. Killer from the bottom of the sea. U.S. Weather Service, NOAA Report, January 1974.

CHAPTER 18

Baumgartner, A. and E. Reichel. *The World Water Balance*, translated by R. Lee. Elsevier, New York, 1975: 179 pp.

Brown, L. R. and E. C. Wolf. Soil erosion: Quiet crisis in the World Economy. World Watch Paper 60, 1984.

Conomos, T. J. and M. G. Gross. Summer mixing of Columbia River and ocean waters. *J. Sanit. Eng. Div., Amer. Soc. Civil Eng.* 94:979–994, 1968.

Czaya, E. *Rivers of the World*, Van Nostrand, New York, 1981: 248 pp.

Defant, A. *Physical Oceanography*, Vol 2. Pergamon Press, Elmsford, N.Y., 1961.

El-Swaify, S. A. and E. W. Dangler. Rainfall erosion in the tropics. In *Soil Erosion and Conservation in the Tropics*. American Society of Agronomy Special Publication 43, Madison, Wis., 1982.

Groves, G. and W. H. Munk. A note on tidal friction. *J. Mar. Res.* 17:199–214, 1958.

Haedrich, R. L. and C. A. S. Hall. Fishes and estuaries. *Oceanus* 19(5):55–63, 1976.

Iles, T. D. and M. Sinclair. Atlantic herring: Stock discreteness and abundance. *Science* 215: 627–633, 1982.

Judson, S. Erosion of the land. *American Scientist*, July–August, 1968.

Ketchum, B. H. and D. J. Keen. The accumulation of river water over the continental shelf between Cape Cod and Chesapeake Bay. *Deep-Sea Res.* 3(suppl):346–357, 1955.

Komar, P. D. *Beach Processes and Sedimentation.* Prentice-Hall, Englewood Cliffs, N.J., 1976.

McGowan, J. A. Personal communication, 1978.

McHugh, J. L. Management of estuarine fisheries. In *A Symposium on Estuarine Fisheries.* American Fisheries Society, Special Publication 3, Bethesda Md.: 133–154, 1966.

Neal, V. T. Physical aspects of the Columbia River and its estuary. In *The Columbia River Estuary and Adjacent Ocean Waters,* Eds. A. T. Pruter and D. L. Alverson. University of Washington Press, Seattle: 19–40, 1972.

Osterberg, C., N. Cutshall, and J. Cronin. Chromium-51 as a radioactive tracer of Columbia River water at sea. *Science* 150:1586, 1965.

Pruter, A. T. Review of commercial fisheries in the Columbia River and in contiguous waters. In *The Columbia River Estuary and Adjacent Ocean Waters,* Eds. A. T. Pruter and D. L. Alverson. University of Washington Press, Seattle, 1972: 868 pp.

Sakou, T. and S. Neshyba. The temporal structure of oceanic motion off the Oregon Coast. *J. Mar. Res.* 30(1):1–14, 1972

Schumacher, J. D., T. H. Kinder, D. J. Pashinski and R. L Charnell. A structural front over the continental shelf of the Eastern Bering Sea. *J. Phys. Ocn.* 9:79–87, 1979.

Shepard, F. P. *Submarine Geology,* Harper and Row, New York, 1963 (2nd ed.): 557 pp.

Small, L. F. and H. C. Curl, Jr. Effects of Columbia River discharge on chlorophyll a and light attenuation in the sea. In *The Columbia River Estuary and Adjacent Ocean Waters,* Eds. A. T. Pruter and D. L. Alverson. University of Washington Press, Seattle: 203–218, 1972.

Stommel, H. Westward intensification of wind-driven ocean currents. Trans. *Amer. Geophys. Union* 29:202–206, 1948.

Sutcliffe, W. H., Jr. Correlations between seasonal river discharge and local landings of Americal lobster *(Homarus americanus),* and Atlantic halibut *(Hippoglossus hippoglossus)* in the Gulf of St. Lawrence. *J. Fish. Res. Bd. Can.* 30(6):856–859, 1973.

CHAPTER 19

Amann, H. The Atlantis II deep project in the Red Sea as a source of technology for the development of marine polymetallic sulfides. In *Proceedings OCEANS '83,* Vol. 2. Marine Technology Society and IEEE, Rockville, Md.: 802–815, 1983.

Ballard, R. D. and J. L. Bischoff. Assessment and scientific understanding of hard mineral resources in the EEZ. *USGS Survey Circular* 929:185–208, 1983.

Bell, F. W. *Food from the Sea: The Economics and Politics of Ocean Fisheries.* Westview Press, Boulder, Colo., 1978:380 pp.

Brooks, R. R. and M. G. Rumsby. The biogeochemistry of trace element uptake by New Zealand bivalves. *Limn. and Ocn.* 10:521–527, 1965.

Brown, L. R. and P. Shaw. Six steps to a sustainable society. Worldwatch Paper 48, March 1982.

Constans, J. *Marine Sources of Energy.* U.N. Department of International Economic and Social Affairs, OST. Pergamon Press, Elmsford, N.Y., 1979: 151 pp.

Crooke, R. C. and L. G. Otteman. Offshore oil and gas technology assessment. *USGS Survey Circular* 929:209–246, 1983.

Duedall, I. W., B. H. Ketchum, P. K. Park, and D. R. Kester. *Wastes in the Ocean.* Vol. 1. Industrial and Sewage Wastes in the Ocean. Wiley-Interscience, New York, 1983: 431 pp.

Food and Agricultural Organization (FAO) Fishery Yearbooks, 1968–1983, U.N. publication, Rome, Italy.

Gulland, J. A. Population dynamics of world fisheries. University of Washington Sea Grant Publication, 296; Seattle, 1972.

Haddad, K., F. T. Manheim, and W. Marr. Waste disposal in the Florida coastal environmental. Unpublished report, Marine Science Program, Dept. of Marine Science, University of South Florida, St. Petersburg.

Halbach, P. Co-rich ferromanganese seamount deposits of the Central Pacific Basin. In *Marine Mineral Deposits—New Research Results and Economic Prospects,* Eds. P. Halbach and P. Winter. Marine Rohstaffe und Meeres technik, Bd 6, Verlag Gluckauf, Essen, Germany: 60–85, 1982.

Hood, D. W. and C. P. McRoy. Uses of the ocean. In *Impingement of Man on the Sea,* Ed. D. W. Hood. Wiley-Interscience, New York, 1971.

Isaacs, John D. Personal communication, 1976.

Japan Fisheries Agency Annual Report, 1980.

Levenspiel, O. and N. deNevers. The osmotic pump. *Science* 183:157–160, 1974.

Manheim, F. T., T. H. Ling and C. M. Lane. An extensive data base for cobalt-rich ferromanganese crusts from the world oceans. In *Proceedings Oceans '83,* Vol. 2. Marine Technology Society and IEEE, Rockville, Md.: 828–831, 1983.

McClain, C. E. Requirements for private sector investment. *USGS Survey Circular* 929:15–54, 1983.

Merlini, M. Heavy-metal contamination. In *Impingement of Man on the Seas,* Ed. D. W. Hood. Wiley-Interscience, New York, 1971:

National Academy of Sciences. *Marine Environmental Quality.* Washington, D.C., 1971: 107 pp.

Oxman, B. H. The new law of the sea. National Oceanographic Data Center, NOAA, Washington, D.C., ABA 69, 1983.

Park, P. K., I. W. Duedall, D. R. Kester, and B. H. Ketchum. *Wastes in the Ocean,* Vol. 3. Radioactive Wastes and the Ocean. Wiley-Interscience, New York, 1983: 5–14 pp.

Peck, D. L. The U.S. geological survey program and plans in the EEZ. *USGS Survey Circular* 929:55–83, 1983.

Pendley, W. P. Importance of the EEZ proclamation. *USGS Survey Circular* 929:3–9, 1983.

Pillay, T. V. R. The role of aquaculture in the fishery development and management. *J. Fish. Res. Bd. Can.* 30:2202–2217, 1973.

Rowland, R. W. and B. A. McGregor. Recommendations from the Department of the Interior EEZ Symposium. In *Exclusive Economic Zone Papers, Oceans '84*, Marine Tehcnology Society and the IEEE, Rockville, Md., 1984: 149 pp.

Scott, S. D. Basalt and sedimentary hosted seafloor polymetallic sulfide deposits and their ancient analogies. In *Proceedings Oceans '83*, Vol. 2. Marine Technology Society and IEEE, Rockville, Md.: 818–824, 1983.

Stowe, K. *Ocean Science*. Wiley, New York, 1983 (2nd ed.).

Templeton, W. L. Artifical radionuclides in the oceans. In *Oceanography: The Past*, Eds. M. Sears and D. Merriman. Springer-Verlag, New York: 420–437, 1980.

Wertenbacker, W. The Law of the Sea: A reporter at large, *New Yorker Mag.*, August 1–8, 1983.

CHAPTER 20

Baumgartner, A. and E. Reichel. *The World Water Balance*, translated by R. Lee. Elsevier, New York, 1975: 179 pp.

Beal, A. B. Bathymetry and structure of the Arctic Ocean (basin). Ph.D. thesis, Oregon State University, Corvallis, 1966.

Bunt, J. S. Microalgae of the Antarctic packice zone. In *Symposium on Antarctic Oceanography*. Published by Scott Polar Research Institute for SCAR, Cambridge, England, 1966.

Dunbar, M. F. Arctic marine ecosystems. In *The Artic Ocean*, Ed. L. Rey and B. Stonehouse. Wiley-Interscience, New York: 233–262, 1982.

Fay, F. H. Marine mammals of the Eastern Bering Sea Shelf. In *The Eastern Bering Sea Shelf: Oceanography and Resources*, Vol. 2, Eds. D. W. Hood and J. A. Calder. Published by NOAA, distributed by University of Washington Press, Seattle, 1981.

Fell, B. *Introduction to Marine Biology*. Harper and Row, New York 1975: 356 pp.

Hood, D. W. *Oceanography of the Bering Sea*. Institute of Marine Science, University of Alaska, Fairbanks, Alaska, 1974: 623 pp.

Horner, R. and G. C. Schrader. Relative contributions of ice algae, phytoplankton and benthic microalgae to primary production in nearshore regions of the Beaufort Sea. *Arctic* 35(4):485–503, December 1982.

Lowry, L. F. and K. J. Frost. Feeding and trophic relationships of Phocid seals and walruses in the Eastern Bering Sea. In *The Eastern Bering Sea Shelf: Oceanography and Resources*, Vol. 2, Eds. D. W. Hood and J. A. Calder. Published by NOAA, distributed by University of Washington Press, Seattle: 807–824, 1981:

Murphy, R. C. The oceanic life of the Antarctic. *Scientific American* 207(3):186–210, 1962.

Nemoto, T. Feeding of baleen whales and krill. In *Symposium on Antarctic Oceanography*. Published by Scott Polar Research Institute for SCAR, Cambridge, England: 240–253, 1968.

NOAA. Ice Limit Charts. Published weekly by the Navy-NOAA Joint Ice Center, Suitland, Md.

Potter, N. *Natural Resource Potentials of the Antarctic*. The American Geographical Society, The Lane Press, Burlington, Vt., 1969.

Rey, L. The Arctic Ocean: A polar Mediterranean. In *The Arctic Ocean*, Ed. L. Rey and B. Stonehouse. Wiley-Interscience, New York, 1982: 433 pp.

CHAPTER 21

Barazangi, M. and J. Dorman. World seismicity maps compiled from ESSA, Coast and Geodetic Survey, Epicenter Data 1961/67. *Bull. Seismol. Soc. Amer.* 59:369, 1969.

Cox, A. Geomagnetic reversals. *Science* 162:237, 1969.

Dewey, J. F. and J. M. Bird. Mountain belts and the new global tectonics. *J. Geophys. Res.* 75(14):2625–2647, 1970.

Drake, E. T. Alfred Wegener's reconstruction of Pangea. *Geology* 4(1):41–44, 1976.

Eicher, D. L. *Geologic Time*, Prentice-Hall, Englewood Cliffs, N.J., 1968: 138 pp.

Fischer, A. G., B. C. Heezen, R. E. Boyce, D. Bukry, R. G. Douglas, R. E. Garrison, S. A. Kling, V. Krasheninnikov, A. P. Lisitzin, and A. C. Pimm. Geologic history of the Western North Pacific. *Science* 168:1210–1214, 1970.

Frank, F. C. Curvature of island arcs. *Nature* 220:363, 1968.

Grow, J. A. Crustal and upper mantle structure of the Aleutian arc. *Geol. Soc. Amer. Bull.* 84(7):2169–2192, 1973.

Hess, H. H. History of ocean basins. In *Petrologic Studies*, Eds. A. E. J. Engel, H. L. James, and B. F. Leonard. Geological Society of America, Boulder, Colo.: 599–620, 1962.

Jacobs, J. A., R. D. Russell, and J. T. Wilson. *Physics and Geology*. McGraw-Hill, New York, 1974 (2nd ed.).

Koppen, W. and A. Wegener. *Die Klimate der geologischen Vorzeit*. Berlin, 1924: 256 pp.

Le Pichon, X. Sea-floor spreading and continental drift. *J. Geophys. Res.* 73:3661, 1968.

Lewis, B. R. The search for longitude. *Oceans*:18–23, Jan. 1982.

Morgan, W. J. Rises, trenches, great faults and crustal blocks. *J. Geophys. Res.* 73:1959, 1968.

Paul, H. A history of geologic time. *Amer. Scientist*:159–166, March–April 1978.

Raff, A. D. and R. G. Mason. Magnetic survey off the west coast of North America. 40°–50°N. *Geol. Soc. Amer. Bull.* 72:1261, 1961.

Schuchert, C. The hypothesis of continental displacement.

In *Theory of Continental Drift*, Eds. W. A. vanderGracht and J. M. Waterschoot. American Association of Petroleum Geologists, Tulsa, Okla.: 104–144, 1928.

Tarling, D. H. and M. P. Tarling. *Continental Drift: A Study of the Earth's Moving Surface*. Anchor Books, G. Bell, London, 1975. (revised ed.): 142 pp.

Valentine, J. W. and E. M. Moores. Plate-tectonic regulation of faunal diversity and sea level: A model. *Nature* 228:657–659, 1970.

Van Andel, T. *Science at Sea*. Freeman, San Francisco, 1977: 186 pp.

Wegener, A. *Die Entstehung der Kontinente und Ozeane*. Vieweg and Sohn, Braunschweig, Germany, 2nd ed., 1929; 3rd ed., 1922, translated into English in 1924; 4th ed., 1924, revised by A. Wegener and translated into English in 1929.

GLOSSARY

AACP   The Antarctic Circumpolar Current.

abiotic matter   Nonliving matter.

absorption   The taking in of one substance by another; incorporation of a substance into a greater structure; capture of radiant energy.

abyssal   Relating to the deep levels of ocean basins and to the features of the ocean floor.

abyssal plain   A flat, almost level area making up the deepest parts of many of the ocean basins.

acceleration   The rate of change of speed of a particle, expressed as a function of time in meters per second squared.

accretion   A process of accumulation through gradual buildup.

accretion prism   An accumulation of material on the margin of a continent, by the process of scraping from basin crust during subduction.

ACPW   Antarctic Circumpolar Water, the water mass intrinsic to the Antarctic Circumpolar Current.

absorption   The adhesion of an extremely thin layer of molecules onto the surface of a solid body or liquid.

advection   The horizontal movement of a mass of fluid.

albedo   The fraction of radiation energy incident upon a surface that is reflected away in all directions.

alga(e)   A simple nonvascular plant, without roots or flowers, reproducing by division; predominantly unicellular in the open ocean, but largely multicellular (kelp) along coastal margins.

alluvial fan   The loosely consolidated material collecting at the base of a mountain (land) or the foot of the continental slope (marine).

ambient   Surrounding on all sides, in the near vicinity.

anadromous   Ascending freshwater rivers from the sea for breeding.

anaerobic   Living in the absence of free oxygen.

andesitic volcano   A volcano derived from magma rich silicates of calcium and sodium.

anion   An ion with a negative electric charge.

anomaly   A deviation in excess of normal variation; usually expressed in units above or below a "norm."

Antarctic Convergence   Name given to the confluence of cold surface waters near the Antarctic continent with the warm surface waters central to the major oceans of the Southern Hemisphere; now called the Polar Front.

anticyclonic   The rotation sense of the large-scale motion of fluid around an atmospheric high-pressure cell (winds) and around a hill of ocean water (currents).

aphotic zone   The vast interior of oceans where the intensity of sunlight is too weak to support photosynthesis.

aquaculture   The seeding, cultivation, and cropping of aquatic flora and fauna.

Arctic (Antarctic) Circle   The parallel located at 66.5° latitude, north or south, at which solar rays are just tangent to the earth's globe during the winter solstice.

arrowworm   Common name for chaetognath, a small, free-swimming marine animal.

anthropods   Animals with a segmented external skeleton and jointed appendages; marine examples are the crab, lobster, and copepods.

asthenosphere   Upper portion of the earth's mantle plastic enough for rocks to flow; extending from the base of the solidified lithosphere to several hundreds of kilometers depth. The material within is believed to yield readily to persistent stress.

Atmospheric high   A zone where the pressure at sea level of an overlying column of relatively dry air is higher than that of standard air (14.7 lb/in.^{2}).

atmospheric low   A zone of relatively low air pressure at sea level established by an overlying mass of air containing relatively large amounts of water vapor; associated with humid climates or storms.

attenuation   Diminution of the intensity of energy propagating through a medium with the distance traveled, through absorption and scattering.

austral   Southern, as in the "austral" (Southern Hemisphere) winter.

autotroph   Organism able to synthesize living protoplasm directly from abiotic material and energy; a plant.

backwash   The flow of water returning to the sea after being cast up the beach by a breaking wave.

baleen   A horny substance growing in two rows of plates suspended from the periphery of the upper jaw of whale species that feed by filtering planktonic and small animals from ingested seawater. The baleen forms a fringelike sieve. In former times called whalebone.

bank   A shallow region, part of the margin of a continent, where the bottom rises to a level that is still safe for navigation but not as deep as that of surrounding areas.

bar   (a) A submerged bank of sand which forms at the entrance to a bay; (b) a unit of pressure, defined as that exerted at sea level by a column of dry air at standard temperature; usually expressed in oceanography and meteorology as 1013 millibars.

barbels   Slender tactile sensory organs in some fish species, usually protruding from the lips.

barrier island Elongate sand bar formed parallel to the shore in areas of considerable sediment flux, but separated from shore by a lagoon, and pierced at intervals by inlets through which the sea communicates with lagoon and river.

barrier reef Coral reef formed along the margin of a continent or island, but eventually separated from the shore by an intervening lagoon.

basalt A fine-grained igneous rock derived from magma, rich in magnesium and iron minerals.

basin The morphological feature formed between continents by an intervening crust of denser material; a great depression lower than the surface of the continents and occupied by an ocean.

basin crust The part of the earth's crust comprised of recently formed basalt rich in magnesium and iron; lithosphere that is noncontinental in origin.

bathymetric Relating to the measurement of depths of water in lakes or oceans.

bathyscaph A navigable submersible ship with a watertight spherical cabin capable of sustaining humans during deep-sea exploration.

bay A body of water in an indentation of a shoreline, smaller than a gulf but larger than a cove.

becquerel A unit of radioactivity, defined as one nuclear disintegration per second.

benthos Living organisms whose life cycles are spent predominantly attached to or in close association with the sea floor.

berm The nearly flat part of a beach or backshore formed of material deposited by breaking waves during highest tide.

bight A long, gradual bend in a shoreline that forms an open bay.

bimodal Relating to the two modes of a process that occur with about equal probability; for example, the two preferred elevations of the earth's lithosphere.

biogenous Having origin by or through living organisms.

biogeography A branch of biology that deals with the geographical distribution of plants and animals.

bioluminescence The emission of visible radiant energy (light) from living organisms.

biomass In the oceans, the amount of living matter that exists within a unit volume of water at a specific instant in time, as in tons per cubic kilometer.

biosphere The total of habitats within a living space of uniform character, as in water or air; the thin shell of the earth's space in which organisms can live.

bitropicality The spatial distribution of one species in a symmetrical pattern about the equator.

bloom The period of the most intense growth in biomass of a plant species; without floral connotations.

bore The response of a body of water when a natural constraint to movement is suddenly overcome; commonly refers to the turbulent tidal flood that rushes with a high abrupt front into restricted inlets.

boundary current The large-scale oceanic flow that sets up parallel to a shoreline.

buffering In chemical oceanography, the process by which the carbon cycle in seawater maintains the acidity–alkalinity ratio near to unity, or at neutral pH.

buoyancy The tendency of a body that is submerged in a fluid to move vertically, depending on its density relative to that of the surrounding fluid.

byssus A tuft of long, tough filaments of proteinaceous material by which certain bivalve mollusks attach themselves to a solid substrate.

caballing A process in which waters of equal density but different T, and S states mix, become denser than either component, and sink.

calcareous ooze An unconsolidated floor sediment consisting primarily of the calcareous shells of dead plants and animals.

calorie A unit of heat energy; changes the temperature of 1 gram of water 1° C.

carapace A chitinous case or shield covering the back or part of the back of species like crab, shrimp, other crustaceans.

carbon cycle The fixing of CO_2 by plants during photosynthesis to form organic nutrients, which then pass through sequential stages of the food chain, and return CO_2 to the air via respiration or decomposition.

carbon-14 A heavy isotope of carbon with 6 protons and 8 neutrons (ordinary carbon has 6 of each); it is radioactive with a half-life of 5700 years. It is used in laboratory experiments to measure the rate of carbon uptake. In the field, the ratio of carbon-14 to carbon-12 atoms indicates the age of calcareous sediments.

catadromous Living in fresh water and migrating from rivers to the sea to spawn, as do some eels.

cation Ion carrying a positive electric charge.

centigrade (Celsius)(°C) Having a temperature scale of 100 divisions between the freezing (0°) and boiling (100°C) points of pure water.

centrifugal force The force acting upon a mass in motion along a curved path and directing it outward from the center of rotation; explains double-wave tide generation.

cephalopods A group of mollusks distinguished by a ring of tentacles around the mouth; includes squids and octopus.

Challenger HMS *Challenger* carried the world's first major, systematic oceanographic expedition, 1872–1876.

chemosynthesis Primary production in the absence of sunlight, by bacteria which gain energy by reducing compounds such as sulfates (SO_4); basis for food webs in the submarine vent communities on the sea floor.

chitin A horny polysaccharide used by many organisms to form an exoskeleton, as for example, crustacean copepods.

chlorinity The chloride concentration in seawater, measured in grams C1/kg of seawater.

chlorophyll A complex green substance of plants which converts the energy available in sunlight into starches and sugars needed for protoplasm construction.

chromatophore A pigment-bearing cell in the skin of animals, exposure of which, through expansion, can alter the color reflectance of the animal and aid in camouflage; common among marine animals such as cephalopods.

chronometer A timepiece of high accuracy, carried aboard ship as a means of calculating longitude of position at sea.

cilia Hairlike processes of surface cells of small organisms, by which they achieve locomotion or move fluids across the body; found in the feeding mechanisms of many marine species.

clay An earthy material composed of extremely small particles of hydrous aluminum silicate, and other minerals, less than 4 micrometers in size.

climax community The last stable stage in community succession achieved through successive adjustment to the environment.

clupeid A member of a large family of finfish including herring, anchovetas, and menhaden, which feed by straining plankton on specialized gill rakers.

Cnidaria Animals which undergo alternation of generations, medusae (floating) and polyp (sessile), and which use stinging cells to capture prey (formerly called Coelenterata).

coastal cell A section of coastline within which the sources and sinks of transportable sediment material just balance.

colonial organism One in which individual animals combine as a unit, with distinct functions performed by each type; an example is the cnidarian Portugese man-of-war.

comber On the open sea, a long curling wave which spills its top as a forward-plunging whitecap; hazardous to small boats.

commensalism A relation between two species, in which one obtains food or other benefits from the other, which is neither damaged nor benefited.

community A population made up of different interacting species sharing a common location.

compensation depth In the photic zone of ever-decreasing light intensity with depth, the depth at which light intensity if just sufficient to sustain photosynthesis at the compensation point. Also, the depth below which the calcareous shells of dead marine organisms are dissolved, a function of pressure, temperature, and the calcium undersaturation level of deep waters.

compensation point A biological concept: the balance point at which the amount of oxygen a plant cell produces through breakdown of CO_2 in photosynthesis just satisfies its oxygen demand for respiration.

compressibility The capability for compression; a physical property of water. Volume decreases with an increase in pressure.

conductivity In an electric field, the ability of a substance to conduct an electric charge.

conservative constituent In seawater, one whose residence time is extremely long, not measureably altered by biological or chemical processes.

continental drift An early hypothesis in the evolution of plate tectonic theory, in which the continents were assumed to drift on the earth's surface, without explaining the cause or energy source needed to accomplish the movement.

continental margin The edge zone between the granitic continents and the basaltic basin crusts.

continental rise An accumulation of sediment at the foot of the continental slope, much like the alluvial fan of the terrestrial counterpart; a gentle, generally smooth-surfaced incline from the abyssal plain.

continental shelf The part of the continental margin presently submerged under the shallow coastal oceans; very flat, with gradient less of 1:500.

continental shelf break Edge of the shelf, where the bottom breaks sharply and slopes rapidly away toward the basin floor.

continental slope The part of the continental margin where the bottom falls off rapidly; gradients are steep, 1:14 (about 4°).

contour line On a chart or graph, the line connecting points of equal value of the quantity being mapped; for example, surface salinity contours.

convection The circulatory vertical motion in a fluid medium, ocean or atmosphere, owing to the variation of its density and the action of gravity.

convection cell A closed circulation pattern of alternate rising and falling motion in a fluid, ocean or atmosphere.

convergence In the oceans, a moving together of water masses, which might result in the buildup of a hill of water on the open sea.

copepods Principal herbivores in the marine biosphere; tiny crustaceans.

coral An animal belonging to the phylum Cnidaria; has modified the alternation of generations so that the dominant form is the sessile polyp. A carnivore, coral captures prey by means of stinging cells.

coralline algae Species of algae which secrete calcareous material during the process of growing over rock surfaces; prime cementing agent for coral reefs.

coral reef A rock like formation consisting of the calcareous structures secreted by coral animals, cemented together through various means such as coralline algae.

core A vertical cylindrical sample of bottom sediment extracted by plunging a pipe into the sediment; now also extracted by special pipe drills.

Coriolis force An apparent force that explains the path deflection of every object moving over spinning, spherical earth; it varies in strength from zero at the equator to a maximum at the poles, causing path deflection to the right of motion in the Northern Hemisphere and to the left of motion in the Southern Hemisphere.

cover A basic factor in promoting stability in an ecosystem.

critical depth The depth below the surface to which mechanical mixing can disperse the phytoplankton population before net production falls to zero.

critical pathway A method for monitoring the potential danger to humans of radioactivity-contaminated seafoods or sea products.

crustaceans Arthropods with bodies of three segments, commonly encased with chitinous exoskeletons; a pair of appendages on each segment, some much modified; and two pairs of antennae. Include copepods, crabs, and shrimps.

crustal plates Segments of the earth's crust which are relatively solid bodies, and which move across its surface as separate entities.

current Horizontal flow of water in an established, defined pattern.

decomposers Organisms such as bacteria and fungi that feed on nonliving protoplasm, causing its breakdown and eventual dissolution in the fluid medium.

deep scattering layer The stratified population of fishes that reflect sound from their air bladders and migrate up and down in the surface layer on a daily cycle; generally found during the day at depths from 200 to 800 in.

deltaic structures The flat, often contorted flatlands created as rivers deposit their load of suspended material; their marshes and tidelands sustain important, productive communities.

demersal fish Bottom-dwelling fish, commonly called groundfish; in particular, the commercially important fishes, such as haddock, sole, flounder.

density field In the oceans, the distribution of mass interior to ocean water, a function of water temperature and salinity.

density current (1) Horizontal water motion caused by nonuniform density fields. (2) Alternate name for turbidity currents caused by sudden changes in fluid density when disturbances resuspend sediments to create high-density muds.

deposit feeder Benthic organism that feeds on the film of nonliving organic detritus settled on the ocean floor; sea cucumbers, brittle stars.

detritus In the oceans, dead organic material that settles to the floor from the biologically productive photic zone.

dew point The temperature of air at which it is moisture-saturated; lower temperature causes condensation.

diatom Microscopic unicellular marine plant possessing a cell wall of overlapping halves impregnated with silica; a principal autotroph.

diatomaceous ooze Unconsolidated sediment in which the dominant material is the siliceous tests of diatom plants; where ancient deposits are exposed at the surface, the material called diatomaceous earth is mined for many industrial applications, including the making of dynamite.

diffusion In the oceans, a mixing process through which a component of seawater (e.g., salt) is transferred from a zone of higher concentration to a zone of lesser concentration; mixing may occur on a molecular scale or on a much larger scale of eddies and gyres.

dinoflagellate Microscopic-size, usually solitary marine plant, enclosed in cellulose plates and bearing two flagella by which the organism achieves limited locomotion.

dispersion Scattering of a material within a fluid medium; differs from simple translation or displacement in that a degree of turbulence is present.

diurnal Cycling once per day.

diurnal inequality The measured difference between the levels of two successive high-water marks over a lunar day.

divergence A horizontal moving apart of fluid masses from a common center; an example is the outward radial flow of surface water when the ocean is subjected to a cyclone (hurricane).

doldrums A part of the ocean near the equator abounding in calms, squalls, and light and variable surface winds.

drift current (*see* wind-driven current.)

dominant species The species in a community or tropic level whose individuals outnumber the individuals of all other species present.

downwelling The opposite of upwelling; the downward transport of seawater.

easterlies Belts of prevailing surface winds coming from the east.

ebb tide, ebb flow The outgoing tide current, with sea level falling.

echinoderms Benthic organisms for the most part, having five-point radial symmetry, a calcareous spiny exoskeleton, and locomoting by means of multiple tube feet: sea stars, sea urchins, crinoids.

ecology Multidiscipline studies of the relations between organisms and a community and the environment and between the organisms themselves.

ecosystem Any part of the natural environment in which both living and nonliving matter is in continuous interchange, with exchanges energized by a unidirectional flow of available energy; ideally defined as a closed system.

eddy A unit of motion in a fluid medium running contrary, usually circularly, to the main current; of any size above the molecular scale; definable, therefore not turbulence.

EEZ (Exclusive Economic Zone) The coastal strip of ocean over whose resources the coastal state claims exclusive jurisdiction; most recently defined as 200-nmi wide, beginning at the beach.

Ekman effect Broadly, the motion of ocean surface water generated by a steady wind; specifically, the principle that net transport of surface water is perpendicular to the direction of the forcing wind.

electromagnetic radiation A succession of electromagnetic waves.

electromagnetic spectrum The entire range of wavelengths or frequencies of electromagnetic radiation from the shortest gamma rays to the longest radio waves and including visible light.

El Niño Aperiodic occurrence of warm surface water off the coast of Peru.

enrichment ratio The degree to which many marine organisms, plant and animal, accumulate chemical elements in far greater concentration than exists in the ambient seawater.

ENSO (El Niño–Southern Oscillation) A global-scale response of the ocean (El Niño) to a driving perturbation in atmospheric circulation in the tropical latitudes of the Pacific Ocean (Southern Oscillation).

entrainment A mixing process between two fluid masses, in which the mixing occurs in the preferential direction of the mass that contains the higher level of turbulent energy available for the mixing; commonly refers to the upward mixing of salt water into the turbulent riverine flow in estuaries.

enzyme Any of numerous complex protein molecules produced by living organisms for the express purpose of catalyzing biochemical reactions without entering chemically into the reaction itself.

epicenter The location projected on the earth's surface corresponding to the focal point of an earthquake in the interior of the lithosphere.

epifauna Animals that live directly on the ocean floor or in close association with it.

epiphyte A plant that derives its moisture and nutrients from the air and grows attached to other plants.

episodic Occurring, appearing, or changing at usually irregular intervals.

equilibrium tide A hypothetical tide wave over an earth covered entirely with oceans, no continents.

estuarine zone The transition domain between fresh river water and the saline water of the open coast.

euphausids Small, shrimplike animals of the open ocean, notable as the first carnivores in the oceanic food web; highly luminescent.

euryhaline Able to live in waters of widely varying salinity.

eurythermal Able to live in waters of widely varying temperature.

evaporite The salt residues remaining after salt or brackish waters have evaporated.

excess volatiles Compounds whose high concentrations in seawater cannot be accounted for as by-products of crust weathering alone; thought to originate in large part in the juvenile waters emanating from geysers and submarine hydrothermal springs.

exoskeleton A covering over the outside of an animal that serves for both body rigidity and protection; common among crustaceans and echinoderms.

fast ice In polar oceans, the sea ice that forms attached to the shoreline and does not drift in coastal currents.

fathometer Device that send and receives sonic signals; used to measure the depth of the ocean floor. A fathom is a unit of length equal to about 6 feet.

fault In geology, a fracture in the earth's crust together with the displacement of crustal blocks on one side relative to those on the other and in a direction parallel to the fracture.

fauna Animals.

feedback loop In a system, the process that reenters a part of the output as a new input signal.

Ferromanganese A mineral that is primarily a mixture of iron and manganese.

fetch In the generation of surface waves by winds, the distance over which the wind travels in a sufficiently coherent manner that it continues to impart energy to the wave.

filter feeder Animal that feeds by straining particulates suspended in water.

fish meal The dry mixture of organic and inorganic residue after whole fish are ground, their oil is removed, and excess moisture is evaporated. Rich in protein, it is a staple component in the preparation of feeds for other animals.

fish ranching The process of releasing fish fry to "graze" freely throughout ocean space before returning to be harvested or otherwise cropped.

fishery conservation zone A political zone set by coastal states for management of fish stocks.

Fishery Management Act of 1976 An executive proclamation by the United States setting limits of national jurisdiction over fisheries.

fjord A drowned valley originally scoured by a glacier, usually with a shallow sill at the entrance, the result of moraine deposits by the glacier.

flagellum A hairlike extension from a living cell; when vibrated it provides a mode of locomotion; characteristic of dinoflagellates.

flocculation The process whereby small particles combine or coagulate to form larger entities; an important process in the deposition of silts and clays from rivers entering the sea.

flood tide Incoming currents during the period of a rising tide.

flora Plants.

flushing time Time interval required for complete replacement of water in a sector of estuary, bay, or gulf; usually measured in numbers of tide cycles.

flux The transport of a material or energy through a known area; expressed per unit area per unit time.

food chain A specific pathway along which the chemical energy created in photosynthesis passes to successive trophic levels.

food web A complex of interlaced food chains.

foraminifers Large protozoans, benthic or pelagic, which secrete a shell of calcium carbonate; important contributors to sediments, hence to paleontology.

forced wave In a fluid, a wave continuously controlled by its generating force, even as it propagates.

fortnight The interval of time between two successive spring tides, 14.75 days.

fracture zone Zone within the earth crust along which there is differential movement of crustal blocks.

free wave In a fluid, a wave that propagates away from the point of generation.

freezing point Temperature at which liquid solidifies.

fringing reef A sea level flat coral reef built out from the shore of an island or continent.

frustule The siliceious shell of a diatom.

fully developed sea For a given wind with its particular speed and direction, the condition at which the wave field reaches a steady-state energy density. Beyond this point, the wave field dissipates energy at the same rate as it acquires energy from the wind field.

GASIMM (Giant Antarctic Sea Ice Mixing Machine) Acronym for the Antarctic sea ice freezing and melting processes which drive the long-term turnover of waters in the total ocean.

gastropods Mollusks which generally construct a spiral shell and move by means of a broad foot; snails, abalones, sea slugs.

GEOSECS Geochemical Ocean Sections Study.

Gondwanaland The part of Pangea south of the Tethys Sea; included what is now South America, India, Australia, Africa.

gradient Change in the value of a quantity per unit distance and specified direction. A topographic gradient is a change in elevation over a measured horizontal distance.

gram Unit of mass, equal to 1 cm^3 of pure water at 4°C.

grazer Herbivore.

greenhouse effect The warming of the earth's surface that results because the atmosphere readily admits the short-wave solar radiation but tends to retain the long-wave heat energy that radiates outward from earth; caused primarily by the gases carbon dioxide and water vapor contained in the atmosphere.

Gyre The large-scale pattern of circulation around the entire ocean basin.

habitat The place or type of site where a plant or animal normally lives and grows.

Hadley cell A pattern of vertical circulation within the atmosphere, usually by three cells in the space between equator and poles. The vertical circulation at the confluence of the low- and middle-latitude cells is downward, that is, subsiding; this is the zone of westerly surface winds. Air rises along the confluence of the middle- and high-latitude cells, the zone of prevailing easterly surface winds and the high-altitude jet stream.

half-life In physics, the time required for half of the atoms of a radioactive substance to become disintegrated to the next substance in the radioactive decay sequence.

halocline Vertical zone of a water column where salinity changes rapidly with depth.

heat The energy associated with the random motion of the molecules, atoms, or smaller structural units of which matter is composed.

heat budget A physics hypothesis, that the heat energy input to a body is balanced by its heat energy losses.

heat capacity A physical property of a substance, whereby the internal energy of its molecular structure, measured on a temperature scale, is directly related to the gain or loss of heat energy per unit mass; for pure water at sea level pressure, 1 cal/(g)/(°C).

heat energy A physics measure of the quanta of energy absorbed or released by a molecular structure, per unit mass.

herbivore Animal that eats plants.

hermatypic coral Species of coral animals (cnidarians) which live in symbiosis with algae (zooxanthellae) in their surface tissue.

heterotroph Organism which needs, for the metabolic synthesis of its own tissues, the complex organic compounds of nitrogen and carbon extracted from the protoplasm of autotrophs or other heterotrophs.

high seas The open ocean. Today, the part of the sea surface not included within the Exclusive Economic Zones of coastal states.

holoplankter A plankter which spends its entire life cylce as plankton, meaning free drifting.

humidity The concentration of water vapor in the air.

hydraulic wave A misnomer; the word ''wave'' implies orderly, defined motion, but hydraulic wave denotes a disordered, turbulent translation of fluid mass, as in a tide bore.

hydrocast In field work, the lowering of a series of sampling bottles into the ocean water column and retrieval of the samples; sometimes called a hydrostation; when executed in sequence along a track, the series is referred to as a hydroline.

hydrogenous sediments Sediments composed of material precipitated chemically from seawater.

hydrologic cycle The cycling of water from reservoir (ocean) to atmosphere (vapor) to land (rain) to ocean (river).

hydrothermal vent Hot-water geyser, either terrestrial or on the basin floor.

hypertonic In a marine organism, having a body fluid more saline, of higher osmotic pressure, than the ambient fluid.

hypotonic In a marine organism, having a body fluid less saline, of lower osmotic pressure, than the ambient fluid.

iceberg A large mass of land ice floating in the ocean; derives from the disintegration of glaciers that reach the sea.

icebreaker A ship with specially shaped and strengthened bow, capable of ramming a path through sea ice.

ice cover The cover of perennial sea ice over polar seas.

ice floe A flat, free mass of floating sea ice; ice formed originally over the liquid ocean.

ice shelf The portion of a glacier system that has moved outward from the land to float on the ocean itself.

index of refraction In physics, the ratio of the speed of light in a vacuum to that in a medium; for seawater, about 1.33.

indicator species An organism whose distribution in the ocean or on land indicates the geographical extent of a living community. A biogeographical term.

infauna Organisms that live within the sediment of the ocean floor.

infrared The part of the electromagnetic radiation spectrum from the sun with wavelengths longer than about 0.750 micrometers and shorter than 1000 micrometers.

inlet A narrow passage leading inland from the shore.

in situ In the natural or original position; a Latin term.

insolation The incoming solar radiation.

interdisciplinary Involving two or more academic or scientific disciplines.

interface A surface forming a common boundary; within a fluid, a discontinuity such as a thermal or density front.

internal wave A disturbance that propagates wavelike along the density interface of two water masses.

intertidal zone The part of the shore exposed between the high- and low-tide levels; see also littoral.

interstitial Relating to the spaces between particles of an unconsolidated substance, for example, beach sand.

invertebrate Animal lacking vertebrae, or backbone.

ion An atom or molecule possessing a positive or negative electric charge from having gained or lost one or more electrons; in seawater, generally ''solvated'' or neutralized by a surrounding cloud of dipolar water molecules.

ionizing radiation The collision of elemental particles with atoms or molecules, converting them into ions.

island arc A sequence of islands or volcanoes geographically arranged in an arc, usually with a deep trench on the convex side. According to plate tectonic theory, they were formed by a collision between segments of a spherical crust, in which one segment was subducted or forced below the other.

iso- In mapping, a curve along which a selected parameter is invariant: *isohaline*—curve of constant salinity; *isopycnic*—curve of constant density; *isotherm*—curve of constant temperature.

isostasy In geophysics, a general equilibrium of flotation of the earth's crust maintained by the yielding of the plastic asthenosphere.

isostatic adjustment In an interglacial period, the slow vertical rebound of a continental mass toward equilibrium flotation over the asthenosphere after its ice cover has melted away.

isotonic In a marine organism, having a body fluid with the same salinity, the same osmotic pressure, as the ambient fluid.

isotope Any of two or more atoms of an element with the same number of protons in their nuclei but differing in the number of neutrons; *see* carbon-14.

ITCZ (Intertropical Convergence Zone) A zone of convergence of tradewinds, north with south, located seasonally within the range of 5° to 10° N, particularly in the Pacific; zone of heavy rainfall on ocean and lands.

jet stream A long, narrow, meandering, high-speed, high-altitude wind blowing west to east, with velocities exceeding 100 km/h.

joule A unit of energy, defined as the work done by a force of 1 newton exerted for a displacement distance of 1 meter.

jurisdiction The authority of a sovereign state to control or legislate.

juvenile water Water derived directly from magma, so released to the earth's surface for the first time.

kilogram Equal to 1000 grams.

kilojoule Equal to 1000 joules.

kilometer A distance equal to 1000 m, or 0.54 nautical mile.

kinetic energy The energy possessed by a moving body, defined as one-half the product of mass times velocity squared. In fluids, it is common to speak of the *Kinetic energy density,* which is a measure of the amount of motional energy possessed by a unit volume of fluid. In describing the mixing of fluids through turbulence, we are actually describing the rate at which kinetic energy density is transformed into heat energy density. That is, the decay of motional energy is offset in part by the rise in the *internal energy* density of the fluid, as measured by its temperature.

knot A unit of speed in maritime navigation; defined as the speed required to cover a distance of 1 nautical mile in 1 hour. The term originated in antiquity. A rope was attached to a ''log,'' which was thrown into the sea at the bow of a moving ship; a sailor would count the knots in the rope as they slipped through his fingers during the time the log passed from bow to stern.

krill In general, euphausids; specifically, the term for the large euphausid, *E. superba,* which dominates the zooplankton of the Southern Ocean.

lagoon Shallow body of water, near or communicating with a larger body of water, for example the lagoons between offshore barrier islands and the mainland.

latent heat The quantity of heat released or absorbed per unit mass when a substance changes state.

Laurasia The part of Pangea north of the Tethys Sea.

Law of the Sea (LOS) Convention signed by 177 nations in 1982, defining the rights and responsibilities of all coastal states and their jurisdiction over coastal waters and resources.

leeward Being in or facing toward the direction in which the wind is blowing; the *leeward side* is therefore the ''protected'' side of an island or sound or bay.

lithogenous Relating to material derived from continental rock.

lithosphere The outer, solid portion of the earth, composed of rock essentially like that exposed at the surface and generally considered 50 miles thick.

littoral the shore area that lies between the highest high-tide and the lowest low-tide levels.

littoral drift The transport of sands in a direction parallel to the shoreline within the zone where waves break, hence within the littoral space. Also called *longshore drift.*

load The quantity of suspended material in seawater or in river water.

longshore current The mean drift of seawater in a direction parallel to the shoreline.

lunar day The period between successive stands of the moon at nadir above a point on the spinning earth. Because the moon orbits earth in the same rotation sense as earth spin about its own axis, the lunar day is longer than the solar day, about 24.8 hours.

magma Molten rock material within the earth; source from which igneous rock comes through cooling.

magnetic anomaly Small variations in the earth's magnetic field strength measured at a point, relative to the average value measured over a chosen area.

magnetic field The composite magnetism of earth that causes orientation in materials having a magnetic dipole sense.

magnetic pole A geographical point toward which all magnetic sensors in the near vicinity will point; has polarity, hence north and south magnetic poles; does not coincide with the poles of the spin axis of earth. The deviation in direction between what points to a magnetic pole and what points to the true geographic pole is called magnetic *declination.*

magnetic reversal Reversal in the polarity of the earth's magnetic field; many reversals have occurred throughout geologic history.

magnetometer Device for measuring strength and orientation of the earth's magnetic field.

manganese nodules Lumpy concretions of minerals widely distributed over the abyssal plains; contain manganese, iron, copper, nickel, and other minerals.

mangroves Multiroot trees found in abundance along

shorelines of brackish waters in the tropics; of great importance ecologically for the cover that exposed roots provide for marine fauna and flora and for preventing erosion.

mantle (1) The interior of the earth that lies beneath the lithosphere and above the core. (2) The tissue of mollusks and gastropods which secretes the shell.

marginal seas Large, semienclosed bodies of water lying between the open sea and continents; examples are the Mediterranean Sea, Bering Sea.

margins The geological structures that form the transition between continental crust and oceanic basin crust.

mariculture Animal husbandry applied to marine organisms, for commercial objectives.

matrix A two-dimensional array of something, such as an array of mucous filaments used by marine filter feeders. In mathematical analysis, a tool for analyzing cause and effect in problems with many variables, such as assessing environmental impact.

maximum permissible concentration (MPC) In the management of human exposure to doses of radioactivity, the largest quantity per unit volume of radioactive material in air, water, and foodstuffs that is not expected to be harmful to human health.

maximum sustainable yield The number or tonnage of individuals of a species that can be harvested from a natural stock each year without a concomitant long-term decline in its population.

mean current In dynamic ocean water with both smooth and turbulent flow, the average flow rate as measured at a point.

meander A convolution in the mean circulation that occupies a large area and is sustained for long periods of time; a perturbation in flow that is too large to be called turbulence and covers too long a period to be a part of mean flow.

mean ocean circulation The average displacement of seawater along streamlines of flow; refers to the large, basin-scale circulation.

mean sea level The average height of the sea at a point over a period of time (usually six months or more), as measured relative to a local datum.

medusa One of the alternate generations of cnidarians; the bell-shaped form of various species of jellyfish.

membrane A thin, pliable sheet, usually of animal or plant origin; generally allows certain substances to pass through while containing others. The covering tissue of a living cell.

meridian A line of constant longitude; the 0° reference is the Greenwich meridian.

meroplankton The portion of the plankton that spends only a part of their life cycles as drifters near the surface; typically, the larval stages of marine animals.

mesenterial Pertaining to threadlike glandular organs attached to the mesenteries of anthozoans.

mesoglea A gelatinous substance that fills the space between the inner and outer skin tissues of certain sponges and cnidarians.

metabolism Chemical reactions in living cells by which energy is provided for vital activities, and new material is assimilated for repair and replacement; the sum of the chemical processes necessary to sustain life.

mica One of several minerals that readily separate into very thin leaves and that are all silicates.

microbial decomposition Process of breakdown and dissolution of organic material by the action of bacteria and fungi.

mid-ocean ridge The characteristic topographic elevation of the sea floor along zones of upwelling magma and sea floor divergence.

milt The reproductive glands of male fishes in breeding condition and filled with secretion; the secretion itself.

mixing As differentiated from "stirring," mixing implies a combining of two components into a uniform blend or mixture.

Moho The Mohorovičić discontinuity, a point, ranging from about 5 km below the ocean floor and about 40 km beneath the surface of continents, at which seismological studies indicate a transition from the earth's crust to the mantle. Seismic waves suddenly increase in velocity.

mole (micromole) The quantity of a substance in grams (micrograms) that has a weight numerically equal to its molecular weight. A mole of N_2 is 28 grams of the gas.

molecular viscosity The internal resistance that a fluid offers to deformation during flow, a consequence of interaction between molecules.

molecular weight The total number of protons plus neutrons in the nuclei of all atoms composing one molecule of a substance.

mollusks A phylum of soft, unsegmented invertebrate animals, usually secreting a calcareous shell; snails, limpets, clams, chitons, squids.

monsoon Strongly prevailing winds which reverse direction from winter to summer; first applied to winds over the Arabian Sea.

morphology The external structure of rocks and their erosional or topographic features; also a branch of biology dealing with the form and structure of plants and animals.

motility The degree of locomotion of which an animal is capable.

mud A slimy fluid-to-plastic mixture of water and finely divided particles, mostly of silts and clays, but often of other materials including organics.

mutualism Mutually beneficial association between two different kinds of organisms.

nannoplankton Extremely small plankton; isolated mainly by centrifuge.

nauplius stage The first larval stage of many crustaceans; the oval, unsegmented body has three pairs of appendages.

nautical mile The distance along the spherical earth's surface between two latitude lines separated by one minute of arc; 1850 m.

neap tide The minimum range of tide, which occurs at fortnightly intervals with the moon in first and third quarters.

nekton Any marine animal with locomotive ability sufficient to override the drift of currents.

nematocyst A stinging cell; a capsule which on contact explodes a barblike proteinaceous hollow thread, which then carries a toxic chemical to the prey. A characteristic feature of cnidarians such as the sea anemone, jellyfish.

neritic Pertaining to the shallow waters over the shelf portion of the continental margins.

niche The "job role" that an organism fulfills in its community.

nitrogen fixation In the oceans, the conversion of dissolved nitrogen gas by algae to a form that can be used for organic compounds; the principal means of supplying nitrogen for primary production in coral reef communities.

node In the oceans, the geographical point at which the range of the standing tide wave is zero.

nutrient depletion In the oceans, the gradual depletion of dissolved nitrates and phosphates in the photic zone as the springtime algal bloom proceeds.

nutrient replenishment In the oceans, the replacement of dissolved plant nutrients by physical processes such as upwelling and mixing by the winds. Waters of deeper layers, not yet depleted of nutrients, are physically forced upward into the photic zone.

nutrients In the oceans, the dissolved nitrates and phosphates that are the principal building blocks used by plants in photosynthesis.

ocean basin A great topographic depression in the earth surface associated with simatic crust and occupied by an ocean.

oceanic Relating to the open sea beyond the continental margin.

oceanic stratosphere Deep waters of the world oceans; separated from the troposphere by the permanent thermocline.

oceanic troposphere The uppermost portion of the oceanic water column; the layer of ocean water above the permanent thermocline.

offshore Directed seaward, away from shore.

omnivore Animal able to feed on either plants or animals.

onshore Directed shoreward, toward the beach.

oolite Sedimentary rock consisting of small rounded grains of calcium carbonate, precipitated out of solution and cemented together; typical of warm tropical beaches.

ooze Fine-grained sediment of clay minerals on the deep-ocean floor, at least 30% of which consists of the tests and shells of marine organisms.

osmosis The process by which a solvent is pressured to move through a semipermeable membrane to equalize the concentrations of solution on both sides of the membrane.

osmotic pressure The pressure on a solvent to move through a semipermeable membrane to a solution of higher solute concentration; in oceans, a direct function of the differences in salinities of two fluids.

OTEC (ocean thermal energy conversion) A physical process that extracts useful electrical or other energy from the temperature difference between the warm surface waters of oceans and the cooler deep waters.

oxygen minimum layer A layer in the ocean column where the concentration of dissolved oxygen is lower than that in water above and below.

$P - E$ The difference between the rates of precipitation (rainfall) and evaporation of seawater, usually on an annual basis; key concept in the study of heat and freshwater budgets for oceans.

pack ice Irregular masses formed of broken pieces of sea ice of varying ages and sizes packed together, interspersed with varying areas of open water.

paleoceanography The science of describing ancient oceans, based on the fossil records found in samples of floor sediments.

paleoclimatology The science of describing climates of the geologic past, based on glacial deposits, fossil records, paleogeography, and sediments.

paleomagnetics The study of the direction and intensity of the earth's magnetic field throughout geologic time.

Pangea Postulated former supercontinent supposedly composed of all the continental structures as these existed about 200 Ma.

parasitism An intimate relationship between organisms of two or more kinds in which one lives on or within the body of another, obtaining food and other benefits and causing some damage.

partial tide A component of the total tide produced by one of the multiple, concurrent tide-generating forces.

pathogen A specific cause of disease, as a bacterium or virus.

pelagic Of, relating to, living in, or occurring in the open sea.

pennate shape   An elongate shape, as in certain species of diatoms.

perennial pack ice   The portion of the ice cover over polar seas that remains through the annual cycle of freezing and melting.

perturbation   In a living community, an abnormal stress; an abnormal event.

pH   A measure of the concentration of hydrogen ions, H^+, in a solution. A pH value of 7 is neutral, neither acidic nor alkaline; the scale is from 1 (acid) to 14 (alkali).

phosphorite   A fibrous, concretionary variety of apatite, a calcium phosphate mineral. It accumulates in shallow upwelling regions where high primary production fixes large quantities of dissolved phosphate, collecting on a variety of substrates, including bones, pebbles, and sand.

photic zone   The sunlit uppermost layer of the ocean receiving enough light to permit photosynthesis. Sometimes defined as ending at the depth at which light intensity falls to 1% of surface intensity, about 100 m.

photoinhibition   A response in which plants reduce photosynthetic activity in strong light.

photon   A quantum of radiant energy having a specific wavelength.

photophore   A specialized organ of deep-sea fishes made up of a lens and a gland capable of an internal chemical reaction that emits light, or bioluminescence; can be triggered active either neurally or chemically.

photosynthesis   The process of constructing carbohydrates (sugars, starches) using energy gained from sunlight together with abiotic materials, in the chlorophyll-containing tissues of plants exposed to sunlight.

physical characteristics of water   Characteristic variables of water, such as its temperature and salinity.

physical properties of water   Properties of the molecule itself; behavior controlled by the structure of water molecules alone or in concert with the structure of adjacent molecules.

physiographic diagram   An illustration of the three-dimensional features of the earth's crust, as they would appear viewed obliquely from the air at an angle of 45°; maplike in nature.

piscivore   Animal that feeds primarily on fishes.

plankton   The passively floating or weakly swimming animal and plant life, usually minute, of a body of water. From the Greek *planktos*, meaning free-drifting.

plate tectonics   A group of theories concerning the global geology of the earth's crust, proposing that it moves in integral segments.

pollutant   Any physical or chemical material that alters an environment from its natural state.

polymetallic sulfides   Concretions of mixed sulfide minerals of various metals: iron, manganese, zinc, copper, and others.

polyp   The sessile generation of cnidarians in which they have a hollow cylindrical body closed and attached at one end and opening at the other with a mouth surrounded by tentacles armed with nematocysts.

population   In biology, all individuals of a species within a community in a definable geographical area.

population dynamics   The processes that determine the size and composition of any population.

potential energy   Energy that an object possesses because of its position in a geopotential field. For example, the potential energy of an initially stratified water column increases as the wind energy mixes deep, saltier water toward the surface.

precipitation   (1) Chemically, the deposition of solid material out of solution. (2) The deposit on earth of water in the form of hail, sleet, snow, rain, or mist.

prevailing winds   Winds that blow predominantly from one direction, whatever their speed.

progressive wave   A moving wave whose form is the transport mechanism propagating energy from one part of the medium to another. The medium itself does not propagate.

protozoans   Minute single-cell animals of varied morphology, physiology, and often complicated life cycles found in almost every habitat. Marine examples are the forminiferans.

pteropod   Free-swimming form of gastropod snail, in which the anterior lobes of the foot are modified into broad flaplike wings for swimming; important member of the pelagic community.

pycnocline   A depth range within a water column where density changes with depth at a greater rate than in other adjacent ranges.

radiant energy   Energy traveling as a wave motion; the energy of electromagnetic waves.

radiation   The process of emitting radiant energy in the form of particles and rays, during disintegrations of radionuclides: in three types, alpha and beta particles and gamma rays.

radiation dose   The amount of ionizing radiation absorbed by living tissue, measured in sievert (Sv) units: 1 sievert = 1 joule of ionizing radiation energy per kilogram of body weight.

radioactive   Having the potential for disintegration and hence radioactivity.

radioactivity   The property possessed by some elements of spontaneously emitting particles or rays during disintegration of the nuclei of their atoms.

radiolarian   A single-celled, microscopic animal of an extensive order of marine protozoans having a siliceous skeleton of spicules and radiating threadlike pseudopodia.

radiometric dating The science of dating materials on the basis of their content of radionuclides of known half-life.

radionuclide An element whose atoms are inherently unstable in structure; the nucleus will disintegrate spontaneously, releasing alpha or beta particles and sometimes gamma rays; the residual atom may be either another radionuclide in the disintegration series, or a stable atom with no potential for further disintegration.

radula In the forward part of the mouth of most mollusks, a filelike ribbon studded with horny teeth on the dorsal surface, by which the animal burrows into substrate or tears up the body of prey and draws it into the mouth.

Reagan Proclamation 1983 Proclamation establishing the sovereignty of the United States over its territorial sea, exclusive economic control of ocean resources seaward to 200 nmi, and jurisdiction over all seabed resources to the limit of the contiguous continental shelf.

red tide Patches of coastal surface waters discolored by the explosive growth of certain species of dinoflagellates. They release a toxin poisonous to many forms of marine life.

reef In nautical terms, a chain of rocks or ridge of sand near the surface of the water that presents a navigation hazard; *see also* barrier reef, coral reef, fringe reef.

relict structures In the stratigraphy of geologic strata, features recognized as formed in past epochs.

remanent magnetism The magnetic effect remaining in rock after original solidification, during which certain mineral crystals having inherent magnetic polar structure were aligned to the then existing magnetic field. They subsequently retained this magnetic orientation.

residence time In the oceans, the interval between the time a salt ion is introduced into the sea for the first time and its removal from the fluid via permanent incorporation in the sediments.

respiration The physical and chemical processes by which an organism supplies its cells with the oxygen needed for metabolism and relieves them of the carbon dioxide formed.

river-induced upwelling Process of upward entrainment of saltwater in an estuary; driven by the turbulent kinetic energy in the river water.

salinity In sea water, the amount of dissolved salts, in grams of salt per kilogram of seawater; the normal ocean value is about 35 parts per thousand.

salt exclusion (1) In the evaporation of seawater, the inability of salts to go into the vapor. (2) In freezing, the exclusion of salts from the crystalline solid, although a small amount remains trapped as pockets of brine. In both exclusions residual salts increase the density of the remaining liquid.

salt wedge The penetration of salt water upstream in an estuary; the flow compensating for the entrainment of seawater into the outflowing river.

Sargasso Sea A central North Atlantic water mass. This large tract of water is comparatively still and named for the seaweed floating there.

Sargassum A branched form of brown alga that sustains itself in the photic zone by means of gas bladders (pneumatocysts); a characteristic plant of the Sargasso Sea.

scarp A line of steep cliffs formed when basin crust blocks undergo faulting. The Mendocino escarpment stretches for hundreds of kilometers westward from Cape Mendocino.

scattering In seawater, the deflection of light rays by suspended particulates.

sea The disordered, highly peaked wave field within a storm-generating area.

sea floor spreading The phenomenon of new magmatic material being inserted into midocean rifts, with subsequent divergence of solidified crust away from the zone of injection.

sea ice Ice formed directly over the surface of the sea.

sea level The average vertical position of the sea surface, measured over an extended period of time and referred to a surveyor's benchmark on the shore; the location of the air–sea interface.

seamount A mountain or volcanic feature on the deep-sea floor, the top of which does not penetrate the sea surface.

sea smoke In arctic regions, vapor evaporated from the sea surface, made visible by its condensation.

sea state A measure of the roughness of the sea surface; a scale of surface wave conditions related to the speed of wind.

seiche The "sloshing" of seawater between the margins of a basin or basin sector for periods varying from a few minutes to several hours. The energy to set up seiches can come from the strong winds and barometric pressure changes of storms, from tides, or from seismic disturbances.

seismic Of or relating to events or processes produced by earthquakes.

seismograph station A setup for recording and monitoring the passage of pressure waves within the earth.

semidiurnal Occurring twice daily; specifically for tides, two events during a lunar day of 24.8 hours, or 12.4 hours per semidiurnal event.

sensible heat transfer In the air and sea, the conduction of heat energy by mass contact between them, as measured by changes in temperature of either fluid.

sessile Immotile, attached to the bottom or other stationary substrate.

set The displacement of a ship; the direction toward which an oceanic current flows.

setae On the appendages of filter-feeding animals, hairlike processes that strain small particulates from the seawater.

shallow-water wave Any water wave that loses energy through friction between moving water particles and the stationary bottom; tides are shallow-water waves everywhere, as are tsunami waves. Wind waves become shallow-water waves only when their approach to a shore allows them to feel the bottom in water about one-half their wavelength or less deep.

shoal A sandbank or sandbar that makes the water shallow and presents a navigation hazard; not to be confused with reefs, which are composed of consolidated rock or coral.

sial The lighter upper portions of the earth's crust composed of rocks rich in silica and alumina; the bulk of the continents is *sialic*.

sievert *See* radiation dose.

sigma-*t* A shortened notation for seawater density: a density of 1.022g/cm^3 is expressed as 22 sigma-*t* units.

silicate Chemically, $Si(OH)_4$, an important constitutent of seawater because of its use by diatoms, radiolarians, and other organisms for forming secreted shells and tests.

sill The relatively shallow ridge on the ocean floor at the entrance to fjords and bays; a submerged ridge at relatively shallow depth separating two ocean basins, or a marginal sea from the open sea.

silt Loose sedimentary material with particles of sizes ranging from sand (0.06 mm) to clay (0.004 mm).

sima The heavier lower portions of the earth's crust composed of rocks rich in silica and magnesia. The lower portion of the continental crust and the bulk of the oceanic crust are *simatic*.

size specificity In oceanic food chains, the "right" size of prey to be taken by a predator and ingested whole.

slack water The period at the turn of the tide, from ebb to flood currents or vice versa, when there is little or no horizontal motion of tidal water; coincides with high- and low-water stands.

SOFAR (1) Acronym for "sound fixing and ranging." (2) The depth range within which acoustic pressure waves can travel extremely long distances.

sonar Acronym for "sound navigation and ranging"; common name for the fathometer instrument which "sounds" the depth of the ocean floor by means of pulsed sonic waves. Used also to detect submerged objects and to survey the waters around a fishing vessel for evidence of fish density, hence potential fish catch.

Southern Oscillation *See* El Niño.

speciation The process by which new species evolve to diversify a community.

species A category of biological classification immediately below the genus, comprising related organisms or populations capable of interbreeding.

specific heat *See* heat capacity.

spectrum For any dynamic process, a graphical display of how its dynamic energy is spread across a range of wavelengths or frequencies of motion.

spit A small point of land usually of sand or gravel running into a body of water.

spring tide The highest range of tide, which occurs at fortnightly intervals with the moon either new or full.

standing crop The number of individuals or biomass existing in a community at one instant in time.

standing wave A type of oscillation in which the wave amplitude oscillates vertically between fixed nodes; the wave does not progress. Examples are seiches in inlets, tide waves in inlets and sounds and even large sectors of ocean basins.

stenohaline Able to live only in seawater of a narrow, specific range of salinities.

stenothermal Able to live only in seawater of a narrow, specific range of temperatures.

storm surge A rise above normal water level on the open coast where the Ekman effect from strong winds causes the shallow waters to pile up against the shore. Exceedingly dangerous, a storm surge accounts for more mortality of humans than any other oceanic factor, owing to their tendency to aggregate in coastal plains most susceptible to sudden and disastrous flooding.

strait A narrow channel connecting two large bodies of water.

stratification In the oceans, the arrangement of different water masses in vertical sequences, with density increasing downward.

subduction In plate tectonic theory, when blocks of the earth's crust converge, the process of the edge of one crustal plate passing beneath the edge of the other.

submarine canyons Erosional features of the ocean floor that cut across both shelf and slope of continental margins; usually steep-sided valleys with dendritic feeder canyons along the shallower, shoreward portions.

substrate In the oceans, any solid or semisolid material between the floor and water, ranging in type from red clay to rock, sand to coral reef; important factor in the distribution of benthic organisms.

surface tension A physical property of water by which its surface seeks to have the least area, that is, a plane.

suspension feeder Animal that feeds directly upon suspended particles; distinguished from filter feeder in the degree of discrimination between particles being ingested. Copepods are examples of suspension feeders, in that they choose between the particles encountered.

suspensoid In seawater, the suite of suspended particulates.

sverdrup A unit of transport of seawater volume, equal to 1 million cm^3/s; named after Harold U. Sverdrup, the "father of oceanography."

swash The turbulent surge of water up the beach face driven by the breaking wave.

swell The long-period, undulating waves that propagate energy to great distances from the point of generation; the source of "breakers" along the beach.

symbiosis The intimate living together of two dissimilar organisms in a mutually beneficial relationship.

system Any mechanism or biological process or chemical reaction in which a signal is sensed and processed according to a defined technique, the resultant of which bears a predicted relation to the input.

tectonics A branch of geology concerned with the broad structural features of the earth's crust, especially folding and faulting.

terrigenous Relating to ocean sediments derived from the destruction of rocks on the continent.

test Shell of an organism.

Tethys Sea Ancient open sea that separated Pangea into Laurasia and Gondwanaland.

thermal expansion coefficient As a physical property of water, the fractional change in volume for a unit change in temperature; accounts for its change in density.

thermocline The depth interval within seawater where its temperature decreases at a maximum rate; gradient of temperature with depth. There is a permanent thermocline and a seasonal, shallower thermocline in summer months.

thermohaline circulation Circulation driven by the unequal heating of earth by the sun, together with surface patterns of evaporation and precipitation.

tidal bore *See* bore.

tidal prism The difference in volume of water in an estuary between the high-water mark after flood tide and the low-water mark after ebb tide; used in calculating the flushing time of an estuary.

tidal wave A long-period wave driven by tide-generating forces; name given as a misnomer to a tsunami wave.

tide The periodic rise and fall in the level of the sea surface; complex wave forced by the composite lunar and solar tide-generating forces.

tide component *See* partial tide.

tide current Flow pattern that accompanies the tide wave; an important dynamic factor in the overall flow pattern in shallow neritic waters but negligible in open-ocean waters.

tide excursion The maximum displacement of a parcel of water during one cycle of the composite tide.

tide flats Sector of coastal plains that is inundated during daily tides or fortnightly tides; important contributor to the productivity of coastal oceans and spawning grounds for some species of marine fishes.

tide-generating force The vector force, sometimes called "tractive force," that is the difference between the gravitational mass attraction force of earth to another planetary body, and the centrifugal force that results because both bodies translate about a common center of mass. Although small, the tractive force is directed horizontally over the earth's sphere and is the cause of the two simultaneous zones of convergence of fluid oceans, one on the side of earth nearest to the other planetary body, the second at the antipode of the first.

tide range The vertical change in the level of sea surface between highest high-water and lowest low-water marks of a tide cycle.

TOGA (tropical ocean, global atmosphere) An experiment worldwide in scope conducted to discover the large-scale dynamic coupling between the atmosphere and the ocean.

torque In the oceans, the force created by the stress pattern of surface winds; responsible for the large-scale spin-up of water. A vorticity-generating force.

trace elements In seawater, dissolved elements whose concentration is minor, less than one part per million.

tracer In seawater, a natural or artificial component whose distribution or behavior suggests the circulation or behavior of the ocean itself. A classic example is the distribution of the radioactive tritium, introduced into the ocean surface layers from the fallout of tests of nuclear weapons during the decade 1950–1960.

trade winds Steady, prevailing surface winds blowing toward the equator from the northeast in the tropical latitudes of the Northern Hemisphere and toward the equator from the southeast in the tropical latitudes of the Southern Hemisphere.

transform fault Geological fault that traverses mid-ocean ridge features; represents the slip plane along which crustal blocks have differential translation during sea floor spreading.

transgression Applied to sea level, the global secular trend of a rising level of seawater. The sea is extending land-ward.

transition zone (1) Geology: the zone of transition between continent and basin crust. (2) Biology: the zone separating distinct biogeographical areas.

trawl A large baglike net used in fishing. Trawls are towed by a ship at specified depths.

trophic level One of the hierarchial strata of a food web in which organisms are the same number of steps removed from the primary producers.

tsunami A long-period wave generated by seismic disturbances to the sea floor. A free wave and a shallow-water wave everywhere, it travels at speeds of about 700 km/h in the open ocean but builds to dangerous heights and energy density when its speed slows in shallow waters. The Japanese word for "wave in bay."

turbidity In seawater, the quality or state of being thick and opaque with roiled sediments and suspended particles; a condition reducing the passage of light.

turbidity current A highly roiled, relatively dense episodic current carrying suspended clay, silt, and sand and flowing down a submarine slope through less dense seawater; speeds of tens of kilometers per hour are common. *See also* density current.

turbulent flow Highly disordered flow in fluids, a source of energy for the mixing of water masses. By definition, not mathematically defined.

typhoon A severe tropical cyclone occurring in the China Sea or the region of the Philippines; analogous to a hurricane.

upwelling The upward motion of seawater anywhere in the oceans; applies specifically to the process that brings cold, nutrient-rich waters into the photic zone to support plant production.

vaporization *See* latent heat.

viscosity *See* molecular viscosity.

water mass A large body of seawater identified by its temperature and salinity, hence a characteristic density, occupying a level in the water column corresponding to its equilibrium density. An example is Antarctic Bottom Water.

wave energy density As applied to surface wind waves, the average number of joules of energy per square meter of sea surface sustaining a wave; computed as $1250\ H^2$ joules/m^2, where H is the height of the wind wave from crest to trough in meters.

wave refraction The process by which the crest of a plane wave is turned as the wave begins to encounter friction with a shoaling bottom. Where the continental shelf is wide, smooth, and regular, the refraction of surface waves is virtually perfect, so that waves enter the surf zone with crests almost parallel to the beach, regardless of their initial approach angle into the coastal waters.

whitecap In open deep waters, the "spillover" of wind waves when their peakedness exceeds a critical value, usually defined as an angle of 120°. The spill is a turbulent, foaming collapse of the wave peak itself, hence the term whitecap. Whitecaps occur whenever the wind speed freshens and waves begin to grow in energy density.

wind-driven current The surface currents that form from the transfer of energy from winds to surface waters. Sometimes called Ekman drift or wind drift, the actual response of surface waters undergoes a transient phase during which the mass of surface water shifts to create a tilt in the sea surface in the crosswind direction. Once accomplished, the surface flow shifts to the downwind direction and remains thus until the wind changes, requiring another shift in the tilt of sea surface, and so on.

zooplankton Animals small enough to be planktonic in the sense of drifting with prevailing ocean currents.

Index